THE HEAVY ELEMENTS

Chemistry, Environmental Impact and Health Effects

by

JACK E. FERGUSSON

Chemistry Department, University of Canterbury, New Zealand

PERGAMON PRESS

OXFORD · NEW YORK · SEOUL · TOKYO

U.K. Pergamon Press plc, Headington Hill Hall, Oxford OX3 0BW, England

U.S.A. Pergamon Press Inc., 395 Saw Mill River Road, Elmsford, New York 10523, U.S.A.

KOREA Pergamon Press Korea, KPO Box 315, Seoul 110-603, Korea

JAPAN Pergamon Press, 8th Floor, Matsuoka Central Building, 1-7-1 Nishi-Shinjuku, Shinjuku-ku, Tokyo 160, Japan

---

First edition 1990
Reprinted 1991 (with corrections)

Library of Congress Cataloging in Publication Data
Fergusson, J. E.
The heavy elements: chemistry, environmental impact and health effects/Jack E. Fergusson.
p. cm.
1. Heavy metals—Environmental aspects.
2. Heavy elements—Environmental aspects.
I. Title.
TD196.M4F47 1990 363.17'91—dc20 90–6783

British Library Cataloguing in Publication Data
Fergusson, J. E. (Jack Erric), *1933–*
The heavy elements.
1. Heavy chemical elements
I. Title
546

ISBN 0–08–034860–2 Hardcover
ISBN 0–08–040275–5 Flexicover

Printed in Great Britain by BPCC Wheatons Ltd, Exeter

Chemistry, Environmental Impact and Health Effects

Related Pergamon Titles of Interest

Books

CLEVER
Mercury in Liquids, Compressed Gases, Molten Salts and other Elements

DICKSON
Fate and Effects of Sediment-bound Chemicals in Aquatic Systems

DIRKSE
Copper, Silver, Gold & Zinc, Cadmium, Mercury Oxides and Hydroxides

FERGUSSON
Inorganic Chemistry and the Earth

GREENWOOD
Chemistry of the Elements

MASSON
Sulfites, Selenites and Tellurites

Journals

Ambio

Applied Geochemistry

Atmospheric Environment

Chemosphere

Environment International

Environmental Toxicology and Chemistry

Geochimica et Cosmochimica Acta

Marine Pollution Bulletin

Urban Atmosphere

Waste Management

Water Quality International

Full details of all Pergamon publications/free specimen copy of any Pergamon Journal available on request from your nearest Pergamon office.

PREFACE

It is the intention in this book to provide a broad survey of the heavy elements, their relevant chemistry, environmental impact and health effects. The particular group of ten elements are the heavier members of the p-block elements, which have a number of features in common, as well displaying periodic trends. Hence it is appropriate to consider them together as a coherent group.

It is clear that to understand the environmental and health effects of the chemical elements we need to know more about their chemistry. Unfortunately this chemistry is frequently avoided in lecture courses, because it is claimed to be too complex or too uninteresting. In addition the best understanding is achieved through the involvement of other science disciplines. Whereas significant advances have been made in research, cooperation and teaching, the whole topic of the environmental and health effects of the heavy elements is in its infancy, but growing fast. It is hoped that this contribution will stimulate some students to investigate further aspects of the heavy elements.

The book is divided into four parts. The first is a brief introduction to the criteria used to select the elements, and the history of the discovery and uses of the elements. The second part is on the chemistry of the elements relevant to the rest of the book. In part three the environmental impact of the elements is reviewed. This includes the concentrations in the environment, sources and chemistry. The final section is a brief introduction to the health effects of the heavy elements.

I wish to thank my colleagues and students who have participated in many helpful discussions. Finally I am grateful for the patience of my wife Beverley, who often saw a back in front of a microcomputer during the several months of writing.

Christchurch
New Zealand
October 1989

Jack Fergusson

CONTENTS

PART I

INTRODUCTION AND HISTORY

CHAPTER 1

INTRODUCTION AND HISTORY

The periodic table is a concise summary of the chemistry of the chemical elements. It is customary to discuss the chemistry of the elements in terms of their position in the table - for example - by group, O, S, Se, Te, Po; or by row, Li, Be, B, C, N, O, F, Ne; or by class e.g. transition elements. One classification is the 'heavy elements', which are the lower members of the periodic groups.

The heavy elements to be considered in this book meet many of the following criteria. (1) The elements are relatively abundant in the earth's crust. (2) They are extracted and used in reasonable amounts. (3) They are used in places where the public may come into contact with them. (4) They are toxic to human beings. (5) They have made significant perturbations to the biogeochemical cycles.

The elements that fall into this group are the heavier members of the p-block elements listed in Fig. 1.1. These elements are unexpectantly abundant as they occur around the three small peaks (atomic number, Z = 35-40, 54-57 and 78-83)

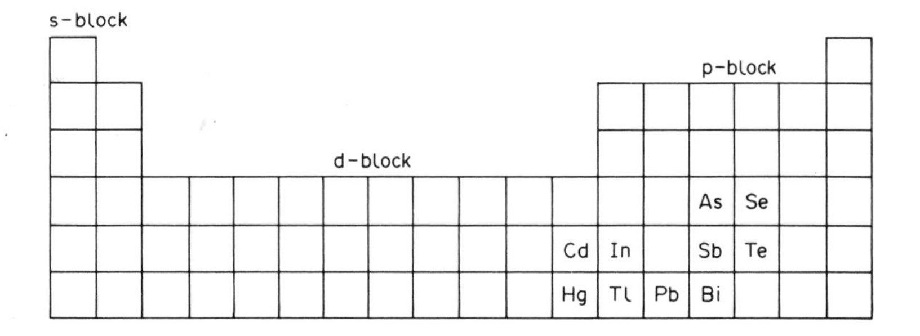

Fig. 1.1 The toxic heavy elements.

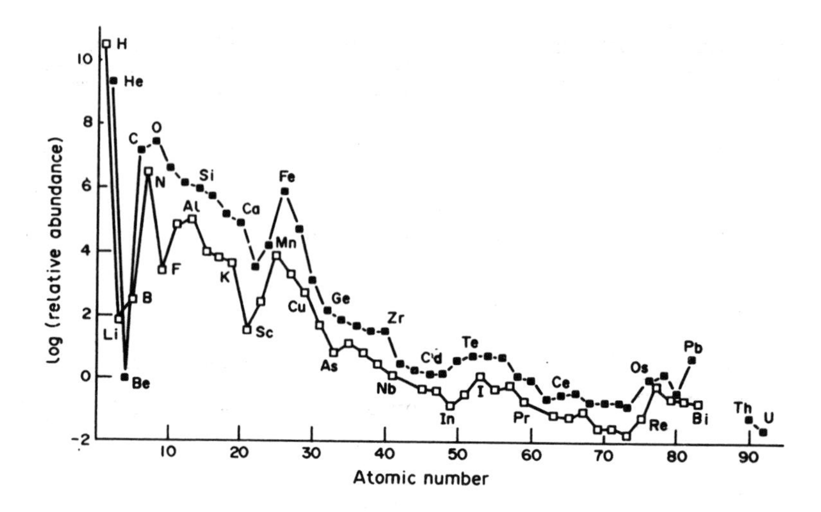

Fig 1.2 The relative abundance of the elements in the universe. (Top curve even atomic numbers, bottom curve odd atomic numbers).

in the relative abundance curves of the elements in the universe (Fig. 1.2). The reason for the small peaks is that some elements in these areas (e.g. Cd, Te and Pb) have low neutron capture cross sections. The elements are produced in stars by neutron capture, and because of the poor capture cross sections of some there is a build up of the elements around them.

The magnitude of the perturbation of the biogeochemical cycles of the ten heavy elements has been estimated [2] using the following criteria; has the cycle been perturbed and on what scale? Is the element mobile in the geochemical process? What is the degree of public health concern and what are the pathways of the metal to man? The results, listed in Table 1.1, indicate that the elements are a semi-coherent group with related environmental effects.

HISTORY

Discovery of the Elements

The discovery of the ten elements spans the period from ancient pre-Roman times to the mid nineteenth century. Reasons for their discovery, and the time of discovery was determined by their chemical and physico-chemical properties, and the state of scientific knowledge and methodology.

The elements known to the ancients and to the alchemists Mercury and lead and probably antimony were known in ancient times. Mercury and lead were isolated because of the relatively low temperatures needed for the reduction of the metal oxides with carbon [1,17]. This is clear from examining the plot of the

TABLE 1. 1 The Pertubation of the Environment by the Heavy Elements

Element	Scale Of Perturbation			Diagnostic environment	Mobility	Health concern	Critical pathway
	Global	Regional	Local				
Pb	+	+	+	a,sd,i,w,h,so	v,a	+	F,A,D
As	(+)	+	+	a,sd,so,w	v,s,a	+	A,W
Cd	(e)	+	+	a,sd,so,w	v,s	+	F, D
Hg	(e)	+	+	a,sd,fish,so	v,a	+ org.	F,(A)
Sb	(e)	+	+	a,sd	v,s	+	F,W,A?
Se	(e)	(+)	+	a	v,s,a	E	F
Tl	?	?	+	em,so	v,s	(+)	A,F?
In	?	?	+	a,so,em	v	(+)	?
Bi	?	?	+	a,so,em	v	(+)	?
Te	?	?	(+)	so	v,a?	(+)	?

Columns 2, 3 and 4: + significant perturbation, (+) possible perturbation, (e) enriched (may not be anthropogenic), ? not enough information.
Column 5: a = air, sd = sediments, so = soil, i = ice cores, w = surface waters, em = emission studies.
Column 6: v = volatile, s = soluble, a = alkylated compound.
Column 7: + toxic in excess, (+) toxic but little data available, E essential but toxic in excess.
Column 8: F = food, W = water, A = air, D = dust
Source; adapted from Andreae 1984 [2].

free energy of formation of the oxides versus temperature (Ellingham diagram) given in Fig. 1.3 [7]. The free energies of formation of CO and CO_2 become more negative than for HgO and PbO at relatively low temperatures. Hence the reactions:

$$MO + C \rightarrow M + CO$$

and

$$MO + CO \rightarrow M + CO_2$$

are achieved for the two metals. Prior to reduction the two ores, HgS (cinnabar) and PbS (galena), are converted to the oxides by roasting in air;

$$2MS + 3O_2 \rightarrow 2MO + 2SO_2$$

A reducing agent is not actually necessary to extract mercury, as the oxide spontaneously decomposes, i.e. $G^o_f > 0$, at temperatures above 500° C.

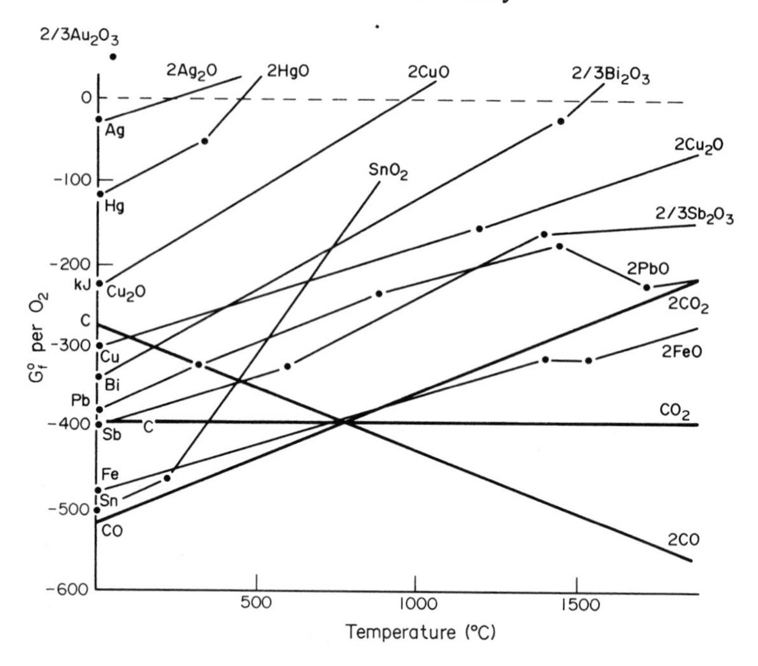

Fig. 1.3 Ellingham diagram for metal oxides. Based on data in Ives 1960 [7].

$$2HgO \rightarrow 2Hg + O_2.$$

Cinnabar was also reduced with iron and copper (or brass). Both metals were often used for holding the sulphide during heating [1,11,15,17],

$$HgS + Fe \rightarrow FeS + Hg,$$

$$HgS + 2Cu \rightarrow Cu_2S + Hg.$$

It is possible that antimony and bismuth were also produced in ancient times, but were not recognized as being different from lead [1,17]. The data given in Fig. 1.3 also indicate that their oxides reduce with carbon at low temperatures. A vase dated around 4000 - 3000 BC is almost pure antimony [17]. The elements were finally identified during the 18th. Century, antimony 1707 and bismuth in 1737 and 1753. The reactions involved in their extraction were;

$$2M_2S_3 + 9O_2 \rightarrow 2M_2O_3 + 6SO_2,$$

$$M_2S_3 + 3Fe \rightarrow 2M + 3FeS$$

and

$$M_2O_3 + 3CO \rightarrow 2M + 3CO_2.$$

Arsenic compounds, in particular arsenic sulphides, were used by the ancients [12,17]. Isolation of the element (attributed to Albert Magnus ca. 1250 AD) was achieved by reduction of the sulphides, orpiment As_2S_3 and realgar As_4S_4, with carbon, or decomposition with lime or soap [1,12,17].

Lead production in ancient times peaked during the Roman period (500 BC to 500 AD) at around 27,000 tons y^{-1}. During the Iron Age, 1200 - 500 BC, production was 12,500 tons y^{-1}, and after the fall of the Roman Empire consumption dropped to 8500 tons y^{-1} during 500 - 1000 AD [11].

Cadmium The discovery of cadmium first required the discovery of zinc (in 1746). Cadmium occurs in association with zinc ores, and is concentrated in flue dusts from the extraction of zinc. In 1817 Stromeyer noticed that the yellow colour of calamine, $ZnCO_3$, was not due to iron and he separated the zinc carbonate from less soluble $CdCO_3$ [9,12,15,17]. The cadmium carbonate was ignited to the oxide, which was then reduced to the metal with lampblack. Cadmium in ZnO was also discovered independently in 1818 by Hermann [12,17]. Significant production of cadmium did not take place until after the first world war, when it was used for electroplating [9,16].

Indium and thallium The two Group III elements indium and thallium were discovered by Reich and Richter in 1863 and by Crookes in 1861 respectively. Their discoveries were possible due to the spectroscope, that had just been discovered by Bunsen and Kirchhoff. The metals were characterized by their atomic line spectra (in flames), and their names originate from the colour of their characteristic lines, indigo blue for indium and green for thallium [5,9,12,15,17]. The actual isolation of thallium metal in 1862 led to a heated debate, between Crookes in England and Lamy in France, over who was first [5].

Selenium and tellurium Tellurium was discovered in 1783 by Muller von Reichenstein, when investigating a gold bearing ore thought to contain native antimony. He concluded that it was not antimony or bismuth but some unknown metal. It took 12 years for the discovery to be confirmed and the element to be named tellurium. Berzelius and Gahn discovered selenium in 1817. They initially thought they had found tellurium in the residue in sulphuric acid, obtained from burning sulphur at their sulphuric acid plant. A year later they decided the material was a new element, which they named selenium [9,12,15,17].

Early Production and Uses

Many of the heavy elements were used prior to this century. Some of their more important early uses will now be considered.

Uses as metals Lead was used as a structural metal in ancient times, and for weather-proofing buildings. The Romans used lead in ducting for water and for cooking vessels. The floor of the Hanging Gardens of Babylon was made of lead, to help retain the moisture in the ground [11].

The Romans report the use of mercury in the recovery of silver and gold. The process was used in Egypt in the 12th. Century, and in Central and South America in the 16th. Century [8].

Cadmium was used in electroplating because of its resistance to corrosion [9,16]. Alloys of Sb, Pb, and Sn with some Bi were used in type metal [12,14]. The antimony causes the alloy to expand on solidification which produces type with sharp edges [14]. Antimony powder was painted on to objects, such as wood, to give a metallic appearance [14].

Arsenic was used by alchemists in efforts to transmutate the base metals. For example, when added to copper arsenic produces a white coloured alloy.

Medicinal and cosmetic uses In ancient times mercuric sulphide (cinnabar) was used to give a red colour to the skin [9]. Hair was darkened, and still is, by treating it with lead compounds, such as lead acetate [11]. White lead was used to hide blemishes on the skin [11]. Galena, PbS, and antimony sulphide, Sb_2S_3, were used to paint around the eyes [3,14].

A number of compounds of these toxic metals were used medicinally (and some continue to be used in this way), even though their toxicity was recognized. Criticisms over the use of such compounds has meant that their popularity has waxed and waned over the years. Arsenic was used by the Greeks, Romans, Arabs and Peruvians both therapeutically and as a poison [3,12]. Mild doses of arsenic were said to improve the complexion. Bismuth was first used around 1889 for treating syphilis [3]. In early times ointments containing mercury compounds were frequently used [9,16]. The Hindus believed in mercury's aphrodisiac properties [8]. Mercury was also used to treat syphilis from the 16th. Century [10]. In fact, mercury was believed to be the panacea for all ills, in spite of its toxicity [3]. One effect of mercury toxicity is excessive salivation. When it was believed that salivation was a useful therapy, people were treated with mixtures containing mercury to produce the effect [4]. Antimony was also used medicinally from early times, and Louis XIV is said to have been cured of typhoid with antimony treatment. Tartar emetic, potassium antimonyl tartrate, was used to produce vomiting as antimony has an irritant effect on the linings of the stomach and intestine [3,14]. Thallium taken orally caused body hair to fall out, and it was used for this purpose, with some obvious tragedies [5].

Mercury amalgams with their putty consistency, which turn hard with time, were used for filling teeth. Tin/silver amalgams are still used today. The use of a cadmium/mercury amalgam last century was discontinued when it was realized it turned the tooth dentine yellow [14].

Toxicity The majority of the elements were recognized as toxic soon after their discovery, and in some cases before the actual element was identified. Arsenic was used by ancient people as a poison, and for killing rats and vermin [3,14]. Because of arsenic poisonings, techniques were developed for the analysis of the element in human tissue, e.g. the Marsh test [12,14]. The Romans knew of the toxic effects of mercury, and used slaves and convicts for mining cinnabar [3]. The life expectancy of miners was as low as 6 months [8]! Thallium was used for killing rodents and pests, but this was discontinued because of its toxicity to humans [3,5,9]. Though selenium was not discovered until 1817, it was known that plants growing in certain areas were poisonous to animals. Marco Polo mentioned poisonous plants in 1295, and in the 19th. Century examples of toxicity in animals grazing in certain areas were reported [3]. It is likely that the poisonous component was selenium, absorbed by the plants from the soil.

Lead and mercury The two metals most frequently used from early times were lead and mercury. In addition to their uses listed above the metals were used in other ways. Lead chromate, $PbCrO_4$, was used as a pigment [14]. Mercury as mercuric nitrate was used in the felting of hats, and this use may be the origin of the term 'mad as a hatter'. The addition of lead to wine, to improve taste and kill bacteria, continued well into the medieval and later periods [3,11]. The fact that the Romans had so much contact with lead, which is a neurotoxin, has prompted the suggestion that lead poisoning among the leaders of the empire may have contributed to the downfall of the Roman Empire [6,10,11,13].

REFERENCES

1. Aitchison, L. A History of Metals (2 Vols.): Interscience; 1960.
2. Andreae, M O. (Reporter). Changing biogeochemical cycles: Group report. in: Nriagu, J. O., Ed. in: Changing metal Cycles and Human Health, Dahlem Konferenzen: Springer-Verlag; 1984: 359-373.
3. Berman, E. Toxic Metals and their Analysis: Heyden; 1980.
4. D'Itri, P. A, and D'Itri, F. M. Mercury Contamination: A Human Tragedy: Wiley-Interscience; 1977.
5. Emsley, J. The trouble with thallium. **New Scientist**; 1978: 392-394.
6. Gilfillan, S. C. Lead poisoning and the fall of Rome. **J. Occup. Med.**; 1965; **7**: 53-60.
7. Ives, D. J. G. Principles of the Extraction of Metals. Monographs for Teachers, No. 3: Chemical Soc. London; 1960.
8. Kaiser, G. and Tolg, G. Mercury. in: Hutzinger, O., Ed. The handbook of Environmental Chemistry: Springer Verlag; 1980; **3A**: 1-58.
9. Kirk-Othmer. Kirk-Othmer Encyclopedia of Chemical Technology. 2nd. Ed.: Wiley-Interscience; 1964.
10. Leninan, J. The Crumbs of Creation: Adam Hilger; 1988.
11. Nriagu, J. O. Lead and Lead Poisoning in Antiquity: Wiley-Interscience; 1983.

12. Partington, J. R. A History of Chemistry.: Macmillan; 1970; Vols. I (Part I), II, III, IV.
13. Patterson, C. C. Lead in ancient bones and its relevance to historical developments of social problems with lead. **The Sci. Total Environ.**; 1987; **61**: 167-200.
14. Roscoe, H. E. and Schorlemmer, C. A Treatise on Chemistry: MacMillan. 1905, 1907.
15. Tylecote, R. F. History of Metallurgy: The Metals Soc. London; 1976.
16. Ullmann, F. Ullmann's Encyclopedia of Industrial Chemistry. 5th.: VCH; 1985.
17. Weeks, M. E. and Leicester, H. M. Discovery of the Elements. 7th. Ed.: J. Chem. Educ.; 1968.

PART II

CHEMISTRY

CHAPTER 2

EXTRACTION AND USES

The extraction of the elements and some of their current uses will now be discussed. Details on the elements' crustal abundance, production and consumption and some of their major uses are listed in Table 2.1, and in Table 2.2 the major minerals and a summary of the extractive metallurgy are presented. All of the elements are strongly chalcophilic (affinity for sulphur) in their mineralogy and chemistry (Fig. 2.1). Their strong affinity for sulphur, compared with oxygen, has a marked influence on the extractive chemistry of the heavy elements.

Of the ten elements, lead is the most widely used industrially, followed by arsenic and antimony. There is a weak relationship between the abundance of the elements in the earth's crust and the amount extracted.

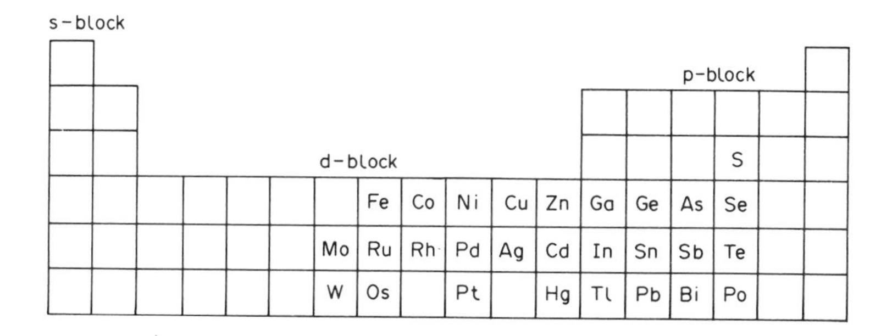

Fig. 2.1 The chalcopile elements.

TABLE 2. 1 Production and Uses of the Elements

Element	Crustal Conc. ppm (1985)	Yearly Prod. x 1000t (1985)	Yearly Consumpt. x 1000t	Uses
Cd	0.2	14	14	Electroplating, anticorrosive coating (22%), pigments (22%), stabelizer in PVC (12%), batteries Ni/Cd (37%), alloys (4%), control rods in nuclear reactors.
Hg	0.08 163,000 flasks	6 200,000 flasks	7	Cloro-alkali cell, paints (mildew proofing), agriculture chemicals, electrical apparatus, dental.
Pb	13	4100	3900	Lead batteries (60%), petrol additive (4%), rolled and extruded products (9%), pigments (13%), alloys (5%), cable sheathing (6%), ammunition (3%).
As	1.8	50		Alloys, storage batteries, agriculture herbicides, wood preservative.
Sb	0.2	53		Alloyed with Pb in batteries (50%), flame retardant, catalyst.
Bi	0.2	4	3	Replace Pb and Cd in enviromental sensitive areas, pharmaceuticals, electronics, catalyst, cosmetics, pigments.
In	0.1	0.075		Alloys, holding of glass lens, electronics.
Tl	0.5	0.005		Alloys, rodenticide, IR windows, electronics.
Se	0.05	1.6	1.7	Electronics, glass, pigments, animal feed, photocopying.
Te	0.01	0.22		Alloys, catalyst, pharmaceuticals, electronics, vulcanizing rubber.

Sources; references 6,11,18,19.

CADMIUM

Sources

Cadmium is concentrated in sulphide ores of zinc, mercury, lead and copper [8]. The principal cadmium mineral is hexagonal CdS, greenockite. Rarer minerals, most of secondary origin, are cubic CdS hawleyite, hexagonal CdSe cadmoselite, CdO monteponite, $CdCO_3$ otavite. Cadmium metacinnabar, or saukevite, is cubic (Hg,Cd)S, and may contain up to 11.7% Cd.

The crustal abundance of cadmium is around 0.2 ppm. Low levels occur in igneous rocks, <0.001 - 1.6 ppm, and in sandstones and limestone. Higher levels occur in sedimentary rocks associated with organic material, such as shales, and in marine manganese nodules and phosphorite deposits. [8,9]. Whereas cadmium is closely associated with zinc, the Zn/Cd concentration ratio varies widely from 27 to 7000. There is no clear association between cadmium and other elements in rocks, but there is some association between cadmium and organic material, as shown by the higher cadmium levels in coal (0.01 - 22 ppm), peat (0.37 - 190 ppm) and crude oil (0.01 - 16 ppm) [8,9,23]. The mean cadmium concentrations in various rocks and similar information for the other elements, are summarized in Table 2.3.

The primary source of cadmium is in zinc, lead and copper sulphide ores. The average levels of cadmium in ZnS (sphalerite) lie in the range 0.02 - 1.4%, (median around 0.3%, and a maximum level of 5%). Another source is the ore tetrahedrite-tennartite $(Cu,Zn)_2(Sb,As)_4S_{13}$, found to have a maximum level of cadmium around 2400 ppm (0.24%) and a median level of 600 ppm [8,23].

Extraction

Cadmium is obtained as a by-product of the extraction of Zn, Pb and Cu from their sulphide ores. The amount produced relates, to some extent, to the production of zinc, about 3 - 3.5 kg per tonne of zinc produced. Production increased in the sixties and seventies, and has evened out at around 14,000 tonne y^{-1} [9,18,23]. Less cadmium products are now being used because of environmental and toxicity concerns [18].

Cadmium accumulates in flue dusts from the pyrometallurgic smelting of Pb, Cu and Zn. It mainly occurs as the metal, because of the reducing environment during smelting. The metal is extracted with sulphuric acid and cadmium sponge is deposited by adding limited amounts of zinc to the solution,

$$Cd^{2+} + Zn \rightarrow Cd + Zn^{2+}.$$

Alternatively NaCl or $ZnCl_2$ are added to the dust, and when heated, $CdCl_2$ (and $PbCl_2$) volatilizes. The lead is removed as $PbSO_4$ and the cadmium obtained as sponge with Zn. [16,28]. Cadmium may also be obtained from the hydrothermal extraction of zinc. Pure metal is obtained from the sponge by smelting, or vacuum distillation or electrolytically, the latter being more common.

Uses

The principal uses of cadmium have been listed in Table 2.1 [1,2,6,8-12,16,18,23,28]. The proportion of metal used in each area varies from year to year, especially as environmental concerns change. Electroplating and pigment use, is decreasing whereas the use in batteries is increasing.

TABLE 2. 2 Major Minerals and Extractive Metallurgy

Metal	Major minerals	Extractive metallurgy
Cd	CdS greenochite associated with Zn ores, e.g.ZnS	As a by-product of zinc production (ca. 3 kg associated with Zn per tonne Zn). Cd/CdO forms in flue dust from Zn smelting and is extracted.
Hg	HgS cinnibar	Roasted in air to form oxide, and then thermal dissociation. $HgS + O_2 \rightarrow Hg + SO_2$ or $HgS + Fe \rightarrow Hg + FeS$ or $HgS + 4CaO \xrightarrow[O_2]{} Hg + 3CaO + CaSO_4$
Pb	PbS galena $PbCO_3$ cerussite $PbSO_4$ anglesite	Roasted, $2PbS + 3O_2 \rightarrow 2PbO + 2SO_2$ reduction, $PbO + C \rightarrow Pb + CO$ $PbO + CO \rightarrow Pb + CO_2$ $2PbO + PbS \rightarrow 3Pb + SO_2$
As	As_4S_4 realgar As_2S_3 orpiment As_2O_3 arsenolite FeAsS arsenopyrite $FeAs_2$ loellingite	$FeAsS \rightarrow As + FeS$ $As_2O_3 + 2C \rightarrow 2As + CO/CO_2$
Sb	Sb_2S_3 stibnite	$Sb_2O_3 + 2C \rightarrow 2Sb + CO/CO_2$ $Sb_2S_3 + 3Fe \rightarrow 2Sb + 3FeS$
Bi	αBi_2O_3 bismite Bi_2S_3 bismuthinite $(BiO_2)CO_3$ bismutite	From flue dust as by-product of Pb/Zn and Cu smelting. Roast and reduce as for antimony.
In	In in ZnS	As for thallium.
Tl	Tl^+ in PbS and with Rb^+ in potassium feldspar and mica	From flue dust from smelting Pb/Zn ores.
Se & Te	Occur with sulphides Cu, Ag, Au, Zn, Cd etc.	Se and Te from anodic slime from electrolytic refining of Cu. Se also in sulphuric acid sludge and flue dust from Cu/Pb smelting. Se, and Te electrolytically. Roast ores to oxyanions and reduce to Se.

Cadmium coatings on metals are produced by electroplating from a Cd^{2+}/CN^- solution. The coatings are anti-corrosive in marine, alkaline and tropical environments, they are ductile and have a low coefficient of friction, they are readily soldered and retain their silver-white lustre. Cadmium coated materials are used widely in car construction, e.g. on bearings, races, nuts, bolts, and disk brakes. Cadmium also protects junctions between different metals.

Alloys containing cadmium are used in bearings (e.g. 99% Cd, 1% Ni), and for soldering, for example Al to Cu and for Al alone (e.g. 95% Cd, 5% Ag; 40% Cd, 60% Zn). Fusible alloys with low melting points contain cadmium and are valuable in fire protection devices, such as sprinklers (e.g. 44.7% Bi, 22.6% Pb, 8.3% Sn, 5.3% Cd). A level of 0.02 - 1.0% cadmium in copper hardens the metal without significantly lowering its electrical conductivity.

Organocadmium compounds are used in PVC, for stabilizing the polymer to heat, light and discolouration. Environmental concerns are restricting its use in this way. Organocadmium compounds are also used as catalysts, in organic hydrogenation and polymerisation reactions.

Pigments containing cadmium are used in plastics, paints, printing inks and ceramics. They have good stability to heat and UV radiation, and do not darken in a H_2S atmosphere, as do lead pigments. The pigments contain CdS, and colours range from yellow to red. Cadmium lithopones are obtained by treating a mixture of $ZnSO_4$ and $CdSO_4$ with BaS to give a coprecipitated product ZnS/CdS/$BaSO_4$. The pigments become darker as the the proportion of Cd to Zn increases. If the precipitation is carried out in the presence of selenium, the incorporation of CdSe also produces darker shades.

Rechargeable batteries are attractive consumer items, and the nickel - cadmium battery (NiCad) with a potential of 1.1 - 1.3 V meets this requirement. The two poles consist of $Ni(OH)_2$ and $Cd(OH)_2$ in the uncharged state, and the electrolyte is KOH. The charging reactions are;

$$Cd(OH)_2 + 2e \rightarrow Cd + 2OH^-, \text{ reduction (-ve pole)}$$

$$2Ni(OH)_2 + 2OH- \rightarrow 2NiO(OH) + 2H_2O + 2e, \text{ oxidation (+ve pole)}$$

i.e.,

$$Cd(OH)_2 + 2Ni(OH)_2 \rightleftharpoons 2NiO(OH) + Cd + 2H_2O.$$

During use the reverse reactions occur. Dioxygen is also produced at the positive pole;

$$4OH^- \rightarrow 2H_2O + O_2 + 4e.$$

The O_2 migrates rapidly to the negative pole and the reaction,

$$O_2 + 2H_2O + 2Cd \rightarrow 2Cd(OH)_2,$$

occurs, otherwise the cell could not be sealed. The less common silver-cadmium battery is also rechargeable,

$$Ag_2O + Cd + H_2O \rightleftharpoons 2Ag + Cd(OH)_2.$$

Cadmium selenide is used in photoconductors, photoelectric cells and phosphors on screens, such as TV screens. Cadmium salts have been used as fungicides and in photographic processes. The metal is also used as the

TABLE 2. 3 Levels of Heavy Elements in Igneous and Sedimentary Rocks (ppm)

Element	Igneous		Sedimentary						
	Mean Basalt	Mean Granite	Mean Shale	Mean Limest.	Mean Sandst.	Mn Nodules	Phos-phorites	Oil	Mean Coal
Cd	0.13	0.09	0.22	0.028	0.05	8	0.01-2.5	0.03	<.01-22
Hg	0.012	0.08	0.18	0.16	0.29	0.5	0.2	0.05	.01-21
Pb	3	24	23	5.7	10	870	2.4	-	2-370
As	1.5	1.5	13	1	1	9-190	30	0.26	0.3-93
Sb	0.2	0.2	1.5	0.3	0.05	4-25	0.2-7	0.05	0.1-
Bi	0.031	0.065	0.48	0.17	0.18	8	<0.05-0.4	-	0.02-4
In	0.058	0.04	0.057	0.009	0.01	0.25	0.03	-	0.01-0.6
Tl	0.08	1.1	1.2	0.14	0.36	100	<0.03-1	-	0.01-2
Se	0.05	0.05	0.5	0.03	<0.01	-	1-300	0.3	0.04-10
Te	-	-	<0.1	-	-	48	-	-	<0.1-0.4

Sources of data; Bowen, 1979 [3], Wedepohl, 1978 [29].

principal material in the control rods of nuclear reactors. The natural occurring isotope ^{113}Cd (12.26% abundance) has a high thermal neutron capture cross section.

MERCURY

Sources

Mercury has a crustal abundance of 0.08 ppm (80 ppb), mainly associated with sulphur. The main ore is cinnabar, red HgO, from which mercury is extracted. Other significant ores are metacinnabar (Hg,Zn,Fe)(S,Se), corderite $Hg_2S_2Cl_2$, livingstonite $HgSb_4S_7$, montroydite HgO, terlinguaite Hg_2OCl, and calomel HgCl [16]. Isomorphous replacement of cations by mercury occurs in pyrite, quartz, calcite, dolomite and stignite. Deposits have formed from hydrothermal solutions, around hot springs or volcanic areas, at relatively low temperatures. The process involves the replacement of sandstone and limestone, entry into fissure veins, breccia filling, stockworks and pore space filling [13,15]. The richest mercury zones lie along plate boundaries, either subduction zones (e.g. west coast of Nth. America), or spreading ridges (e.g. Mid-Atlantic ridge) [5,14].

The average levels of mercury in basaltic rocks is 12 ppb, in granitic rocks 80 ppb, and in igneous rocks a range of 5 - 26 ppb (Table 2.3). Levels are higher in shales, reaching 400 ppb (0.4 ppm) with a mean of 0.18 ppm. In general levels are lower in sandstone and limestone, 30 and 16 ppb respectively. In metamorphic rocks a range of 0.002 - 2.5 ppm has been reported [1,3,29].

Extraction

Cinnabar is crushed and concentrated by flotation and roasted at 500-600°C in air to give mercuric oxide and then mercury.

$$2HgS + 3O_2 \rightarrow 2HgO + 2SO_2$$

$$2HgO \rightarrow 2Hg + O_2$$

The net reaction is;

$$HgS + O_2 \rightarrow Hg + SO_2.$$

Reduction of Hg(II) can also be achieved with iron or heating with lime;

$$HgS + Fe \rightarrow Hg + FeS$$

$$4HgS + 4CaO \rightarrow 4Hg + 3CaS + CaSO_4.$$

The metal is purified from solids by allowing it to pass through perforated paper or leather. Dissolved metals are removed by passing a thin stream of the metal through dilute nitric acid, and then either distilling or electrolysing [6,11,15]. Mercury production is measured in flasks, and one flask contains 34.5 kg. Production increased until the early 1970's, but has dropped off significantly (Fig. 2.2) [1,16].

Uses

The production figures (Fig. 2.2) indicate a declining use of mercury, which arises from concern over its toxicity to human beings, and declining mineral resources. The main use of mercury, though declining, is as the mobile cathode in the chloroalkali cell, for the production of NaOH and Cl_2 from brine. Two separate cell reactions occur [4,20,26].

Electrolyser cell:

at carbon cathode: $Cl^-(aq) \rightarrow 1/2Cl_2\,(g) + e$, $-E^\circ = 1.358$ V

at mercury anode: $Na^+(aq) + e + xHg \rightarrow NaHg_x$, $E^\circ = -1.868$ V

Decomposer cell:

at $NaHg_x$ cathode: $NaHg_x \rightarrow Na^+(aq) + e + xHg$

at carbon anode: $H_2O + e \rightarrow 1/2H_2 + OH^-(aq)$

Caustic soda is produced in the last cell, and the mercury recycled back to the first cell. In the past large quantities of mercury were lost to the environment from these cells. The cell is being replaced by diaphragm and membrane cells, which are less polluting and run at lower voltages.

Mercury is used in electrical and measuring apparatus, such as mercury discharge lamps, power rectifiers, mercury batteries, thermometers, barometers and electrical switches [15,16,26]. In the Zn/Hg cells the cathode is HgO

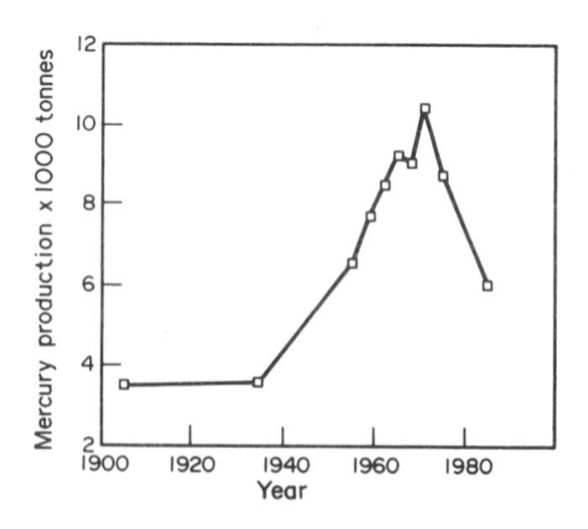

Fig. 2. 2 Production of mercury. Sources of data; references 1,16.

and the anode is Zn, and the cell reaction is:

$$Zn + HgO \rightarrow ZnO + Hg,$$

with KOH saturated with ZnO as the electrolyte.

A number of mercury compounds have fungicidal activity, and have been used in paints and as seed dressings. This use of mercury is now banned in many countries, especially the use of methylmercury compounds. Phenylmercury acetate, oleate or dodecenylsuccinate are used in paints to prevent the growth of fungi and mildew. The compounds are however subject to photochemical decomposition. Similar compounds are used in the pulp and paper industry as slimicides [15,16,26].

Mercury salts are used as catalysts in the industrial production of vinyl chloride, vinyl acetate and acetaldehye from acetylene [15,16,26].

$$C_2H_2 + HCl \xrightarrow[HgCl_2]{} CH_2{=}CHCl$$

$$C_2H_2 + H_2O \xrightarrow[HgSO_4]{} CH_3CHO$$

In the last reaction Hg(II) is reduced to Hg, but is regenerated with iron(III). The serious outbreaks of mercury poisoning, resulting from effluents from factories using these reactions, will be considered in Chapter 15. Mercury is also used as a catalyst in the dyestuff industry, e.g.

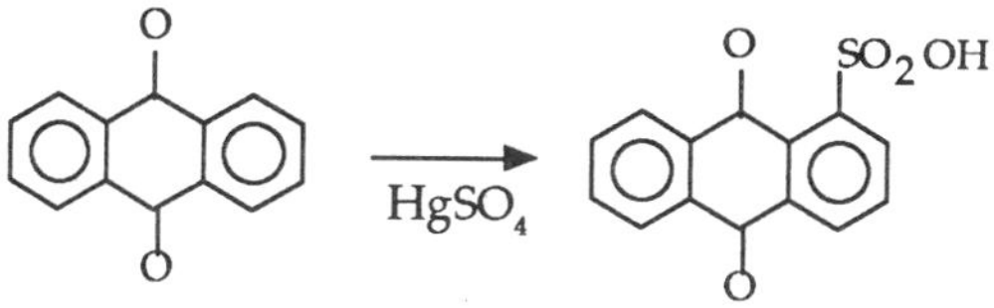

O O SO_2OH $HgSO_4$ O O

Many people carry mercury around in their mouths as dental fillings [27]. A dental alloy of Ag (66.7-74.5%), Sn (25.3-27.0%), Cu(0.0-6.0%) and Zn (0.0-1.9%) is mixed with elemental mercury (approximately 1:1 weight ratio) in the dentists' rooms to give a dental amalgam, a paste which soon hardens in the tooth cavity. The dental alloy Ag_3Sn forms two non-stoichiometric phases when mixed with mercury,

$$(6+x)Ag_3Sn + 29Hg \rightarrow 9Ag_2Hg_3 + 2Sn_3Hg + xAg_3Sn$$

The two amalgam phases gradually form as the paste hardens, Ag_2Hg_3 being the harder phase. The final solid matrix is strong, resistant to abrasion, adheres strongly to the tooth, is of very low solubility and impermeable to saliva.

The antiseptic and preservative qualities of some mercury compounds ($HgCl_2$, HgO, $Hg(CN)_2$, $HgNH_2Cl$, HgI_2, HgCl and organo-compounds such as o-chloromercuriphenol and mercury oleate) are still made use of in pharmaceuticals and cosmetics, even though the toxicity of mercury is well recognized [5,15,16,26]. Mercuric oxide is used in eye ointments to treat irritation, and HgI is used to treat skin diseases. Organomercury compounds have also been employed as diuretics and in the treatment of syphilis, though the latter has been superseded by antibiotics.

LEAD

Sources

There are over 200 lead minerals known, but only three are significant, galena, PbS, cerussite, $PbCO_3$ and anglesite, $PbSO_4$, and of these galena is used for the extraction of lead. Galena crystallized from residual molten magma in fissure veins and in replacement bodies. The carbonate and sulphate ores have been formed from the sulphide by weathering processes. Lead also occurs with zinc in sphalerite, and with copper in calcopyrite, as well as by isomorphous replacement of K, Sr, Ba, Ca and maybe Na ions in various lattices. The average levels of lead in rocks are given in Table 2.3, the high levels in coal and organic shales are presumably due to complexes formed with organic compounds [1,16,22].

Extraction

The three main steps in the extraction of lead are; concentration of the sulphide ore by flotation, roasting to give the oxide, followed by reduction. After Al, Cu, and Zn, lead is the fourth most important non-ferrous metal produced [1].

Roasting is carried out at around 600° C, and numerous reactions occur, including;

$$2PbS + 3O_2 \rightarrow 2PbO + 2SO_2,$$ (the main reaction)

$$2SO_2 + O_2 \rightarrow 2SO_3,$$

$$PbO + SO_3 \rightarrow PbSO_4,$$

$$PbS + 2PbO \rightarrow 3Pb + SO_2,$$

$$PbS + PbSO_4 \rightarrow 2Pb + 2SO_2,$$

$$PbO + SiO_2 \rightarrow PbSiO_3,$$

Smelting commences around 400° C, the main reaction being:

$$PbO + CO \rightarrow Pb + CO_2,$$

but if any iron is present the two reactions below also occur:

$$PbS + Fe \rightarrow Pb + FeS,$$

and

$$PbO + Fe \rightarrow Pb + FeO.$$

The crude lead produced contains varying amounts of Cu, Ag, Au, Zn, Sn, As and Sb. Copper is removed by liquefying the lead and the solid copper floats to the top. The three metals Sn, As and Sb are removed as solid oxides after oxidation with O_2, called softening of the lead. A mixture of NaOH and $NaNO_3$ also achieves the oxidation. If the melt is cooled slowly from 480 to 420° C the Au and Ag dissolve preferentially in the zinc in the lead which then crystallizes out from the melt. Zinc residue may then be removed by treatment with Cl_2 or by vacuum distillation. Final purification of the lead is achieved electrolytically in $PbSiF_6$ as electrolyte, and if required, by zone refining [11,16,22].

Uses

The consumption of lead has risen dramatically during this century and the data presented in Fig. 2.3 [24], shows a significant increase over the last 30 years. The automobile industry uses a large proportion of the lead, in car batteries, petrol additives and solder.

The principal use of lead is in the lead accumulator battery. Even though the battery has a poor current density relative to its weight, this is not a limiting factor for a car [7,11,17,22]. The anode is lead and the cathode is a paste of lead oxides, PbO litharge, Pb_3O_4 red lead and PbO_2 grey oxide. The paste is set into

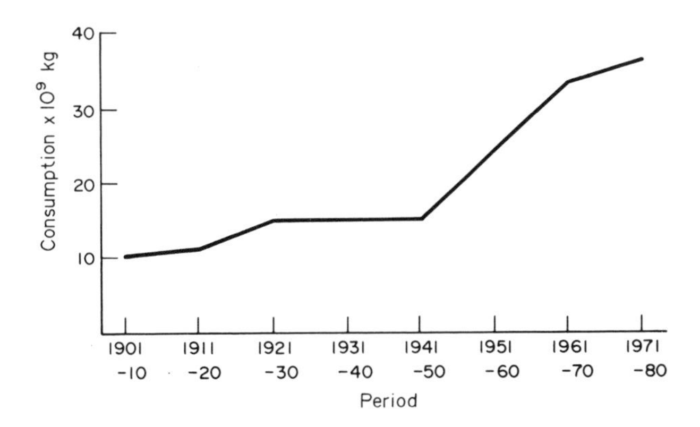

Fig. 2. 3 Lead consumption 1901-1980. Source of data; Nriagu, 1979 [24].

a grid of a lead/antimony alloy (91% Pb, 9% Sb) and the electrolyte is sulphuric acid. The cell reactions when being used are:

$$\text{anode}\ Pb(s) + SO_4^{2-} \rightarrow PbSO_4(s) + 2e$$

$$\text{cathode}\ PbO_2(s) + 4H^+ + SO_4^{2-} \rightarrow PbSO_4(s) + 2H_2O - 2e$$

the net reaction being:

$$Pb(s) + PbO_2(s) + 4H^+ + 2SO_4^{2-} \rightarrow 2PbSO_4(s) + 2H_2O$$

which gives 2.01 volts [7,11,16,22].

Tetraethyl- and/or tetramethyllead have been added to petrol from the mid 1920's to increase the octane rating of the fuel. This is a declining use of lead, as more countries reduce or eliminate the additive. If the octane level of petrol is not sufficiently high, a process called knocking occurs in the engine, which can damage the cylinders. Just before the end of the compression stroke remaining gas can ignite rapidly giving the knock, and at high speeds the knocking can be dangerous, as it may not be heard, and will damage the engine. The pre-ignition arises from free radicals in the mixture. The free radicals are reduced by using hydrocarbons with a significant amount of branched chains, or scavenged by the use of tetraalkyllead compounds.

$$\text{e.g.}\ PbO_2 + RCH_2\cdot \rightarrow PbO + RCH_2O\cdot \xrightarrow{O_2} RCHO$$

The use of lead alkyls increases the octane rating without the need to use expensive branched chain hydrocarbons. The lead is converted to PbO and PbO_2 which is then transformed into volatile $PbCl_2$, $PbBr_2$, and PbBrCl by the addition of dichloro- or dibromoethane to the petrol [17,25].

Lead has a number of properties, such as corrosion resistance, sound absorption, radioactive radiation absorption, anti-friction and low melting that make it useful as the metal. It is used in the building industry for roofing and flashing, and for sound proofing. Lead water pipes, are now only used under special circumstances. The metal is used for cable sheathing, for ammunition, and it is an effective shield against high energy radiation [11,16,17,22].

Lead is used in a number of alloys, particularly low melting alloys such as solder. Solder contains lead (27 - 97%, typical 67%) and tin (3 - 63%, typical 33%), it is widely used, though not now in food containers. Type metal is lead (62 - 94%), tin (3 - 14%) and antimony (3 - 24%), the proportions depending on its application. Other alloys of lead that have uses are; Pb/Cu, Pb/Ca, Pb/Sr, Pb/Sb, Pb/Ag, Pb/Te, babbitt Pb/Sn/Cu or Pb/Sn/Sb/Cu a heavy duty bearing metal and Pb/Sn/Sb pewter [11,16,17,22].

Another declining use of lead, because of its toxicity, is in paint. Lead paints have good covering and anticorrosive properties. Some lead compounds used as pigments are red lead Pb_3O_4, white lead $2PbCO_3.Pb(OH)_2$ in rust resistant paint, $PbCrO_4$, a yellow pigment used in road markings, Ca_2PbO_4, for painting on galvanized iron, $PbMoO_4$ and PbO. Because of the high levels of lead in the dried paint film it is necessary to ensure that the paint is not used where children may consume it. Unfortunately, this has not been the situation in the past, and a great deal of lead paint, on and in houses, is still around. Lead pigments have also been used in coloured pictures in magazines and children's books [10,11, 16,17,22].

Lead is used in ceramic glazes, in glass (crystal), as a PVC stabilizer, for lead toys and leaded windows. Lead containing insecticide are used much less today, PbO removes sulphur from petroleum during refining, and lead compounds (e.g. lead acetate) are used in mixtures to darken the hair.

ARSENIC, ANTIMONY AND BISMUTH

The three elements arsenic, antimony and bismuth have a long history, and have been used in a variety of ways over the centuries.

Sources

A list of the more common minerals of the three elements is given in Table 2.4. Deposits of the native elements also occur. The elements are associated with gold, silver, cobalt, nickel, lead, zinc and copper in their ores [6,11,28]. The highest levels of the elements in rocks occur in sedimentary materials especially shale, coal, phosphorites and Mn nodules (Table 2.3) [3,29], because of the organic content of some of the materials.

Extraction

The elements are mainly obtained as by-products of the extraction of lead, zinc and copper. Only in a few cases are ores of the elements used directly for their extraction.

Arsenic Arsenic can be obtained from two of its ores, arsenopyrite and lollingite by smelting in the presence of air around 650-700° C [11,16,28],

$$4FeAsS \rightarrow 4FeS + As_4,$$

$$2FeAs_2 + O_2 \rightarrow 2FeO + As_4.$$

However, the principal source is the As_2O_3 in the flue dust from the extraction of lead and copper [6,11,16,28]. The addition of galena or pyrite prevent the formation of arsenites. The oxide is used to make either arsenic compounds or, by reduction with charcoal at 500-800° C, elemental arsenic.

TABLE 2.4 Minerals of Arsenic, Antimony and Bismuth

Arsenic	Antimony	Bismuth
Arsenopyrite, FeAsS	Stibinite, Sb_2S_3	Bismuth glance, Bi_2S_3
Lollingite, $FeAs_2$	Valentinite, Sb_2O_3	Bismite, Bi_2O_3
Orpiment, As_2S_3	Senarmontite, Sb_2O_3	Bismuth ocher, $Bi_2O_3.3H_2O$
Realgar, As_4S_4	Cervantite, Sb_2O_4	Bismutosphaerite, $Bi_2(O_2/CO_3)$
Chloanthite, NiAs2	Stibiconite, Sb2O4.H2O	
Niciolite, NiAs	Kermesite, $2Sb_2S_3.Sb_2O_3$	
Smalite, $CoAs_2$	Jamesonite, $Sb_2S_3.2PbS$	
Cobaltite, CoAsS	Livingstonite, $2Sb_2S_3.HgS$	
Gersdorffite, NiAsS	Ullmanite, NiSbS	
Tennantite, $4Cu_2S.As_2S_3$	Tetrahaedrite, Cu_3SbS_3	
Proustite $3Ag_2S.As_2S_3$	Wolfsbergite, $CuSbS_2$	
Enargite $3Cu_2S.As_2S_5$		

$$2As_2O_3 + 3C \rightarrow As_4 + 3CO_2.$$

Arsenic(III) oxide is purified by sublimation or recrystallisation from water as it has a steep increase in solubility with temperature [16,28]. A purity of <1 in 10^6 is necessary for semiconductor use. The arsenic is sublimed from a solution in lead which retains the sulphur. Arsenic is then crystallized under pressure and converted to arsine (AsH_3) which is scrubbed with alkali, dried and pyrolysed at 600 °C to give back the arsenic [6,28].

Antimony The extraction method for antimony depends on the concentration of the element in the source. If the concentration is in the range 5 - 25%, the Sb ores are roasted and Sb_2O_3 collects in the flue dust. The reaction approximates to;

$$2Sb_2S_3 + 9O_2 \xrightarrow[500\,^\circ C]{} 2Sb_2O_3 + 6SO_2.$$

The oxide is reduced using alkali metal carbonates or sulphates as fluxes. If the antimony concentration is in the range 25 - 40%, the ores are smelted and the metal (and oxide) collected in the flue dusts. Finally in the concentration range 40 - 60%, the ore body is liquefied at around 550-600 °C from which the sulphide Sb_2S_3 crystallizes out. The metal is then obtained by treating with scrap iron,

$$Sb_2S_3 + 3Fe \rightarrow 2Sb + 3FeS.$$

In more complex systems the ores are leached with reagents such as HCl or alkali metal sulphides. In the latter case the antimony is removed as Na_3SbS_4, from which the metal is obtained by electrolysis [6,11,16,28].

The impurities in antimony (such as Fe, As, Sn, Zn, Cu, Pb and Bi) need to be removed if it is to be used in semiconductors. Lead is difficult to remove and is done by electrolysis, or treatment with chloride or sulphide. However, this may not be necessary if the antimony is to be used in lead alloys. Iron, arsenic, tin, copper and sulphur can be removed by oxidation in alkali, the antimony remaining unoxidised. Very pure antimony is obtained by subliming the metal from an alloy of antimony containing 1-2% Al and 0.5-4% Mn to get rid of Sn, S, Pb, and As, which remain behind. Antimony containing 0.1% Al may be zone refined, and a ternary phase of Sb/Al/As lowers the melting point so that arsenic remains in the liquid phase. An intermediate step is to produce an antimony compound, e.g. $HSbCl_6.9H_2O$, which can be purified and reduced in pure dihydrogen to give the metal [6,11,16,28].

Bismuth As for antimony, the extraction method for bismuth depends on the source. The principal source occurs in flue dusts, from the smelting of lead, zinc or copper. It is also obtained from the anode slime from the electrolytic processing of copper. Bismuth rich concentrates are leached with hydrochloric acid for extracting the metal.

$$Bi_2S_3 + 6HCl \rightarrow 2BiCl_3 + 3H_2S.$$

The chloride is treated with iron to obtain bismuth, i.e.

$$2BiCl_3 + 3Fe \rightarrow 2Bi + 3FeCl_3,$$

or hydrolysed to give $Bi(OH)Cl_2$, from which BiOCl is obtained. This material and bismuth metal are then smelted with sodium sulphide. In smelting lead the bismuth stays with the lead. In order to concentrate the bismuth a Ca/Mg alloy is added to give a ternary alloy Bi/Ca/Mg, which has a higher melting point than lead and crystallizes out. Chlorination of the ternary material liberates the bismuth, but still contains some lead. Electrolytic separation of the two metals deposits bismuth in the anode slime. Treatment of the Pb/Bi alloy with molten caustic soda removes arsenic and tellurium, and silver and gold are removed by desilverization. Chlorination, of the remaining material, removes zinc and lead as chlorides leaving pure bismuth. Very pure bismuth for use in semiconductors is produced by reduction of bismuth oxide obtained from recrystallized $Bi(NO_3)_2$. The metal is then vacuum distilled and zone refined [6,11,16,28].

Uses

One of the main uses of the three elements is in non-ferrous alloys, especially alloyed with lead. Around 0.5% As and 2.5-3% Sb are added to lead in the grids in the lead accumulator battery. The antimony imparts fluidity, creep resistance, fatigue strength and electrochemical stability to the alloy, whereas the arsenic minimizes self-discharge, gassing and the poisoning of the negative electrode. Arsenic (0.5%) added to copper improves its high temperature stability, corrosion resistance and recrystallization temperature. The spherical shape of lead shot is improved by the addition of 0.5-2% arsenic, and antimony increases the hardness of the lead ammunition. All three elements are usually in type metal, arsenic improves the hardness and castability, antimony increases the hardness, minimizes shrinkage, gives a sharp definition, and lowers the melting point, whereas bismuth has the effect of increasing the volume of the type metal when it solidifies, which means the alloy casts with sharp edges. Arsenic and antimony are in solder used on the bodies of automobiles (92% Pb, 5% Sb, 2.5% Sn and 0.5% As). Antimony gives alloys the property of precision in duplication, durability and metallic lustre, as found in pewter. Antimony is added to copper in naval brass, admiralty metal and muntz metal. Low temperature fusible alloys (melting range from 47 to 262° C) which contain bismuth (In/Cd/Pb/Sn/Bi), are used in fire sprinkler systems and other safety devices, and for holding lens while being ground [11,16,28].

The three metals of very high purity are used in III-V semiconductors. These binary compounds, such as GaAs, InAs, AlSb, GaSb, and InSb will be discussed in more detail in Chapter 3 [6,11,16].

The low absorption cross section of bismuth for thermal neutrons is made use of in liquid-metal fission reactors, and as windows for neutron irradiation in medicine [16].

Arsenic compounds The use of arsenate pigments (e.g. copper arsenate, green) and pesticides (calcium and lead arsenates) have now been discontinued because of the toxicity of the element. Some compounds however, (e.g. $NaMeHAsO_3$, Na_2MeAsO_3 and $Me_2AsO(OH)$) are still used in agriculture for insect and weed control, in cotton, coffee and rice growing areas. The pentavalent compound $AsO(OH)_3$ is used as a desiccant for the defoliation of cotton bolls prior to harvesting, and as a wood preserver. Sodium arsenate is used as a sheep and cattle dip, and for aquatic weed control [1,6,11,16,28].

The oxide As_2O_3 decolourises bottle (green) glass, and As_4S_4 gives a red colour to glass, it is also used as a depilatory in the production of fine leather. Organoarsenic compounds are used in the treatment of trypanosomal diseases (sleeping sickness), amebiasis and non-parasitic skin diseases in animals [6,16,28].

Antimony compounds Antimony trichloride is used as a catalyst in some organic chlorinations and polymerisations [28]. The halide, oxide and organoantimony compounds are used for making textiles flame resistant [10,28]. The antimony(III) oxide is used as a white pigment in glass, and the trisulphide in pryotechnics, for certain types of matches and in camouflage paint. When burnt

Sb_2S_3 gives a dense smoke and is used for markers and signalling [28]. The pentasulphide is used in the vulcanizing of rubber [28]. Some antimony compounds have application in medicine, such as potassium antimony tartrate in treating schistosomiasis, ethylstibamine for treating oriental sore, and antimony(V) sodium gluconate as an antileishmanial agent [16].

Bismuth compounds Some bismuth compounds have uses in pharmaceuticals, for the treatment of amoebic dysentery, diarrhea, gastric disturbances and gastric and duodenal ulcers [16].

INDIUM AND THALLIUM

Sources

Indium and thallium are associated with the ores of other elements, e.g. indium with zinc sulphide and thallium with lead sulphide. The ionic radius of thallium(I) is similar to that of rubidium and it is concentrated with rubidium in late magmatic potassium minerals, such as feldspars and micas. The few thallium minerals are rare, e.g., crookesite, $(Cu, Tl, Ag)_2Se$; lorandite, $TlAsS_2$; urbaite, $TlAs_2SbS_5$; and hutchinsonite, $(Tl, Cu, Ag)_2S.PbS.2As_2S_3$ [6,11].

Extraction

Both indium and thallium are obtained from the flue dusts produced from the roasting of zinc and lead sulphides. Thallium is also obtained when sulphides are roasted for the production of sulphuric acid. Indium is leached out of the dusts with sulphuric or hydrochloric acids, and the addition of zinc or aluminium produces indium sponge, which may be refined electrolytically from a halide solution. Volatile materials may be removed by heating under vacuum. The thallium in flue dusts occurs as the oxide or sulphate, and the thallium is leached out with water or acid, and then precipitated as TlCl. Pure thallium is obtained from the electrolysis of Tl(I) sulphate. The production of the two elements is not high, around 75 tonnes y^{-1} of indium and 5 tonnes y^{-1} of thallium [1,6,11,16].

Uses

Indium alloys, which have good resistance to alkali, are used for joining metals to non metals, and in low melting devices, e.g., heat regulators, sprinkler systems, surgical casts, in pattern making and lens blocking. The metal increases hardness, tarnish and corrosion resistance, and anti-seizure properties of alloys. Therefore indium alloys are used for heavy duty bearings and high speed devices, and indium coatings protect metals against wear and tear. Indium has a high thermal neutron capture cross section, and is alloyed with Cd and Ag for making control rods for nuclear reactors (5% In, 15% Ag, 80% Cd). An alloy of 24% In and 76% Ga has a melting point of 16° C, and is used in nuclear reactors to circulate γ-ray activity [6,11,16].

Indium III-V semiconductors (see Chapter 3) are used in low temperature transistors, thermistors and photoconductors (InAs, and InSb), whereas InP is used for high temperature transistors [6,11,16].

Thallium confers to alloys high endurance, a very low coefficient of friction and good resistance to acids (e.g. 72% Pb, 15% Sb, 8% Tl and 5% Sn, and Ag/Tl and Au/Al/Tl). An alloy containing 70% Pb, 20% Sn and 10% Tl has negligible solubility and is used as the anode in the electrolytic extraction of copper. A Tl/Hg alloy (8.7 wt % Tl) has a melting point of -60° C and is a substitute for mercury in switches and seals in polar regions [1,6,11,16].

Thallium(I) bromide and iodide are transparent to long wavelength infrared radiation, and are used in infrared spectroscopy. Aqueous solutions of thallium formate and malonate have high densities (4.3 g ml $^{-1}$) and are used in the density separation of minerals. Incompletely oxidized Tl_2S is a better photocell than selenium in the low wavelength region. Thallous oxide confers on glass a high refractive index [1,6,11,16].

SELENIUM AND TELLURIUM

Sources

Both elements occur in the sulphide minerals of Ag, Cu, Pb, Hg, and Ni, in which the sulphur is replaced by selenium or tellurium. Primary sources of selenium are associated with volcanic activity, and metallic sulphides linked with igneous activity. There are around 50 selenium minerals, e.g. clausthalite, PbSe; crookesite $(Cu, Tl, Ag)_2Se$; eucairite, $(Cu, Ag)_2Se$; and naumannite, (Ag, Pb)Se, and also minerals containing oxidized selenium, e.g. $MSeO_3.2H_2O$ (M = Ni, Cu, Pb). Tellurium is less abundant than selenium, and examples of its minerals are calaverite, $AuTe_2$; nagyagite, $Au_2Sb_2Pb_{10}Te_6S_{12}$; petzite, Ag_3AuTe_2; and sylvanite, $(Au, Ag)Te_2$ [1,6,11,16].

Extraction

The two elements are obtained from the anode slime from the electrolytic refining of copper, and selenium from the sludge in sulphuric acid plants, and dusts from Cu and Pb smelting. Recovery from their minerals is not possible because they are so rare [1,6,11,16].

The selenium and tellurium in the anode slimes occur mainly as copper and silver selenides or tellurides, e.g. CuAgSe, Ag_2Se and Cu_2Te. Roasting of these materials with sodium carbonate around 650 - 700° C oxidizes the elements to selenite and tellurite,

$$Ag_2Se + Na_2CO_3 + O_2 \rightarrow 2Ag + Na_2SeO_3 + CO_2,$$

$$Cu_2Te + Na_2CO_3 + 2O_2 \rightarrow 2CuO + Na_2TeO_3 + CO_2.$$

The selenite and tellurite are extracted with alkali followed by acidification which precipitates the tellurium as $TeO_2.nH_2O$, and the selenium which remains in solution is obtained by reduction with sulphur dioxide,

$$H_2SeO_3 + 2SO_2 + H_2O \rightarrow Se + 2H_2SO_4.$$

Crude selenium is converted to hydrogen selenide by heating in dihydrogen at 650° C, and the reaction is reversed at 1000° C. Hydrogen sulphide does not decompose at the higher temperature, whereas the hydrides of the elements Te, P, As and Sb, do not form at the lower temperature. Pure tellurium is obtained similarly if the tellurium hydride is allowed to decompose at 300° C. Other purifications of selenium are achieved by subliming SeO_2, followed by reduction with ammonia,

$$3SeO_2 + 4NH_3 \rightarrow 3Se + 2N_2 + 6H_2O,$$

or distillation of $SeBr_4$ from concentrated HBr solution. The latter method separates selenium from tellurium, but not from arsenic or germanium. The reversible reaction,

$$Se + Na_2SO_3 \rightleftharpoons Na_2SeSO_3,$$

gives pure selenium on cooling the solution or adding sulphuric acid. Finally both elements can be purified by zone refining [1,6,11,16].

Uses

Selenium Selenium counteracts the coloration of glass due to iron, and if more is added the colour will become pink. Solid particles of Cd(S/Se) in glass give ruby red glass. Selenium increases the heat and wear resistance of rubber, and the rate of vulcanization. Selenium is also used as an anti-oxidant in inks, mineral oils and lubricants [1,6,11,16,21].

Selenium is used widely in the electronics industry, in photoelectric cells, and in xerography. Change in light intensity causes a variation in the electric current in selenium, the current increasing with light intensity. Crystalline grey selenium, which has an infinite helical chain structure, with no side branches, is the allotrope of selenium with this property. It is obtained from heating other modifications, or slowly cooling a selenium melt, or cooling a saturated solution of amorphous selenium in aniline, or condensing selenium vapour close to its melting point. The same allotrope is used in ac rectification, making use of asymmetric conduction of thin layers of the element.

The interaction of selenium and light is made use of in photocopying. A thin film of selenium on aluminium is a photoreceptor, and is sensitized with an electrostatic charge by a corona discharge. When exposed to light the charge is reduced, but remains in dark areas (i.e.the image of what is being copied). The image of charge is developed by passing a toner (black particles) over the receptor, the particles become attached to the charged area. The particles are then transferred to charged paper and the image fused into the paper with heat. Excess toner is removed and the image on the photoreceptor removed by uniform flooding with light [1,6,11,16,21].

Selenium is added to metals to improve their stability and machine-ability. It also colours copper alloys, and increases corrosion resistance of stainless steel, and decreases the surface tension of liquid steel [1,6,11,16,21].

Low levels of selenium are essential for animals and probably human beings. It is added to livestock feeds at a level of 0.1 - 2 ppm to prevent muscular dystrophy and white muscle disease. Selenium sulphides are used in the treatment of dandruff, and fungal infection. Selenium dioxide is a catalyst in the preparation of some drugs, such as hormones and vitamins [16,21].

Tellurium Most tellurium is used in steel to improve its machine-ability, and in non-ferrous alloys. It is used as a secondary vulcanizing agent, and like selenium is an anti-oxidant in lubricating oils and greases [6,11,16].

REFERENCES

1. Adriano, D. C. Trace Elements in the Terrestial Environment: Springer Verlag; 1986.
2. Berman, E. Toxic Metals and their Analysis: Heyden; 1980.
3. Bowen, H. J. M. Environmental Chemistry of the Elements: Academic Press; 1979.
4. Brooks, W. N. The chloroalkali cell: from mercury to membrane. **Chem. in Brit.**; 1986; **22**: 1095-1098.
5. D'Itri, P. A, and D'Itri, F. M. Mercury Contamination: A Human Tragedy: Wiley-Interscience; 1977.
6. Editorial Board. Comprehensive Inorganic Chemistry: Pergamon Press; 1973.
7. Fergusson, J. E. Inorganic Chemistry and the Earth: Pergamon Press; 1982.
8. Fleisher, M., Sarofim, A. F., Fassett, D. W., Hammond, P., Shacklette, H. T., Nisbet, I. C. T. and Epstein, S. Environmental impact of cadmium: a review by the panel on hazardous trace substances. **Environ. Health Perspec.**; 1974: 253-323.
9. Forstner, U. Cadmium. in: Huntzinger, O., Ed. Handbook of Environmental Chemistry: Springer Verlag; 1980; **3A**: 59-107.
10. Funk, W. Inorganic pigments. in: Hutzinger, O., Ed. Handbook of Environmental Chemistry: Springer Verlag; 1980; **3A**: 217-229.
11. Greenwood, N. N. and Earnshaw, A. Chemistry of the Elements: Pergamon Press; 1984.
12. Hem, J. D. Chemistry and occurrence of cadmium and zinc in surface water and ground water. **Water Resources Res.**; 1972; **8**: 661-679.

13. Jensen, M. L. and Bateman, A. M. Economic Mineral deposits. 3rd. Ed.: Wiley; 1981.
14. Jonasson, I. R. and Boyle R. W. Geochemistry of mercury and origins of natural contamination of the environment. **Canad. Mining Met. Bull.**; 1972; **65**: 32-39.
15. Kaiser, G. and Tolg, G. Mercury. in: Hutzinger, O., Ed. The handbook of Environmental Chemistry: Springer Verlag; 1980; **3A**: 1-58.
16. Kirk-Othmer. Kirk-Othmer Encyclopedia of Chemical Technology. 2nd. Ed.: Wiley-Interscience; 1964.
17. Landsdown, R. and Yule, W. The Lead Debate: The Environment, Toxicology and Child Health: Croom Helm; 1986.
18. Mason, B and Moore, C. B. Principles of Geochemistry. 4th. Ed.: J. Wiley; 1982.
19. Mining Annual Review; 1986.
20. Moore, J. W. and Moore, E. A. Environmental Chemistry: Academic Press; 1976.
21. Newland, L. W. Arsenic, beryllium, selenium and vanadium. in: Hutzinger, O., Ed. Handbook of Environmental Chemistry: Springer Verlag; 1982; **3B**: 27-68.
22. Newland, L. W. and Dawn, K. A. Lead. in: Hutzinger, O., Ed. Handbook of Environmental Chemistry: Springer Verlag; 1982; **3B**: 1-26.
23. Nriagu, J. O. Global inventory of natural and anthropogenic emissions of trace metals. **Nature**; 1979; **279**: 409-411.
24. Nriagu, J. O. Production, uses and properties of cadmium. in: Nriagu, J. O., Ed. Cadmium in the Environment, Part I, Ecological Cycling: Wiley; 1980.
25. Pyle, J. L. Chemistry and the Technological Backlash: Prentice Hall; 1974.
26. Saito, N. Uses of mercury and its compounds in industry and medicine. in: Mercury Contamination in Man and his Environment, **Tech. Report**: IAEA; 1972; No. 1335-42.
27. Treptow, R. S. Amalgam dental filling, Part II The chemistry and a few problems. **Chemistry**; 1978; **51**: 15-19.
28. Ullmann, F. Ullmann's Encyclopedia of Industrial Chemistry. 5th. Ed.: VCH; 1985.
29. Wedepohl, K. H., Ed. Handbook of Geochemistry: Springer Verlag; 1978.

CHAPTER 3

CHEMISTRY

The ten elements considered in this text are p-block elements and we will discuss some of their chemical features relevant to this study. Some properties of the elements are given in Table 3.1, and these will be discussed at the appropriate place. There are numerous sources of information on the chemistry of the elements, some of the more recent monographs are: [4,7,9,12,16,18,21,24,27,33,35].

THE ELEMENTS

Electronic Structure

The elements are members of Groups IIB (Cd, Hg), IIIB (In, Tl), IVB (Pb), VB (As, Sb, Bi) and VIB (Se, Te), or in the new IUPAC system for labelling the periodic groups 12, 13, 14, 15 and 16 respectively. Arsenic and selenium are in the first long period, cadmium, indium, antimony and tellurium in the second long period whereas the rest are in the third long period (Fig. 3.1). The electronic structure of the elements in the first and second long periods is represented by the symbol $[IG](n\text{-}1)d^{10}ns^2np^x$, where [IG] is the electronic structure of the inert gases argon and krypton for n = 4 and 5 respectively. The value of x is the Group

IIB (12)	IIIB (13)	IVB (14)	VB (15)	VIB (16)	Long Period	n
Zn	Ga	Ge	**As**	**Se**	First	4
Cd	**In**	Sn	**Sb**	**Te**	Second	5
Hg	**Tl**	**Pb**	**Bi**	Po	Third	6

Fig. 3.1 The heavy metals of the p-block.

TABLE 3.1 Some Properties of the Heavy Elements

Property	Cd	Hg	In	Tl	Pb	As	Sb	Bi	Se	Te
Atomic number	48	80	49	81	82	33	51	83	34	52
Valence shell structure	$5s^2$	$6s^2$	$5s^25p^1$	$6s^26p^1$	$6s^26p^2$	$4s^24p^3$	$5s^25p^3$	$6s^26p^3$	$4s^24p^4$	$5s^25p^4$
Atomic weight	112.41	200.59	114.82	204.37	207.2	74.922	121.75	208.98	78.96	127.6
Density (g cm^{-3})	8.65	13.534	7.31	11.85	11.342	5.78	6.68	9.80	4.82	6.25
Melting point (°C)	320.8	-38.9	156.6	303.5	327	816a	630.7a	271.4a	217	452
Boiling point (°C)	765	357	2080	1457	1751	615sub	1587	1564	685	990
Ionisation energy 1st.	868	1007	558	589	716	947	834	703	941	869
(kJ $mole^{-1}$) 2nd.	1631	1810	1821	1971	1450	1798	1595	1610	2045	1790
3rd.	3616	3300	2750	2878	3082	2736	2440	2466	2974	2698
Electronegativity	1.69	2.00	1.78	2.04	2.33	2.18	2.05	2.02	2.55	2.1
Atomisation energy (kJ $mole^{-1}$)	112	61	243	182	196	302	262	207	227	197
Oxidation states	2	1,2	1,3	1,3	2,4	-3,3,5	-3,3,5	-3,3,5	-2,2,4,6	-2,2,4,6
No. stable isotopes	8	7	2	2	4	1	2	1	6	8
E°, M^{n+} + ne Æ M (volt) (n)	-0.40 (2)	0.79 0.85 (1,2)	-0.34 (3)	0.72 -0.34 (3,1)	-0.13 (2)					
Atomic radius (pm)	154	157	166	171	175	139	159	170	140	160
Covalent radius (pm)	144	147	144	155	154	121	141	148	117	137
Ionic radius (pm) (n+) (6 coord.)	95(2)	119(1) 102(2)	80(3)	150(1) 89(3)	119(2) 78(4)	58(3)	78(3)	103(3)	198(-2)	221(-2)
Elemental structure	hcp	liquid	ccp (dist.)	hcp	ccp	As_4, polymer	Sb_4, polymer	Bi_4, polymer	Se_8, chains	chains

number (Roman numeral) - 2. For the elements of the third long period the electronic structure is [Xe]$4f^{14}5d^{10}6s^26p^x$. The elements of the first and second long periods have a filled d-shell just below the valence shell, whereas the elements of the third long period have filled f- and d-shells. These filled f- and d-shells have an influence on the chemistry of the elements.

Elemental Structure

The structures of the elements are listed in Table 3.1. The elements of Groups 12, 13 and 14 have typical metallic structures, (some with distortion). Mercury is liquid at STP, but the usual form of solid mercury (α) is distorted cubic close packed, making it rhombohedral. The distortions in the metallic structures produce less dense materials. This may be due to weak metallic bonding involving the valence s- and p-electrons, compared with the transition metals which also use their d-electrons. The low atomization energies of the five metals (Cd, Hg, In, Tl and Pb) reflect this bond weakness [12,35].

The Group 15 (VB) and 16 (VIB) elements have elemental structures more typical of non-metals. In the vapour state the Group 15 elements occur as tetrameric molecules As_4, Sb_4 and Bi_4. The 'metallic' form of the elements are puckered layers, each atom has three close neighbours arranged pyramidally which are intra-layer, and three neighbours further away which are inter-layer. The distinction between the two groups of neighbours is less for Sb and Bi, as shown by the ratio of the two distances (in pm).

M	3 at	3 at	Ratio
As	252	312	1.24
Sb	291	336	1.15
Bi	307	353	1.15

Arsenic has two other polymorphs, one polymeric form is similar to black phosphorus. Antimony has five other polymorphs, and bismuth a number of ill defined forms. Grey 'metallic' selenium consists of infinite helical chains, which is isostructural to the normal hexagonal form of tellurium. Three red forms of selenium differ in the packing of the Se_8 rings. The usual form of selenium, vitreous black selenium, has irregular packing of large selenium rings [12,35].

Oxidation States

A feature of the heavy elements of the p-block is their variable oxidation states as listed in Table 3.1. The highest oxidation state corresponds to the number of valence electrons, i.e. np-electrons + two ns-electrons (= N). The next oxidation state is two less than the number of valence electrons, N - 2, i.e. In(I) and Tl(I); Pb(II); As(III), Sb(III) and Bi(III) and Se(IV) and Te(IV). For mercury and cadmium this oxidation state is Hg(0) and Cd(0). Selenium and tellurium also have a divalent oxidation state. The elements of Groups 15 and 16 also have a -III and -II oxidation state respectively, when bonded to elements which are less electronegative, e.g. hydrogen and metals.

The N - 2 oxidation state, which only exists for the heavier elements of the p-block, becomes more stable down a group and across the periods. The existence of the N - 2 oxidation state has been called an 'inert pair' effect. The term is not helpful, and is not an explanation. The inference that the ns ^{2} electrons are inert is not correct. The valence electrons (s and p) of the p-block elements in the 1st. 2nd. and 3rd. long periods are expected to be more firmly held than the valence electrons of the elements of the 1st. and 2nd. short periods. This is because of the greater nuclear charge of the heavier elements and their contracted size as a result of the influence and shielding by the preceding d- and f-series of elements. This is not borne out however, from a consideration of ionization potentials, i.e. comparing the energy $I_{N-1} + I_N$ (in kJ mol^{-1}) to convert oxidation state N - 2 into N.

M	$I_2 + I_3$	M	$I_3 + I_4$
B	6090	C	10820
Al	4550	Si	7580
Ga	4940	Ge	7710
In	4520	Sn	6870
Tl	4840	Pb	7160

There is little difference in the relative stability of the N and N - 2 oxidation states for the two series Al to Tl and Si to Pb. Yet there is a definite chemical trend in increasing stability of the lower oxidation states for the heavier elements, as indicated by the following reactions.

$GeCl_2 + Cl_2 \rightarrow GeCl_4$, very rapid at 25°C

$SnCl_2 + Cl_2 \rightarrow SnCl_4$, slow at 25°C

$PbCl_2 + Cl_2 \rightarrow PbCl_4$, when forced.

This trend is also clear when comparing the E°(M^{3+}/M) and E°(M^+/M) reduction potentials (in volts);

E°	B	Al	Ga	In	Tl
E°, M^{3+}/M	-0.87	-1.66	-0.56	-0.34	+1.26
E°, M^+/M	-	+0.55	-0.79	-0.14	-0.34

As the atomic number increases M^{3+} becomes more difficult to form and M^+ easier. The stability of elemental mercury (N - 2 = 0) is another example of the increasing stability of the N - 2 oxidation state (E° (Hg^{2+}/Hg) = 0.85V).

The observed trend in the stability of the N - 2 oxidation state is therefore the result of a balance of a number of energy terms in compound formation, and not just ionization energies, but also, covalent bond energies, $s^2p^n \rightarrow s^1p^{n+1}$ promotion energies and lattice energies. The M-X bond energies for the heavy p-block elements decrease as the metal M, gets heavier and as the oxidation state of M increases, as shown in the following data for chlorides (in kJ mol^{-1}).

M	MCl_2	MCl_4
Si	-	380
Ge	385	354
Sn	386	323
Pb	304	240

The electronegativity of the heavy elements are similar, and therefore it is likely that the ionic component of the bonds are comparable. Hence the differences given above may reflect the covalent component. The promotion energy to involve the s^2 electrons in the covalent bonding, may not be compensated for by the additional two bonds, especially as the atoms get heavier and orbital overlap becomes more diffuse. Therefore, a point is reached where the lower oxidation state becomes the more stable. The same may apply for the ionic component of the bonds. As the atoms get larger the lattice energy falls, and it is possible that the lattice energy difference between the two oxidation states no longer favours the higher state, due to the energy required in the reaction;

$$M^{n+} - 2e \rightarrow M^{(n+2)+}.$$

It is clear that it is difficult to be precise over the reasons for the stability of the N - 2 oxidation state. Regardless of the reasons the existence of the lower oxidation state does have a significant influence on the chemistry of the elements, and their reactions in the environment [4,6,12,16,27].

Coordination Numbers and Stereochemistry

A feature of the chemistry of the heavy elements is that they can have high coordination numbers because of their large size. Summaries of the coordination numbers and stereochemistry of the ten heavy p-block elements are given in Tables 3.2, 3.3 and 3.4 [4,9,12,16,27,35]. The 'inert pair' is stereochemically active in compounds such as MX_3 (M = As, Sb and Bi, X = H and halogen), but not in species such as MX_6^{2-} (M = Se and Te, X = halogen). For covalent species, their stereochemistry can be interpreted in terms of the valence state electron pair repulsion concept.

In ionic compounds (e.g. compounds of Pb(II)) high coordination numbers occur, because of the large radius of the ions. In PbX_2 and $PbCO_3$ the lead is nine coordinate with a tripyramid structure, whereas in $PbSO_4$ the lead is twelve coordinate. The covalent radii and ionic radii [29] of the elements are listed in Tables 3.5 and 3.6 respectively.

TABLE 3.2 Coordination Numbers and Stereochemistry of Selenium and Tellurium

Coord. No.	Stereochemistry	Total Pairs	Bond Pairs	Lone Pairs	Examples	
					Selenium	Tellurium
1		4	1?	3?	$NCSe^-$	COTe
2	Bent	4	2	2	H_2Se, R_2Se	H_2Te, R_2Te
3	Trigonal planar	3	3	0	$SeO_3(g)$	$TeO_3(g)$
3	Pyramidal	4	3	1	$SeOCl_3$ $SeO_2(s)$	$Te(CH_3)_3^+$ TeO_3^{2-}
4	Tetrahedron	4	4	0	SeO_4^{2-} SeO_2Cl_2	
4	Distorted Tetrahedron	5	4	1	R_2SeX_2	TeX_4, TeO_2 R_2TeX_2
5	Square pyramid	6	5	1	$SeOCl_2(py)_2$	TeX_5^- $TeI_4(CH_3)^-$
5	Planar pentagon	6	5	1		$Te(S_2COEt)_3^-$
6	Octahedron	6	6	0	SeF_6	TeF_6, $TeO_3(g)$
	Octahedron	7	6	1	SeX_6^{2-}	TeX_6^{2-}
6	Trigonal prism				VSe, MnSe	VTe, MnTe
8	Cubic	8	8	0		TeF_8^{2-}?

Plots of the six-coordinate ionic radii of the M^{3+} ions of Group 13 (IIIB) and the covalent radii of the Group 15 (VB) atoms are given in Fig. 3.2. These demonstrate the influence of the contraction effect of the first row transition elements on the sizes of Ga and As, and the influence of the lanthanide contraction and relativistic effects on Tl and Bi. An important consequence is that the second and third members of each group have a similar chemistry, as do the last two members, because of the comparable size of the members of each pair.

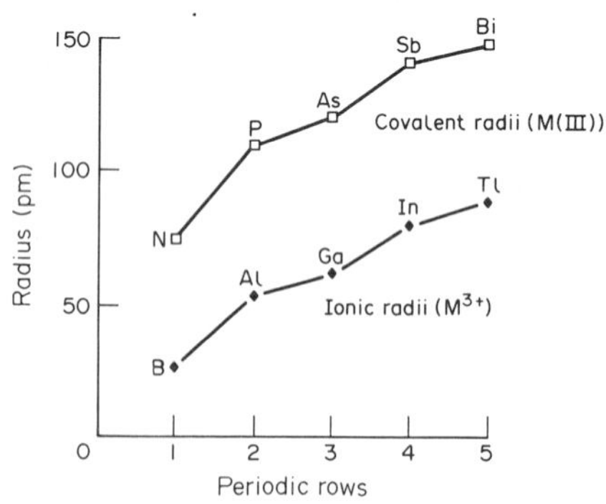

Fig. 3. 2 Ionic and covalent radii of the elements of groups III and V respectively.

TABLE 3.3 Coordination Numbers and Stereochemistry of Arsenic, Antimony and Bismuth

Coord. No.	Stereochemistry	Total Pairs	Bond Pairs	Lone Pairs	Examples		
					Arsenic	Antimony	Bismuth
1		4	1	3	$SiAs_4^{8-}$		
2	Bent	4	2	2	As_7^{3-}	Sb_7^{3-}	
3	Pyramidal	4	3	1	AsH_3 AsX_3 $AsMe_3$	SbH_3 SbX_3	$BiCl_3(g)$ $BiMe_3$
4	Tetrahedron	4	4	0	$AsPh_4^+$		
4	Distorted Tetrahedron	4	3	1		Sb in Sb-tartrate	
5	Trigonal Bipyramid	5	5	0	AsF_5 $AsPh_5$	SbF_5 $SbMe_3Cl_2$	BiF_5 $BiPh_3Cl_2$
5	Square pyramid	6	5	1		SbF_5^{2-}	$BiCl_5^{2-}$
6	Octahedron	7	6	1		$SbBr_6^{3-}$	
6	Octahedron	6	6	0	AsF_6^-	SbX_6^-	BiF_6^- ?
6	Trigonal prisim				NiAs	NiSb	NiBi

Bonding

The heavy p-block elements, in their high oxidation states, are predominantly covalent in character, whereas in the lower oxidation states they are predominantly ionic. The tendency towards covalency is in part a result of the filled inner d- and f-shells, which makes the elements polarisable. This property of the heavy p-block elements means they have less affinity for the more electronegative elements, and more affinity for polarisable ligands such as S^{2-} [4,12,27].

CHEMISTRY OF THE p-BLOCK ELEMENTS

The chemistry of the ten heavy p-block elements to be outlined, is not exhaustive in coverage, but selective and relevant to the following chapters on the environmental chemistry of the elements. The emphasis is on reaction rather than structural chemistry. Topics covered include the aqueous chemistry of the elements, their acid-base chemistry, their redox chemistry, solution chemistry as it relates to solubility, pertinent environmental chemistry and organometallic chemistry. The chemistry of cadmium, mercury and lead will be considered first, as they are the three significant environmental elements. The rest will be discussed in relation to their position in the periodic table.

Cadmium

Cadmium has one principal oxidation state Cd(II), but evidence exists for Cd(I) in Cd_2^{2+} species [7]. The chemistry, but not the biochemistry, of cadmium

TABLE 3.4 Coordination Numbers and Stereochemistry of Cadmium, Mercury, Indium, Thallium and Lead

Coord. No.	Stereo-chemistry	Cadmium	Mercury	Indium	Thallium	Lead
2	Linear	$CdEt_2$	$Hg(NH3)_2^{2+}$ HgO Hg_2X_2	$Tl(CH_3)^+$		
2	Bent					$PbX_2(g)$ $PbPh_2$
3	Planar		HgI_3^- HgBr3-	$In(CH_3)_3$	$Tl(CH_3)_2$- $(py)^+$	
3	Pyramidal					PbI_3^-
4	Tetrahedron	CdX_4^{2-}	$Hg(SCN)_4^{2-}$	InX_4^- $In_2Cl_6^{2-}$	TlX_4^- TlS	$PbCl_4$ PbR_4
4	Pyramidal					PbO
4	Square planar		HgCl2- 1,4 dioxane			PbO in Pb_2O_3
5	Trigonal bipyramid	$CdCl_5^{2-}$	$HgCl_2$- (terpy)	InBr $In(CH_3)_3$		$Pb(CH_3)_3OH$
5	Square pyramid	Cd- $(S_2CNEt_2)_2)_2$	$Hg(SMe)_2$	$InCL_5^{2-}$	TlI	
6	Octahedron	$Cd(NH_3)_6^{2+}$ CdO	$Hg(en)_3^{2+}$	$InCl_3$ $InCl_6^{3-}$	TlF $TlCl_6^{3-}$	PbO_2 $PbCl_6^{2-}$ PbS
7	Pentagonal bipyramid	$Cd(quin)_2$- $(NO_3)2H_2O$		$InF_4(H_2O)^-$		$PbCl_2$- $(thiourea)_2$
8	Dodecahedral	$Cd(NO_3)_2$- $4H_2O$	$Hg(NO_2)_4$		Tl+ in $Tl^+TlBr_4^-$	$Pb(O_2CMe)_4$
8	Cube				$TlNO_3$- $(thiourea)_4$	$PbWO_4$
9	Tripyramid					Pb(OH)Cl $PbX_2(s)$
12						$PbCO_3$ $PbSO_4$

follows closely that of zinc [2]. The cadmium species of particular relevance to environmental processes are listed in Table 3.7 at the end of the chapter, with some of their thermodynamic and solubility data. All the species have relatively high free energies of formation and the compounds of low solubility in water are CdO, CdS, $Cd(OH)_2$, $CdCO_3$ and $CdSiO_3$, however this depends on the pH.

Oxy-species Cadmium oxide is produced by oxidation of the metal, and is the product when the metal tarnishes in air. The colour of the oxide, which varies, from green/yellow to red and nearly black, is a function of the particle size and lattice defects. The oxide has the NaCl structure and defects occur (loss of

TABLE 3.5 The Single Bond Covalent Radii of the p-Block Elements (in pm)

H	C	N	O	F
37	77	74	74	72
	Si	P	S	Cl
	117	110	104	99
	Ge	As	Se	Br
	122	121	117	114
	Sn	Sb	Te	I
	140	141	137	133
	Pb	Bi		
	154	148		

TABLE 3.6 Ionic Radii of the p-Block Elements (in pm)

Ion	Coord. no.	Radius	Ion	Coord. no.	Radius
Cd^{2+}	4	78	As^{3+}	6	58
	6	95	Sb^{3+}	4	76
	8	110		6	76
Hg^{+}	3	97	Bi^{3+}	6	103
	6	119		8	117
Hg^{2+}	2	69	O^{2-}	2	135
	4	96		4	138
	6	102	OH^{-}	2	132
In^{3+}	4	62		4	135
	6	80	S^{2-}	6	184
	8	92	Se^{2-}	6	198
Tl^{+}	6	150	Te^{2-}	6	221
	8	159	F^{-}	2	129
Tl^{3+}	4	75		4	131
	6	89		6	133
	8	98	Cl^{-}	6	181
Pb^{2+}	4	98	Br^{-}	6	196
	6	119	I^{-}	6	220
	9	135			
Pb^{4+}	4	65			
	6	78			

Source of data; Shannon, 1976 [29].

oxygen) when it is heated. The oxide is soluble in acids to give the Cd^{2+} ion, but because it tends to be basic, it has low solubility in water and alkali.

The solid hydroxide precipitates from solution when hydroxide ions are added to Cd^{2+} ions, but redissolves in strong alkali.

$$Cd^{2+} + 2OH^- \rightarrow Cd(OH)_2$$

$$Cd(OH)_2 + 1 \text{ or } 2OH^- \rightarrow Cd(OH)_3^- \text{ or } Cd(OH)_4^-$$

The monohydroxide species $Cd(OH)^+$ is also known in solution, and is probably the initial product in the hydrolysis of the cadmium ion,

$$Cd^{2+} + H_2O \rightarrow Cd(OH)^+ + H^+.$$

The hydroxide and the basic salt CdClOH have the CdI_2 layer lattice structure, and in the hydroxychloride the OH^- and Cl^- ions occur in alternate layers.

The structure of cadmium carbonate is rhombohedral, of the $NaNO_3$ type. It has a low solubility in water, with $\log K_{sp} = -13.7$. It is soluble in acids and only exists above a pH of 8. The carbonate is an important species in fresh water, but converts to the cadmium chloro-species in sea water [20].

Cadmium halides The four halides of cadmium are known, but of these the chloride is the only one of environmental importance. The structure of $CdCl_2$ is cubic close packing of chloride ions with cadmium occupying the octahedral cavities between pairs of layers. The halides are all soluble in water, and form complex ions, such as CdX^+, CdX_3^- and CdX_4^{2-} in solution in excess halide ion. The distribution of the chloro-species in water (i.e. fraction of each species versus concentration of Cl^-) is given in Fig. 3.3. This shows that at around 10^{-3} moles l^{-1} chloride ions, the chloro-complexes start to become significant [34], and that at a concentration of 0.06 moles l^{-1} the main species are Cd^{2+}, $CdCl^+$ and $CdCl_2$. In sea water the main species are $CdCl^+$, $CdCl_2$ and $CdCl_3^-$. This diagram is helpful in understanding the environmental chemistry of the elements, and we will refer to others in the following pages.

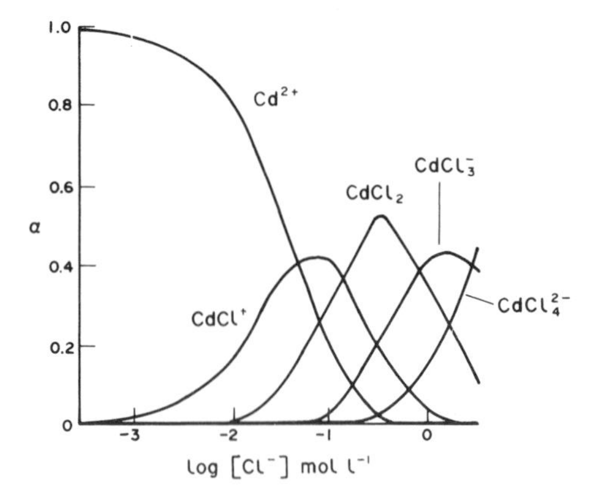

Fig. 3.3 Distribution diagram for cadmium-chloro species. Sources; references 15,30,31,34.

We will first consider the method of its construction [13,34]. The total concentration of cadmium (C), in the system is given by;

$$C = [Cd^{2+}] + [CdCl^{+}] + [CdCl_2] + [CdCl_3^{-}] + [CdCl_4^{2-}],$$

i.e the sum of the concentration of all the cadmium species present. The concentration of each species can be expressed in terms of $[Cd^{2+}]$, $[Cl^{-}]$ and the cumulative formation constants β_i [1,34].

$$Cd^{2+} + Cl^{-} \rightarrow CdCl^{+}, \quad \beta_1 = 21.0$$

$$Cd^{2+} + 2Cl^{-} \rightarrow CdCl_2, \quad \beta_2 = 166.0$$

$$Cd^{2+} + 3Cl^{-} \rightarrow CdCl_3^{-}, \quad \beta_3 = 204.0$$

$$Cd^{2+} + 4Cl^{-} \rightarrow CdCl_4^{2-}, \quad \beta_4 = 71.5.$$

Hence $C = [Cd^{2+}]\{1 + \beta_1[Cl^{-}] + \beta_2[Cl^{-}]^2 + \beta_3[Cl^{-}]^3 + \beta_4[Cl^{-}]^4\}$

The fraction (α_i) of each species X, is defined as $[X] = \alpha_i C$, e.g. $[Cd^{2+}] = \alpha_0 C$, and $[CdCl^{+}] = \alpha_1 C$ etc., where,

$$\alpha_0 + \alpha_1 + \alpha_2 + \alpha_3 + \alpha_4 = 1$$

$$\text{Since } \alpha_1 = \frac{[CdCl^{+}]}{C},$$

α_1 can be re-written using the following procedure;

$$\beta_1 = \frac{[CdCl^{+}]}{[Cd^{2+}][Cl^{-}]}, \text{ hence}$$

$$\alpha_1 = \beta_1 \frac{[Cd^{2+}][Cl^{-}]}{C} = \beta_1 \alpha_0 [Cl^{-}], \text{ where}$$

$$\alpha_0 = \frac{[Cd^{2+}]}{C} = \frac{1}{\{1 + \beta_1[Cl^{-}] + \beta_2[Cl^{-}]^2 + \beta_3[Cl^{-}]^3 + \beta_4[Cl^{-}]^4\}}$$

Similar expressions can be obtained for α_2, α_3 and α_4 viz., $\alpha_2 = \beta_2\alpha_0[Cl^{-}]^2$, $\alpha_3 = \beta_3\alpha_0[Cl^{-}]^3$ and $\alpha_4 = \beta_4\alpha_0[Cl^{-}]^4$. It is now possible to calculate the values for different chloride ion concentrations, provided the cumulative formation constants are available. The results are given in Fig. 3.3. Compilations of the various constants are available [15,30,31]. A scan of these tables shows that different studies give somewhat different values, therefore it is important to give the conditions under which the constants were measured. The values used for the Cd/Cl system were obtained at 25°C in a 4.5 mole l^{-1} $NaClO_4$ solution.

The cumulative formation constants may also be obtained from the successive formation constants, as they are related by the equations $\beta_1 = K_1$, $\beta_2 = K_1K_2$, $\beta_3 = K_1K_2K_3$ etc.

Cadmium sulphide Cadmium sulphide is produced by passing H_2S through a solution of Cd(II). Two structural forms exist, a cubic (zinc blende) structure and a hexagonal (wurtzite) structure. It has a very low solubility in water, $\log K_{sp} = -27.2$. Cadmium has a strong affinity for sulphur, and the two ions Cd^{2+} and S^{2-} are mutually polarisable which assists in the formation of a strong lattice. In reducing environments it is the stable species, however, in an atmosphere of dioxygen it is unstable because of the oxidation of the sulphide ion [8].

$$S^{2-} + 3H_2O - 6e \rightarrow SO_3^{2-} + 6H^+, \ -E° = -0.59V,$$

$$SO_3^{2-} + H_2O - 2e \rightarrow SO_4^{2-} + 2H^+, \ -E° = -0.17V,$$

$$O_2 + 4H^+ + 4e \rightarrow 2H_2O, \ E° = 1.23V.$$

A question may be raised as to how stable cadmium sulphide pigments are, when exposed to a moist and oxidising environment.

Both CdS and $CdCO_3$ can occur together in the environment, and it is possible that the reaction,

$$CdCO_3(s) + H_2S(g) \rightleftharpoons CdS(s) + H_2O + CO_2(g), \ \log K = 12.24,$$

takes place. A stability field diagram for this system is given in Fig. 3.4, which indicates that the carbonate is relatively unstable with respect to the sulphide, at even small partial pressures of H_2S. The two other reactions represented in the diagram are [32];

$$CdCO_3(s) + 2H^+ \rightleftharpoons Cd^{2+} + CO_2(g) + H_2O, \ \log K = 6.44,$$

$$CdS(s) + 2H^+ \rightleftharpoons Cd^{2+} + H_2S(g) \ \log K = -5.8.$$

The diagram is constructed from a knowledge of the equilibrium constants for the three reactions.

$$12.24 = \log\left(\frac{p_{CO_2}}{p_{H_2S}}\right).$$

$$6.64 = \log[Cd^{2+}] + \log(p_{CO_2}) + 2pH,$$

and

$$-5.8 = \log[Cd^{2+}] + \log(p_{H_2S}) + 2pH,$$

For a value of -3.5 for $\log(p_{CO_2})$ the last two equations can be rewritten as;

$$\log[Cd^{2+}] = 9.94 - 2pH, \text{ and}$$

$$\log[Cd^{2+}] = -2.3 - 2pH + \log\left(\frac{p_{CO_2}}{p_{H_2S}}\right)$$

respectively. The three equations have been plotted ($\log(p_{CO_2}/p_{H_2S})$ versus pH), in Fig. 3.4 for two cadmium concentrations 10^{-4} and 1 mole l^{-1}.

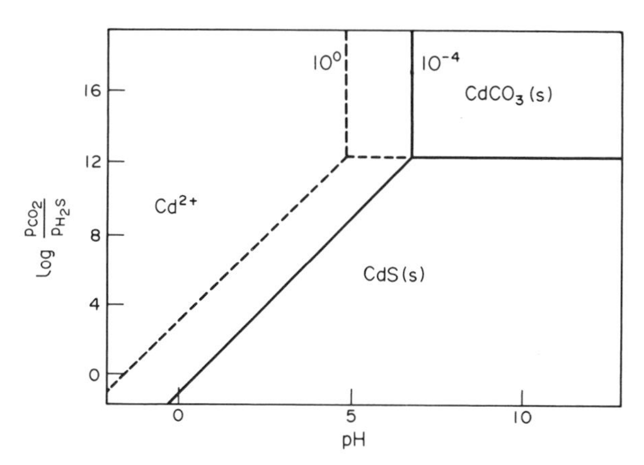

Fig 3. 4 Stability diagram for the system $Cd^{2+}/H_2S/CO_2/H_2O$. Source of data; Stumm and Morgan, 1970 [32].

In a chloride environment (e.g. sea water), the sulphide is the preferred species at a depth where bacterial reduction of SO_4^{2-} to HS^- can occur.

$$2\{CH_2O\} + SO_4^{2-} + H^+ \rightarrow 2CO_2 + HS^- + 2H_2O,$$

$$CdCl^+ + HS^- \rightarrow CdS + H^+ + Cl^-.$$

Therefore in a reducing environment the concentration of cadmium in solution should decrease [25].

Cadmium aqueous system The cadmium ion in water undergoes hydrolysis,

$$Cd^{2+} + H_2O \rightarrow Cd(OH)^+ + H^+, \log K = -10.2.$$

A useful description of the aqueous chemistry of the elements is in terms of a diagram that relates E_h and pH. This summarizes the redox potential/pH characteristics of a number of species on the same diagram, and indicates the areas in which certain species are more stable. The construction of these diagrams has been described in some detail previously [3,8,11,32]. We will go through the procedure for a simple Cd/H_2O system, containing the species Cd, Cd^{2+} and $Cd(OH)_2$. The chemical equations relating the three species are;

$$Cd^{2+} + 2e \rightarrow Cd(s),\ E° = -0.403V,$$

$$Cd(OH)_2(s) + 2H^+ \rightarrow Cd^{2+} + 2H_2O,\ \log K = 13.6$$

$$Cd(OH)_2(s) + 2H^+ + 2e \rightarrow Cd(s) + 2H_2O,\ E° = 0.002V.$$

If an E° value is unknown it may be calculated from $E° = -\Delta G/nF$, provided the free energy, ΔG_f, data are available. For the last reaction, $\Delta G_f(H_2O) = -237$ and $\Delta G_f(Cd(OH)_2) = -473.6$ kJ mole^{-1}, hence $\Delta G = -0.4$ kJ mole^{-1} and $E° = 0.002V$. In order to construct the diagram equations relating E_h and pH have to be developed. For the first equation the Nernst equation is;

$$E_h = E^0 - \frac{RT}{nF} \ln\left(\frac{1}{[Cd^{2+}]}\right)$$

$$\text{i.e. } E_h = -0.403 + \frac{0.059}{2} \log([Cd^{2+}]).$$

For the second equation the equilibrium constant is;

$$\log K = \log\left(\frac{[Cd^{2+}]}{[H^+]^2}\right)$$

$$\text{i.e. } 13.6 = \log[Cd^{2+}] + 2pH$$

For the third equation the Nernst equation is;

$$E_h = E^0 - \frac{RT}{nF} \ln\left(\frac{1}{[H^+]^2}\right)$$

$$\text{i.e. } E_h = 0.002 - 0.059pH.$$

The three E_h-pH lines are plotted (Fig. 3.5) for two cadmium ion concentrations of 1×10^{-5} and 8.9×10^{-8} moles l^{-1}, i.e. 10 ppb or 10 $\mu g\ l^{-1}$. The diagram shows that the predominant species is Cd^{2+}, even at a pH of 8-9. The dashed lines in the figure are the lines representing the oxidation (top line) and reduction (bottom line) of water. The area between the dashed lines is the region where water is stable.

A more detailed E_h-pH diagram, which includes the compounds $CdCO_3$ and CdS, is given in Fig. 3.6 [10,14,20]. Again the important species up to a pH of 7-8 is the cadmium ion. At higher pH's, $CaCO_3$(s) controls the cadmium concentration in solution in oxidising and slightly reducing conditions, whereas CdS(s) is the controlling species in more strongly reducing environments [14,20].

The distribution diagram for the cadmium/oxy/hydroxy species Cd^{2+}, $Cd(OH)^+$, $Cd(OH)_2$, $HCdO_2^-$ and CdO_2^{2-} given in Fig. 3.7 [8,13,34] portrays the

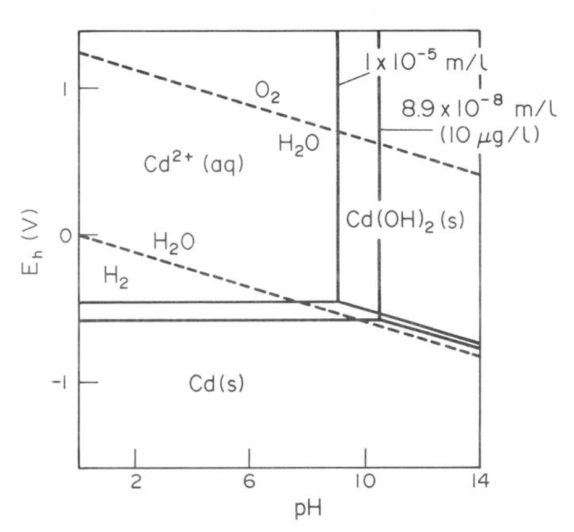

Fig. 3.5 E_h-pH diagram for Cd/H_2O system.

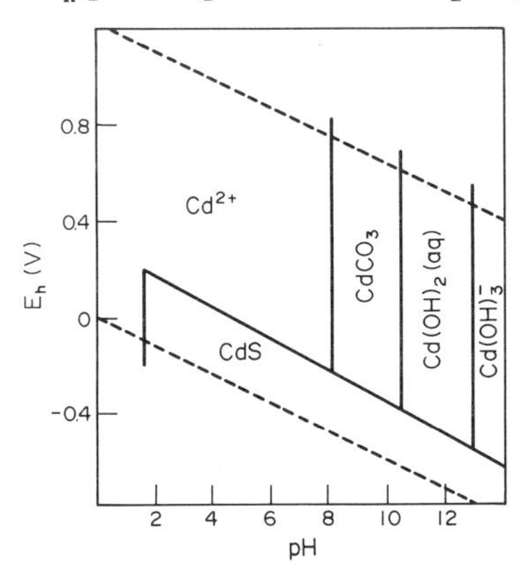

Fig. 3.6 E_h-pH diagram for cadmium species. Sources of data; Förstner, 1980 [10], Hem, 1972 [14].

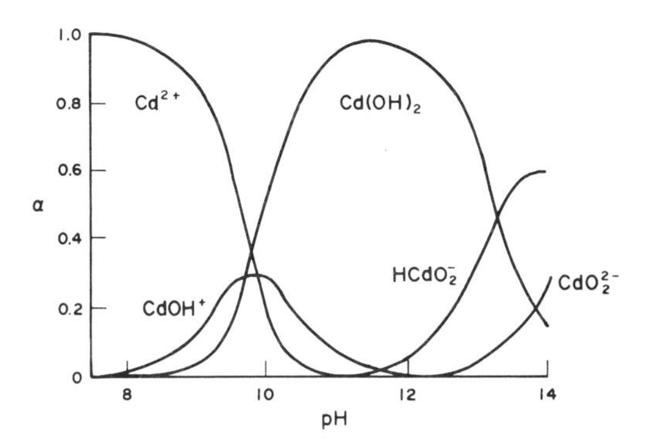

Fig. 3. 7 Distribution diagram for cadmium/hydroxide system.

pH dependence of the cadmium species. The diagram is determined as outlined above for the cadmium/chloro-species, where $\log\beta_i$ values are 4.30, 7.70, 10.30 and 12.00, i = 1 to 4 [13]. The distribution and E_h/pH diagrams have some features in common, but complement each other.

The oxidation or corrosion of cadmium metal occurs in acid, the reduction of dioxygen being the energy source [28];

$$Cd(s) - 2e \rightarrow Cd^{2+}, -E° = 0.403V,$$

$$O_2 + 4H^+ + 4e \rightarrow 2H_2O, E° = 1.23.$$

Hence,

$$2Cd + O_2 + 4H^+ \rightarrow 2H_2O + 2Cd^{2+}, E° = 1.63V,$$

at a pH = 0. The potential as a function of pH is given by the equation, E = 1.63 - 0.059pH [28], i.e. the reaction is more difficult as the pH increases, and at a pH = 10 the potential is 1.04V.

Organocadmium compounds Compounds of the type R_2Cd are known and are prepared from the appropriate Grignard reagents. Dimethylcadmium boils at 105.7° C and diethylcadmium at 64° C at 19 mm pressure. The volatile compounds are moderately reactive, they do not normally catch fire in air, and do not react vigorously with water. The production of dimethylcadmium by biomethylation is not well demonstrated [5], and it is presumed that if formed would decompose in an aqueous environment.

Mercury

Mercury is the only metallic element that is liquid at normal temperatures and pressures, it has a melting point at -38.9° C. Mercury has three oxidation states, Hg(0), Hg(I) and Hg(II). Mercury is unusual for a p-block element in that the monovalent oxidation state is not expected. Mercury(0) is the state with the inert pair, and Hg(I) and Hg(II) are formed by loss of one or two electrons respectively. Mononuclear Hg(I) would be paramagnetic, an unusual situation for a p-block element. Relevant thermodynamic and solubility data are presented in Table 3.8 at the end of the chapter.

Mercury (I) The experimental evidence strongly suggests that two mercury atoms are bonded together to give the dimeric Hg(I) species Hg_2^{2+}. Presumably the bond, which is 250 pm (short compared with 310 pm in the metal) is formed by the overlap of the 6s orbitals achieving electron pairing and a diamagnetic species [27]. The size of the 6s orbitals allow for a metal-metal bond to form, which would be less possible for the smaller 5s and 4s orbitals. The diamagnetism, Raman spectra and chemical equilibria can only be explained by a dimeric species.

The Hg_2^{2+} ions are moderately stable towards disproportionation in solution,

$$Hg_2^{2+} \rightarrow Hg + Hg^{2+}.$$

The potential for the reaction is -0.115V, and the equilibrium constant is 1.15 x 10^{-2}. Any reagent that lowers the concentration of Hg(II) in solution (i.e. a precipitate or complexing agent) will drive the reaction to the right. Hence reagents such as OH^-, S^{2-}, CN^- and R_2S produce mercury(II) compounds from a solution containing Hg(I). Most Hg(I) compounds are insoluble except the nitrate, chlorate and perchlorate. The reduction potentials $E°(Hg^{2+}/Hg)$ and $E°(Hg_2^{2+}/Hg)$ are 0.854V and 0.793V respectively, therefore any oxidising agent that oxidises mercury to give Hg(I) will also produce Hg(II). In fact there is no oxidising agent with a potential that lies within the narrow range. To obtain Hg(I) it is necessary to have an excess (at least 50%) of mercury in the reaction [4,7,12,19].

The four halides, Hg_2X_2, are known and have a linear structure. The chloride, bromide and iodide are insoluble and precipitate from solution by adding an alkali metal halide to mercurous nitrate. The chloride (calomel) is used in the saturated calomel secondary reference cell,

$$Hg_2Cl_2 + 2e \rightarrow 2Hg(l) + 2Cl^-,$$

which has a potential E = 0.2415 - 0.00076(t-25)V, in a saturated KCl solution, and t = °C. The potential varies with the KCl concentration. Mercurous sulphate is used in the standard Weston cell.

$$Cd/Hg \mid 3CdSO_4 8H_2O(s), Hg_2SO_4(s) \mid Hg.$$

The cell reaction is;

$$Cd/Hg + Hg_2SO_4 + 8/3H_2O \rightarrow CdSO_4 8/3H_2O + 2Hg(l),$$

and the emf is $E = 1.01845 - 4.05 \times 10^{-5}(t - 20) - 9.5 \times 10^{-7}(t - 20)^2$, and t is the temperature, °C.

Definite evidence for Hg(I) oxide and hydroxide does not exist.

Mercury (II) Mercury(II) is readily produced by oxidation of mercury with nitric acid;

$$3Hg + 8H^+ + 2NO_3^- \rightarrow 3Hg^{2+} + 2NO + 4H_2O.$$

The ion is polarisable and as a soft acid associates strongly with soft bases, such as the S^{2-} ion.

Mercuric oxide The oxide HgO, occurs in a red or yellow form, the difference arising from the particle size. The oxide is prepared by heating mercury in oxygen at 360° C, or heating mercuric nitrate, or anodic oxidation of mercury, or addition of hydroxide or carbonate to Hg^{2+}. It has a chain structure, with a linear O-Hg-O bond system, with four next nearest oxygen atoms. It is a weak base and dissolves in concentrated alkali.

Mercuric halides The four halides are known and are prepared (X = Cl, Br and I) by direct action of the halogen on the metal or by the following reactions.

$$HgSO_4 + 2NaX \rightarrow Na_2SO_4 + HgX_2$$

$$HgO + 2HX \rightarrow HgX_2 + H_2O.$$

The chloride structure is linear, and the units are stacked in sheets which are themselves stacked. Aqueous solutions of $HgCl_2$ (in the Cl^- concentration range 0.01-1 x 10^{-6} mol l^{-1}) contain mainly the dichloride, but small amounts of Hg^{2+}, $HgCl^+$, $HgCl_3^-$ and $HgCl_4^{2+}$ may also be present, depending on the chloride concentration (Fig. 3.8). The distribution diagram was calculated using the cumulative formation constants

$$\beta = \frac{[MX_n^{(-n+2)}]}{[M^{2+}][X^-]^n},$$

where $\log\beta_n$ = 6.74, 13.22, 14.12 and 15.12 and n = 1 to 4 [13,27]. Solid complexes containing the chloro-species, as well as more complex chloro-anions can be isolated.

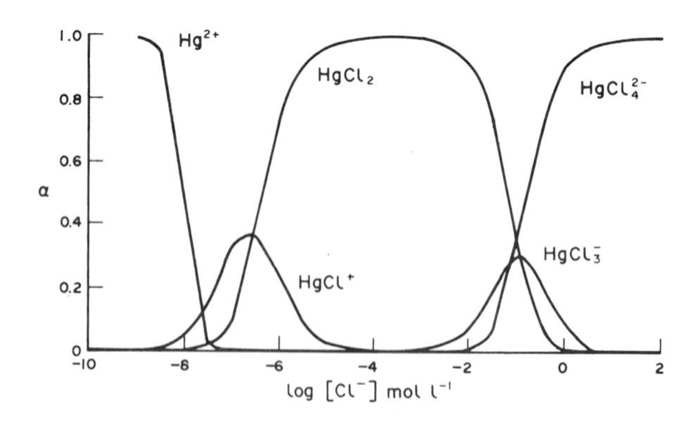

Fig. 3. 8 Distribution diagram for mercury-chloro species.

Partial hydrolysis of $HgCl_2$ produces a hydroxy chloride,

$$HgCl_2 + H_2O \rightarrow Hg(OH)Cl + Cl^- + H^+,$$

whereas in alkali the oxide is formed.

$$2HgCl_2 + 2OH^- \rightarrow HgO + H_2O + HgCl_3^- + Cl^-.$$

Mercuric sulphide The reaction of mercury with sulphur gives the red HgS, as does the addition of H_2S to a Hg(II) solution. Two forms exist, red hexagonal HgS (cinnabar) with the HgO structure and black cubic HgS (metacinnabar) with the zinc blend structure. The black form is produced by bubbling H_2S through an acidic solution of Hg(II) ions. The compound is very insoluble, $\log K_{sp} = -53$. However, owing to the possibility of forming complexes in solution, such as HgS_2^{2-} in alkaline solution, and $Hg(SH)_2$ in acid solution, the sulphide may be more soluble than the solubility product indicates. For example the following equilibria may occur;

$$HgS(s) \rightleftharpoons Hg^{2+} + S^{2-}, \log K_{sp} = -53,$$

$$HgS(s) + H^+ \rightleftharpoons Hg^{2+} + H_2S, \log K = -30.8,$$

$$HgS(s) + H_2S(g) \rightleftharpoons Hg(SH)_2, \log K = -6.2,$$

$$HgS(s) + S^{2-} \rightleftharpoons HgS_2^{2-}, \log K = -1.5.$$

The concentration of S^{2-} must be $>10^{-2}$ mole l^{-1}, to have a significant effect on the solubility of HgS.

Mercury aqueous system The mercury(II) ion hydrolyses, in the pH range 2 to 6, probably giving $Hg(OH)_2$.

$$Hg^{2+} + H_2O \rightarrow HgOH^+ + H^+, K = 2.6 \times 10^{-4},$$

$$HgOH^+ + H_2O \rightarrow Hg(OH)_2 + H^+, K = 2.6 \times 10^{-3}.$$

The solid hydroxide has not been isolated, because in alkaline solution the oxide forms. No further hydroxy species of mercury appear to exist, and the Hg^{2+}/OH^- system, $Hg(OH)_2$, is the dominant and only species at pH's > 4 - 5 [13].

The E_h-pH diagram for the system $Hg/Hg_2^{2+}/Hg^{2+}/HgO$ (Fig. 3.9) confirms the dominance of HgO down to a pH of 3 - 4 in an oxidising environment. In a reducing environment metallic mercury is the dominant species [3,7]. The relevant chemical equations are;

$$Hg_2^{2+}(aq) + 2e \rightarrow 2Hg(l),\ E° = 0.793V,$$

$$2Hg^{2+}(aq) + 2e \rightarrow Hg_2^{2+}(aq),\ E° = 0.908V,$$

$$HgO(s) + H_2O \rightarrow Hg^{2+}(aq) + 2OH^-,\ K = 3 \times 10^{-26},$$

$$2HgO(s) + 4H^+ + 2e \rightarrow Hg_2^{2+}(aq) + 2H_2O,\ E° = 1.06V.$$

Chloride and sulphide in particular, influence the speciation of mercury in aqueous systems. The stability regions for mercury sulphides, oxide and chlorides are given in the E_h-pH diagrams in Figs. 3.10 and 3.11 [23]. In an acidic and oxidising environment the chlorides dominate, whereas the oxide/hydroxide dominate in an alkaline and oxidising environment, and the sulphides in a reducing environment. A marked feature, is the stability of metallic

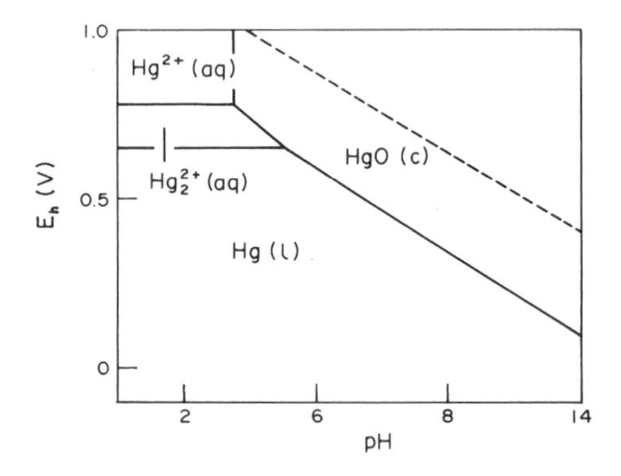

Fig. 3. 9 E_h-pH Diagram for the Hg-H_2O system.

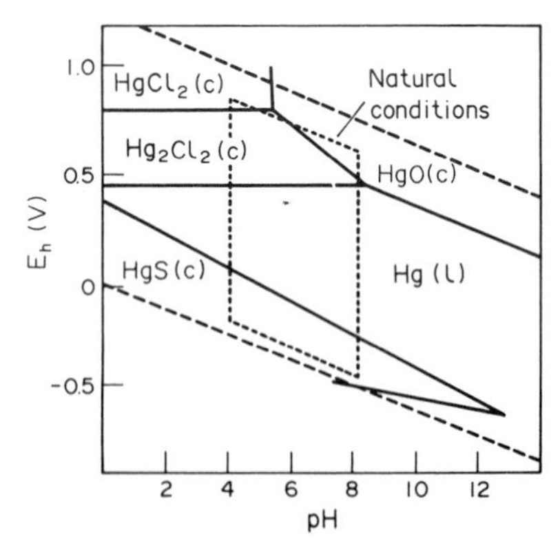

Fig. 3.10 E_h-pH Diagram for Mercury-chloro-sulphur system. Source of data; Leckie and James, 1976 [23].

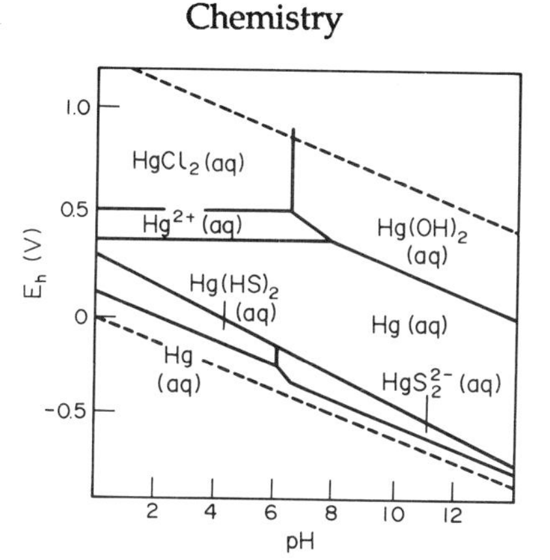

Fig. 3.11 E_h-pH Diagram for mercury-chloro-sulphur-aqueous system. Source of data; Leckie and James, 1976 [23].

mercury over a wide range of conditions. The relative existence of chloro- and hydroxy-species of mercury, in relation to pH, are demonstrated in a log concentration versus pH plot (Fig. 3.12 [23]), at a total chloride ion concentration of 0.1 mole l^{-1}. The concentrations of $HgCl_2$ and $Hg(OH)_2$ (the two major species) are similar at a pH around 8 - 9. The concentration of $HgCl_2$ falls off at higher pH, whereas the concentration of $Hg(OH)_2$ drops at lower pH. The pH at which the two species have the same concentration depends on the chloride concentration, and at 0.001 mole l^{-1} it is around a pH 6 - 7.

Organomercury compounds Organomercury compounds are historically important, and CH_3Hg^+ was discovered in 1851 by Frankland, from the reaction;

$$Hg + CH_3I \rightarrow CH_3Hg^+I^-.$$

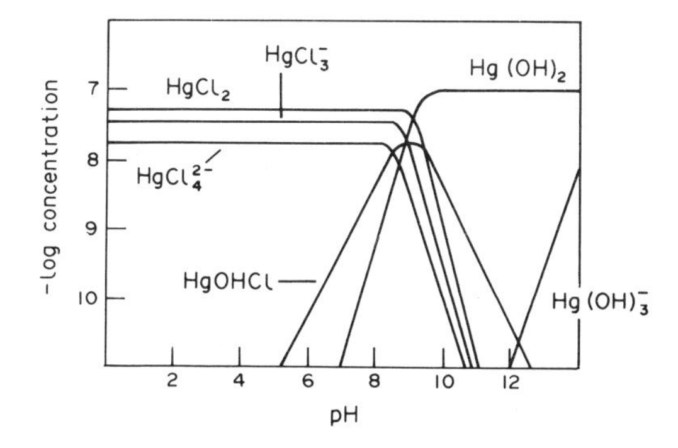

Fig. 3.12 Log concentration-pH diagram for mercury salts. Source of data; Leckie and James, 1976 [23].

The Hg-C bond is not strong (ca. 60 kJ mole^{-1}), but is stronger than the Hg-O bond. This helps to explain why organomercury compounds are stable in air and water, and therefore persist in the environment. This is one of the reasons for concern over the toxicity of mercury. Organomercury compounds, because of the weakness of the Hg-C bond, have been used for the preparation of more stable organometallic systems [12].

Dialkylmercury compounds, R_2Hg, are prepared by reacting a Grignard reagent, or alkyllithium or aluminiumalkyls with HgX_2. The compounds are typically covalent, and $(CH_3)_2Hg$ (b.pt. 92.5° C) and $(C_2H_5)_2Hg$ (b.pt. 159° C) are liquids. The chloro-compounds are solids e.g. CH_3HgCl (m.pt. 170° C) and C_2H_5HgCl (m.pt. 193° C). The CH_3Hg^+ species is a soft acid, and has been likened to a proton, as both CH_3Hg^+ and H^+ form one-coordinate species with a range of groups, and the compounds range from ionic to covalent, e.g. $RHg^+ClO_4^-$ is ionic. Also an ion $(CH_3Hg)_3O^+$, like the oxonium ion H_3O^+ exists.

The interconversion of mercury species Inorganic and organomercury species interconvert in the environment, and in some cases bacteria are involved. Some of the important environmental interconversions given in Fig 3.13 [5,8,17,19], will be discussed briefly.

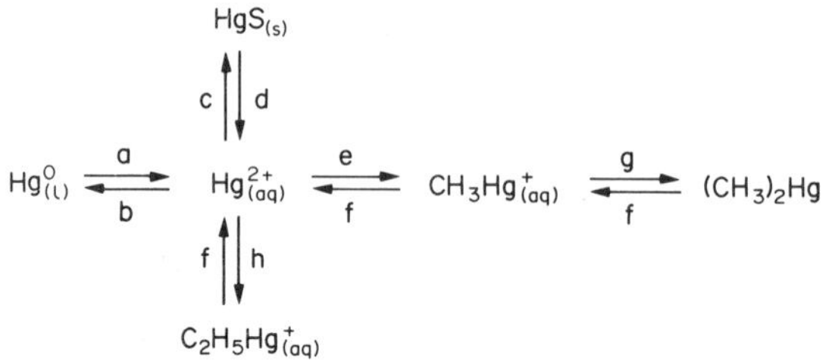

$$HgS_{(s)} \underset{d}{\overset{c}{\rightleftharpoons}} Hg^{2+}_{(aq)}$$

$$Hg^0_{(l)} \underset{b}{\overset{a}{\rightleftharpoons}} Hg^{2+}_{(aq)} \underset{f}{\overset{e}{\rightleftharpoons}} CH_3Hg^+_{(aq)} \underset{f}{\overset{g}{\rightleftharpoons}} (CH_3)_2Hg$$

$$Hg^{2+}_{(aq)} \underset{h}{\overset{f}{\rightleftharpoons}} C_2H_5Hg^+_{(aq)}$$

Fig. 3.13 The interconversion of mercury species in the environment.

Reaction (a), $Hg^o \rightarrow Hg^{2+}$. The oxidation of mercury to Hg(II) can be achieved by dioxygen.

$$Hg - 2e \rightarrow Hg^{2+}, \ -E^\circ = -0.854V,$$

$$O_2 + 4H^+ + 4e \rightarrow 2H_2O, \ E^\circ = 1.23V.$$

Reagents that coordinate to the Hg^{2+} ion assist the reaction.

Reaction (b), $Hg^{2+} \rightarrow Hg$. The reduction (E° = 0.854V) is achieved in aerobic conditions by bacteria, such as the *Pseudomonas* genus.

Reaction (c), $Hg^{2+} \rightarrow HgS$. The very low solubility of HgS drives the reaction to the right, especially in anaerobic conditions where H_2S, SH^- and S^{2-} exist.

$$Hg^{2+} + H_2S \rightarrow HgS(s) + 2H^+.$$

This is an important reaction of mercury in anaerobic conditions but, as described above, excess sulphide takes some mercury back into solution in the complex HgS_2^{2-}.

Reaction (d), $HgS \rightarrow Hg^{2+}$. This reaction appears to be of little consequence because of the low solubility of HgS. However, in aerobic conditions, the reaction is moved progressively to the right by removal of sulphide by its oxidation to sulphite and sulphate. This will occur under the conditions of natural water, as shown by the E_h-pH diagram for sulphur species in water (Fig. 3.14). Dioxygen and bacteria are involved in the oxidations.

Reactions (e) to (h), $Hg^{2+} \rightleftharpoons RHg^+/R_2Hg$. Methylation of mercury is achieved in the environment by methylcobalamine, a methyl derivative of vitamin B_{12}. The degradation of methylmercury is also achieved by bacteria. The chemistry of these reactions will be discussed in greater detail in Chapter 12.

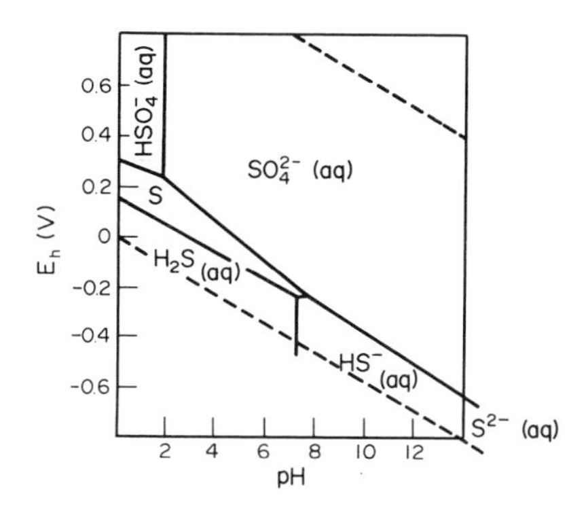

Fig. 3.14 E_h-pH diagram for sulphur species. Sourceof data; Garrels and Christ, 1965 [11]

Lead

Lead, the most widely used of the heavy metals, has two oxidation states Pb(II) and Pb(IV) as well as the elemental state Pb(0). The tetravalent state is a powerful oxidising agent, and the divalent state is the more stable oxidation level as seen from the following standard reduction potentials,

$$E°(Pb^{4+}/Pb^{2+}) = 1.7V, \text{ and } E°(Pb^{2+}/Pb) = -0.126V.$$

Hence Pb(IV) is associated with the more electronegative ligands, e.g. O^{2-}, F^- and Cl^-, and the bonding is essentially covalent, whereas Pb(II) is also associated with reducing ligands such as Br^-, I^- and S^{2-}. Bond strengths are also relevant, and for Pb(IV) organo-compounds the energy associated with the two additional Pb-C bonds (bond energy ca. 544 kJ mole $^{-1}$) is sufficient to offset the promotional energy from Pb(II) to Pb(IV). The relative stability of lead compounds is generally in the order: inorganic Pb(II) > organolead Pb(IV) >

inorganic Pb(IV) > organolead Pb(II) [26]. Hence the major environmental inorganic forms of lead are divalent, whereas the organic forms are tetravalent. Relevant thermodynamic and solubility data for compounds of environmental interest are given in Table 3.9 at the end of the chapter.

Lead oxy-species Lead monoxide, PbO, exists in two forms, a stable red form, which is tetragonal, and a yellow orthorhombic form, produced at temperatures above 488° C. The red oxide made by heating lead in air, has four oxygen atoms attached to each lead atom, all on one side giving a square pyramidal structure. The yellow form of PbO is a distorted form of this structure.

Lead dioxide, PbO_2, has the rutile structure, and is made from the hydrolysis of lead(IV) tetraacetate, or oxidation of lead(II) acetate with hypochlorite, or oxidation of lead(II) in Pb_3O_4 with nitric acid. The compound is a good oxidising agent, with applications in organic chemistry.

The oxide Pb_3O_4 is lead plumbate, $2Pb^{2+}PbO_4^{4-}$, in which the Pb(IV) atom is six coordinate and the Pb(II) ion in a three coordinate pyramidal environment [7]. When PbO_2 is heated a number of non-stoichiometric oxide phases form (temperatures in °C);

$$PbO_2 \xrightarrow[293]{} Pb_{12}O_{19} \xrightarrow[351]{} Pb_{12}O_{17} \xrightarrow[374]{} Pb_3O_4 \xrightarrow[605]{} PbO$$

The species Pb_2O_3 is also formed around 550-620° C.

Lead(II) hydrolyses in water to give initially $PbOH^+$ (logK = -7.9). Hydroxide produces basic materials, but even at a pH of 12 there is no sign of solid $Pb(OH)_2$, though it does occur in solution. The Pb^{2+}/OH^- distribution diagram (Fig. 3.15 [13]) is based on the cumulative formation constants log β_i = 7.82, 10.88, 13.94 and 16.30 for i = 1-4, for the species $PbOH^+$, $Pb(OH)_2$, $Pb(OH)_3^-$ and

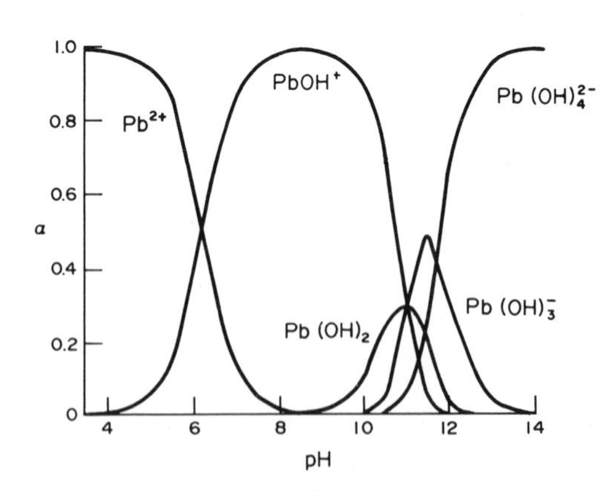

Fig. 3.15 Distribution diagram for lead hydroxy species.

$Pb(OH)_4^{2-}$. Other hydroxy species also exist in solution, including $Pb_4(OH)_4^{4+}$, (the main one), Pb_2OH^{3+}, $Pb_3(OH)_4^{2+}$ and $Pb_6(OH)_8^{4+}$. They contain clusters of lead atoms. Metaplumbates are salts of $Pb(OH)_6^{2-}$, obtained by treating PbO_2 with KOH. The compound $3PbO.H_2O$ (i.e. $Pb_6O_8H_4$) is obtained from the hydrolysis of lead(II) acetate, or evaporation of a solution of PbO at low temperature. It contains an octahedral cluster of lead atoms, surrounded by a cube of oxygen atoms.

Metallic lead has a reasonable solubility in soft and acidic water to give soluble hydroxy species. This is called the 'plumbosolvency' of lead.

$$Pb - 2e \rightarrow Pb^{2+}, -E° = 1.26V$$

$$O_2 + 4H^+ + 4e \rightarrow 2H_2O, E° = 1.23V.$$

Increasing the pH and/or the hardness of water reduces the plumbosolvency, hence the addition of lime to water is a method of reducing the lead content in water that passes through lead pipes.

Lead carbonate which has the $CaCO_3$ aragonite structure, is obtained by treating lead(II) acetate or nitrate with an alkali metal carbonate in the cold. Basic carbonates i.e. $3PbCO_3.2Pb(OH)_2$ and $2PbCO_3.Pb(OH)_2$, come from hot solutions. The second compound is the essential component of white lead, used in some lead paints. The compound turns black in air due to the slow formation of PbS from sulphide species in the atmosphere. A chloro-carbonate, $PbCl_2.PbCO_3$, occurs as a mineral phosgenite.

Lead halides The four lead(II) halides are known. The dichloride has a tricapped trigonal prism structure, with nine chloride ions around each lead. Mixed halides also exist, such as PbBrCl and PbFCl. Only PbF_4 and $PbCl_4$ occur in the tetravalent state, both compounds are covalent. Chloro-species of lead in solution only occur when the chloride concentration is > 0.1 mole l^{-1} (Fig. 3.16).

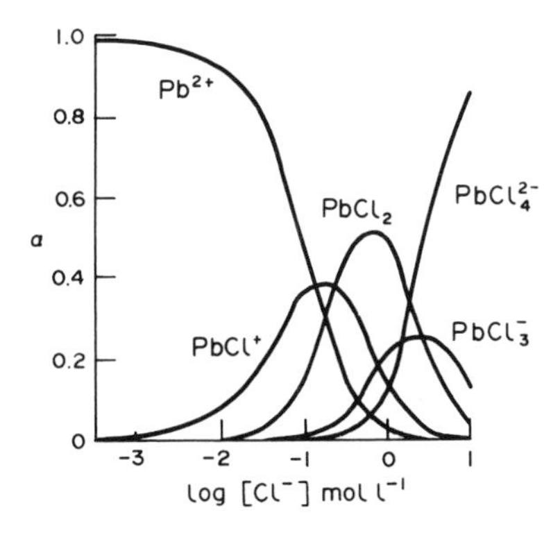

Fig. 3.16 Distribution diagram for lead chloro species.

Log of the cumulative formation constants ($\log\beta_i$) used in constructing the diagram, are 0.88 ($PbCl^+$), 1.49 ($PbCl_2$), 1.09 ($PbCl_3^-$) and 0.94 ($PbCl_4^{2-}$). The last species only becomes significant in solution at [Cl^-] > 1.8 mole l^{-1} [13].

A number of oxychlorides of lead exist, some as minerals, e.g. Pb_2OCl_2 ($PbCl_2.PbO$) matlockite, $Pb_3O_2Cl_2$ ($PbCl_2.2PbO$) mendipite, Pb_3OCl_4 ($2PbCl_2.PbO$) penfieldite and PbOHCl laurionite.

Lead sulphide Lead sulphide PbS (galena) has the NaCl structure. It can be produced in a lead rich form or sulphur rich form, which are p- and n-type semiconductors respectively.

Lead(II) aqueous systems Many lead(II) compounds are either of low solubility in water (e.g. $PbCl_2$) or almost insoluble (e.g. $PbCO_3$, PbS). The lead water system containg the species Pb, Pb^{2+}, PbO, $Pb(OH)_3^-$ and PbO_2 is summarized in the E_h-pH diagram in Fig. 3.17 [32]. Eight chemical equations are involved in

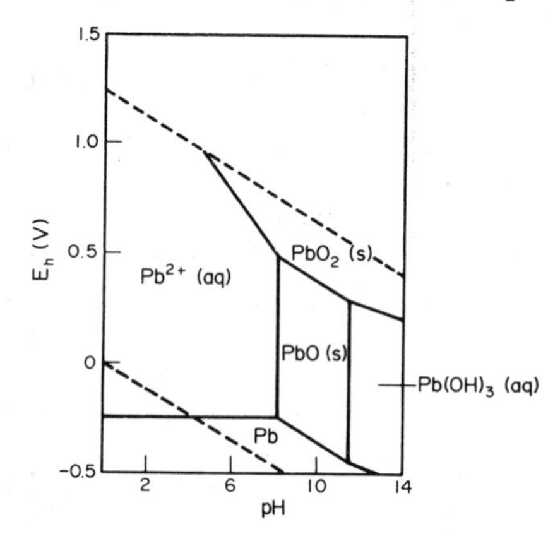

Fig. 3.17 E_h-pH diagram for $Pb/H_2O/OH^-$ system. Source of data; Stumm and Morgan, 1970 [32].

the diagram (one for each line), which has been calculated for a lead concentration of 1 x 10^{-4} mole l^{-1}. The dioxide is only stable at high +E_h and pH values, whereas the dominant species in acid and over a range of E_h values is the dipositive ion, and in alkaline solution the species is PbO.

If carbon dioxide is added to the system, lead carbonate species form, and reactions such as;

$$PbCO_3 \rightleftharpoons Pb^{2+} + CO_3^{2-}, \log K_{sp} = -12.83, \text{and}$$

$$2PbCO_3.Pb(OH)_2 + 2H^+ \rightarrow 3Pb^{2+} + 2CO_3^{2-} + 2H_2O, \log K = -18.8$$

are possible. In this situation, lead carbonate species (see Fig. 3.18 [32]) are dominant in natural environmental conditions. The interrelationship between

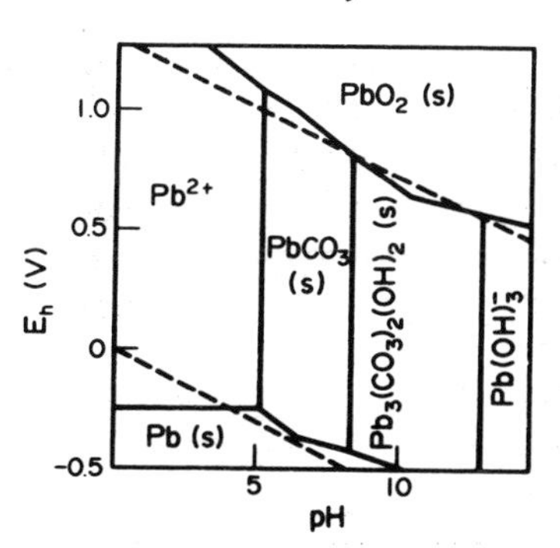

Fig. 3.18 E_h-pH diagram for Pb-H_2O-CO_2 system. Source of data; Stumm and Morgan, 1970 [32].

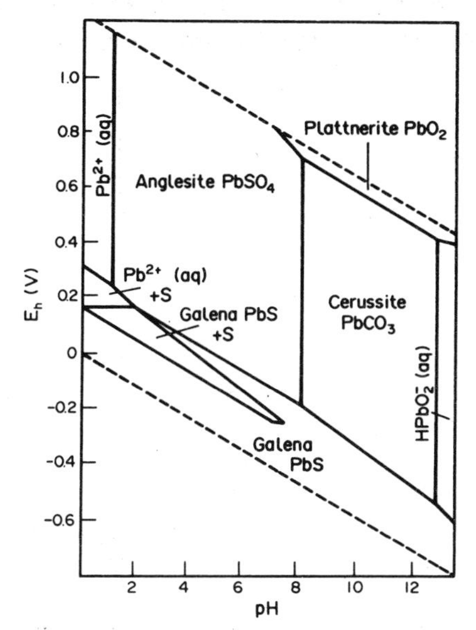

Fig. 3.19 E_h-pH diagram for some lead minerals and species. Source of data; Garrels and Christ, 1965 [11].

a number of lead minerals in an aqueous environment is shown in the E_h-pH diagram in Fig. 3.19 [11].

The solubilization of low soluble lead compounds can be achieved by complex formation, (such as the chloro- and hydroxo-complexes described above), and by organic ligands. The tribasic reagent nitrilotriacetic acid $N(CH_2COOH)_3$, NTA has been, and still is, used in some detergents to replace phosphate. Over the pH range 3.5 to 9.5 [25] the main form of the acid TH_3 is HT^{2-}. We will consider the effect of NTA on the solubility of lead carbonate as a typical example of solubilizing lead compounds. In the pH range 7 to 10 the principal carbonate species is the bicarbonate ion. The relevant equations are;

$$PbCO_3 \rightleftharpoons Pb^{2+} + CO_3^{2-}, K_{sp} = 1.48 \times 10^{-13},$$

$$HT^{2-} \rightleftharpoons T^{3-} + H^+, K_a = 5.25 \times 10^{-11},$$

$$Pb^{2+} + T^{3-} \rightleftharpoons PbT^-, K_f = 2.45 \times 10^{11}, \text{ and}$$

$$CO_3^{2-} + H^+ \rightleftharpoons HCO_3^-, K_b = 2.13 \times 10^{10}.$$

Combining the four equations gives;

$$PbCO_3 + HT^{2-} \rightleftharpoons PbT^- + HCO_3^-,$$

which has the equilibrium constant;

$$K = \frac{[PbT^-][HCO_3^-]}{[HT^{2-}]} = K_{sp}K_aK_fK_b = 4.05 \times 10^{-2}.$$

Hence

$$\frac{[PbT^-]}{[HT^{2-}]} = \frac{K}{[HCO_3^-]} = \frac{4.05 \times 10^{-2}}{1 \times 10^{-3}} = 40.5.$$

for $[HCO_3^-] = 1 \times 10^{-3}$. Therefore most of the lead is complexed, and this will increase as the concentration of the bicarbonate ion decreases.

Different cations may compete for the complexing agent however, so the solubilization of lead will depend on the presence of other cations. For example if calcium cations are present they will also compete for the NTA [25].

$$Ca^{2+} + HT^{2-} \rightleftharpoons CaT^- + H^+, K = 7.75 \times 10^{-3},$$

where

$$K = \frac{[CaT^-][H^+]}{[Ca^{2+}][HT^{2-}]}.$$

Hence

$$\frac{[CaT^-]}{[HT^{2-}]} = K\left(\frac{[Ca^{2+}]}{[H^+]}\right) = 7.75 \times 10^{-3}\left(\frac{1 \times 10^{-3}}{1 \times 10^{-7}}\right)$$

$$= 77.5,$$

for $[Ca^{2+}] = 40 \text{ mg l}^{-1} = 1 \times 10^{-3} \text{ mole l}^{-1}$ at a pH of 7.

The competition reaction between lead and calcium for the ligand is;

$$PbCO_3(s) + CaT^- + H^+ \rightleftharpoons Ca^{2+} + PbT^- + HCO_3^-, \text{ i.e.}$$

$$K = \frac{[Ca^{2+}][HCO_3^-][PbT^-]}{[CaT^-][H^+]},$$

$$= \frac{[HCO_3^-][PbT^-]}{[HT^{2-}]} \times \frac{[Ca^{2+}][HT^{2-}]}{[CaT^-][H^+]},$$

$$= \frac{4.05 \times 10^{-2}}{7.75 \times 10^{-3}} = 5.24.$$

Hence

$$\frac{[PbT^-]}{[CaT^-]} = K \frac{[H^+]}{[Ca^{2+}][HCO_3^-]}$$

$$= \frac{5.24 \times 1 \times 10^{-7}}{1 \times 10^{-3} \times 1 \times 10^{-3}},$$

$$= 0.524.$$

Less than half of the NTA is complexed with the lead. Therefore what actually occurs depends on a number of factors, such as the relative stability of the metal complexes, concentration of the complexing agent, pH of the solution, nature of the insoluble precipitates and the number of competing cations.

Organolead compounds As described above organolead compounds contain tetravalent lead. The compounds R_4Pb are insoluble in water, but soluble in non-polar solvents, and stable in air. The compounds R_3PbX and R_2PbX_2 are white solids, stable in air and soluble in water. Tetramethyllead boils at 110° C and melts at -27.5° C, and tetraethyllead boils at 78° C at 10 mm pressure. These compounds are made industrially, for the addition to petrol (Chapter 2), by treating a sodium/lead alloy with alkyl chloride;

$$4Na/Pb + 4RCl \rightarrow R_4Pb + 4NaCl + 3Pb,$$

at 80-100° C and in an autoclave. They may also be made by electrolysis of RMgX, using a sacrifical lead anode,

$$Pb + 4R. \text{ or } 4R^- \rightarrow R_4Pb.$$

In the laboratory the treatment of PbX_2 with RMgX, RLi or R_3Al may be used.

Indium and Thallium

Indium and thallium, the heavy elements in Group 13 (IIIB) have two oxidation states, III and I. The chemistry of indium tends to be more similar to gallium rather than thallium. This is because of the dominance of the mono-valent oxidation state for thallium. The monovalent chemistry of thallium resembles that of the alkali metals, and is a reason why thallium is toxic. It can replace monovalent, biochemically active metals, in particular potassium, but it does not have the same biochemical activity. Thermodynamic and solubility data of some indium and thallium compounds are listed in Table 2.10 at the end of the chapter.

Oxy-species The oxide In_2O_3 is obtained by igniting In(III) hydroxide to 850° C to constant weight and then heating to 1000° C for thirty minutes. The compound is soluble in acids, but not in alkalis. When heated in vacuo to more than 700° C it loses dioxygen and give In_2O. Analogous oxides exist for thallium, as well as Tl_4O_3 ($3Tl_2O.Tl_2O_3$) and TlO_2. The Tl(I) oxide is obtained by heating TlOH to 100° C, or Tl_2CO_3 to 370° C. The sesquioxide is produced by oxidation of $TlNO_3$ with chlorine, or bromine, and then treating the solution with hydroxide to give a hydrated material which is desiccated to the anhydrous oxide. The oxide is insoluble in water, but dissolves in mineral acids.

Aqueous solutions of In(III) salts undergo hydrolysis;

$$In^{3+} + H_2O \rightleftharpoons InOH^{2+} + H^+, \; pK = 4.43$$

$$InOH^{2+} + H_2O \rightleftharpoons In(OH)^{2+} + H^+, \; pK = 3.9.$$

Polynuclear indium cations occur at higher concentrations. Indium hydroxide $In(OH)_3$, which has the ReO_3 structure, begins precipitating at a pH of 3.4. It is sparingly soluble in water and in low concentrations of alkali, the solubility decreases with aging. Crystalline hydroxy salts can be obtained, $In(OH)_6^{3-}$, $In(OH)_5^{2-}$, which decompose in water to give $In(OH)_3$. When the hydroxide is heated to around 245-435° C in an atmosphere of water at a pressure of 1000-25,000 psi the compound InO(OH) is produced.

Solid Tl(III) hydroxide does not exist, but it and hydroxy species, such as $TlOH^{2+}$ and $Tl(OH)_2^+$ exist in solution. Their distribution with respect to pH (Fig. 3.20) are based on the cumulative formation constants log β_i = 12.82, 25.27, 37.46, i = 1 to 3. A similar diagram for the In^{3+}/OH^- system is given in Fig. 3.21

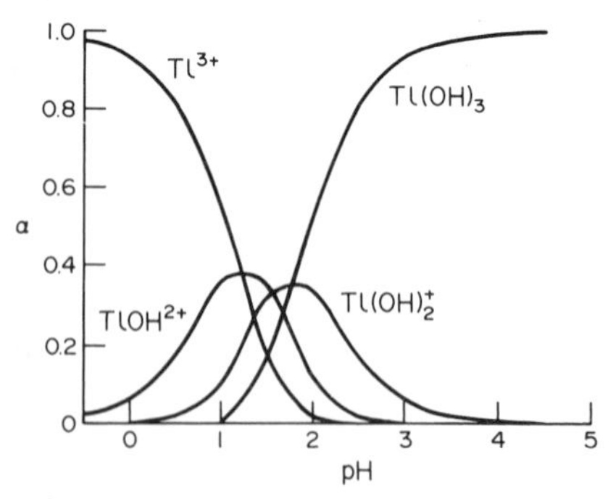

Fig. 3.20 Distribution diagram for thallium hydroxy species.

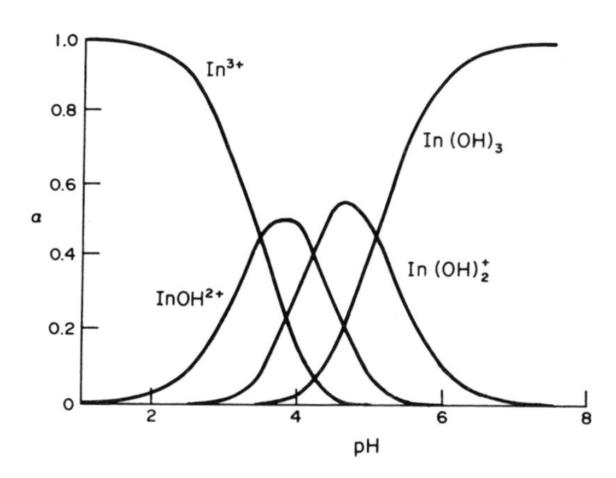

Fig. 3.21 Distribution diagram for indium hydroxy species.

using the constants log β_i = 10.52, 20.32, 29.26, i = 1 to 3 [15,30,31]. It is clear that the trivalent ions M^{3+}, only exist at low pH, and that hydroxy species commence forming in acid solution.

The monovalent hydroxide of thallium, TlOH, is obtained as a solid from the reaction of thallium with ethanol in the presence of dioxygen. In solution it is obtained from the 1:1 reaction of thallous sulphate and barium hydroxide. The hydroxide is a strong base and slowly attacks glass.

Thallous oxyanion salts, (SO_4^{2-}, NO_3^-, CO_3^{2-}) are similar to the salts of the alkali metals, in particular rubidium and caesium. The ionic radius of Tl(I) is close to that of Rb^+ (in pm),

Tl^+	Rb^+	Cs^+	
150	152	167	6 coordination
159	156	176	8 coordination

Thallous carbonate Tl_2CO_3, which is like the carbonates of sodium and potassium, is prepared by adding CO_2 to TlOH, and further additon of CO_2 gives the bicarbonate. Solutions of the carbonates are alkaline owing to hydrolysis, and the compounds are stable in air. Thallous sulphate is obtained from the reaction of sulphuric acid with thallous hydroxide or nitrate.

Halides The eight halides of In(III) and In(I) all exist, as well as In_2X_4 which is $In^+[InX_4]^-$. The monohalides are obtained from heating the elements together, or, for the chloride, reacting indium with Hg_2Cl_2. The chloride has a deformed NaCl structure, with an In-In interaction and it hydrolyses rapidly in water. The trichloride tetrahydrate, $InCl_3.4H_2O$, crystalizes from a solution prepared from the metal in hydrochloric acid. The anhydrous compound is obtained from the reaction of In_2O_3 with hydrogen chloride at 100 ° C, or with $SOCl_2$ in a sealed tube at 300° C. The structure is layer lattice with each indium surrounded by six chlorine atoms. In chloride solutions chloro-anions form, such as $InCl_4^-$, $InCl_5^{2-}$ and $InCl_6^{3-}$, which can also be isolated in solid salts. The distribution diagram for indium(III) chloro-species given in Fig. 3.22, is based on the cumulative formation constants log β_i = 2.58, 3.84, 4.20, i = 1 to 3 [15,30,31]. An oxychloride, InOCl, forms from the reaction of $InCl_3$ with dioxygen, and a thiochloride, InSCl, can also be prepared.

The trihalides of thallium are thermally unstable, losing halogen on heating, and are hydrolysed by water, as are other thallic salts. However, stable Tl(III) chloro-species in solution and as solid salts, as well as complexes with N-, O- and P-donors exist. The Tl(III)/Cl^- system in solution is represented by the distribution diagram in Fig 3.23 based on the cumulative formation constants log β_i = 7.04, 12.32, 15.30, 17.36, for i = 1 to 4 [15,30,31]. Comparison of the In(III) and Tl(III) diagrams (Figs.3.22, 3.23) shows that thallium species exist at a lower chloride concentration. The thallous halides all exist, the fluoride is soluble in water whereas the others are only sparingly soluble. The compounds are principally ionic, and the chloride has the CsCl structure. The halides are neutral in water, and in the presence of chloride ions Tl(I) chloro-species form (Fig. 3.24). The relavant cumulative formation constants are logβ_i = 0.50, -0.80, -1.68, -2.64 for i = 1 to 4 [15,30,31]. Thallous chloride forms from the addition of chloride ions to Tl(I) ions. Chlorination of it gives Tl_2Cl_4, whereas in acetonitrile the trichloride is produced. The oxychloride is produced from TlCl and Cl_2O at 4° C.

Sulphides Heating stoichiometric proportions of indium and sulphur gives dark coloured sulphides, red InS and near black In_6S_7, the former crystallizes at 600-675° C, and the latter at 740-760° C. A higher temperature sulphide also occurs, In_3S_4. The addition of H_2S to a solution of In(III) in a weak acid or alkaline solution gives orange In_2S_3, which has two forms, α resembling Al_2O_3 and, β

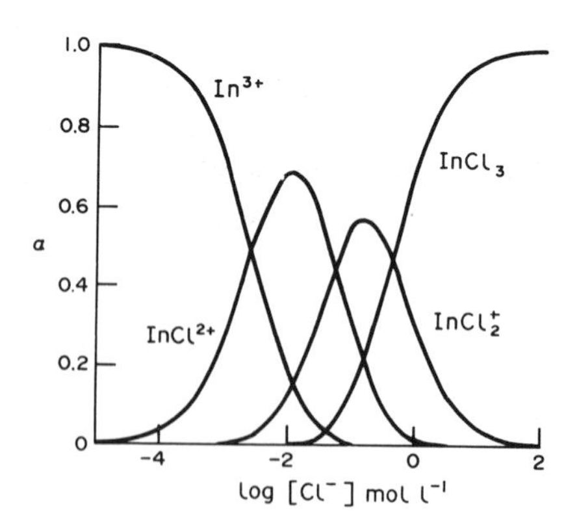

Fig. 3.22 Distribution diagram for indium(III)-chloro species.

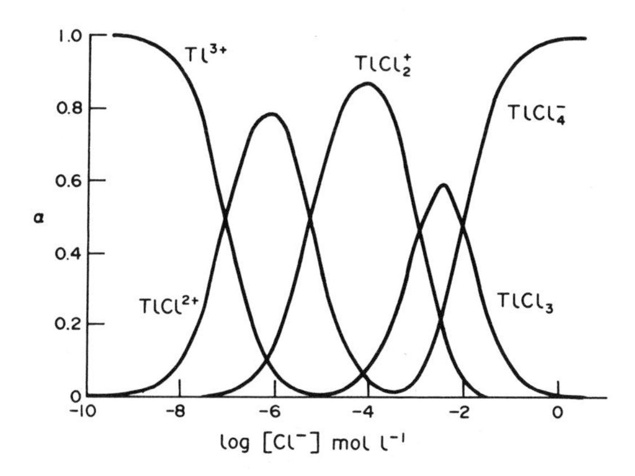

Fig. 3.23 Distribution diagram for thallium(III)-chloro species.

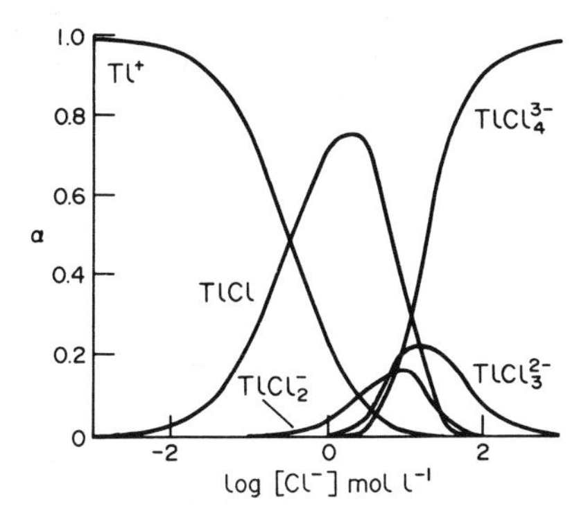

Fig. 3.24 Distribution diagram for thallium(I)-chloro species.

which is a cation deficient spinel structure. No simple thallium(III) sulphide has been made, though TiS contains the thallic disulphide ion, $Tl(I)[Tl(III)S_2]$. The sulphide Tl_2S is obtained from Tl(I) solutions when treated with hydrogen sulphide, or from heating the elements together. The compound is ionic and oxidises in air, however, the sulphide ion oxidises first followed by the thallium. Polysulphides can also be produced, such as TlS_2 and Tl_2S_5.

Aqueous chemistry The monovalent oxidation state of thallium is distinctly more stable than that of the trivalent state, whereas the reverse is true for indium. The E° reduction potentials $M^{3+}/M^{+}$are -0.34V and 1.26V for In and Tl respectively. The difference is clearly demonstrated from their M/H_2O E_h-pH diagrams given in Figs. 3.25 and 3.26. The principal indium species are In(III) and $In(OH)_3$, whereas for thallium they are Tl(I) and $Tl(OH)_3$. The thallous ion is not sensitive to pH but the thallic ion undergoes extensive hydrolysis. The

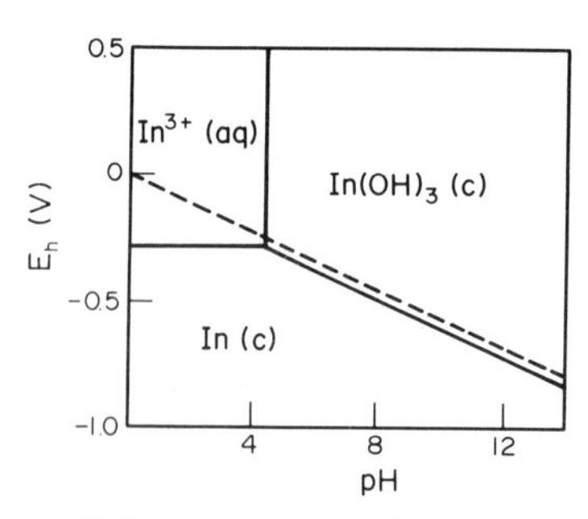

Fig. 3.25 E_h-pH diagram for indium- water system.

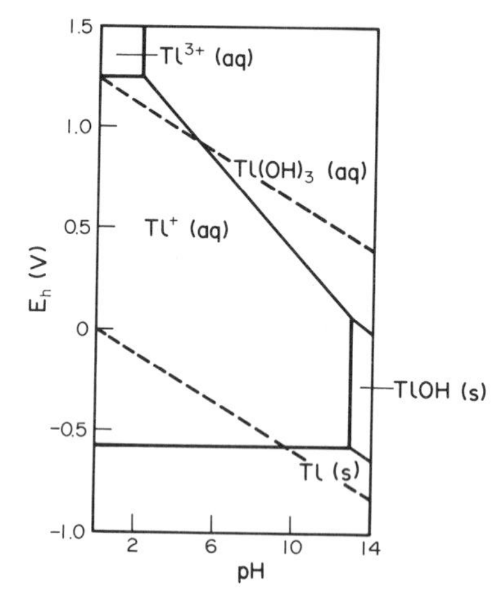

Fig. 3.26 E_h-pH diagram for a thallium-water-hydroxide system.

reduction potential of Tl(III) is affected by pH and complexing agents, e.g. in 1 mole l^{-1} HCl, the potential drops to 0.77V, because the ion is strongly complexed with chloride, compared with Tl(I).

Organo-compounds Both metals form compounds of the type R_3M, R_2MX and RMX_2 in the trivalent state. The reactivity of the M-C bond lies in the order Al > Ga = In > Tl, and the indium and thallium R_3M compounds are air and moisture sensitive. The lower alkyl compounds inflame in air, whereas other compounds are more stable, and for thallium are salt like, i.e. $R_2M^+ X^-$. Some preparative reactions are:

$$2In + 3R_2Hg \rightarrow 2R_3In + 3Hg,$$

$$R_3In + HX \rightarrow R_2InX + RH,$$

$$R_3In + X_2 \rightarrow R_2InX + RX,$$

$$2MeLi + MeI + TlI \rightarrow Me_3Tl + 2LiI,$$

$$TlX_3 + 2RMgX \rightarrow R_2TlX + 2MgX_2.$$

Binary intermetallic compounds Binary compounds formed between elements of groups 13 (IIIB) and 15, (VB) called III-V compounds, have important semiconductor properties. Some of the properties of the six compounds GaP, GaAs, GaSb, InP, InAs and InSb, are listed in Table 3.11. They all have the zinc blende structure, and are made mainly by heating, in sealed tubes, stoichiometric amounts of the elements. Care is necessary over the regulation of the amounts of impurity in the products. The range of compounds and therefore range of energy gaps (Table 3.11) give a wide choice of semiconductors. Some of their uses are: in power rectifiers (e.g. GaP), transistors (e.g. GaAs), galvanomagnetic devices (e.g. InSb, InAs), photosensitive and photovoltaic cells (e.g. GaAs, CdS, CdSe) and injection lasers (e.g. GaAs). A major use of GaAs, as $GaAs_{1-x}P_x$, is in light emitting diodes (LED), used in calculator displays. The colour of the emitted radiation relates to the energy gap. For GaAs the gap is 138 kJ mole^{-1} which increases to 184 kJ mole^{-1} when $x = 0.4$ and the light is red (650 nm), but when $x > 0.4$ the gap increases to 218 kJ mole^{-1} and the light becomes green (550 nm).

TABLE 3.11 Some Properties of Binary Intermetallic Compounds

Property		GaP	GaAs	GaSb	InP	InAs	InSb
Band gap (kJ mole^{-1})		218	138	69	130	34	17
Mobility	n	150	8500	4000	5000	23,000	30,000
(cm^{2}s^{-1}V^{-1})	p	150	400	700		240	1000
Lattice const. (pm)		1030.0	1068.4	1156.1	1109.0	1140.6	1224.2
M. Pt (°C)		1465	1238	712	1070	942	525

Arsenic, Antimony and Bismuth

The chemistry of the three elements, arsenic, antimony and bismuth of Group 15 is similar, with the expected periodic trends. The lower oxidation state M(III) becomes more stable, and ionic in character, as the elements get heavier. The two main oxidation states are pentavalent and trivalent, and in the hydrides MH_3 the oxidation state is -III. Arsenic is in the first long period, and follows the first transition metal series. The contraction in size across the transition metals means the arsenic atom is smaller than expected. The influence of the lanthanide contraction on bismuth is to reduce its radius, bringing its size and chemistry closer to that of antimony. These two effects are apparent in the plot of the covalent radii of the Group 15 elements versus Period number, in Fig. 3.2. The covalent bond strength between the three elements and other elements lie in the order As > Sb > Bi. For example BiH_3 and organobismuth compounds are unstable.

The binary compounds of the three elements, either react with water, or are insoluble, therefore the elements are more non-metallic in character. For example in water, oxyanions or complexes exist rather than cationic species, such as $M(H_2O)_x^{n+}$. Bismuth does form cations, but only in strong acid, the bismuthyl ion BiO^+ is an example. In general arsenic form oxyanions e.g. arsenates and arsenites, antimony forms oxyanions and oxycations, and bismuth forms oxycations and maybe hydrated cations. This trend down the group is typical of a change towards more metallic and basic character.

The reduction potentials for the three elements are similar, except for bismuth(V), as shown in Fig. 3.27, and they are quite distinct from those of phosphorus. The Bi(V)/Bi(III) couple of 2.03V, is strongly oxidising, and oxidizes water to dioxygen, whereas the couples for arsenic and antimony are weaker. Thermodynamic and solubility data for some species for each element are listed in Table 3.12 at the end of the chapter.

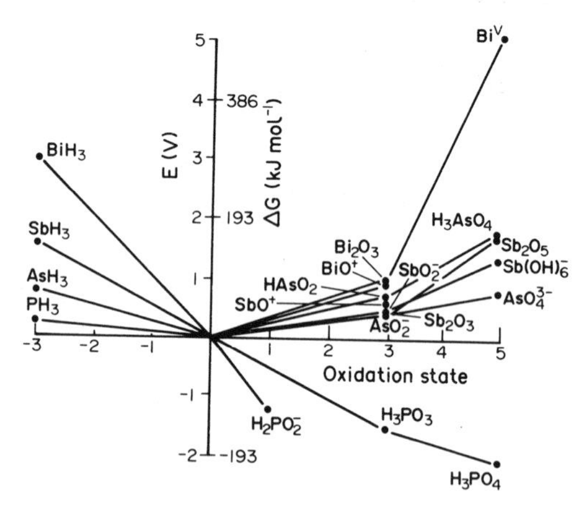

Fig. 3.27 Oxidation state diagram for phosphorus, arsenic, antimony and bismuth.

Oxy-species The oxides of arsenic, antimony and bismuth are:

As_4O_6	Sb_4O_6	Bi_2O_3	M(III)
As_2O_4?	Sb_2O_4		M(V), M(III)
As_2O_5	Sb_2O_5	Bi_2O_5?	M(V)

The bascity of the oxides increases in the order As < Sb < Bi, and M(V) < M(III). The trivalent oxides are made by heating the elements in dioxygen, or hydrolysis of the trichlorides. Industrially, As_4O_6 is obtained from roasting FeAsS. The oxides all display polymorphism, and the cubic form of M_4O_6 has the P_4O_6 structure. High temperature forms are polymeric, and the bismuth oxide can have a defect lattice. The compounds are not very soluble in water, but the arsenic(III) and antimony(III) oxides are soluble in basic solutions giving oxy/hydroxy-species, and bismuth(III) oxide is soluble in acid giving cationic bismuth species.

The pentavalent arsenic oxide is obtained by oxidation of the trivalent oxide using nitric acid, followed by dehydration of the arsenic acid produced. The compound has arsenic in both octahedral and tetrahedral cavities. Antimony(V) oxide is produced by hydrolysis of the pentachloride with aqueous ammonia, and both oxides (As and Sb), are obtained from high pressure and temperature oxidations of M_2O_3 with dioxygen. The bismuth(V) oxide is poorly defined, and is extremely unstable, acting as a powerful oxidising agent. Heating Sb_4O_6 or Sb_2O_5 around 900° C gives Sb_2O_4, which contains Sb(V) in octahedral cavities and $Sb(III)O_4$ pyramidal units.

Arsenic(III) oxide in basic solution gives the species $As(OH)_3$, $AsO(OH)_2^-$, $AsO_2(OH)^{2-}$ and AsO_3^{3-}. The arsenite ion which may be isolated in salts, is pyramidal with a lone pair occupying the fourth position. Arsenious acid is a weak acid:

$$H_3AsO_3 \rightleftharpoons H^+ + H_2AsO_3^-, K = 6 \times 10^{-6}.$$

Alkali metal arsenites are soluble in water, but many other arsenites are not, for example Cu(II) arsenites have been used as pigments, such as Paris green $Cu_2(MeCO_2)AsO_3$ and Scheele's green $CuHAsO_3$. Antimony(III) oxide in base gives $Sb(OH)_3$, $Sb(OH)_4^-$ and SbO^+, and less well characterised antimonites can be obtained from the solution. For bismuth(III), complex hydroxy species are produced in solution, such as $Bi_6(OH)_{12}^{6+}$, $Bi_6O_6^{6+}$ and $Bi_6O_6(OH)_3^{3+}$. They all contain an octahedral cluster of bismuth atoms, and the first species has the $Ta_6Cl_{12}^{2+}$, structure with the hydroxides bridging between two metals across each edge of the Bi_6 octahedron.

Arsenic(V) oxide gives in solution the arsenate ion AsO_4^{3-}, which can be isolated as arsenic acid and arsenate salts. The anion has a tetrahedral arrangement of oxygen atoms around the arsenic. Most arsenates have low solubility in water, and the chromic and cupric salts have $\log K_{sp}$ = -20.1 and -35.1 respectively. This is the reason why they are useful wood preservatives as they do not leach out of the wood. The acid is tribasic and $\log K_i$ = -2.2, -6.9, -11.5 for i = 1 to 3.

The relationships between the oxides and various oxy/hydroxy species for each element are shown in the E_h-pH diagrams given in Fig 3.28 a, b, c [3]. A comparison of the three diagrams reveals the increasing stability of the lower oxidation state from arsenic to bismuth.

Bismuth and antimony form oxy-salts in their trivalent oxidation state, e.g. perchlorates, nitrates, sulphates and phosphates. The trend in stability is As << Sb < Bi. Two nitrates of bismuth are $BiO(NO_3)$ and $(Bi_6O_6)_2(NO_3)_{11}(OH).6H_2O$. Potassium tartratoantimonite(III) (tartar emetic), $K_2(Sb_2(C_4H_2O_6)_2).nH_2O$ is a good source of antimony(III) in solution.

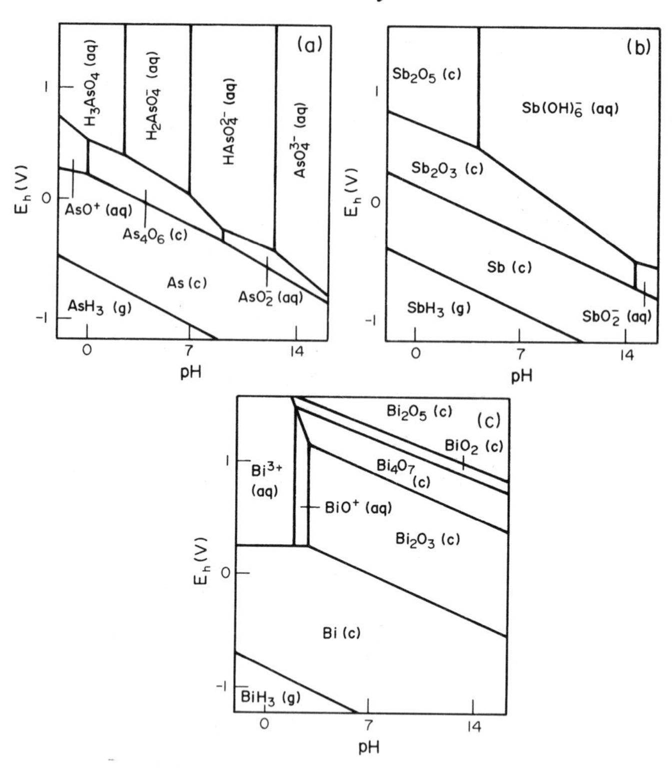

Fig. 3.28 E_h-pH diagrams for (a) arsenic, (b) antimony, and (c) bismuth in aqueous systems.

Halides The four trihalides are known for the three elements, and they are made from reaction of the elements, or from the M(III) oxides and HX. In the solid the halogen atoms form close packed layers, with the metal pyramidal and bond angles close to 100°. The halides hydrolyse in water, which is reversed by the addition of HX. Arsenic trichloride is immisicble with water at a $AsCl_3$:H_2O mole ratio > 1:10, but at lower ratios a monohydrate forms and then hydroxy-species and finally at a ratio 1:20 arsenious oxide precipitates out. Antimony trichloride gives a clear solution in water in a limited amount of water, but on dilution SbOCl and $Sb_4O_5Cl_2$ precipitate out. Bismuth trichloride forms the oxychloride, BiOCl, in water. When the compound is heated > 600° C a compound with stoichiometry $Bi_{24}O_{31}Cl_{10}$ forms, which is $(Bi_2O_3)_{12}$ with 5 oxygen atoms replaced by 10 chlorine atoms. A chloride $BiCl_{1.167}$ contains three bismuth species viz. $(Bi_9^{5+})_2$.$(BiCl_5)_4^{2-}$.$(Bi_2Cl_8)^{2-}$. The trihalides react with alcohols to give alkoxides $M(OR)_3$. The chloro-species for antimony(III) and bismuth(III) range from MCl^{2+} to MCl_6^{3-}, and the distribution diagrams given in Figs. 3.29 and 3.30 are based on cumulative formation constants log β_i = 2.26, 3.49, 4.18, 4.72, 4.72 and 4.11 for antimony [30] and 3.7, 5.5, 6.9, 7.9, 7.0 and 7.3 for bismuth [15] for i = 1 to 6.

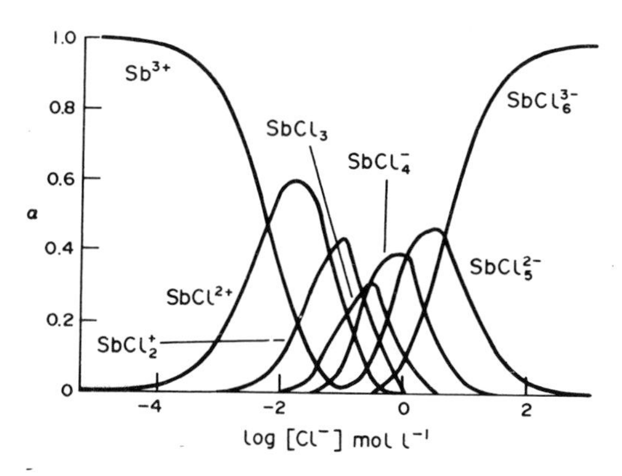

Fig. 3.29 Distribution diagram for antimony(III)-chloro species.

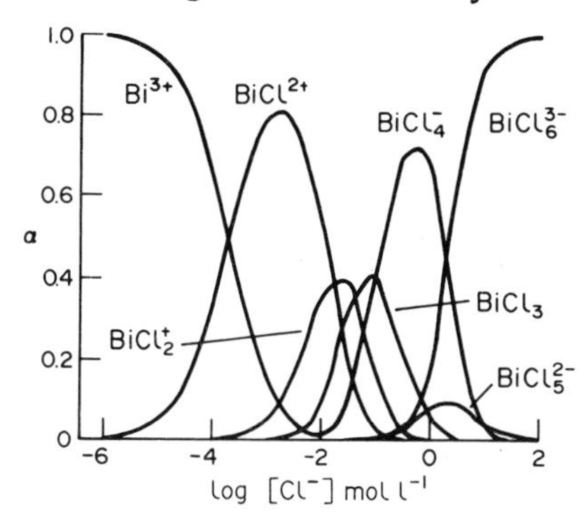

Fig. 3.30 Distribution diagram for bismuth(III)-chloro species.

Only the pentafluorides are known for all three elements, and $SbCl_5$ is the only stable chloride. It is a fuming liquid, but in the solid is a dimer with two chlorine bridges. The absence of stable $AsCl_5$ is unexpected as it is thermodynamically feasible, and As(V) chloro-complexes exist when associated with large cations, e.g. $AsCl_4^+AsF_6^-$, and $AsCl_4^+SbCl_6^-$. Photochlorination of $AsCl_3$ at -105° C does give $AsCl_5$, which decomposes above -50° C. The relative stability of MCl_5 compounds is P >> As << Sb. The non-existence of $BiCl_5$ has been attributed to the high ionization energy of the trihalide i.e.

$$MCl_3 \underset{-2e}{\rightarrow} MCl_3^{2+} \underset{+Cl^-}{\rightarrow} MCl_4^+ \underset{+Cl^-}{\rightarrow} MCl_5$$

For phosphorus, the ionization energy is 1012 kJ $mole^{-1}$, whereas for arsenic, it is 1128 kJ $mole^{-1}$. The extra promotional energy for arsenic may be the reason for the low stability of $AsCl_5$. The stronger attachment of the electrons to As(III), may be a result of stabilization of the orbitals of arsenic, due to the contraction effect of the preceding transition metals. A pentavalent arsenic oxychloride $AsOCl_3$ which exists below -25°C, is made by the ozonization of $AsCl_3$ at -78° C.

Sulphides The trivalent sulphides of arsenic, antimony and bismuth are prepared by heating the elements together or by reaction of the trivalent oxides with sulphur. The structures of the arsenic sulphides are based on the tetrahedral structure of As_4 (Fig. 3.31). The stoichiometries are, As_4S_3, As_4S_6 (orpiment), and As_4S_4, of which there are a number of modifications, one of which is realgar. Some of the As-As bonds in As_4 are retained in the structures. The same structural unit also occurs in thio-anions. For the compounds M_2X_3 the trend in

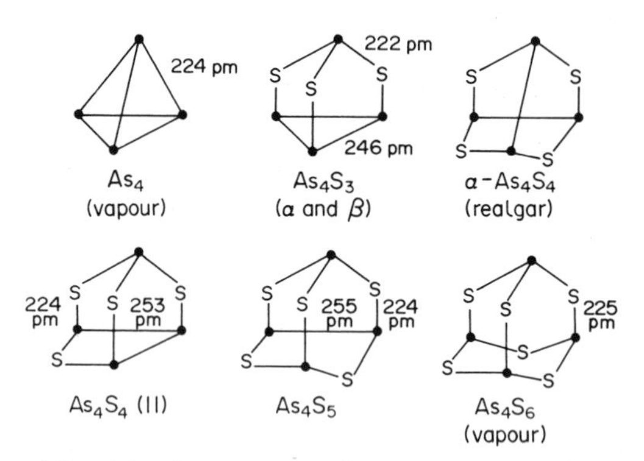

Fig. 3.31 Structures of arsenic sulphides.

the band gap for doped materials is, for M: As > Sb > Bi, and for X: S > Se > Te giving n- and p-semiconductors. The antimony and bismuth sesquisulphides have complex network structures. The compounds are stable in air, but are oxidised when heated. They are sparingly soluble in water, but in excess sulphide form thio-anions.

The pentavalent sulphides are not well characterized, both the arsenic and antimony compounds are produced from an ice cold solution of the M(V) oxyanion in HCl through which H_2S is bubbled. The products appear to contain some M(III).

Organo-compounds The three elements have a well developed organic chemistry in both oxidation states, such as $R_{3-n}X_nM$ and $R_{5-n}X_nM$, where X is a monovalent species. When X is oxygen the stoichiometry changes. The R_3M compounds are pyramidal in the vapour and solid states, and they are readily oxidised in air. Some of the oxy-compounds of arsenic(III) are $R_2As(OH)$, $(R_2As)_2O$, $RAs(OH)_2$ and RAsO. The stability of the organo-compounds lie in the order As > Sb > Bi. This may be seen from the M-C bond dissociation energy data (in kJ $mole^{-1}$).

M	$M(CH_3)_3$	MPh_3
As	238	280
Sb	224	267
Bi	140	200

Selenium and Tellurium

The elements selenium and tellurium of Group 16 (VI) are typical non-metals, with an extensive oxy/hydroxy chemistry like that of sulphur. They have three oxidation states VI, IV, II, and in compounds like H_2M also -II. The tetravalent oxidation state is the most common. The thermodynamic and solubility data for some selenium and tellurium compounds are given in Table 3.13 at the end of the chapter.

Oxy-species The divalent oxides SeO and TeO do not exist in the solid state, but there is evidence that they occur in the gaseous state. The oxides SeO_2 and TeO_2 are formed from heating the elements, or from dehydration of selenous or tellurous acids. Selenium dioxide is very soluble in water and acts as a weak acid, whereas tellurium dioxide is amphoteric, and has a minimum solubility at a pH 3.8-4.2. The trioxide of selenium is hygroscopic, and thermodynamically unstable with respect to SeO_2. The tellurium(VI) oxide is a powerful oxidising agent. A 'pentoxide' is in fact $TeO_2.TeO_3$, i.e. tellurium(IV) tellurate.

Selenous and tellurous acids, H_2SeO_3 and H_2TeO_3, are obtained from the crystallization of SeO_2 in water, and the hydrolysis of $TeCl_4$ respectively. The selenium compound can also be produced by nitric acid oxidation of selenium.

$$3Se + 4HNO_3 + H_2O \rightarrow 3H_2SeO_3 + 4NO$$

The acids occur as white solids that dehydrate to the oxides. Alkali metal salts are soluble, whereas other salts are less soluble. The first and second acid dissociation constants are 3.5×10^{-3} and 5×10^{-8} for selenous acid and 3×10^{-3} and 2×10^{-8} for tellurous acid.

Selenic acid H_2SeO_4, is obtained by oxidation of selenium or SeO_2 or selenous acid with hydrogen peroxide. The compound is white and deliquesent, it decomposes to SeO_2 on heating and is a strong oxidising agent, markedly different to sulphuric acid. The Te(VI) oxy-acids $Te(OH)_6$ and $(H_2TeO_4)_n$ do not resemble the selenium and sulphur analogues. The acid $Te(OH)_6$ is soluble in water and obtained by oxidation of tellurium with chloric acid or hydrogen peroxide or with permanganate. Oxidation of the dioxide also produces the acid, which is weak and dibasic. Alkali and alkaline earth salts are soluble in water, whereas other metals salts are sparingly soluble. The hexavalent oxy-species are oxidising agents;

$$SeO_4^{2-} + 4H^+ + 2e \rightarrow H_2SeO_3 + H_2O, \quad E° = 1.15V,$$

$$Te(OH)_6 + 2H^+ + 2e \rightarrow TeO_2 + 4H_2O,\ E° = 1.02V.$$

though the reactions are slow. The relative oxidising powers of Se(VI) and S(VI) is shown by the fact that selenium is oxidised to selenous acid by nitric acid (see equation above), whereas sulphur is oxidised to sulphuric acid. The relative stability of the Se and Te oxidation states with respect to pH are shown in the E_h-pH diagrams given in Fig. 3.32 a and b [3].

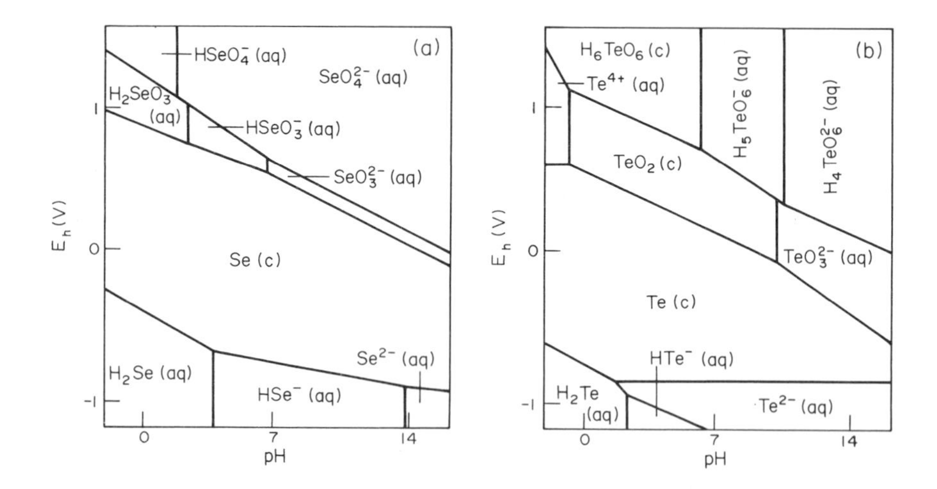

Fig. 3.32 E_h-pH diagrams for selenium and tellurium aqueous systems.

Halides The halides of selenium and tellurium occur mainly in the divalent and tetravalent oxidation states. The selenium(I) compound Se_2Cl_2 is a pungent oily liquid, which decomposes at its boiling point. It is obtained from the stoichiometric reaction of the elements in CCl_4 (before the tetrahalide forms), or from reduction of the tetrahalide. No equivalent tellurium compound has been made, though compounds Te_2Cl and Te_3Cl_2 have been reported.

The dichloride of selenium is unknown, but may occur in the vapour of $SeCl_4$. On the other hand $TeCl_2$ is a black solid melting at 208° C. It is prepared from the reaction of Te and CF_2Cl_2, further chlorination gives the tetrachloride. The compound disproportionates in water to Te and Te(IV).

All tetrahalides are known except the iodide of selenium. The chlorides $SeCl_4$ and $TeCl_4$ are prepared from the elements, or by oxidising Se or Te or treating the dioxides with chlorinating agents such as CCl_4, PCl_5 and $SOCl_2$. The selenium tetrachloride dissociates in the vapour state, whereas gaseous $TeCl_4$ is undissociated and has a distorted tetrahedral structure as a result of the lone pair in the valence shell. In the solid the chloride is ionic $TeCl^{3+}Cl^-$, and also has a cubane type structure Te_4Cl_{16}. The compounds are soluble in non-polar organic solvents, and form adducts with electron donors. In water the compounds hydrolyse, and chloro-species do not occur in solution The only hexavalent halides are the fluorides.

The oxyhalides $SeOX_2$ (X = F, Cl, Br) are less stable than their sulphur analogues, and hydrolyse readily. They have the same pyramidal structure. The oxychloride is a useful solvent, with a high dielectric constant, high dipole moment, and an appreciable electrical conductivity due to self ionization,

$$2SeOCl_2 \rightleftharpoons SeOCl^+ + SeOCl_3^-$$

TABLE 3.7 Cadmium Species and Some Physical Properties

Species	State	M Pt °C	H_f kJ	G_f kJ	S_f J deg^{-1}	Solubil. g/100g	Log K_{sp}
Cd^{2+}	aq		-72.4	-77.7	-61.1		
CdO	c	1559 subl.	-258.2	-228.4	54.8	insol	
$Cd(OH)^+$	aq 1M			-261.1			
$Cd(OH)_2$	c	~175 decomp.	-560.7	-473.6	96.2	2.6x10^{-4}	-13.8
$Cd(OH)_2$	aq 1M		-535.9	-393.2	-94..6		
$Cd(OH)_3^-$				-600.8			
$Cd(OH)_4^{2-}$	aq 1M			-758.6			
$CdCl^+$	aq 1M		-240.6	-224.4	43.5		
Cd(OH)Cl	aq			-106.7			
$CdCl_2$	c	568	-391.5	-344.0	115.3	140	
$CdCl_2$	aq 1M		-410.2	-340.1	39.7		
$CdCl_3^-$	aq 1M		-561.1	-487.0	202.9		
$CdCO_3$	c	~340 decomp.	-750.6	-669	4.92.5		-13.7
$CdSO_4$	c	1000	-933.5	-823.0	123.1	75.5	
$CdSO_4$	aq 1M		-985.3	-822.2	-53.1		
CdS	c	1000 subl.	-161.9	-156.5	64.9	1.3x10^{-4}	-27.2
$CdSiO_3$	c	1246	-1189	-1105	95.5		

Sources; references 7,11,14.

TABLE 3.8 Mercury Species and Some Physical Properties

Species	State	M Pt. °C	H_f kJ	G_f kJ	S_f J deg^{-1}	Solubil. g/100g	Log K_{sp}
Hg_2^{2+}	aq			152.1			
Hg^{2+}	aq		174.0	164.8	-22.6		
HgO	c	350 decomp.	-90.7	-58.5	70.2	$5.2x10^{-3}$	
$Hg(OH)_2$	aq			-274.9			
$HgCl_2$	c	277	-230.1	-184.1	144.3	6.6	
$HgCl_2$	aq 1 M		-216.7				
$HgCl_3^-$	aq 1 M		-392.0				
$HgCl_4^{2-}$	aq			-450.6			
Hg_2Cl_2	c		-264.9	-210.5	192.5	$2x10^{-4}$	-17.88
HgS	c subl.	583.5	-58.2	-48.8	77.8	$1x10^{-6}$	-53
HgS_2	aq				48.5		
$HgSO_4$	c	decomp	-704.2	-589.9	136.4		

Sources; references 7,11,12,22

TABLE 3.9 Lead Species and Some Physical Properties

Species	State	M Pt °C	H_f kJ	G_f kJ	S_f J deg^{-1}	Solubil. g/100g	Log K_{sp}
Pb^{2+}	aq		1.6	-24.3	21.3		
Pb^{4+}	aq			302.5			
PbO	c	897	-219.2	-227.0	67.8	0.0017	
PbO_2	c	decomp	-277.0	-219.0	76.6		
Pb_3O_4	c	830	-734.7	-617.6	211.3		
$HPbO_2^-$	aq			-338.9			
PbO_3^{2-}	aq			-277.6			
PbO_4^{4-}	aq			-282.1			
$Pb(OH)_2$	c		-514.6	-420.9	87.9	0.0155	-19.8
$PbCl_2$	c	500	-359.2	-314.0	136.3	0.99	
PbS	c	subl.	-94.3	-92.7	91.3	$8.6x10^{-5}$	-28.2
$PbSO_4$	c		-918.4	-811.2	147.3	0.00425	-7.9
$PbO.PbSO_4$	c		-1182	-1083	203.8		
$PbCO_3$	c		-700.0	-626.3	131.0		-12.8
$PbO.PbCO_3$	c		-920.5	-818.4	202.9		
$2PbO.PbCO_3$	c		-1142	-1013	272		
$Pb_3(OH)_2$-$(CO_3)_2$	c			-1699	-18.8		
$PbCrO_4$	c		-942.2	-851.9	152.7	$6x10^{-6}$	-15.7
$PbSiO_3$	c		-1083	-1000	113.0		
Pb_2SiO_4	c		-1308	-1195	180		

Sources; references 7,11,12,22

TABLE 3.10 Indium and Thallium Species and Some Physical Properties

Species	State	M Pt °C	H_f kJ	G_f kJ	S_f J deg^{-1}	Solubil. g/100g	Log K_{sp}
In^{3+}	aq		-133.9	-99.2	-259.4		
In_2O_3	c	2000 decomp	-925.8	-830.7	121		
$InOH^{2+}$	aq		-232.2				
$In(OH)_3$	c		-895	-761	105	-33	
InCl	c	225	-186.2	-164.0	96		
$InCl_3$	c	498 subl.	-537.2	-463.2	138	66.1	
$InCl_3$	aq		-636.0				
$In_2(SO_4)_3$	c		-2098	-2566	281		
In_2S_3	c	1098					
Tl^{3+}	aq		195.8	209.2	-176		
Tl_2O	c	596	-175.3	-136.0	99.6		
Tl_2O_3	c	716	-394.6				
TlOH	c		-238.1	-190.4	72.4	3.9	
$Tl(OH)_3$	c		-653	-514.6	102		-43.8
TlCl	c	430.8	-205.0	-184.9	108.4	0.325	-3.7
$TlCl_3$	c	155					
$TlCl_2^+$	aq		-42.7				
Tl_2S	c	448.5	-97.1	-93.7	163	0.022	-23.9
Tl_2CO_3	c	272	-700.0	-614.6	155.2	5.2	
Tl_2SO_4	c	632	-931.8	-830.4	230.5	4.3	
Tl_2SO_4	aq		-898.6	-809.6	271.1		

Sources; references 7,11,12,22

TABLE 3.12 Arsenic, Antimony and Bismuth Species and Some Physical Properties

Species	State	M Pt °C	H_f kJ	G_f kJ	S_f J deg^{-1}	Solubil. g/100g	Log K_{sp}
As_2O_5	c		-914.6	-772.4	105.4	230	
As_2O_5	aq		-939.7				
AsO_4^{3-}	aq		-870	-636	-144.6		
$HAsO_4^{2-}$	aq		-898.7	-707	3.8		
$H_2AsO_4^-$	aq		-904.6	-748.5	117.2		
H_3AsO_4	c		-900.4				
H_3AsO_4	aq		-898.7	-769.0	206.3		
As_4O_6	c	278	-1314	-1152	214.2	2.16	
AsO^+	aq			-163.6			
AsO_2^-	aq			-350.2			
$HAsO_2$	aq		-456.1	-402.7	126.8		

Table 3.12 continued

Species	State	M Pt °C	H_f kJ	G_f kJ	S_f J deg^{-1}	Solubil. g/100g	Log K_{sp}
$H_2AsO_3^-$	aq		-712.5	-587.4			
H_3AsO_3	aq		-741.8	-639.9	196.6		
$AsCl_3$	liq	-16.2	-305.0	-256.9	207.5		
As_2S_2	c	320	-133.5	-134.5	138		
As_2S_3	c		-169.0	-168.6	163.6		
Sb_2O_5	c		-980.7	-838.9	125.1		
Sb_2O_5	aq		-947.3				
Sb_2O_4	c		-808.8	-694.1			
Sb_4O_6	c	655	-1408	-1247	246.0	8.7×10^{-4}	
SbO^{2-}	aq			-345.2			
$HSbO_2$	aq			-407.9			
$SbCl_5$	liq	4	-440	-350	301.2		
$SbCl_3$	c	73	-382.0	-322.8	184.1		
SbOCl	c		-379.9				
Sb_2S_3	c	546	-174.9	-173.6	182.0		-93
SbS_2^-	aq			-54			
SbS_3^{2-}	aq		-17	-134			
Bi_2O_4	c			-456.1			
Bi_2O_3	c	824	-574.0	-493.8	151.5		
BiO	c		-208.6	-182.0	71		
BiO^+	aq			-144.5			
BiO(OH)	g			-369.9			
$Bi(OH)_3$	c		-709.6	-573.2	102.9		-30
$BiCl_3$	c	233.5	-379.1	-314.6	189.5		
$BiCl_3$	aq		-379	-425.1			
BiOCl	c	575 dec.	-367	-322	-121		-34.9
$BiCl_4^-$	aq			-477.8			
Bi_2S_3	c	850	-143.1	-140.6	200.4		-96

Sources; references 7,11,12,22

TABLE 3.13 Selenium and Tellurium Species and Some Physical Properties

Species	State	M Pt °C	H_f kJ	G_f kJ	S_f J deg^{-1}
Se^{2-}	aq		126.8	178.2	0
SeO_2	c	340	-230.1	-173.6	57
SeO_2	aq		-226.2		
SeO_3	c	118	-184.1		
SeO_3^{2-}	aq		-512.1	-373.8	16.3

Table 3.13 continued

Species	State	M Pt °C	H_f kJ	G_f kJ	S_f J deg^{-1}
SeO_4^{2-}	aq		-607.9	-441.1	23.8
HSe^-	aq		70.3	98.3	92
H_2Se	g	-65.7	85.8	71.1	221.3
H_2Se	aq		75.7	77.0	166.9
$HSeO_3^-$	aq		-516.7	-411.3	127.2
$HSeO_4^-$	aq		-598.7	-452.7	92.0
H_2SeO_3	c		-529.3		
H_2SeO_3	aq		-512.1	-425.9	191.2
H_2SeO_4	c	58	-538.1		
H_2SeO_4	aq		-607.9	-441.1	23.8
$SeCl_4$	c	305	-188.3	-97.5	184
Se_2Cl_2	liq	-85	-83.7	-48.5	188
$SeOCl_2$	liq	10.9			
Te^{2-}	aq			220.5	
HTe^-	aq			157.7	
H_2Te	g	-51	154.4	138.5	234.3
H_2Te	aq			142.7	
TeO_3	c		-348.1		
TeO_2	c	732.6	-325.1	-270.3	71.1
TeO_3^{2-}	aq		-594.5	-451.9	-8
$Te(OH)_6$	c	136		-702.5	
H_2TeO_3	c		-605.4	-484.1	199.6
H_2TeO_3	aq		-605.4		
$TeO(OH)^+$	aq			-258.5	
$TeCl_4$	c	224	-323.0	-237.2	209.2
$TeCl_6^{2-}$				-574.9	

Sources; references 7,11,12,22

REFERENCES

1. Bjerrum, J. G., Schwarzenbach, G. Sillen, L. G. Stability Constants: The Chemical Society, London; 1957.
2. Bryce-Smith, D. Heavy metals as contaminats of the human environment: Chem. Soc. Chem. Cassetts; 1975.
3. Campbell, J. A. and Whiteker, R. A. A periodic table based on potential-pH diagrams. **J. Chem. Educ.**; 1969; **46:** 90-92.
4. Cotton, F. A. and Wilkinson, G. Advanced Inorganic Chemistry. 5th. ed.: Wiley; 1988.
5. Craig, P. J. Metal cycles and biological methylation. in: Huntzinger, O., Ed. Handbook of Environmental Chemistry: Springer Verlag; 1980; **1A:** 169-227.
6. Drago,. Thermodynamic evaluation of the inert pair effect. **J. Chem. Phys.**; 1958; **62:** 353-357.

7. Editorial Board. Comprehensive Inorganic Chemistry: Pergamon press; 1973.
8. Fergusson, J. E. Inorganic Chemistry and the Earth: Pergamon Press; 1982.
9. Fergusson, J. E. Stereochemistry and Bonding in Inorganic Chemistry: Prentice Hall; 1974.
10. Förstner, U. Cadmium. in: Huntzinger, O. Handbook of Environmental Chemistry: Springer Verlag; 1980; **3A:** 59-107.
11. Garrels, R. M. and Christ, C. L. Solutions, Minerals and Equilibria: Harper Row; 1965.
12. Greenwood, N. N. and Earnshaw, A. Chemistry of the Elements: Pergamon Press; 1984.
13. Hahne, H. C. H. and Kroontje, W. Significance of pH and chloride concentration and behaviour of heavy metal pollutants: mercury(II), cadmium(II), zinc(II) and lead(II). **J. Environ. Quality;** 1973; **2:** 444-450.
14. Hem, J. D. Chemistry and occurrence of cadmium and zinc in surface water and ground water. **Water Resources Res.;** 1972; **8:** 661-679.
15. Hogfeldt, E., Compiler. Stability Constants of Metal-Ion Complexes, Part A, Inorganic Ligands: Pergamon Press, IUPAC; 1982.
16. Huheey, J. E. Inorganic Chemistry. 2nd. ed.: Harper Row; 1978.
17. Jensen, S. and Jernelov, A. Behaviour of mercury in the environment. in: Mercury Contamination in Man and his Environment, Tech. Report.: IAEA; 1972; **No. 137:** 43-47.
18. Jolly, W. L. Modern Inorganic Chemistry: McGraw Hill; 1984.
19. Kaiser, G. and Tolg, G. Mercury. in: Hutzinger, O., Ed. The Handbook of Environmental Chemistry: Springer Verlag; 1980; **3A:** 1-58.
20. Khalid, R. A. Chemical mobility of cadmium in sediment-water systems. in: Nriagu, J. O., Ed. Cadmium in the Environment. Part I. Ecological Cycling: Wiley; 1980: 257-304.
21. Kirk-Othmer. Kirk-Othmer Encyclopedia of Chemical Technology. 2nd.Ed.: Wiley-Interscience; 1964.
22. Latimer, W. M. Oxidation Potentials. 2nd. ed.: Prentice Hall; 1952.
23. Leckie, J. O. and James, R. O. Control mechanisms for trace metals in natural waters. in: Aqueous Environmental Chemistry of Metals: Ann Arbor Science; 1976: 1-76.
24. Mackay, K. M. and Mackay, R. A. Introduction to Modern Inorganic Chemistry. 2nd. ed.: Intertext; 1972.
25. Manahan, S. E. Environmental Chemistry. 3rd. Ed.: Willard Grant Press; 1979.
26. Newland, L. W. and Dawn, K. A. Lead. in: Hutzinger, O., Ed. Handbook of Environmental Chemistry: Springer Verlag; 1982; **3B:** 1-26.
27. Phillips, C. G. S. and Williams, R. J. P. Inorganic Chemistry (2 Vols.): Oxford U. Press; 1966.

28. Posselt, H. S. and Weber, W. J. Studies on the aqueous corrosion chemistry of cadmium. in: Aqueous Environmental Chemistry of Metals: Ann Arbor Science; 1976: 291-315.
29. Shannon, R. D. Revised effective ionic radii and systematic studies of interatomic distances in halides and chalcogenides. **Acta Cryst.**; 1976; **A32:** 751-767.
30. Sillen, L. G. and Martell, A. E., Compilers. Stability Constants of Metal-Ion Complexes (Spec. Pub.): Chemical Soc. London; 1964.
31. Sillen, L. G. and Martell, A. E., Compilers. Stability Constants of Metal- Ion Complexes Suppl. No. 1: Chemical Soc. London; 1971.
32. Stumm, W. and Morgan, J. J. Aquatic Chemistry: Wiley-Interscience; 1970.
33. Ullmann, F. Ullmann's Encyclopedia of Industrial Chemistry. 5th.Ed.: VCH; 1985.
34. Weber, W. J. and Posselt, H. S. Equilibrium models and precipitation reactions for cadmium(II). in: Aqueous Environmental Chemistry of Metals: Ann Arbor Science; 1976: 291-315.
35. Wells, A. F. Structural Inorganic Chemistry. 5th. ed.: Oxford U. Press; 1984.

CHAPTER 4

ANALYSIS

A significant aspect of our understanding of the chemistry and properties of the chemical elements comes from chemical analysis. Because of the rapid growth in interest in pollution and health effects, the need for analysis has increased enormously. Unfortunately skill in the use of the various techniques, and appreciation of the reliability of the analytical data has not kept up with the demand. Hence unreliable analytical data gets into the literature, and it is sometimes difficult to separate these from reliable data, because estimates of accuracy and precision are not given.

It is not possible to give an exhaustive survey of the analytical chemistry of the ten heavy elements. The objective is to highlight problem areas, especially as they relate to the validity of analytical results. Where appropriate, examples will be given for the heavy elements.

An example of the inaccuracy of analytical results can be demonstrated by the trend in the levels of chromium in blood over more than thirty years [137,211]. The typical level of chromium in blood 1950-1970 was reported as being 30-100 ppb, whereas during the 1970's the levels reported steadily dropped. It now appears that the level is more likely to be around 1 ppb. Therefore a person with a genuine high level of Cr in blood may not have been detected prior to 1970, as all levels were found to be incorrectly high. The reason appears to be the use of stainless steel needles in the earlier work. The same situation applies to lead levels in the ocean, see Fig. 8.15. The levels have decreased over the years, because of improved contamination control, not reduced lead inputs [54]. These examples highlights the need for careful, accurate and reliable analysis so safe levels can be ascertained.

INTRODUCTION

There are four major steps in analytical chemistry, (a) sample selection, (b) preparation for analysis, (c) analysis, and (d) interpretation of the results. Of these areas (c) has received most attention, and should not normally be a problem, provided analysts are aware of the limitations of the various instrumental methods, and problems relating to the matrix and interferences. Sampling (a) and interpretation of the results (d), have received least attention. There is little value in obtaining precise results if the sampling method has been flawed, or contamination occurred during handling, or the statistical analysis of the results mishandled

We will consider the four areas, using examples from the ten heavy metals. The literature on analysis of the heavy elements is extensive. Sources are Analytical Abstracts, Citation Index, Chemical Abstracts and annual reviews published in Analytical Chemistry and by the Royal Society of Chemistry (UK). Some of the more important journals are; Analytical Chemistry, Analytical Chimica Acta, Talanta, Analyst (London), Journal of the Association of Official Analysts, Zh. Anal. Khim., Fresnius' Zeit. Anal. Chem., Bunseki Kagaku and Mikrochim. Acta.

SAMPLING

Sample selection

The reason for analysing a material will determine to some extent the method of sampling [112,121,166,167,200,315]. The sampling must be such that the analytical results obtained give the information required, and that the sample is representative of the material being investigated [167,194,200, 262,264]. Hence a sampling regime needs to be set up that achieves this end as closely as possible. Obtaining a representative sample is easiest for air and water compared with solid samples (e.g. rock, dust, soil). It is important to ensure the sampling is representative and random. Wandering around, taking occasional samples is not random sampling, in fact such sampling may be biased, as frequently the obvious samples are taken, such as rock outcrops [167,262].

The samples that are encountered in environmental studies are: gases (e.g. air), water, solids (such as aerosol, particulate in water, soil, sediment, rock and dust), plant material, human tissue and fluids (including blood, urine, hair, teeth) and animal tissue. In addition samples may be taken from a host of other materials, such as metals, food, consumer materials (e.g. carpets, bricks, paint) and industrial effluent [166].

Each material has its own sampling problems and general techniques do not apply, for example sampling soil pore water is quite different to sampling river water. Some of the difficulties in sampling are problems over contamination during sampling, loss or gain of trace elements from the sample during storage and preservation, and sampling heterogeneous materials.

Contamination

Contamination of samples can occur at each stage in an analytical study. Two areas of concern are during sampling and during storage. This problem is especially relevant the lower the concentration of the analyte, and the smaller the sample being analysed. The range of sample sizes and concentration levels that can be expected for environmental samples are given in Fig. 4.1. The problem of contamination during sampling, increases the higher up the two scales. Each sample has its own contamination problems, for example the materials used for constructing soil samplers, the cleanliness of containers used in water sampling [156,316], the purity of filters used in the collection of aerosols [76,283] and the type of needles used in blood collection [344]. Surface water in the ocean is subject to contamination from the vessel and sampler and sampling must be carried out at a reasonable distance from the ship [203,261,279]. Travelling upstream in rivers minimizes the risks of contamination from the vessels and sampler. The influence of the sampler is seen from the following data [279].

Sampler type	Measured lead ng kg^{-1}	Approx. true lead* ng kg^{-1}
A	20	4
B	10	1.7
C	40	4
D	5.7	2.9
E	1.0	1.0

* All collected with sampler E

The collection of soil moisture on to blotting paper is another procedure where extremely clean conditions are required [80].

The grinding of a solid sample prior to or after sub-sampling in the laboratory can add impurities from the apparatus, for example the grinding of ultrapure graphite in an agate pestle and mortar added 0.3 μg g^{-1} of cadmium [121]. Dust can be a source of Cd, As, Se, and Pb, paint of Cd and Pb, sweat of Cd, tobacco ash of Cd, Sb, cosmetics of Pb and PVC of Pb [121]. Various plastic, rubber and paper materials have been analysed for Cd, Hg and Pb by extracting with acetic acid, and reasonably high levels of Pb were found [284]. Mercury vapour can pass through polyethylene or linear polyethylene or FEP-teflon containers into water. The transfer increases if an oxidizing agent (e.g. permanganate or nitric acid) is used to preserve the mercury in the water. Glass containers, or freezing prevent this contamination [62].

The determination of the concentrations of lead in blood is frequently carried out [69]. For children, a capillary method of sampling has been used, where only a few millilitres of blood are removed. The method is prone to severe contamination, unless great care is taken, such as shaving the skin around the

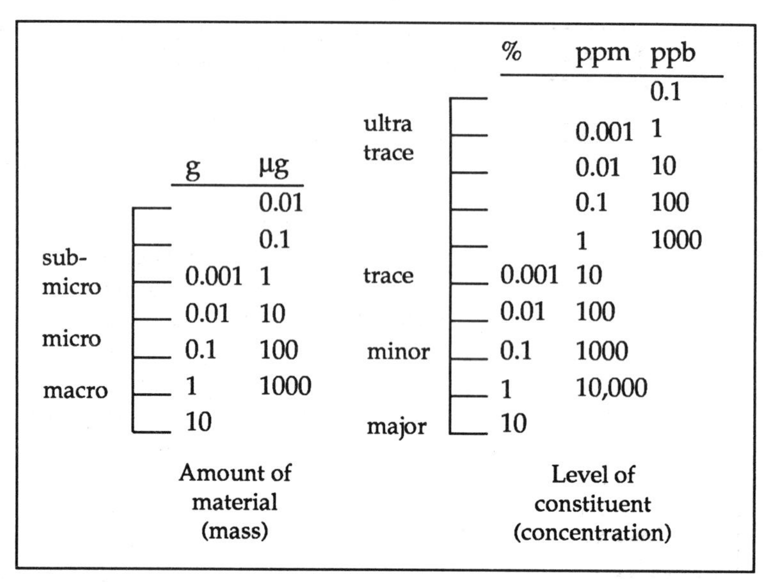

Fig 4.1 Scale of amounts and concentrations in trace element analysis (log scale).

sampling area (e.g. arm or finger) washing the area thoroughly (e.g. successively with water, acetone, 1M HCl and water) and protecting the cleaned area from the rest of the body [87]. Careful work has meant a significant fall in the lead level in plasma (most of the lead is in the red cells) i.e. 0.02 $\mu g\ l^{-1}$ (adults) and 2 $\mu g\ g^{-1}$ (lead workers) compared with earlier values sometimes many orders of magnitude greater [48,87]. The historical drop in the concentrations of the heavy metals reflect more the improvement in the control of contamination during sampling, rather than reduced inputs to the environment [201].

Losses and Preservation

A difficulty in the collection and storage of gas and water samples is loss (or gain) of material onto (or from) the surface of the container. Acidification of water to a pH of 1.5-2 is necessary, to prevent the sorption of trace metals on to the surface of containers. Details for Cd, Hg, In, Pb and Sb are given in Table 4.1, where it is clear that acidification is necessary during storage [52,158,262,269, 317]. Cooling water samples to 4° C or deep freezing also helps keep them, and preservation with CH_2Cl_2 or $HgCl_2$ can be used [112,262,307,309]. Losses from sea water in containers cleaned ultrasonically are negligible [156]. The loss of mercury from aqueous systems is avoided by adding oxidizing agents e.g. HNO_3 or $KMnO_4$ or $K_2Cr_2O_7$ to keep the metal oxidized and non-volatile [112].

For gases, there are problems of contamination of the sample with air and reaction of the gas sample with some of the constituents of air. Also losses by

TABLE 4.1 Sorption of Ions onto the Surface of Containers

Element	Conc. $\mu g\ l^{-1}$	pH	Time	% Absorption	
				Polethylene	Borosilicate
Cd	1.0	6.0	20 days	ND	20
	1.0	2.0	20 days	ND	ND
Hg	trace	5.5	1 week	82	35
	trace	1.0	2 weeks	ND	ND
In	trace	8.0	20 days	90	20
	trace	1.5	20 days	ND	-
Pb	10	6.0	4 days	76	95
	10	2.0	4 days	5	1
Sb	trace	8.1	75 days	ND	ND
	trace	1.5	75 days	ND	-

ND = not detected. Sources of data; Reeves and Brooks, 1978 [262], Struempler, 1973 [317].

sorption on to the walls of the container or gains through desorption from the walls can alter the sample composition [200,264].

Sample Homogeneity

There is, generally, little problem over sample homogeneity for air and water, but this is not the situation for many solid materials. For substances, such as rocks, soil and sediments, it is necessary to ensure that the samples taken represent the bulk, and that any sub-samples taken in the laboratory are homogeneous and are representative of the field sample [200]. Grain size within rocks needs to be considered when sampling so that the sample contains a representative selection of the different grains. The larger the grains the bigger the sample that has to be taken. When a subsample is taken for analysis, it may be so small that the chance of the larger grains being included is reduced. A sampling constant $K = R^2w$, (where R = relative standard deviation (%) and w = weight of the sub sample), should be determined to ensure the correct amount of material (w) is taken [141,167].

In the laboratory samples may be cleaned, dried, sorted, crushed (without destroying natural particles) and sub-sampled. [262,264]. Unless an adequate description be given of the sample, e.g. of a sediment core, it is possible to sub-sample incorrectly, and end up in mixing different strata in the same sample [194,349]. The particulate material in water creates an inhomogeneous sample unless the water is first filtered [112,309]. Also the spatial and temporal variation of samples can add a further inhomogeneity factor [75,97].

PREPARATION FOR ANALYSIS

The next step is preparation of the sample for analysis. This may involve a number of processes, including physical manipulation, separation of trace elements from the matrices, enrichment of trace elements and separation of the trace elements. The steps are outlined in Fig. 4.2 [217,328] for solid materials. A similar scheme would apply to liquids with the omission of some steps. At each step contamination or loss can occur and clearly the less handling the better, i.e. the further up the scheme the analysis is done the less chance there is for contamination and/or losses.

Sources of error	Analytical process		Analytical method
Inhomogeneity Contamination	Sampling ↓		
Alteration of sample Contamination Losses	Storing ↓		
Contamination handling from apparatus	Sample preparation crushing grinding surface cleaning ↓	→	Direct methods* (solid state)
Inhomogeniety	Weighing ↓	→	Direct methods* (solid state)
Contamination from air, dust reagents vessels	Dissolution, Decomp. ↓ Separation of matrices / Separation of traces	→	Direct multi-element methods* (solution, melts)
Losses from adsorption volatilization	↓ Enrichment of traces	→	Direct multi-element methods* (solution, melts)
	↓ Separation of traces	→	Single element determination* (solution, melts)

Fig. 4.2 Steps in the analysis of solids. * Also has a measurement error. Source; adapted from Tolg, 1974 [328].

Clean Environment

In order to achieve as little contamination as possible a clean laboratory is essential for analysis at the ppm and ppb levels. A class 100 clean room is defined as an area containing ≤100 particles ft^3 (≤3.5 particles l^{-1}) which are ≥0.5 μm in diameter and no particles >5μm (see Fig. 4.3) [121,165,182,202,215,217, 221,222,236,245,270,327,328,370,371]. If certain manipulations, such as wet chemistry, cannot be carried out in a clean room there needs also to be a separate clean area, or a closed working space [202,221]. A major problem in the chemical laboratory is contamination from deposited and aerosol dust. The concentration of some of the trace elements in aerosol particles are: Pb, 10^2-10^3 ng m^{-3}; Pb, As, Sb, Cd, 10-10^2ng m^{-3}; and Pb, Hg, Cd, Sb, Se, As, 1-10 ng m^{-3} [4,217,245]. High

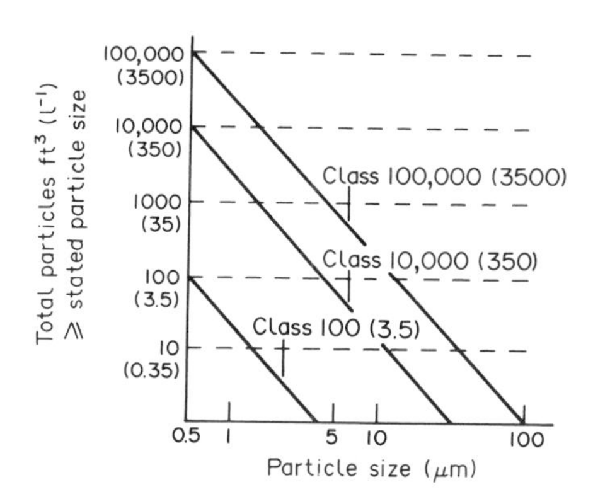

Fig 4.3 Clean roon classification, permissible particle size and number of particels per unit volume. Source of data; Moody, 1982 [221].

efficiency particulate filters (HEPF) will remove much of the dust and aerosol problem, but not gases.

A further source of contamination in the laboratory lies in the reagents and apparatus used. Containers need to be cleaned and soaked so leachable materials are removed. For example, glass may contain the elements As, Bi, Pb, Sb, and Se which contaminate solutions stored in them [202,217,242]. Attack on the glass may occur [3]. Whereas plastic containers are probably less of a problem for some of the elements they are still a problem for Pb and maybe Cd [164]. Reagent purity can be checked from the manufacturers specifications (e.g. AnalaR and Aristar) and in the case of acids and water they can be distilled by a sub-boiling technique [121,168,202,215,217,222,371]. The impurity levels in sub-boiled water and acids for the elements Pb, Tl, Te, In, Cd and Se are given in Table 4.2 [168,217,222]. In order to allow for these impurities reagent blanks must be used in any analytical study. Purification of other analytical reagents may be necessary, using techniques such as ion exchange, mercury cathode electrolysis, isopiestic distillation, low temperature sublimation, solvent ex-

TABLE 4.2 Purity of Sub-Boiled Acids and Water (ng g^{-1})

Element	HCl 31%	HNO_3 70%	$HClO_4$ 70%	H_2SO_4 96%	HF 48%	H_2O
Pb	0.07	0.02	0.2	0.6	0.05	0.008
Tl	0.01		0.1	0.1	0.1	0.01
Te	0.01	0.01	0.05	0.1	0.05	0.004
In	0.01	0.01				
Cd	0.02	0.01	0.05	0.3	0.03	0.005
Se	0.03	0.06	0.5		0.21	

Sources of data; Kuehner et al., 1972 [168], Moody, 1982 [222].

traction, zone refining, direct synthesis, crystallization and gas liquid chromatography [214]. Rubber is often used in the laboratory (e.g. gloves) and can contain Hg, Cd, Pb, Sb added in various compounds as accelerators, stabilizers, fillers, driers and pigments [270].

Decomposition and Ashing

It is frequently necessary, in the case of solids to alter the physical state (e.g. dissolve) so analysis can be carried out. Methods of decomposition depend on the materials and what analysis is required, e.g. total or extractable levels. Mineral acid dissolution is often used (e.g. HCl, HNO_3, H_2SO_4, HF, $HClO_4$, and various mixtures of these) or fusion of the sample in bisulphate, pyrosulphate, carbonate, peroxide, borates or hydroxides. Other decomposition methods that have been used are, reduction in hydrogen, oxidation and chlorination [21,200, 207,217,262,291,325,327,328]. Organic material may be removed by either wet ashing, with an oxidising acid, or dry ashing at high temperatures (400-500°C) in a furnace, or at low temperatures in an oxygen gas plasma or using radio- or microwaves [277].

A problem with various decomposition and ashing methods is that elements may be lost from the system. This is particularly a problem for most of the heavy elements, which form volatile compounds, such as chlorides [5,267] or hydrides [327]. There is also the possibility of loss of the free elements during decomposition and ashing procedures [15,277,327], because of their low melting points. The ten elements all have relatively low melting and boiling points (see Table 3.1) and in a reducing environment (during dry ashing of organic material) the free elements may be produced. Low temperature ashing or ashing in a sealed container, such as the O_2 combustion flask [116,327,336], avoids losses. Wet ashing in the open is feasible in an acidic oxidizing medium [118]. Some examples of losses during dry ashing are listed in Table 4.3 [277]. Loss of mercury is high because of its volatility. The best method is to use, either a sealed combustion chamber and then taking up the material in a solution containing an oxidizing agent (e.g. $KMnO_4$), or in an open system in the

TABLE 4.3 Loss of Elements During Dry Ashing

Element	Matrix	Temp. °C	Time (hours)	Loss (%)
As	Ox blood	850	16	35
	Rat blood	450	16	44
Cd	Tuna (food)	500	24	7
Hg	Fish (whole)	110	24	81
Pb	Animal kidney	450	?	<12
	Human rib	710	16	40

Source of data; Sansoni and Panday, 1983 [277].

presence of excess oxidizing agent (e.g. $KMnO_4$, HNO_3) to maintain the mercury in the oxidized state [327,371].

Once a material has been taken into solution there is the possibility of loss of analytes by sorption onto the walls of the container. To avoid this, it is usual to lower the pH of the solution to ≤2pH [317,327,371]. The concentration of the elements Pb, Bi, In and Sb in borosilicate glass, fall when the pH rises above 3.0, whereas for Cd and Tl the fall does not occur until the pH had risen to >6.5 [371]. In some cases (e.g Pb) the sorption is almost complete, whereas for the other ions the drop in concentration is gradual and incomplete, and for antimony its concentration increases again at high pH [371]. Sorption becomes a significant problem when the trace metal concentration <10^{-3} mole l^{-1}, and sorption on the container walls falls in the order: borosilicate glass > soda glass > platinum > quartz > polyethylene > polypropylene > teflon, but is dependent on the pH, metal ion and what other ions are present [327].

Enrichment and Separation

When the level of an analyte is low, or the amount of sample is small it may be necessary to enrich the analyte by a physical, or chemical process. It may also be necessary to separate elements from the matrix or from each other, to avoid interferences [74,181,217]. Both these processes are often achieved simultaneously.

Volatilization Enrichment can be achieved by transfer of the analyte into the vapour phase, followed by direct analysis or condensation for further treatment. Suitable volatile compounds of the heavy elements (<1000° C) are: chlorides (Cd, Hg, In, Tl, Pb, As, Sb, Bi, Se, Te), fluorides (Hg, As, Sb, Bi, Se, Te), oxides (Se, Te, As, Cd, Hg), hydrides (As, Se, Pb, Te, Sb, Bi), and free elements (Se, Te, As, Sb, Bi, Pb, Tl, Cd, Hg) [15].

Hydrides generated using acidic borohydride, can be analysed directly using atomic absorption. [132,138,147,198,216,217,262,266,268,327,355]. The analysis of mercury can be carried out similarly, but in this case free mercury is produced and flushed into the light path [61,65,77,78,192,217]. The elements, Sb, As and Se, can be distilled as bromides from $HBr/HCl/H_2SO_4$ or $HClO_4$ solutions [217,262]. The volatile compounds may be analysed using inorganic gas chromatography, or by thermochromatography which utilizes a temperature gradient in the column [366]. Alternatively volatilization of the matrix may be used. Acidic solutions of trace metals may be enriched by evaporation, without boiling, in quartz or teflon vessels [34]. Freeze drying and removal of water by dehydration are also pre-concentration techniques [120,224].

Liquid extraction Chelation of metal ions and then liquid-liquid extraction is a common method for enrichment and separation of analytes. the resulting solution may be directly analysed, (e.g. colorimetrically or by atomic absorption spectrophotometry), or treated further. A wide variety of organic agents are available for chelation [53,121,144,217,262]. Four of the more common reagents used for the heavy metals are; 8-hydroxyquinoline (oxine) for Bi, Cd, Hg, In, Pb, Sb and Tl, diphenylthiocarbazone (dithizone) for Bi, Cd, Hg, In, Pb, Te and Tl, sodium diethyldithiocarbamate (DDTC) for As, Bi, Cd, Hg, In, Pb, Sb, Se, Te and Tl, and ammonium pyrrolidinedithiocarbamate (APDC) for As, Bi, Cd, Hg, In, Pb, Sb, Se, Te and Tl [53,121,144,217,231,262]. Typical examples of extraction data for DDTC complexes are given in Table 4.4 [53]. The elements extracted at pH > 11 are, Bi, Cd, Hg(II), Pb, Tl(I), Tl(III) and those at pH >9 are, In(III), Sb(III) and Te(IV) and those extracted at pH >6 are, As(III) and Se(IV) [53]. In order to selectively extract a certain element it is often necessary to mask other elements, i.e. react the other elements with reagents that prevent them being chelated and/or extracted. Many of the chelating agents, as well as inorganic species, can be used for masking. Masking may involve precipitation, or forming stable compounds that do not react with the chelating agent. Some common masking agents for the ten heavy elements are listed in Table 4.5 [217].

Selective dissolution Acids generally dissolve extractable ions from a solid matrix, and if HF is used, the silicate matrix of rocks, soils and dusts can be destroyed, e.g.;

$$SiO_2 + 4HF \rightarrow SiF_4 + 2H_2O,$$

$$3SiF_4 + 4H_2O \rightarrow 2H_2SiF_6 + H_4SiO_4.$$

Selective dissolution of the matrix or trace elements is used for enriching trace

TABLE 4.4 Extraction of Heavy Metals with DDTC

Metal ion	Stoichio-metry	$\log K_{sp}$	Solubility mol l^{-1}	pH range for extraction	
				$CHCl_3$	CCl_4
As(III)				3 - 6	4 - 5.8
Bi	ML_3		2.1×10^{-10}	2 - 9	4 - 11
Cd	ML_2	-22.0	1.8×10^{-7}	0 - 9.5	4 - 11
Hg(II)*	ML_2	-43.8		2 - 14	4 - 11
In(III)	ML_3	-27.1			4 - 10
Pb(II)*	ML_2	-21.7	3.4×10^{-8}	3 - 9.5	4 - 11
Sb(III)				2 - 9	4 - 8.5
Se(IV)				3 - 5	4 - 6.2
Te(IV)				3 - 5	4 - 8.8
Te(VI)				3 - 5	3 - 5
Tl(I)*	ML	-10.1		2 - 14	4 - 11
Tl(III)				2 - 14	4 - 11

* Stability constants $\log K_{MLx}$ = 22.2 (Hg), 4.3 (Tl), $\beta_2 = (\log K_{ML})(\log K_{ML2})$ = 38.1 (Hg), 18.3 (Pb), 5.3 (Tl). Source of data; Cheng et al., 1982 [53].

TABLE 4.5 Masking Agents

Element	Masking Agents*
As	S^{2-}, DMP, citrate, tartrate, TGA, DDTC
Bi	X^- (F, Cl I), SCN^-, S^{2-}, OH^-, DDTC, DMP, dithizone, TGA, cysteine, thiourea, citrate, tartrate, oxalate, tiron, SSA, NTA, EDTA, TEA, DHG, triphosphate, ascorbic acid
Cd	I^-, CN^-, S^{2-}, SCN^-. cysteine, DDTC, DMP, dithizone, TGA, citrate, tartrate, glycine, DHG, NTA, EDTA, NH_3, tetren, phen
Hg	X^-(Cl, I), CN^-, SCN^-, S^{2-}, SO_3^{2-}, cysteine, DDTC, DMP, TGA, thiourea, tartrate, citrate, NTA, EDTA, TEA, DHG, trien, tren, tetren, penten, phen
In	X^-(F, Cl), SCN^-, TGA, thiourea, tartrate, EDTA, TEA
Pb	OH^-, X^-(F, Cl, I), SO_4^{2-}, S^{2-}, PO_4^{3-}, triphosphate, DDTC, DMP, TGA, acetate, citrate, tartrate, tiron, NTA, EDTA, TEA, DHG.
Sb	X^-(F, Cl, I), OH^-, S^{2-}, citrate, tartrate, oxalate, TEA, DMP, TGA
Se	X^-(F, I), S^{2-}, SO_3^{2-}, tartrate, citrate
Te	X^- (F, I), S^{2-}, SO_3^{2-}, tartrate, citrate
Tl	Cl^-, CN^-, citrate, tartrate, oxalate, TEA, DHG, NTA, EDTA, TGA

*DDTC, diethyldithiocarbamate; DHG, N,N-dihydroxyethylglycine; DMP, 2,3-dimercaptopropanol; EDTA, ethylenediaminetetraacetic acid; NTA, nitrilotriacetic acid; penten, pentaethylenehexamine; phen, 1,10-phenanthroline; SSA, sulphosalicylic acid; TEA, triethanolamine; tetren, tetraethylenepentamine; TGA, thioglycollic acid; tren, triaminotriethylamine; trien, triethylenetetramine, tiron. Source of data; Mizuiki, 1983 [217].

elements in the analysis of pure metals or alloys [217]. A variety of reagents are used for selective removal of trace elements from solids. The reagents are selected to remove the elements from certain chemical components of the matrix, such as exchangeable ions, materials soluble in weak acid i.e. carbonates, metal ions bound to organic material, metal ions attached to the oxides/ hydroxides of iron and manganese and residual silicate material. A list of some of the reagents used are given in Table 4.6 [103,136]. There is, however, some concern over this procedure, because once a reagent extracts a certain proportion of a metal ion the ion may then redistribute itself among the remaining phases. The extraction procedure therefore, interferes with the trace metal distribution in the solid [155].

Precipitation and coprecipitation Enrichment and separation can be achieved by precipitating a trace element from a matrix, or vice-versa. If the level of the trace element is low coprecipitation with a carrier precipitate may be necessary [121,217,225,262,300,327]. Some of the common precipitates, or carrier precipitates are hydroxides or oxides, e.g. $Fe(OH)_3$, $Al(OH)_3$, MnO_2, $MnO(OH)_2$, $Bi(OH)_3$, phosphates, oxalates, sulphates, sulphides e.g. CuS, HgS, CdS, PbS, carbonates, chromates, iodates, metal chelates or free elements e.g. Te, As, Au, Se [217,262,327]. The iron(III) hydroxide, $Fe(OH)_3$ enriches As, Cd, Pb and Se,

TABLE 4.6 Some Reagents used in Sequential Extractions

Species	Reagents
Exchangeable	Solutions of NaOAc or NH_4OAc or $MgCl_2$ or $BaCl_2$ at a pH around 7 - 8.
Carbonates	Weak acid solutions, e.g. HOAc or HOAc/NaOAc at a pH around 4 - 5.
Reducible	Solutions of oxalate or dithionite/citrate or NH_2OH-HCl/acid.
Organics	Most often solutions of hydrogen peroxide in acid.
Sulphides	Hydrogen peroxide or acid
Detrital silicates	HF or HF/$HClO_4$/HNO_3 or borate fusions.

Sources of data; Förstner, 1986 [103], Horowitz, 1885 [136] and see Table 8.32.

whereas $MnO(OH)_2$ enriches Tl, Sb, Bi and Pb. The sulphide CuS is used for enriching Bi, Cd, Hg, Sb, In, Tl and Pb, gold for enriching Se and Te, and CuAPDC for enriching Bi, Cd, In, Pb and Tl [217,262,327]. Sorption on to active carbon has also been used [139]. A good summary of enrichment on solid phases is given in reference [225].

Electrochemical deposition Electrolytic deposition of trace elements on to electrodes is an enrichment process. The method is selective, as it depends on the electrode potential. For example Bi, Cd, Hg, Pb, Sb and Te can, under the appropriate conditions, be depositedon to a Pt cathode as a metal coating . Mercury is often used as the enriching cathode, and the heavy elements that quantitatively deposit into mercury are Bi, Cd, Hg, In and Tl, whereas those quantitatively separated, but not into the mercury, are As, Pb, Se and Te [191,217,262]. Anodic stripping voltammetry is a sensitive analytical procedure, whereby the elements are first enriched in the mercury (as a cathode) as an amalgam, and then stripped quantitatively from the mercury (now as an anode).

Ion exchange Various cation and anion exchange resins and chelating resins have been developed for enriching and separating trace species. For example Hg, Cd, and Pb can be sorbed from water as chloro-anions on to an anion exchange resin, and the same metals are enriched from water, by use of chelating resins such as Chelex-100 [121,217,225,246,262,319,327]. Activated charcoal is also a useful material for sorbing and enriching trace elements e.g. Bi, Cd, In, Pb, Hg, Tl and methylmercury from water, with or without added chelating agents [139,217,225].

Gas and liquid chromatography Volatile species (e.g. alkyllead and mercury compounds) and species soluble in organic solvents can be separated and enriched using gas or liquid chromatography [121,217,257,262,327]. Gas chromatographs can be attached to an atomic absorption spectrometer, and separation and enrichment carried out on the GC prior to quantitative estimation using the AAS. This has been successfully used for separation and analysis of alkyllead compounds [257].

Flotation Enrichment can be achieved by using a carrier precipitate to absorb a trace element (see above), and then separate it from other solids by flotation. The required precipitate collects at the top of a liquid with the aid of a stream of air bubbles. The heavy metals As, Sb, Bi, Se, Cd, Pb and Hg have been enriched in this manner [217].

Mechanical methods For solid samples (rock, soil, dust) some separation and enrichment can be achieved by mechanical processes. These include magnetic and electrostatic separations, and fractionation according to density and particle size [103,262]. Frequently the heavy metals occur in the heavy and small size fractions [95].

ANALYSIS

After the sample has been prepared for analysis the next step is the actual analysis. A range of instrumental analytical techniques are available, and this aspect of analysis is not generally considered a problem as a lot of effort has been put into developing new and sensitive methods. It is not intended in this book to give a description of the various instrumental analytical techniques, as they are well described in a number of texts [20,29,88,252,262,265,271,282,293, 299,323].

The objective of the analysis is to obtain a result which is accurate, i.e. as close to the true value as possible. This is not always possible because of a number of factors which include contamination, inhomogeneity and interferences. The results should be precise, i.e. as consistent as possible, with a good agreement between values, even if they are not accurate. The measure of precision is given by, the range or magnitude of the standard deviation σ, or the coefficient of variation, $CV(\%) = (\sigma/\text{mean}) \times 100$. The lower the concentration of the analyte the analytical results becomes less precise [127,150,186,190,20,299]. It has been suggested that a concentration of lead at 1 ng kg^{-1} in water needs to be analysed successfully before embarking on a programme of reliably analysing lead in the marine ecosystem [44].

Sensitivity and Detection Limits

The concentrations of analytes that can be measured in various materials are decreasing, as sensitivity and detection limits of analytical techniques improve [90]. The sensitivity of a method is given by the slope of the relationship between concentration and the analytical signal. The relative sensitivities of flame AAS, for the ten elements, is demonstrated in Figs. 4.4 a, b. Cadmium is the most sensitive (i.e. shows the biggest change in absorbance with change in concen-

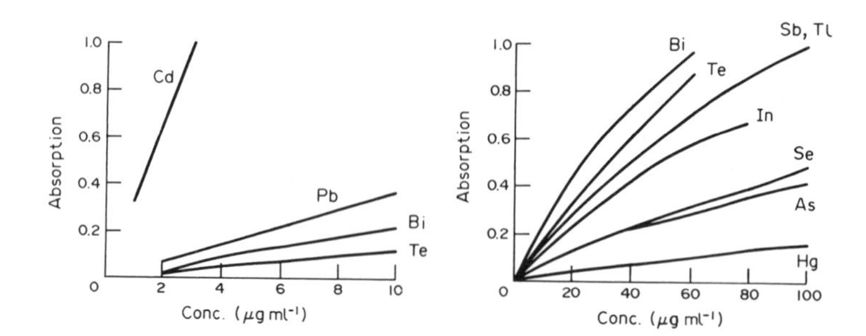

Fig. 4.4 Sensitivity of heavy metals in flame AAS, (a) Cd, Pb, Bi, Te, (b) Bi, Te, Sb, Tl, In, Se, As, Hg. Source of data; Varian, 1979 [343].

tration) and mercury is the least sensitive [343]. The sensitivity portrayed is the instrumental sensitivity, however, preconcentration of the analyte will increase the overall sensitivity.

The detection limit for an analytical method is expressed, either as an absolute limit (in ng) or as a relative limit (μg ml^{-1}) [21,76,150,186,200,299, 300,361]. The limit of detection is the lowest amount or concentration that can be reliably estimated, and has been defined as;

$$(x - x_b)_{min} = k\sqrt{(2\sigma_b)},$$

where x = sample mean, x_b = blank mean, k is a constant = 1.96 for the 95% confidence level and ≈ 3 for the 99% confidence level and σ_b = standard deviation for the blank. For k = 1.96,

$$(x - x_b)_{min} \approx 3\sigma_b,$$

which gives the minimum level of detection as $x = x_b + 3\sigma_b$ [186,200,299]. A problem in emphasising the detection limit is, that it is sometimes mistakenly believed, that the limit represents the level that can be accurately or precisely measured. The absolute detection limits for a number of techniques are listed in Table 4.7 [76,186,328] and the detection limits for specific elements are listed in Table 4.8. When a range of values are given in Table 4.8, it is because different values are given in the literature. The table is a guide to the most sensitive techniques to use for the various elements. For example AAS is good for cadmium, and NAA is good for indium. The limits represent the instrumental capability, but overall detection can be improved by prior preconcentration.

Interferences

In any analytical method it is necessary to be aware of possible chemical or instrumental interferences in the analysis. Matrix effects may be overcome by

TABLE 4.7 Absolute Detection Limits (ng)

Method	Detection limit
X-ray fluoresence	1
Colorimetry	$1 - 10^{-2}$
Anodic stripping voltammetry	$1 - 10^{-2}$
D.C. Arc emission	$1 - 10^{-2}$
Fluorimetry	$0.1 - 10^{-2}$
AAS flame	$0.1 - 10^{-3}$
AAS electrothermal	$10^{-3} - 10^{-5}$
Mass spectrometry	$10^{-3} - 10^{-6}$
Neutron activation analysis	$10^{-6} - 10^{-8}$

Sources; references 186,328.

ensuring that the calibration standards are in a similar matrix, or as close to it as possible. This is achieved exactly by the method of standard additions. It is necessary to first check if interferences occur, how they influence the result, and if necessary how to remove the effect. A significant problem in multielement techniques, such as NAA and ICP is spectral interference, where signals from two or more elements may overlap [60,85,131,148,178,197,205,206,304,353,360].

Analytical Methods

The main analytical methods, and the processes involved in trace analysis are listed in Table 4.9. Some of the more common methods are; colorimetry, atomic absorption, ICP, polarography, stripping voltammetry, XRF and NAA. Probably AAS is the most widely used method.

Quality Control

It is very necessary in analytical chemistry to use some means of quality control [2,32,40,79,83,107,127,142,145,189,209,244,255,298,299,310,312,314,316, 336,338,367]. This is most important, in order to be certain that the analytical results are as close to being accurate as possible. Quality control can be achieved by periodically analysing international standards, whose trace element levels have been determined in an interlaboratory study. Such standards are available from different sources, such as the National Bureau of Standards, USA, and the International Atomic Energy Agency, Vienna. Laboratories may also prepare their own internal standards, which can be checked against international standards. Involvement in an interlaboratory study, is helpful for a laboratory. The results of two such studies are given in Figs. 4.5 and 4.6. One study is of cadmium in a sediment [142] and the other is of lead in tooth powder [298] It is clear that a wide range of results have been obtained for the same sample.

The development of standard analytical methods has been advocated, for example, for the analysis of lead in blood and teeth [83], and standardized

TABLE 4.8 Detection Limits

Element	Colori-metry	Fluori-metry	Flame AAS	Electr0-thermal AAS	Flame emission	Polaro-graphy	ASV	DC arc emission	XRF	NAA	Mass spect.	ICP	Atomic fluores.
As	1[a]		25	0.008-0.03	100			3	$1x10^4$	0.1-0.6	0.06		
			100-300[b]	0.6-2	$1x10^4$	0.02-20	100	$1x10^5$				30-40	100
Bi	60		1	0.007-0.03	150			0.8	600	50	0.2		
			25-50	0.4	$\approx 1x10^4$	0.02-20	0.01-0.2	4000				50-1000	5
Cd	0.3	200	0.5	0.0005-0.003	100		0.2	770	5-30	0.3			
		10	1-5	0.02-0.1	$1x10^4$	0.02-20	0.01-0.2	$1.5x10^4$				0.001-2	0.001
Hg	0.5		10	0.35	35			8	$1x10^5$	1-6	0.6		
		100	50-500	2-6	$2x10^4$		40	$1x10^6$				0.01-200	0.02
In	5	4	3	0.03	2			3	1000	0.005-0.01		0.1	
		40	50	0.4-2	0.4-3	0.02-20	0.01-0.2	$1x10^4$				0.01-30	100
Pb	0.6	0.01	5	0.003-0.005	50		0.5	400	40-1000	0.3			
			10-100	0.2	100-200		$1x10^4$				2-8	10	
Sb	0.4		10	0.01	10			3	1000	0.5-1	0.2		
			30-100	0.5	600	0.02-20	0.01-0.2	$5x10^4$				30-200	50
Se	20	0.2	25	0.025-0.2	150			500		40-500	0.1		
		1-2	100-500	1	$5x10^4$	0.02-20	10	1000				30-40	40
Te	40		20	0.02-0.1	30			250	$1x10^5$	4-5	0.3		
			30-300	0.1-1	20-200	0.02-20		$4x10^5$				80-200	5
Tl	2	3	10	0.025	5			0.2	1000	40-500	0.2		
		10	30-200	1	20	0.02-20		$5x10^4$				0.01-200	8

[a] First row; absolute detection limit in ng, [b] Second row; relative detection limit in ng ml^{-1} Sources from references 148, 149, 006, 150, 180, 181, 214, 153, 206, 272.

TABLE 4.9 Some Methods of Chemical Analysis

Process	Method	Comments	References
Quantity			
Mass	Gravimetric	For major and minor constituents	
Volume	Titrimetric		
Absorption of radiation	Colorimetry (UV - visible)	Versatile for trace levels	20,88,122,250, 262,265,271,293
	Atomic absorption Flame Electrothermal Cold vapour	Simple, sensitive technique for many elements	17,20,21,37,88,122, 130,159,161,182, 183,205,206212,239, 243,247,250,262,271, 282,293,306,313,315, 318,332,334,360
Emission of radiation	Flame emission	Mainly used for Group I and II elements	20,122,161,250, 262,293
	Atomic fluoresence	Wide range of elements can be analysed	20,250,262,265,
	Spark emission		271,293
	Inductively coupled plasma		21,122,188,250, 271
	Fluorimetry	Limited; species that fluoresce/quench fluoresence	20,88,247,250, 262,265,293
Electrical	Polarography	For certain elements at trace levels	20,29,43,88,122, 250,262,265,293
	Anodic or cathodic stripping voltammetry	For selected elements that amalgamate with mercury	20,29,88,100,102, 146,165,182,235, 236,250,265,293
	Ion selective electrode	For certain elements in the ppm range	20,122,250,262, 293,337
Chromatography	Ion exchange	Pre-concentration of ions	20,265
	Liquid	Good for organometallic & volatile compounds & pre-concentration	68,88,234,265, 293
	Gas		15,271
X-rays	X-ray fluoresence	Non-destructive for major & minor elements	20,88,122,161, 250,262,265,271, 293,323
	Electron microprobe Particle induced X-rays	Sensitive technique	228,265,293
Other	Spark source mass spectrometry	Sensitive multielement technique	20,121,161,262, 265,293
	Isotope dilution mass spectrometry	Needs isotopes	20,122,293,331
	Neutron activation analysis	Multielement, sensitive for some elements	20,24,88,115,122, 161,247,250,265, 293,321,371
	Electron spin resonance	Limited to paramagnetic species	20,293

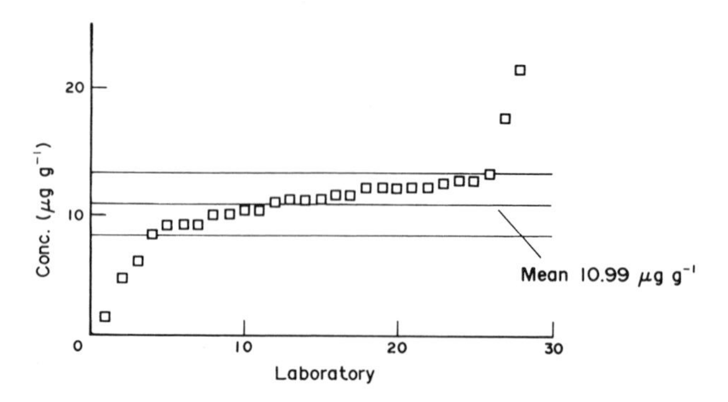

Fig. 4.5 Interlaboratory study of cadmium concentrations in a sediment. Sourceof data; IAEA, 1984 [142].

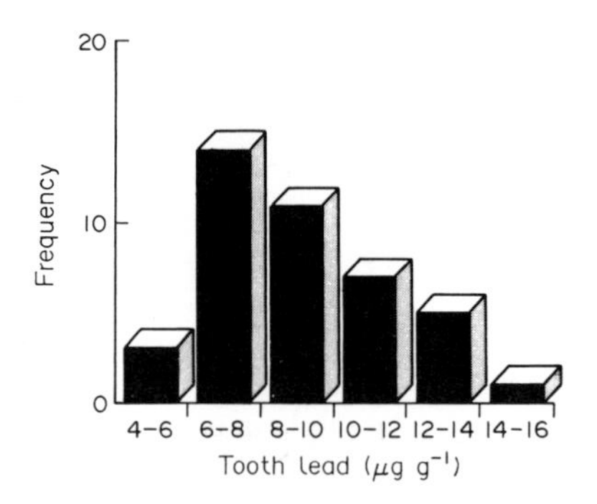

Fig. 4.6 External quality control for lead levels in teeth. Source of data; Smith et al., 1983 [298].

analytical and extraction techniques for sediments [66]. There is also value, however, in using different analytical methods, and sample preparation, to cross check analytical methods and results.

Analysis of the Elements

The analytical chemistry literature on the individual elements is vast and it is not possible to cover it in any adequate way in a short space. However, to indicate the types of analyses done recently details on a few studies are listed in Tables 4.10 to 4.15 given at the end of the chapter. The selection of references has been arbitrary and far from complete. It is necessary to consult the analytical abstracting literature to locate papers relevant to a particular study.

Of the ten elements, the most widely studied are cadmium, mercury and lead, followed by arsenic, then selenium. However, this does depend on the area of interest and the above comment applies mainly to environmental chemistry. Review articles on some of the elements are; cadmium [23,31,106,

200,297,312], mercury [23,82,173,200,321,335,363], arsenic [8,38,177,339], lead [23,31,58,59,182,200,312], selenium [11,23,200,247] and antimony [23,31,200]. A range of analytical methods and sample preparation techniques are available for most elements. It is not usually difficult to find a method which can be adapted for a particular problem. Crucial factors are, the reliability of the analytical method and its application to a range of materials.

Analysis of Materials

The type of material being analysed can determine the analytical method, though again, a range of methods are available, provided certain pre-treatments are carried out. A few of the materials that have been analysed for the heavy metals are; water [36,195,294,339], food [21,134,290,336], organometallic compounds [22,45,46,50,51,68,123,124257,278], aerosols (particulates) [18,26, 28,67,110,125,154,179,227,233,301,345,369], hair [16,49,135,180,195,294] or particular compounds [46,101,117]. Specific examples of materials are listed in the Tables 4.10 - 4.15.

To summarize the problems in trace analysis that have been discussed we will consider in more detail two examples, the analysis for lead in teeth [93], and mercury in food materials.

Lead in Teeth

Human teeth are a readily accessible biological tissue, and keep well. They have been analysed for lead in various studies, in order to classify people in terms of their lead exposure and absorption. Teeth provide an integrated historical record of a person's lead exposure and therefore have some attractive features as biological indicators compared with blood and urine [93].

Three significant problem areas exist; the lead is not homogeneously distributed throughout a tooth, the lead levels vary with tooth type, (which relates to the age of a tooth), and there are significant variations in results from different laboratories, which reflect problems with contamination, pretreatment and analytical methods (see Fig. 4.6).

Teeth consist of two main zones of hydroxyapatite, $Ca_5(PO_4)_3(OH)$. Enamel, composed of tightly packed hexagonal prisms of apatite, is formed by the time a tooth erupts into the mouth. The inner zone of teeth, called dentine, grows during the lifetime of the tooth, the apatite is packed less densely than the enamel, and has a 20-23% organic content. The area in contact with the pulpal cavity, called circumpulpal or secondary dentine, has a higher organic content as it includes the cellular membrane responsible for the formation of dentine.

Cleaning teeth It is essential to remove indigenous non-tooth material, and any contamination, prior to analysis [93,274]. Organic material can be removed with hydrogen peroxide, papain (a proteolytic enzyme that destroys protein) and hypochlorite [70,93,160,169,193,254,289,298,308]. Grease may be removed with organic solvents, or with detergents [12,70,84,98,223,298]. Abrasive methods, ultrasonic washing and a brief soak in dilute acid, have been used for surface cleaning [84,96,308,341]. Water is frequently used as a washing agent, but is probably inadequate by itself. In a number of studies, cleaning methods have not been reported, and some doubt must remain as to the reliability of the analytical data.

Unwanted material (e.g. fillings and caries) can be removed with dental drills [152,169,193,223,311], or with a prophylactic paste [25,286], or the tooth can be discarded. Many studies do not mention whether fillings or damaged areas are present or absent.

Teeth sections The teeth sections, surface enamel, bulk enamel, bulk dentine and secondary or circumpulpal dentine, as well as cementum, (the enamel-dentine junction) and plaque each present their own analytical problems. The analysis of surface enamel is subject to the greatest sources of error because the sample is small and prone to contamination. The bulk enamel is probably the material most reliably analysed because of its greater mass and low porosity. Dentine is subject to contamination problems because of its porosity, but is probably a better material than enamel for indicating lead exposure, as it forms during the life of the tooth. Secondary dentine, which is intimately associated with the blood, is a thin section around the pulp cavity and is difficult to sample. Teeth crowns are relatively easy to sample but consist of both enamel and dentine, and the relative proportion of each tissue could vary from sample to sample. Whole teeth provide a sample of sufficient size to reduce the problems of contamination, however, as for crowns, and especially for deciduous teeth where resorption of some of the root (dentine) can occur, the enamel to dentine ratio may vary from tooth to tooth.

Teeth sections may be obtained by cutting with a dental diamond saw, or grinding with a dental burr drill, or chipping, chiseling, splitting and powdering [93,96,98,113,175,176,219,229,288,287,298,352]. The risk of contamination is high especially when the tooth is powdered. These operations need to be carried out in an clean environment [83,96].

Analysis of teeth In wet methods of analysis the teeth are either dissolved in acid, most commonly nitric acid or a mixture of nitric and perchloric acids, or first dry ashed at around 400-500°C and then dissolved in acid. The lead may be measured directly in the solution or complexed with reagents such as APDC and dithizone and extracted into a solvent, such as methylisobutylketone before analysis.

For analysis with flame AAS it is necessary to preconcentrate and separate the lead, because the signal is suppressed by both the calcium and phosphate ions, reducing the detection limit. To achieve a detection limit of 1 $\mu g\ g^{-1}$ (Pb in a tooth) the calcium concentration in solution will be around 10% w/v, which is a high concentration of interfering ions. The detection limit is reduced by extraction processes, or use of a tantalum-boat in the flame, or use of electro-thermal-AAS (where the detection limit is reduced to 1-5 ng ml^{-1}).

Anodic stripping voltammetry, the next most popular analytical technique, has problems from interference from organic material in the teeth, which may be overcome by first dry ashing the sample. X-ray emission techniques such as X-ray fluorescence (XRF), and proton induced X-ray emission (PIXE) have been used. The techniques are not as sensitive as ET-AAS or ASV, but do allow for measurement of the distribution of lead across a solid sample. The XRF method (as well as acid etching) has also been used for measurements in-situ. Mass spectrometric (MS) methods offer a multielement technique of high sensitivity. For example the isotope dilution MS method has been used to measure lead levels down to 0.04 μg g^{-1} in enamel [84].

Lead levels in teeth It is not possible to arrive at a standard value, or range of concentrations, for lead in teeth, because of different teeth type and materials within a tooth, as well as different lead exposures. The data presented in Table 4.16 lists some typical levels of lead. It is difficult to discern any trends, in the overall data, because of the large variations found. A rather poorly defined drop in lead levels over the last 20 years, may suggest improved sampling and analytical methods. High lead levels in teeth of >100 μg g^{-1} generally reflect lead poisoning. However, some results may appear elevated because of the low detection limits of the instrumental technique. For example, the lead in 250 mg of tooth in a 25 ml solution (1% w/v), analysed by flame AAS, with a detection limit of 0.5 μg ml^{-1}, would only be detected above background if the concentration of lead in the tooth was >50 μg g^{-1}.

Two intrinsic sources of variation in the lead levels in teeth are the tooth type and the tooth material. Between 15% to 20% or the variance in dentine lead values has been attributed to to random errors of measurement, of which most (80%), is due to the inherent variability in the sample due to its heterogenity [96]. In order to study any trends and reduce other variations (e.g. exposure) the study of teeth from the same mouth is the best approach. Lead concentrations are found to vary with tooth type which relates to the tooth age, i.e. levels increase with age. The difference in the lead concentrations between incisors and molars appears to increase with an increase in the concentration of the lead [93]. The levels are not too different at around 5 μg g^{-1}, which may explain conflicting results in the literature.

For both permanent and deciduous teeth the lead concentrations are reported to decrease in the order; secondary dentine > dentine > enamel [10,57, 93,98,104,105,249,287,341,342], whereas in some reports the levels are greater in enamel than dentine, or there is little difference [57,119,238]. The health of the tooth [57,276,329], level of lead, age of the person [196] and contamination all have a bearing on the actual concentration obtained. Lead levels are highest in surface enamel, presumably because of surface absorption. The concentration quickly falls a few μm into the enamel [27,41,42,63,93,184]. The rapid fall off in

TABLE 4.16 Typical Lead Levels in Teeth

Sample	Range of means $\mu g\ g^{-1}$	Typical median values $\mu g\ g^{-1}$
Deciduous teeth		
Whole teeth	<0.1-50	6
incisors	3-56	6
canines	4-17	10
molars	2-20	7
Incisor enamel	4-39	
dentine	5-143	5-8
Canine enamel	37*	
dentine	7-726	
Molar enamel	29*	
dentine	4-730	
Surface enamel	31-10250	50
Enamel	<0.1-82	6
Dentine	<0.1-55	6
Secondary dentine	35-607	100
Permanent teeth		
Whole teeth	2-300	10
incisors	328*	
canines	254*	
premolars	4, 18	
molars	38*	
Surface enamel	2-2500	500
Enamel	1-80	10
Dentine	3-60	10
Secondary dentine	18-600	100
Ancient teeth	0.1-80	7

Samples with known high lead exposures are not included.
* One study only. Source of data; Fergusson and Purchase, 1987 [93].

lead levels with depth into the enamel is consistent with either a surface absorption effect, or surface contamination, or both.

The complexity of teeth as biological material for analytical studies, and the difficulty in ensuring the teeth are clean and remain free of contamination, means considerable planning must go into any analytical study. In a number of investigations one or more of these factors have been forgotten, reducing the value of the work, (not necessarily from internal comparisons but from comparison with other investigations). Even with careful planning however, sampling and maintaining the integrity of the sample during handling and analysis remain as a problem. There is also a need for an international quality control reference material, and investigation into a suitable cleaning regime.

Mercury in Food

As mercury is the most toxic of the heavy metals it is necessary to have reliable analytical methods for its determination in foods. The most serious problem in its analysis is the potential for formation of the metal (and/or other volatile species) and its loss from the system due to its volatility (M. Pt. -38.9°C, B. Pt. 357° C). The mercuric ion is readily reduced to the free element,

$$Hg^{2+} + 2e \rightarrow Hg, \quad E^{\circ} = 0.854 \text{ V}.$$

The reduction could occur when ashing samples to remove organic material, especially if there is a deficiency of dioxygen or an oxidizing agent. The oxide, HgO, which may form in oxidizing conditions, decomposes to the elements around 400-500° C.

Contamination with mercury is a problem when the levels to be determined are low, because of the wide occurrence of mercury in our environment, especially in chemical and analytical laboratories. Gaseous mercury can be absorbed into solutions in plastic containers when an oxidizing agent present is in the solution [33,62,112]. Presumably the metal absorbed through the plastic is oxidized, providing a pathway for its continuing transfer into the solution.

Organomercury compounds are the most serious toxic form of the element and they accumulate in marine organisms. Therefore it is important to establish the organomercury content of foods in addition to the total mercury content. Unlike the determination of lead in teeth the determination of mercury in foodstuffs has an extensive literature. A number of reviews on different aspects of mercury analysis are available [71,149,185,359].

Storage of samples During sample collection and storage for mercury analysis it is important to avoid losses and contamination gains. Freezing of foodstuffs prior to analysis is probably the best way to maintain the integrity of the sample [62,82,112,149,200]. Care to avoid contamination, and loss of mercury during sample preparation is also necessary [82,149,200]. It is possible mercury may be lost [19] by reduction of Hg(II) to Hg or conversion to organomercury that by bacterial activity. There is evidence that the Hg(II) ion can be reduced to the free element in teflon containers [187]. Hence samples may need to be stored in the presence of oxidizing agents. Loss of mercury during freeze drying (lyophilization) has been reported [187], but also no losses have been reported [172].

Decomposition of the Sample Except for Instrumental NAA (INAA) it is normally necessary to decompose the sample for mercury analysis, producing the Hg^{2+} ion in solution. This approach gives total mercury content. At this stage extreme care is necessary to ensure no loss of the metal occurs. Either a closed system is used, such as a Schöniger flask [116,173,280], or an open system is used in which sufficient oxidizing agent is present to oxidize the organic material and keep the mercury oxidized. In the Schöniger flask the sample is ashed in O_2 in a vessel which allows for any increase in pressure and the products taken up in an acidic oxidizing agent (e.g. $H_2SO_4/KMnO_4$). Dry ashing of a sample for mercury analysis is not feasible due to losses, even at low temperatures, see Table 4.3, [277].

Most wet digestions (ashings) are carried out in acid solution with, if necessary, an added oxidizing agent. Provided the sample does not contain a large amount of organic material, and provided sufficient oxidizing agent is added, and the ash temperature is low i.e. 50-60°C then there should be no loss of mercury [335]. Some of the oxidants used are; HNO_3, $KMnO_4$, H_2O_2, $HClO_4$, V_2O_5, $K_2Cr_2O_7$, Cl_2, $HBrO_3$, and $K_2S_2O_8$, [21,23,82,92,108,149,173,187,200,335, 354, and see references in Table 4.12]. Other methods of decomposition include digestion by heating under pressure in a sealed unit using acids or dioxygen [108,354], hydrolysis of fish at room temperature in 20% KOH [200] and irradiation with ultraviolet which releases Hg from its matrix [7].

Separation of Mercury The volatility of mercury can be made use of to separate it from its matrix. Mercury being collected for analysis after pyrolysis of samples at high temperature [149,159]. Reduction of the mercuric ion to Hg(0) is another method [61,82,192,200,218,230,335,354]. Many of the methods of preconcentration and separation discussed earlier are applicable to mercury. One particular method for mercury, is to make use of its reversible formation of amalgams with gold and silver. These metals, in the form of a wire, or on asbestos, or as wool, collect the mercury vapour which may then be released for analysis by heating [72,78,82,170,241,285,335,368].

Analysis for Mercury The two principal methods for the analysis of total mercury are AAS and NAA. Other methods such as XRF [55,149], electroanalytical [149], atomic emission [149], colorimetric [21,71,82,149,173,335,359] and mass spectrometric [173] have also been used.

The most frequently used reagent for mercury in colorimetric analysis, and in its separation, using liquid extraction, is dithizone (H_2Dz) [144,200]. It forms complexes with mercury of the type $Hg(HDz)_2$, $Hg_2(HDz)_2$, RHg(HDz), and HgCl(HDz). The bis-complex is quite stable and exists at low pH values, as the following analysis indicates.

$$Hg^{2+} + 2H_2Dz \rightleftharpoons Hg(HDz)_2 + 2H^+,$$

The Hg^{2+} ion is in the aqueous phase and $Hg(HDz)_2$ can be extracted into an organic phase (e.g. in CCl_4). The equilibrium constant K, for the reaction is given by;

$$K = \frac{[Hg(HDz)_2][H^+]^2}{[Hg^{2+}][H_2Dz]^2},$$

$$\text{i.e. } \log K = \log\frac{[Hg(HDz)_2]}{[Hg^{2+}]} - 2\log[H_2Dz] - 2pH$$

hence

$$\log\frac{[Hg]_{org}}{[Hg]_{aq}} = 2pH + 2\log[H_2Dz] + \log K.$$

Since the equilibrium constant, using CCl_4 as the solvent, is large (logK = 26.6) [200], the pH can be quite low, and still maintain a high ratio $[Hg]_{org}/[Hg]_{aq}$ corresponding to an efficient separation of the Hg(II) ion into the complex and organic phase. The reaction conditions need to be adjusted however, so that other complexes of mercury with dithizone do not form [173].

The principal atomic absorption spectrophotometric method for the analysis of mercury is the cold vapour technique, whereby free mercury is produced from an aqueous solution by reduction with Sn(II), or hydroxylamine or ascorbic acid. The mercury is flushed out with an inert gas into a cell in the AAS. The method is well documented and used widely [21,23,65,71,78,82,99,108,127, 149,185,200,335,338,354,359]. Sometimes the mercury vapour is collected using Au, Ag and Pt, which is then heated to vaporize the element into the cell of an AAS. [61,72,77,78,335]. The cold vapour method is specific for mercury, and most interferences have been removed.

Neutron activation is used either non-destructively (INAA) or destructively. Mercury has seven stable isotopes, the most abundant being ^{202}Hg, ^{200}Hg, ^{199}Hg, and ^{201}Hg and less abundant ^{198}Hg, ^{204}Hg and ^{196}Hg. They undergo (n, γ) reactions producing the five radioactive isotopes ^{197m}Hg, ^{197}Hg, ^{199m}Hg, ^{203}Hg and ^{205}Hg. The most sensitive of these for measurement by γ-ray spectroscopy is ^{197}Hg ($t_{1/2}$ = 65 hr, detection limit = 0.0003 μg), the next being ^{197m}Hg ($t_{1/2}$ = 24 hr detection limit = 0.002 μg). However, they are produced from the least abundant isotope ^{196}Hg (abundance = 0.15%), but by increasing the neutron flux or radiation time good sensitive analyses can be obtained. The problem with using INAA is that γ-ray spectral interferences from other elements in the sample are possible, and caution is needed in analysing the γ-ray spectrum. In destructive NAA the mercury is separated after irradiation using first wet digestion (maybe with a mercury carrier) and then a variety of separation methods such as, distillation, electrolysis and ion exchange [21,23,71,78,82,115,127,149,185,187, 242,335,338,356,359].

Organomercury determination A knowledge of the amount of organomercury in a sample is important because of its high toxicity. Regulatory bodies often specify the maximum permissible levels of both organomercury and inorganic mercury intake. For example for an intake of no more than 0.3 mg of mercury per week no more than 0.2 mg should be methylmercury [359].

The organomercury compounds are normally removed from a homogenized sample by extraction with benzene or toluene, with or without the addition of HCl. The different compounds may then be separated chromatographically using TLC or GC and analysed using a mass spectrometer or AAS as detectors [23,71,82,99,149,185,200,322,335,357,358,359]. Inorganic mercury can be converted into organomercury, for example with $SnMe_4$, extracted with benzene and analysed in the same way [99,354].

Some typical concentrations of mercury reported in foodstuffs are listed in Table 4.17. The concentration range can be quite wide depending on the circumstances.

TABLE 4.17 Some Levels of Mercury in Food

Foodstuff	Concentration range $\mu g\ g^{-1}$
Canned meats	0.001-0.01
Ham	0.001-0.05
Eggs (total Hg)	0.02-0.06
(MeHg)	0.02
Fish	
Canned salmon & tuna	0.04-0.44
Pike	0.19-0.59
Canned sea food	0.009-0.73
Shellfish	0.01-0.23
Fresh vegetables	0.001-0.05
Baby foods	nd-0.02
Grain flours	0.004-0.018
Cereals	0.004-0.008

Sources of data; Fricke et al., 1979 [108], Kaiser and Tolg, 1980 [149].

STATISTICAL TREATMENT OF DATA

The last stage in an analytical study is the interpretation of the results, often with the aid of statistics. It is not intended to give a review of statistics as applied to analytical work here, as there are a number of monographs available [56,186, 210,213,220,262,302,303,326]. A list of common statistical tests and notation is given in Table 4.18. The important thing for analysts to be aware of is, that they either become competent with statistics or make use of a statistician in their work. It is not uncommon to find inaccurate interpretations published because of the incorrect application of statistics, or authors trying to infer too much from the data.

TABLE 4.18 Statistical Notation and some Tests

Notation/tests	Equation
Notation	
Population mean	$\mu = \frac{1}{N}\sum_{i=0}^{N} x_i$
Sample mean	$\bar{x} = \frac{1}{n}\sum_{i=0}^{n} x_i$
Sample variance	$s^2 = \frac{1}{n-1}\sum_{i=0}^{n}(x_i - \bar{x})^2$
Standard deviation	$+\sqrt{s^2}$
Coefficient of variation %	$CV = \frac{s}{\bar{x}} 100$
Range	(lowest value - highest value)
Confidence limits (95%)	$x \pm 1.96 \frac{s}{\sqrt{n}}$
Confidence limits (99%)	$x \pm 2.58 \frac{s}{\sqrt{n}}$
Tests of significance	
F	Comparison of the variability of two sets of data $F = \frac{s_A^2}{s_B^2}$
Student t	Comparison of the means of two sets of data $t = \frac{(x_1 - x_2)}{\sigma_p \sqrt{1/n_1 - 1/n_2}}$ where $\sigma_p = \frac{(n_1-1)s_1^2 + (n_2-1)s_2^2}{n_1 - n_2 - 2}$
χ^2	Comparison of variation of a sample with theoretical $\chi^2 = \sum \frac{(f_0 - f_e)^2}{f_e}$ f_0 = frequencies in sample, f_e = frequencies expected
Relationship between variables	
Regression analysis	Relation between sets of data, one set without error
Correlation coefficient	Testing association between two or more random variables
Multiple regession analysis	Expressing one variable in terms of more than one independent variable
Factor analysis	Interpretation of the relationship between variables in a set of multivariate data
Cluster analysis	Classifying variables into groups or subgroups

TABLE 4.10 Some Examples of Analysis for Cadmium

Sample	Sample treatment	Analytical method*	Concentration	Refs.
Aerosol	Acid digest, take up in water.	ASV	1.5-41 ng m^{-3}	154
Aerosol	Acid digest	AAS		110
Aerosol	None	ET-AAS		37
Animal, urine, plant		ASV	0.57 μg l^{-1} urine	5
Aerosol, soil, dust	Acid digest and chlorate	AAS		138
Biol. material	Acid peroxide digest	ASV	ng g^{-1} level	292
Biol. material	Acid digest, APDC extraction	ET-AAS	dl 1.2 ng g^{-1}	305
Biol. sludge	Electrodeposition on to Pt	IDMS, ET-AAS, ASV		348
Blood	Add KI absorb on column and desorb	ET-AAS		259
Blood	Dilute blood	ET-AAS, AAS	0.002-0.009 μg ml^{-1}	351
Blood, teeth	Acid digest	ASV	2-18 μg dl^{-1} 5.2 μg g^{-1}	153
Blood, urine	Ashing, low temp and wet	ASV	0.3-2.0 μg l^{-1} blood	235
Eye tissue	Direct	INAA	<1 μg g^{-1}	365
Human tissue	Irradiate, wet digest, chromatography	NAA destructive		35
Kidney		NAA, Cf-252		86
in-vivo		Pu-238, Be		81
Saliva	Gel filtered to give protein fraction	ET-AAS	3.5 ppb	174
Seawater	Acidified	ET-AAS	dl 0.013 μg l^{-1}	253
Seawater	Acidified	ASV	0.042 μg l^{-1}	114
Seawater	APDC/MIBK extraction, acid extraction, ion exchange	ET-AAS	0.043 μg l^{-1}	333
Sewage sludge	Acid digest and dry ashing	ET-AAS, AAS		47
Urine	Direct	ET-AAS	0.2-15 μg l^{-1}	109
Water		ASV cellulose acetate coated mercury	dl 1.3x10^{-9}M	350
Water	Anion exchange (Cl^-, Br^-, SCN^-)	AAS		162
Water	Sorption on Fe/OH collect by flotation	NAA		364
Water effluent	APDC/MIBK extraction	AAS	<0.2-4.0 μg l^{-1}	30
Water, river	Acid/persulphate digest, APDC/MIBK extraction	AAS	dl 0.2 μg l^{-1}	324
Water, urine blood	Water, urine acidified blood deproteinated	AAS, slotted quartz tube	dl 0.004 μg l^{-1}	39
	Sorb onto Chelex-100 form XRF pellet	XRF		55

APDC Ammonium pyrrolidinedithiocarbamate, MIBK Methylisobutylketone. * ASV, anodic stripping voltammetry, AAS, atomic absorption spectroscopy, ET-AAS, electrothermal atomic absorption spectroscopy, IDMS, isotope dilution mass spectrometry, INAA, instrumental neutron activation analysis, NAA, neutron activation analysis, XRF, X-ray fluoresence, GC-MS, gas chromatographic mass spectrometry, GC-AAS, gas chromatographic atomic absorption spectroscopy

TABLE 4.11 Some Examples of Analysis for Lead

Sample	Sample treatment	Analytical method*	Concentration	Refs.
Aerosol	Acid digest and take up in water	ASV	16-550 ng m^{-3}	154
Aerosol	None	ET-AAS		37
Aerosol	Collect directly into furnace	ET-AAS		330
Aerosol	Acid digest	AAS		110
Aerosol, soil dust	Acid/ chlorate digest	AAS		138
Biological	Acid peroxide digest	ASV	ng g^{-1} level	292
Biol. sludge	Electrodeposition onto Pt	IDMS, ET-AAS, ASV		348
Blood	Dilute	ET-AAS, AAS	0.09-0.32 μg ml^{-1}	351
Blood	Solubilise with Me_4NOH	ET-AAS		14
Blood	Acid digest	ASV		204
Blood	Direct	ET-AAS	dl 0.02 μg ml^{-1}	13
Blood, teeth	Acid digest	ASV	1.8-18 μg dl^{-1} 5.2 μg g^{-1}	153
Blood, urine	Ash low temp, & wet ash	ASV	70-190 μg l^{-1} blood	235
Blood, urine	Dilute blood, acidify urine	ET-AAS	dl 3ppb	251
Bone	Nitric acid digest, La to suppress matrix	ET-AAS	14-61 μg g^{-1} ash	362
Fish, shell	Powder and direct	AAS (Ta cup)		91
Hair, teeth	Nitric acid digest	AAS	4-5 μg g^{-1} teeth 1-10 μg g^{-1}hair	289
Organo	Sorb on column and desorb	AAS (various modes)	dl 0.03 ng	258
Organo		GC-AAS, GC-MS		46
Organo	Sorb on column and desorb	AAS	dl 1 ng l^{-1}	272
Organo		GC-AAS	dl 0.1 ng	51
Organo, aerosol	Organo in ICl, aerosol in acid ash and take up in acid	ET-AAS	8-10 ng m^{-3}	67
Plant	Make into a disc	XRF	150-1100 μg g^{-1}	157
Plant, animal urine		ASV	129 μg l^{-1} urine	5
Sea water	APDC/MIBK extraction, acid extraction	ET-AAS	0.28 μg l-1	333
Sea water	Acidification	ASV	0.095 μg l^{-1}	114
Sewage sludge	Acid digest, dry ash	ET-AAS, AAS		47
Soil, paint plastic	Acid digest, ppte as chromate redissolve	AAS	Range	199
Teeth	Acid digest, dry ash, APDC extraction	AAS	1-27 μg g^{-1}	70
Water	Anion exchange (Cl^-, Br^-)	AAS		162
Water		ASV cellulose acetate coated mercury	$7x10^{-10}$ M	350
Water	Hydride generated	AAS (hydride), ASV ET-AAS	8-567 μg l^{-1}	347
Water effluent	APDC/MIBK extraction	AAS	<10 μg l^{-1}	30
Water, organo	Separate organolead by extr. react with ICl	AAS, GLS-MS		232
Water, river	Acid/persulphate digest APDC/MIBK extraction	AAS	dl 2.5 μg l^{-1}	324

Table 4.11 continued

Sample	Sample treatment	Analytical method*	Concentration	Refs.
Water, river	Extr. 8-hydroxyquinoline	AAS		111
Water, urine, blood	Water, urine acidified, blood deproteinated	AAS with slotted quartz tube	dl 0.03 $\mu g\ l^{-1}$	39

APDC Ammonium pyrrolidinedithiocarbamate, MIBK Methylisobutylketone. * See Table 4.10

TABLE 4.12 Some Examples of Analysis for Mercury

Sample	Sample treatment	Analytical method	Concentration	Refs.
Air	Collect on gold & release at 500° C	AAS		285
Air	Sorb on gold and desorb	AAS cold vapour	dl 0.1 ng	77
Blood	Combust in oxygen, sorb onto Cu & direct analysis	ET-AAS	3-11 $ng\ l^{-1}$	73
Fish	Acid peroxide digest	AAS cold vapour	0.1-1 $\mu g\ g^{-1}$	65
Fish, grain	None	INAA	$4x10^{-6}$ µg	115
Fish, blood, urine	Interlab.	NAA, AAS	0.03-0.21 $\mu g\ g^{-1}$ fish	338
Flour	Interlab.	INAA, NAA destruct.	30-301 ppb	127
Fungicides		MS		140
Human tissue	Sequential precipitation	NAA destructive	1.3-1.8 $\mu g\ l^{-1}$	64
Human tissue	Irradiate, wet digest, chromatography	NAA destructive		35
Marine organ.	Acid digest, ion exchange	NAA	0.43-8.8 $\mu g\ g^{-1}$	242
Organo		GC, TLC	> 0.5 ng	322
Organo		GC-AAS, GC-MS		46
Tobacco prods.	Direct	INAA	0.02-0.08 ppm	9
Vapour	Collect on gold, heat to desorb	ET-AAS		170, 368
Various	Pyrolyse, collect with gold	AAS cold vapour	dl 5 ppb	78
Various	Add Hg carrier, wet digest distill, electrolyse	NAA destructive	0.3-1 $ng\ g^{-1}$	356
Water	Reduction with Sn(II)	AAS cold vapour	100-1200 $ng\ l^{-1}$	192
Water	Sorb on Fe/OH + ADPC collect by flotation	NAA		364
Water	Redn. with Sn(II), collect on Au	AAS cold vapour	0.8-75 $ng\ l^{-1}$	61
Water	Collect on Ag, reduce organoHg	AAS cold vapour		72
Water	Sorb onto carbon, ash collect on gold	AAS cold vapour	1-3 $ng\ l^{-1}$	241
Water	Separate org. and inorg. using borohydride and Fe(III)	AAS	0.2-8 ppb	218
Water, fish sediment, plant	Acid/permanganate/persulphate	AAS cold vapour		1
Water, waste effluent	Acid digest, Sn(II) reduction	AAS cold vapour		230
	Sorb onto Chelex-100 and make XRF pellets	XRF		55
	Sorb onto 8-hydroxyquinoline resin and direct analysis	ET-AAS	1-50 ng	295

TABLE 4.13 Some Examples of Analysis for Arsenic, Antimony and Bismuth

Sample	Sample treatment	Analytical method*	Concentration	Refs.
ARSENIC				
Biological		NAA destructive		216
Biological	Generate hydride, sorb on C	NAA	dl 5 ppb	133
Human tissue	Irradiate, acid digest, chromatography	NAA destructive		35
Marine organ.	Acid digest, ion exchange	NAA destructive	0.27-133 $\mu g\ g^{-1}$	242
Plant, water	Hydride evolution	AAS, hydride	0.164 $\mu g\ g^{-1}$ water	8
Sediment	Heat in HCl to give $AsCl_3$, trap	Arc emission	15-760 $\mu g\ g^{-1}$	216
Sediment	Direct	INAA	18-474 $\mu g\ g^{-1}$	89
Soil, dust, aerosol	Acid/chlorate digest	AAS, hydride		138
Syn-fuel		ET-AAS,AAS hyd. ICP	1-30 ng l^{-1}	198
Tobacco prods.	Direct	INAA	0.25-2.7 ppm	9
Urine	Hydride generation	AAS, hydride		355
Water	Sorb onto Fe/OH + APDC collect by flotation	NAA		364
Water	Inorg As reduced to As(III), org + inorg ash and reduce	Polarography	dl 1 $\mu g\ l^{-1}$	
	Analyse in presence of $CuCl_2$	CSV	0.2-20 ppb	128
	Sorb onto thiol resin, direct	ET-AAS	dl 0.32 ng ml^{-1}	296
ANTIMONY				
Biological	Generate hydride & sorb on C	NAA	dl 5 ppb	133
Blood	Dry ash, extract with DDTC	Emission spectr.		340
Organic, TiO_2	Acid/peroxide digest	ET-AAS	0.02-0.6 $\mu g\ g^{-1}$	281
Tobacco prods.	Direct	INAA	0.03-0.22 ppm	9
Water	Sorb onto Fe/OH + APDC collect by flotation	NAA		364
	Sorb onto thiol resin, direct	ET-AAS	dl 1.3 ng ml^{-1}	296
BISMUTH				
Blood	Acid digest	ASV		204
Rock	Acid digest, extr. as iodide, form EDTA complex	ET-AAS	> 10 ng g^{-1}	151
Tea, leaves	Acid digest, extr. with DDTC	ET-ASS		143
	Sorb onto thiol resin. direct	ET-AAS	dl 0.28 ng ml^{-1}	296
	Extr with DDTC, acid digest	AAS		74

APDC Ammonium pyrrolidinedithiocarbamate, DDTC Diethyldithiocarbamate.
* See Table 4.10

TABLE 4.14 Some Examples of Analysis for Indium and Thallium

Sample	Sample treatment	Analytical method*	Concentration	Refs.
INDIUM				
Rock	Acid digest, extr. elements	AAS	0.08 $\mu g\ ml^{-1}$	248
	Extract In(III)/Br	Polarography	0.3-1.6 μM	226
THALLIUM				
Biol. sludge	Electrodeposition onto Pt	NAA		364

Table 4. 14 continued

Sample	Sample treatment	Analytical method*	Concentration	Refs.
Plant	Form WC discs	XRF	dl 0.3-0.6 ppm	263
Water	Form Br^-, sorb, elute, acid digest	AAS	> 0.85 ppm	163
Water	Anion exchange (Cl^-, Br^-)	AAS		162

* See Table 4.10

TABLE 4.15 Some Examples of Analysis for Selenium and Tellurium

Sample	Sample treatment	Analytical method*	Concentration	Refs.
SELENIUM				
Biol. material	Acid digest or combust in oxygen, reduce to Se	XRF	dl 50 ng	260
Blood	Form selenadiazole & extr.	GLC	> 2ng	171
Blood serum	Dry ash in AAS, use Ni^{2+}, Ag^+	ET-AAS	dl 5 ppb	275
Eye tissue	Direct	INAA	0.1-2.5 $\mu g\ g^{-1}$	365
Human tissue	Sequential precipitation	NAA destructive	2.2-2.6 $\mu g\ g^{-1}$	64
Marine organ.	Acid digest, ion exchange	NAA destructive	0.3-2.6 $\mu g\ g^{-1}$	242
Plant, animal tissue, urine		ASV	23.6 $\mu g\ l^{-1}$ urine	5
Sea water	Form Se(VI) compound, for Se(VI) and total Se	GC	dl 0.8 ng l^{-1}	208
Sea water, urine	Preconc. on Pt by electrolysis	AAS	dl 5 ppb	191
Tobacco prods.	Direct	INAA	0.11-3.15 ppm	9
Water	Reduce to Se, sorb onto C	XRF	dl 0.05 $\mu g\ l^{-1}$	256
Water, biol.	Hydride generation	AAS, hydride		346
	Hydride generation	AAS, hydride	dl 20 pg	266
	Preconc on Hg drop	CSV		129
TELLURIUM				
Blood	Dry ash, Extr. with DDTC	Emission spectr.		6
Water	Sorb onto Fe/OH + APDC collect by flotation	NAA		364
	Te(IV) and Te(VI) separated with $TiCl_3$ and borohydride	AAS, hydride	dl 8 ng	147
	Preconc. on Hg drop	CSV		129

APDC Ammonium pyrrolidinedithiocarbamate, DDTC Diethyldithiocarbamate. * See Table 4.10, CSV cathodic stripping voltammetry.

REFERENCES

1. Abo-Rady, M. D. K. Determination of mercury in water, fish, plant and sediment samples by atomic absorption spectroscopy. **Fresenius' Z. Anal. Chem.**; 1979; **299:** 187-189.
2. Ackermann, F., Bergmann, H. and Schleichert, U. On the reliability of trace metal analysis results of intercomparison analysis of a river sediment and an estuarine sediment. **Fresenius' Z. Anal. Chem.**; 1976; **296:** 270-276.

3. Adams, P. B. Glass containers for ultrapure solutions. in: Zief, M. and Speights, R., Eds. Ultrapurity Methods and Techniques: Dekker; 1972.
4. Adeloju, S. B. and Bond, A. M. Influence of laboratory environment on the precision and accuracy of trace element analysis. **Anal. Chem.**; 1985; **57**: 1728-1733.
5. Adeloju, S. B., Bond, A. M. and Briggs, M. H. Multielement determination in biological materials by differential pulse voltammetry. **Anal. Chem.**; 1985; **57**: 1386-1390.
6. Adriano, D. C. Trace Elements in the Terrestial Environment: Springer Verlag; 1986.
7. Agemain, H. and Chau, A. S. Y. Automated method for the determination of total dissolved mercury in fresh and saline waters by ultraviolet digestion and cold vapour atomic absorption spectrometry. **Anal. Chem.**; 1978; **50**: 13-16.
8. Aggett, J. and Aspell, A. C. The determination of arsenic(III) and total arsenic by atomic absorption spectroscopy. **Analyst**; 1976; **101**: 341-347.
9. Ahmad, S., Chaudhry, M. S. and Qureshi, I. H. Determination of toxic elements in tobacco products by instrumental neutron activation analysis. **J. Radioanal. Chem.**; 1979; **54**: 331-3412.
10. Al-Naimi T., Edmonds M.I. & Fremlin, J.H. The distribution of lead in human teeth using charged particle analysis. **Phys. Med. Biol.**; 1980; **25**: 719-726.
11. Analytical Methods Committee. Determination of small amounts of selenium in organic matter. **Analyst**; 1979; **104**: 778-787.
12. Annegarn H.J., Jodaikin A., Cleaton-Jones P.E., Sellschop J.P.F., Madiba C.C. PIXIE analysis of caries related trace elements in tooth enamel. **Nucl. Instr. Methods**; 1981; **181**: 323-326.
13. Auermann, E., Heidel, G., Cumbrowski, J., Jacobi, J. and Meckel, U. Critical observations on blood lead estimations. **Dtsch. Gesundheitswes.**; 1978; **33**: 1769-1772.
14. Aungst, B. J. , Dolce, J. and Ho-Leung Fung. Solubilisation of rat whole blood and erythrocytes for automated determination of lead using atomic absorption spectrophotometry. **Anal. Lett. Part B**; 1980; **13**: 347-355.
15. Bachmann, K. Separation of trace elements in solid samples by formation of volatile inorganic compounds. **Talanta**; 1982; **29**: 1-25.
16. Bagliano, G., Benischek, F. and Huber, A. A rapid and simple method for the determination of trace metals in hair samples by atomic absorption spectrometry. **Anal. Chim. Acta**; 1981; **123**: 45-56.
17. Bailey, P., Norval, E., Kilroe-Smith, T. A., Skikne, M. I. and Roellin, H. B. Application of metal coated graphite tubes to the determination of trace metals in biological materials. I. Determination of lead in blood using a tungsten coated graphite furnace. **Microchem. J.**; 1978; **60**: 107-116.

18. Baiulescu, G. E. and Marinescu, D. M. Effect of sampling parameters on total suspended particulates and their lead content estimation. **Strab-Reinhalt. Luft.**; 1979; **39:** 79-82.
19. Baler, R. W., Wijnowich, L. and Petrie, L. Mercury losses from culture media. **Anal. Chem.**; 1975; **47:** 2464-2467.
20. Bauer, H. H., Christian, G. D. and O'Reilly, J. E. Instrumental Analysis: Allyn Bacon; 1978.
21. Benton Jones J. Developments in the measurement of trace metal constituents in food. in: Gilbert, J., Ed. Analysis of Food Contaminants: Elsevier.; 1984: 157-205.
22. Berg, S. and Jonsson, A. Analysis of airborne organic lead. in: Grandjean and Grandjean, Eds. Biological Effects of Organolead Compounds: CRC Press; 1984: 33-42.
23. Berman, E. Toxic Metals and their Analysis: Heyden; 1980.
24. Berry, D. L. Evaluation of some procedures relevant to the determination of trace elemental components in biological materials by destructive neutron activation analysis. U. S. Dept. Energy Report, IS-T-821; 1979.
25. Bloch P., Garavaglia G., Mitchell G. & Shapiro I.M. Measurement of lead content of children's teeth in situ by X-ray fluoresence. **Phys. Med. Biol.**; 1976; **20:** 56-63.
26. Bloom, H. and Noller, B. N. Application of trace analysis techniques to the study of atmospheric metal particulates. **Clean Air Symp. (May)**; 1977.
27. Bodart F., Deconninck G. & Martin M. Large scale study of tooth enamel. IEEE Trans. Nucl. Sci.; 1981; **28:** 1401-1403.
28. Boiteau, H. L. and leblois-Remond, D. Use of SF filters for monitoring atmospheric pollution by lead. **Pollut. Atmos.**; 1978; **20:** 120-122.
29. Bond, A. M. Modern Polarographic Methods In Analytical Chemistry: Dekker; 1980.
30. Bone, K. M. and Hibbert, W. D. Solvent extraction with ammonium pyrrolidenethiocarbamate and 2,6-dimethyl-4-heptanone for the determination of trace metals in effluents and natural waters. **Anal. Chim. Acta**; 1979; **107:** 219-229.
31. Boniface, H. J. Analytical Notes: a Summary of Inorganic Methods of chemical Analysis: Sigma Tech. Press; 1978.
32. Boone, J., Hearn, T. and Lewis, S. Comparison of interlaboratory results for blood lead with results from a definitive method. **Clin. Chem.**; 1979; **25:** 389-393.
33. Bothner, M. H. and Robertson, D. E. Mercury contamination of sea water samples stored in polyethylene containers. **Anal. Chem.**; 1975; **47:** 592-598
34. Boutron, C. Concentration of dilute solutions at ppb by non-boiling evaporation in quartz and teflon. **Anal. Chim. Acta**; 1972; **61:** 140-143.
35. Brandone, A., Borromi, P. A. and Genova, N. Determination of arsenic, cadmium and mercury in biological samples by neutron activation

analysis. **Radiochem. Radioanal. Lett.;** 1983; **57:** 83-94.

36. Brezonik, P. L. Analysis and speciation of trace metals in water supplies. Aqueous Environmental Chemistry of Metals: Ann Arbor Science; 1976: 167-191.
37. Brodie, K. G. and Routh, M. W. Trace analysis of lead in blood, aluminium and manganese in serum and chromium in urine by graphite furnace atomic absorption spectrometry. **Clin. Biochem.;** 1984; **17:** 19-26.
38. Brooks, R. R., Ryan, D. E. and Hanfei Zhang. Atomic absorption spectrometry and other instrumental methods for quantitative measurement of arsenic. **Anal. Chim. Acta;** 1981; **131:** 1-16.
39. Brown, A. A., Milner, B. A. and Taylor, A. Use of slotted quartz tube to enhance the sensitivity of conventional flame atomic absorption spectrometry. **Analyst;** 1985; **110:** 501-505.
40. Bruaux, P. and Svartengren, M. Assessment of human exposure to lead: comparison between Belgium, Mexico and Sweden: UNEP/WHO; 1985.
41. Brudevold F., Aasenden R., Srinivasian B.N. and Bakhos Y. Lead in enamel and saliva, dental caries and the use of enamel biopsies for. **J. Dent. Res.;** 1977; **56:** 1165-1171.
42. Brudevold F., Reda A., Aasenden R. and Bakhos Y. Determination of trace elements in surface enamel of human teeth by a new biopsy procedure. **Arch. Oral Biol.;** 1975; **20:** 667-673.
43. Buldini, P. L., Ferri, D. and Zini, Q. Differential pulse polarographic determination of inorganic and organic arsenic in natural waters. **Mikrochim. Acta;** 1980; **1:** 71-78.
44. Burnett, M. and Patterson, C. C. Analysis of natural and industrial lead in the marine ecosystems. in: Branica, M. and Konrad. Z., Eds. Lead in the Marine Environment: Pergamon; 1980: 15-30.
45. Bye, R. and Paus, P. E. Determination of alkylmercury compounds in fish tissue with an atomic absorption spectrometer used as a specific gas chromatographic detector. **Anal. Chim. Acta;** 1979; **107:** 169-175.
46. Cantillo, A. Y. and Segar, D. A. Metal species identification in the environment major challenge to the analyst. in: Int. Conf. Heavy Metals in the Environment, Toronto; 1975; 1: 183-204.
47. Carronds, M. J. T., Perry, R. and Lester, J. N. Comparison of electrothermal atomic absorption spectrometry of the metal content of sewage sludge with flame atomic absorption spectrometry in conjunction with different treatments. **Anal. Chim. Acta;** 1979; **106:** 309-317.
48. Cavalleri, A, Minoia, C., Pozzoli, L. and Baruffini, A. Determination of plasma lead levels in normal subjects and in lead exposed workers. **Brit. J. Ind. Med.;** 1978; **35:** 21-26.
49. Chatt, A., Saijad, M. DeSilva, K. N. and Secord, C. A. Human scalp hair as an epidemiologic monitor of environmental exposure to elemental pollutants. Health related monitoring of trace element pollution using

nuclear techniques. IAEA TECDOC 330: IAEA; 1985: 33-50.

50. Chau, Y. K., Wong, P.T.S. and Kramar, O. The determination of dialkyllead, trialkyllead, tetraalkyllead and lead(II) ions in water by chelation/ extraction and gas chromatography/atomic absorption. **Anal Chim. Acta**; 1983; **146:** 211-217.
51. Chau, Y. K., Wong, P. T. S. and Goulden, P. D. A gas chromatographic atomic absorption spectrophotometer system for the determination of volatile alkyl lead and selenium compounds. in: Int. Conf. Heavy Metals in the Environment, Toronto; 1975; **1:** 295-302.
52. Cheam, V. and Agemain, H. Preservation of inorganic arsenic species at microgram levels in water samples. **Analyst (London)**; 1980; **105:** 737-743.
53. Cheng, K. L., Ueno, K. and Imamura, T. Handbook of Organic Analytical Reagents: CRC Press; 1982.
54. Chow, T. J. Lead in natural waters. In: Nriagu, J. O., Ed. The Biogeochemistry of Lead in the Environment, Part A, Ecological Cycling: Elsevier/ Nth Holland; 1978: 185-218.
55. Clanet, F. and Deloncle, R. Compressible chromatography column for the determination of trace elements by X-ray fluoresence spectrometry on chelating resins. **Anal. Chim. Acta**; 1980; **117:** 343-348.
56. Cochran, W. G. Sampling Techniques. 3rd. Ed.: Wiley; 1977.
57. Cohen D.D., Clayton E. & Ainsworth T. Preliminary investigations of trace element concentrations in human teeth. **Nucl. Instr. Methods Phys. Res.**; 1981; **188:** 203-209.
58. Committee on Biological Effects of Atmospheric Pollution. Lead: Airborne Lead in Perspective: Nat. Acad. Sci.; 1972.
59. Committee on Lead in the Human Environment. Lead in the Human Environment: Nat. Acad. Sci.; 1980.
60. Cornelis, R., Hoste, J. and Versieck, J. Potential interferences inherent in neutron activation analysis of trace elements in biological materials. **Talanta**; 1982; **29:** 1029-1034.
61. Courau, P., Laumond, F. and Hardstedt-Romeo, M. Improvement of the cold-vapour atomic absorption method for the determination of mercury in sea water. in: Albaiges, J., Ed. Analytical Techniques in Environmental Chemistry: Pergamon; 1980: 607-613.
62. Cragin, J. H. Increased mercury contamination of distilled water and natural water samples caused by oxidising preservatives. **Anal. Chim. Acta**; 1979; **110:** 313-319.
63. Cutress T.W. A preliminary study of the microelement composition of the outer layer of dental enamel. **Caries Res.**; 1979; **13:** 73-79.
64. Czauderna, M. Simultaneous determination of some trace elements in biological materials by neutron activation analysis. **Int. J. Appl. Radiat. Isot.**; 1984; **35:** 681684.

65. Davidson, J. W. Improved method for determination of mercury in fish tissue using 50% hydrogen peroxide and a hot block. **Analyst (London)**; 1979; **104:** 683-687.

66. de Groot, A. J., Zschuppe, K. H. and Salomons, W. Standardization of methods of analysis for heavy metals in sediments. **Hydrobiologia**; 1982; **92:** 689-695.

67. De Jongh, W. and Adams, F. Determination of organic and inorganic lead compounds in urban air by atomic absorption spectrometry with electrothermal atomization. **Anal. Chim. Acta**; 1979; **108:** 21-30.

68. De Jonghe, W. R. A. and Adams, F. C. Measurement of organic lead in air: a review. **Talanta**; 1982; **29:** 1057-1067.

69. Delves, H. T. Analytical techniques for blood lead measurements. **J. Anal. Toxicol.**; 1977; **1:** 261-264.

70. Delves, H. T., Clayton, B. E., Carmichael, A., Bubear, M. and Smith, M. An appraisal of the analytical signifiance of tooth-lead measurements as possible indices of environmental exposure of children to lead. **Ann. Clin. Biochem.**; 1982; **19:** 329-337.

71. D'Itri, F. M. The Environmental Mercury Problem: CRC Press; 1972.

72. Dogan, S. and Haerdi, W. Pre-concentration on silver wool of volatile organomercuty compounds in natural waters and air, the determination of mercury by flameless atomic absorption spectrometry. **Int, J. Environ. Anal. Chem.**; 1978; **5:** 157-162.

73. Dogan, S. and Haerdi, W. Rapid separation on copper powder of total mercury in blood and determination of mercury by flameless atomic absorption spectrometry. **Int. J. Environ. Anal. Chem.**; 1979; **6:** 327-334.

74. Donaldson, E. M. Determination of bismuth in ores concentrates and nonferrous alloys by atomic absorption spectrophotometry after separation by diethyldithiocarbamate extraction or iron collection. **Talanta**; 1979; **26:** 1119-1123.

75. Duggan, M. Temporal and spatial variations of lead in air and in surface dust - implications for monitoring. **The Sci. Total Environ.**; 1984; **33:** 37-48.

76. Dulka, J. J. and Risby, T. A. Ultratrace metals in some environmental and biological systems. **Anal. Chem.**; 1976; **48:** 640A-653A.

77. Dumarey, R., Heindryckx, R., Dams, R. and Hoste, J. Determination of volatile mercury compounds in air with the MAS-50 mercury analyser system. **Anal. Chim. Acta**; 1979; **107:** 159-167.

78. Dumarey, R., Heindryckx, R. and Dams, R. Determination of mercury in environmental standard reference materials by pyrolysis. **Anal. Chim. Acta**; 1980; **118:** 381-383.

79. Dybczynski, R. Relative accuracy precision and frequency of use of neutron activation analysis and other techniques as revealed by the results of some recent IAEA intercomparisons. Quality assurance in biomedical

neutron activation analysis, IAEA-TCHDOC-323: IAEA; 1984: 39-51.

80. Elias, R. W., Hirao, Y. and Patterson, C. C. The circumvention of the natural biopurification of calcium along nutrient pathways by atmospheric inputs of industrial lead. **Geochim. et Cosmochim. Acta;** 1982; **46:** 2561-2580.
81. Ellis, K. J., Vartsky, D. and Cohn, S. H. A. A mobile prompt gamma in-vivo neutron activation facility. Nuclear Activation Techniques in the Life Sciences 1978: IAEA; 1979: 733-744.
82. Environmental Studies Board Panel on Mercury. An assessment of mercury in the environment: Nat. Acad. Sci.; 1978: 139-169.
83. Epidemiological study protocol on biological indicators of lead neurotoxicity in children: WHO; 1983; ICP/RCE 903.
84. Ericson J.E., Shirahata, H. & Patterson C.C. Skeletal concentrations of lead in ancient Peruvians. **New Eng. J. Med.;** 1979; **300:** 946-951.
85. Erspamer, J. P. and Niemczyk, T. M. Effect of graphite surface type on determination of lead and nickel in a magnesium chloride matrix by furnace atomic absorption spectrometry. **Anal. Chem.;** 1982; **54:** 2150-2154.
86. Evans, C. J., Cummins, P., Dutton, J., Morgan, W. D., Sivyer, A. and Ghose, R. R. A californium-252 facility for in-vivo measurement of organ cadmium. Nuclear Activation Techniques in the Life Sciences 1978: IAEA; 1979: 719-731.
87. Everson, J. and Patterson, C. C. Ultra-clean isotope dilution/mass spectrometric analysis for lead in human blood plasma indicate that most reported results are artifically high. **Clin. Chem.;** 1980; **26:** 1603-1607.
88. Ewing, G. W. Instrumental Methods of Chemical Analysis. 4th. Ed.: McGraw Hill; 1975.
89. Farmer, J. G. and Cross, J. D. The determination of arsenic in Loch Lomond sediment by instrumental neutron activation analysis. **Radiochem. Radioanal. Lett.;** 1979; **39:** 424-440.
90. Farrer, K. T. H. Tangled in the traces. **Food Tech. in Aust.;** 1978: 312-316.
91. Favretto, G.L., Pertoldi, M. G. and Favretto, L. Determination of lead in mussels by atomic absorption spectrophotometry and solid microsampling. **At. Spectro.;** 1980; **1:** 35-37.
92. Feldman, C. Perchloric acid procedure for wet ashing organics for the determination of mercury (and other metals). **Anal. Chem.;** 1974; **46:** 1606-1609.
93. Fergusson, J. E. and Purchase, N. G. The analysis and levels of lead in human teeth - a review. **Environ. Pollut.;** 1987; **46:** 11-44.
94. Fergusson, J. E. Inorganic Chemistry and the Earth: Pergamon Press; 1982.
95. Fergusson, J. E. Lead: petrol lead in the environment and its contributions to human blood lead levels. **The Sci. Total Environ.;** 1986; **50:** 1-54.
96. Fergusson, J. E., Kinzett, N., Fergusson, D. M. and Horwood, L. T. Measurement of dentine lead. **The Sci. Total Environ.;** 1989; **80:** 229-241.

97. Fergusson, J. E. The signifance of the variability in analytical results for lead, copper, nickel and zinc in street dust. **Canad. J. Chem.;** 1987; **65:** 1002-1006.

98. Fergusson J.E., Jansen M.L. & Sheat A.W. Lead in deciduous teeth in relation to environmental lead. **Environ. Techn. Lett.;** 1980; **1:** 376-383.

99. Filippelli, M. Determination of trace amounts of organic and inorganic mercury in biological materials by graphite furnace atomic absorption spectrometry and organic mercury speciation by gas chromatography. **Anal. Chem.;** 1987; **59:** 116-118.

100. Florence T. M. Electrochemical approach to trace element speciation in waters: a review. **Analyst;** 1986; **111:** 489-505.

101. Florence, T. M. and Batley, G. E. Determination of the chemical forms of trace metals in natural waters with special reference to copper, lead, cadmium and zinc. **Talanta;** 1977; **24:** 151-158.

102. Florence, T. M. Recent advances in stripping analysis. **J. Electroanal. Chem.;** 1984; **168:** 207-218.

103. Förstner, U. Chemical forms and environmental effects of critical elements in solid-waste materials: combustion residues. In: Bernhard, M., Brinckman, F. E. and Sadler, P. J., Eds. The Importance of Chemical "Speciation" in Environmental Processes, Dahlem Konferenzen: Springer Verlag; 1986: 465-491.

104. Fosse G. and Justesen N-P.B. Lead in deciduous teeth of Norwegian children. **Arch. Environ. Health;** 1978; **33:** 166-175.

105. Fremlin J.H. and Edmonds, M.I. The determination of lead in human teeth. **Nucl. Instr. Methods;** 1980; **173:** 211-215.

106. Friberg, L., Piscator, M., Nordberg, G. F. and Kjellstrom, T. Cadmiun in the Environment: CRC Press; 1971.

107. Friberg, L. and Vahter, M. WHO/UNEP pilot project on assessment of human exposure to pollutants through biological monitoring, metals component. in: Facchetti, S., Ed. Analytical Techniques for Heavy Metals in Biological Fluids: Elsevier; 1983: 1-15.

108. Fricke, F. L., Robbins, W. B. and Caruso, J. A. Trace element analysis of food and beverages by atomic absorption spectrometry. **Prog. Anal. Atom. Spectros.;** 1979; **2:** 185-286.

109. Gardiner, P. E. Ottaway, J. M. and Fell, G. S. Accuracy of the direct determination of cadmium in urine by carbon furnace atomic absorption spectrometry. **Talanta;** 1979; **26:** 841-847.

110. Geladi, P. and Adams, F. The determination of cadmium, copper, iron, lead and zinc in aerosols by atomic absorption spectrometry. **Anal. Chim. Acta;** 1978; **96:** 229-241.

111. Gohda, S., Nishikawa, Y. and Mitani, A. Preconcentration of heavy metals, as 8-hydroxyquinoline-5-sulphonic acid complexes, from water and their determination by atomic absorption spectrophotometry. **Bunseki Kagaku;** 1979; **28:** 845-890.

112. Goulden, P. D. Environmental Pollution Analysis: Heyden; 1978.

113. Grandjean, P., Hansen, O. N. and Lyngbyi, K. Analysis of lead in circumpulpal dentine in deciduous teeth. **Annals Clin. Lab. Sci.**; 1984; **14:** 270-275.

114. Green, D. G., Green, L. W., Page, J. A., Poland, J. S. and van Loon, G. The determination of copper, cadmium and lead in sea water by anodic stripping voltammetry with a thin film mercury electrode. **Canad. J. Chem.**; 1981; **59:** 1476-1486.

115. Guinn, V.P. Determination of Mercury by instrumental neutron activation analysis. Mercury Contamination in Man and his Environment, (Tech. Report): IAEA; 1972; 137: 87-97.

116. Gutenmann, W. H. and Lisk, D. J. Rapid determination of mercury in apples by modified Schöniger combustion. **J. Agric. Fd. Chem.**; 1960; **8:** 306.

117. Guy, R. D. and Chakrabarti, C. L. Analytical techniques for speciation of trace metals. Int. Conf. on Heavy metals in the Environment, Toronto; 1975: 275-294.

118. Haas, H. E. and Krivan, V. Open wet ashing of some types of biological materials for the determination of mercury and other toxic elements. **Talanta**; 1984; **31:** 307-309.

119. Haavikko, K., Antilla, A., Helle, A., and Vuori, E. Lead concentrations of enamel and dentine teeth of children from two Finnish towns. **Arch. Environ. Health**; 1984; **39:** 78-84.

120. Hall, A. and Godinho, M. C. Concentration of trace metals from natural waters by freeze-drying prior to flame atomic absorption spectrometry. **Anal. Chim. Acta**; 1980; **113:** 369-373.

121. Hamilton, E. I. Analysis for trace elements I. Sample treatment and laboratory quality control. in: Davies, B. E., Ed. **Applied Soil Trace Elements:** Wiley; 1980: 21-68.

122. Hamilton, E. I. Analysis for trace elements II. Instrumental analysis. in: Davies, B. E., Ed. **Applied Soil Trace Elements:** Wiley; 1980: 69-130.

123. Harrison, R. M., Perry, R. and Slater, D. H. An adsorption technique for the determination of organic lead in street air. **Atmos. Environ.**; 1974; **8:** 1187-1194.

124. Harrison, R. M. and Perry, R. The analysis of tetraalkyl lead compounds and their significance as urban air pollutants. **Atmos. Environ.**; 1977; **11:** 847-852.

125. Harrison, R. M. and Sturges, W. T. The measurement and interpretation Br/Pb ratios in airborne particles. **Atmos. Environ.**; 1983; **17:** 311-328.

126. Heinonen, J. Assurance and control of quality in trace element analysis. Nuclear Activation Techniques in the Life Sciences 1978: IAEA; 1979: 7-25.

127. Heinonen, J., Merten, D. and Suschny, D. The reliability of mercury analysis in environmental samples: results of an intercomparison

organised by IAEA. Mercury Contamination in Man and his Environment, (Tech. Report): IAEA; 1972; No. 137: 137-141.

128. Henze, G., Joshi, A. P. and Neeb, R. Determination of arsenic in the sub-ppb range by differential pulse cathodic stripping voltammetry. **Fresenius' Z. Anal. Chem.;** 1980; **300:** 267-272.

129. Henze, G., Tolg, G., Umland, F. and Wessling, E. Simultaneous determination of selenium and tellurium in the low ppb range by cathodic stripping voltammetry. **Fresenius' Z. Anal. Chem.;** 1979; **295:** 1-6.

130. Herber, R. F. M. Some instrumental improvements in electrothermal atomization - atomic absorption spectrometry application in biomedical research. **Spectrochim. Acta;** 1983; **38B:** 783-789.

131. Heydorn, K. and Damsgaard, E. Gains or losses of ultratrace elements in polyethylene containers. **Talanta;** 1982; **29:** 1019-1024.

132. Hinners, T. A. Arsenic speciation:limitations with direct hydride analysis. **Analyst (London);** 1980; **105:** 751-755.

133. Hoede, D. and Van der Sloot, H. A. App;ication of hydride generation for determination of antimony and arsenic in biological material by activation analysis. **Anal. Chim. Acta;** 1979; **111:** 321-325.

134. Holak, W. Analysis of foods for lead, cadmium, copper, zinc, arsenic and selenium using closed system sample digestion: collaborative study. **J. Assoc. Official Anal. Chem.;** 1980; **63:** 485-495.

135. Holzbecher, J. and Ryan, D. E. Some observations on the interpretation of hair analysis data. **Clin. Biochem.;** 1982; **15:** 80.

136. Horowitz, A. J. A primer on trace metal sediment chemistry. **US Geol. Survey, Water supply paper, 2277;** 1985.

137. Horwitz, W. The problem of utilizing trace analysis in regulatory analytical chemistry. **Chem. in Aust;** 1982; **49:** 56-63.

138. Hubert, J., Candelaria, R. M. and Applegate, H. G. Determination of lead, zinc, cadmium and arsenic in environmental samples. **Atom. Spectros.;** 1980; **1:** 90-93.

139. Huess, E. and Lieser, K. H. Separation of trace elements from sea-water by absorption, and their determination by neutron activation analysis. **J. Radioanal. Chem.;** 1979; **52:** 2276-2280.

140. Hutzinger, O., Jamieson, W. D. and Safe, S. Mass spectra of some fungicidal organomercury compounds. in: Frei, R. W. and Hutzinger, O., Eds. Analytical Aspects of Mercury and Other Heavy Metals in the Environment: Gordon Breach; 1975: 55-67.

141. Ingamells, C. O. and Switzer, P. A proposed sampling constant for the use in geochemical analysis. **Talanta;** 1973; **20:** 547-568.

142. Intercomparison of trace metal measurements in marine sediment sample SD-N-1/2: IAEA; 1984; **No. 22.**

143. Inui, T., Fudagawa, N. and Kawase, A. Extraction and atomic absorption spectrometric determination of bismuth with electrothermal atomisation. **Fresenius' Z. Anal. Chem.;** 1979; **299:** 190-193.

144. Irving, H. M. N. H. The Analytical Applications of Dithizone: CRC; 1980.
145. Iyengar, G. V. Preservation and preparation of biological materials for trace element analysis. Quality assurance in biomedical neutron activation analysis, IAEA-TECDOC-323: IAEA; 1984: 83-105.
146. Jagner, D. and Aren, K. Potentiometric stripping analysis for zinc, cadmium, lead and copper in sea water. **Anal. Chim. Acta;** 1979; **107:** 29-35.
147. Jin, K., Taga, M., Yoshida, H. and Hikime, S. Differential determination of tellurium(IV) and tellurium(VI) by atomic absorption spectrophotometry after hydride generation. Combined use of titanium(III) chloride as pre-reductant and sodium borohydride solution. **Bull. Chem. Soc. Japan;** 1979; **12:** 935-950.
148. Kabil, M. A. and Mostafa, M. A. Atomic absorption spectrophotometric determination of cadmium. **Bull. Chem. Soc. Japan;** 1985; **58:** 3667-3668.
149. Kaiser, G. and Tolg, G. Mercury. in: Hutzinger, O., Ed. The handbook of Environmental Chemistry: Springer Verlag; 1980; **3A:** 1-58.
150. Kaleman, G. and Pijpers, F. W. Quality Control in Analytical Chemistry: Wiley; 1981.
151. Kane, J. S. Determination of nanogram amounts of bismuth in rocks by atomic absorption spectrometry with electrothermal atomisation. **Anal. Chim. Acta;** 1979; **106:** 325-331.
152. Kelsall M.A. and Hunter R.E. Biological indicator of summational exposures to lead. US EPA, EPA 600/1-78-053, PB 286 196.; (1978).
153. Khandekar, R. N. and Mishra, U. C. Determination of lead, cadmium, copper and zinc in human tissues by differential pulse anodic stripping voltammetry. **Fresenius' Z. Anal. Chem.;** 1984; **319:** 577-580.
154. Khandekar, R. N., Dhaneshwar, R. G., Palrecha, M. M. and Zarapkar, L. R. Simultaneous determination of lead, cadmium and zinc in aerosols by anodic stripping voltammetry. **Fresenius' Z. Anal. Chem.;** 1981; **307:** 365-368.
155. Kheboian, C. and Bauer, C. F. Accuracy of selective extraction procedures for metal speciation in model aquatic sediments. **Anal. Chem.;** 1987; **59:** 1417-1423.
156. Kinsella, B. and Willix, R. L. Ultrasonic bath in container preparation for storage of sea water samples in trace metal analysis. **Anal. Chem.;** 1982; **54:** 2614-2616.
157. Klopfenstein, D. and Nussbaum, E. Analysis of grass samples for lead content by X-ray fluoresence. **Proc. Indiana Acad. Sci.;** 1976; **85:** 339-342.
158. Knechtel, J. R. Utilisation of the mercury(II) chloride complex for the long term storage of samples containing parts per billion levels of mercury. **Analyst (London);** 1980; **105:** 826-829.

159. Koirtyohann, S. R. and Kaiser, M. L. Furnace atomic absorption - a method approaching maturity. **Anal. Chem.**; 1982; **54:** 1515A-1524A.
160. Kollmeier, H., Seeman, J., Wittig, P. Thiele, H. and Schach, S. Altersabhnagige Kumulation von Blei in den Zahnen. **Klin. Wochenschrift;** 1984; **62:** 826-831.
161. Kopp, J. F. Current status of analytical methodology for trace metals. in: Int. Conf. on Heavy metals in the Environment, Toronto; 1975: 261-274.
162. Korkisch, J. Application of ion-exchange techniques to the analysis of natural waters for toxic metals. in: Albaiges, J., Ed. Analytical Techniques in Environmental Chemistry: Pergamon; 1980: 449-461.
163. Korkisch, J. and Steffan, I. Determination of thallium in natural waters. **Int. J. Environ. Anal. Chem.**; 1979; **6:** 111-118.
164. Kosta, L. Contamination as a limiting parameter in trace analysis. **Talanta;** 1982; **29:** 985-992.
165. Kramer, C. J. M., Yu Guo-Hui, and Duinber, J. C. Possibilities for misinterpretation in ASV-speciation studies of natural waters. **Fresenius' Z. Anal. Chem.**; 1984; **317:** 383-384.
166. Kratochvil, B., Wallace, D. and Taylor, J. K. Sampling for chemical analysis. **Anal. Chem.**; 1984; **56:** 113R-129R.
167. Kratochvil, B. Sampling for chemical analysis in the environment: statistical considerations. In: Kurtz, D. A., Ed. Trace Residue Analysis, ACS Symposium Series: ACS; 1985; **284:** 5-23.
168. Kuehner, E. C., Alvarez, R. Paulsen, P. J. and Murphy, T. J. Production and analysis of special high purity acids purified by sub-boiling distillation. **Anal. Chem.**; 1972; **44:** 2050-2056.
169. Kuhnlein H.V. and Calloway D.H. minerals in human teeth: differences between preindustrial and contemporary Hopi Indians. **Amer. J. Clin. Nutr.**; 1977; **30:** 883-886.
170. Kunert, I., Komarek, J. and Somer, L. Determination of mercury by atomic absorption spectrometry with cold-vapour and electrothermal techniques. **Anal. Chim. Acta;** 1979; **106:** 285-297.
171. Kurahashi, K., Inoue, S., Yonekura, S., Shimoishi, Y. and Toei, K. Determination of selenium in human blood by gas chromatography with electron capture detection. **Analyst (London)**; 1980; **105:** 690-695.
172. La Fleur, P. D. Retention of mercury when freeze drying biological materials. **Anal. Chem.**; 1973; **45:** 1534-1536.
173. Lamm, G. G. and Ruzicka, J. The determination of traces of mercury by spectrophotometry, atomic absorption, radioisotope dilution and other methods. Mercury Contamination in Man and his Environment, (Tech. Report): IAEA; 1972; No. 137: 111-130.
174. Langmyhr, F. J. and Eyde, B. Determination of the total content and distribution of cadmium, copper and zinc in human parotid saliva. **Anal. Chim. Acta;** 1979; **107:** 211-218.
175. Lappalainen R. and Knuuttila M. The concentrations of Pb, Cu, Co, and Ni

in extracted permanent teeth related to donor's age and elements in the soil. **Acta Odontol. Scand.**; 1981; **39:** 163-167.

176. Lappalainen R. and Knuuttila M. The distribution and accumulation of Cd, Zn, Pb, Cu, Co, Ni, Mn and K in human teeth from five different geological areas of Finland. **Arch. Oral Biol.**; 1979; **24:** 363-368.
177. Lauwerys, R. R., Buchet, J. P. and Roels, H. The determination of trace levels of arsenic in human biological materials. **Arch. Toxicol.**; 1979; **41:** 239-247.
178. Lawson, S, R. and Woodriff, R. A method for reduction of matrix interferences in a commercial electrothermal atomizer for atomic absorption spectroscopy. **Spectrochim. Acta**; 1980; **35B:** 753-763.
179. Lech, J. F., Siemer, D. and Woodriff, R. Determination of lead in atmospheric particulates by furnace atomic absorption. **Environ. Sci. Tech.**; 1974; **8:** 840-844.
180. Lenihan, J. Hair analysis prospects and pitfalls. **MLW** ; 1984: 15-19.
181. Leyden, D. E., Wegscheider, W. Bodnar, W. B. Sexton, E. D. and Nonidez, W. K. Comparison of methods of trace element enrichment for XRF determination. in: Albaiges, J., Ed. Analytical Techniques in Environmental Chemistry: Pergamon; 1980: 469-476.
182. Leyendecker, W. Blood lead determinations by AAS and anodic stripping voltammetry techniques, comparative results: AAS-DC, AAS-ETA, ASV. in: Facchetti, S., Ed. Analytical Techniques for Heavy Metals in Biological Fluids: Elsevier; 1983: 233-254.
183. Liddle, P., Athanasopoulous, N. and Grey, R. The effect of background correction speed on the accuracy of atomic absorption measurements. AA Application: GBC; 1986; 3.
184. Lindh U. & Tveit, A.B. Proton microprobe determination of fluorine depth distributions and surface. **J. Radioanal. Chem.**; 1980; **59:** 167-191.
185. Lindstedt, G. and Skerfving, S. Methods of analysis. in: Friberg, L. T. and Vostal, J. J., Eds. Mercury in the Environment: CRC Press; 1972.
186. Liteanu, C. and Rica, I. Statistical Theory and Methodology of Trace Analysis: Ellis Horwood; 1980.
187. Litman, R., Finston, H. L. and Williams, E. T. Evaluation of sample pretreatment for mercury determinations. **Anal. Chem.**; 1975; **47:** 2364-2369.
188. Locke, J. The application of plasma source atomic emission spectrometry in forensic science. **Anal. Chim. Acta**; 1980; **113:** 3-12.
189. Long, S. J., Suggs, J. C. and Walling, J. F. Lead analysis of ambient air particulates: interlaboratory evaluation of EPA lead reference method. **J. Air Pollut. Control Assoc.**; 1979; **29:** 28-31.
190. Lucas, J. M. Effect of analytical variability on measurements of population blood lead levels. **Amer. Ind. Hyg. Assoc. J.**; 1981; **42:** 88-96.
191. Lund, W. and Bye, S. Flame atomic absorption analysis for selenium after electrothermal pre-concentration. **Anal. Chim. Acta**; 1979; **110:** 279-

284.
192. Lutze, R. L. Implementation of a sensitive method for determining mercury in surface waters and sediments by cold vapour atomic absorption spectrophotometry. **Analyst;** 1979; **104:** 979-982.
193. Mackie A.C., Stephens R., Townshend A. and Waldron H.A. Tooth lead levels in Birmingham children. **Arch. Environ. Health;** 1977; **32:** 178-185.
194. Macpherson, J. M. and Lewis, D. W. What are you sampling? **J. Sediment. Petrology;** 1978; **48:** 1077-1079.
195. Maier, D., Sinemus, H. -W. and Wiedeking, E. AAS Bestimmung geloster Spurenelemente im Bodenseewasser des Uberlinger Sees. **Fresenius' Z. Anal. Chem.;** 1979; **296:** 114-124.
196. Malik S.R. and Fremlin J.H. A study of lead distribution in human teeth using charged particle activation. **Caries Res.;** 1974; **8:** 283-292.
197. Maney, J. P. and Luciano, V. J. Time resolution of interferences in electrothermal atomic absorption spectrometry. **Anal. Chim. Acta;** 1981; **125:** 183-186.
198. Manning, D. C., Ediger, R. D. and Hoult, D. W. Determination of arsenic in synthetic-fuel process waters. **At. Spectrosc.;** 1980; **1:** 52-54.
199. Markunas, L. D. Barry, E. F. Guiffre, G. P. and Litman, R. An improved procedure for the determination of lead in environmental samples by atomic absorption spectroscopy. **J. Environ. Sci. Health;** 1979; **A14:** 501-506.
200. Marr, I. L. and Cresser, M. S. Environmental Chemical Analysis: Int. Textbook Co.; 1983.
201. Mart, L., Nurnberg, H. W. and Valenta, P. Comparative base line studies on Pb-levels in European coastal waters. in: Branica, M. and Konrad, Z., Eds. **Lead in the Marine Environment:** Pergamon; 1980: 155-179.
202. Mart, L. Minimization of accuracy risks in voltammetric ultratrace determination of heavy metals in natural waters. **Talanta;** 1982; **29:** 1035-1040.
203. Mart, L. Prevention of contamination and other accuracy risks in voltammetric trace metal analysis of natural waters. **Fresenius' Z. Anal. Chem.;** 1979; **299:** 97-102.
204. Martin, M., Pelletier, M and Haerdi, W. Micro-determination of traces of lead in blood by differential pulse anodic stripping voltammetry. **Mitt. Geb. Lebensmittelunters. Hyg.;** 1980; **71:** 260-262.
205. Matousek, J. P. Interferences in electrothermal atomic absorption spectrometry, their elimination and control. **Prog. Anal. Atom. Spect.;** 1981; **4:** 247-310.
206. May, T. W. and Brumbaugh, W. G. Matrix modifier and L'vov platform for elimination of matrix interferences in the analysis of fish tissues for lead by graphite furnace atomic absorption spectrometry. **Anal. Chem.;** 1982; **54:** 1032-1037.

207. McGrath, S. P. and Cunliffe, C.H. A simplified method for the extraction of the metals Fe, Zn, Cu, Ni, Cd, Pb, Cr, Co and Mn from solis and sewage sludge. **J. Sci. Food Agric.**; 1985; **36:** 794-798.
208. Measures, C. I. and Burton, J. D. Gas chromatographic method for the determination of selenite and total selenium in sea water. **Anal. Chim. Acta;** 1980; **120:** 177-186.
209. Meeting Report. Interlaboratory lead analysis of standardised samples of seawater. **Marine Chem.;** 1974; **2:** 69-84.
210. Mendenhall, W. Introduction to Probability and Statistics. 5th. Ed.: Duxbury Press; 1979.
211. Mertz, W. Trace element nutrition in health and disease: Contribution and problems of analysis. **Clin. Chem.;** 1975; **21:** 468-475.
212. Michel, R. G., Ottaway, J. M., Sneddon, J. and Fell, G. S. Preparation and operation of selenium electrodeless-discharge lamps for use in atomic-fluoresence flame spectroscopy. **Analyst (London);** 1979; **104:** 697-.
213. Miller, J. C. and Miller, J. N. Statistics for Analytical Chemistry: Wiley; 1984.
214. Mitchell, J. W. Purification of analytical reagents. **Talanta;** 1982; **29:** 993-1002.
215. Mitchell, J. W. Ultrapurity in trace analysis. **Anal. Chem.;** 1973; **45:** 492A-500A.
216. Miyazaki, A., Kimura, A. and Umezaki, V. Determination of arsenic in sediments by chloride formation and DC plasma arc emission spectrometry. **Anal. Chim. Acta;** 1979; **107:** 395-398.
217. Mizuiki, A. Enrichment Techniques for Inorganic Trace Analysis: Springer Verlag; 1983.
218. Mizunuma, H., Morita, H., Sakurai, H. and Shimomura, S. Selective atomic absorption determination of inorganic and organic mercury by combined use of iron(III) and sodium tetrahydroborate. **Bunseki Kagaku;** 1979; **28:** 695-699.
219. Moller B., Carisson L.E., Johansson G.I., Malmqvist K.G., Hammarström L. and B. Lead levels determined in Swedish permanent teeth by particle-induced X-ray emission. **Scand. J. Work Environ. Health;** 1982; **8:** 267-272.
220. Mood, A. M., Graybill, F. A. and Boes, D. C. Introduction to the Theory of Statistics. 3rd. ed.: McGraw Hill; 1974.
221. Moody, J. R. NBS clean laboratories for trace element analyses. **Anal. Chem.;** 1982; **54:** 1358A-1376A.
222. Moody, J. R. and Beary, E. S. Purified reagents for trace metal analysis. **Talanta;** 1982; **29:** 1003-1010.
223. Moore M.R., Campbell B.C., Meredith P.A., Beattie A.D., Goldberg A. and Campbell. The association between lead concentrations in teeth and domestic water lead concentrations. **Clin. Chim. Acta;** 1978; **87:** 77-83.

224. Mor, E. D. and Beccaria, A. M. A dehydration method to prevent loss of trace elements in biological samples. in: Branica, M. and Konrad, Z., Eds. Lead in the Marine Environment: Pergamon; 1980: 53-59.
225. Murthy, R. S. S., Holzbecher, J. and Ryan, D. E. Trace element preconcentration from aqueous solutions on a solid phase. **Rev. Anal. Chem.;** 1982; **6:** 113-150.
226. Nagaosa, Y. Polarographic determination of indium(III) after salting out extraction of the bromide complex into acetonitrile. **Talanta;** 1979; **26:** 987-990.
227. Nakahara, T. Application of hydride generation techniques in atomic absorption, atomic fluoresence and plasma atomic emission spectroscopy. **Prog. Anal. Atom. Spectros.;** 1985; **6:** 163-223.
228. Natusch, D. F. S. Recent advances in surface analysis of environmental particles. in: Albaiges, J., Ed. Analytical Techniques in Environmental Chemistry: Pergamon; 1980: 501-514.
229. Needleman H.L., Gunnoe C., Leviton A., Reed R., Persie H., Maher C. and Barret. Deficits in psychologic and classroom performance of children with elevated dentine lead levels. **New Engl. J. Med.;** 1979; **300:** 689-695.
230. Nelson, K. H., Brown, W. D. and Staruch, S. J. Determination of trace levels of mercury in effluents and waste waters. in: Frei, R. W. and Hutzinger, O., Ed. in: Analytical Aspects of Mercury and Other Heavy Metals in the Environment: Gordon Breach; 1975: 77-88.
231. Newland, L. W. and Clements, R. G. Variations in extraction efficiency of aqueous cadmium(II) using APDC/MIBK procedure. in: Albaiges, J., Ed. Analytical Techniques in Environmental Chemistry: Pergamon; 1980: 621-628.
232. Noden, F. G. The determination of tetraalkyllead compounds and their degradation products in natural waters. in: Branica, M. and Konrad, Z., Eds. Lead in the Marine Environment: Pergamon; 1980: 83-91.
233. Noller, B. N. and Bloom, H. The application of graphite furnace atomic absorption spectrometry to the determination of metals in air particulates. **Clean Air;** 1980: 9-15.
234. Nriagu, J. O. Global cadmium cycle. In: Nriagu, J. O., Ed. Cadmium in the Environment Part 1. Ecological Cycles: Wiley; 1980: 1-12.
235. Nurnberg, H. W. Potentialities and applications of voltammetry in analysis of toxic trace metals in body fluids. in: Facchetti, S., Ed. Analytical Techniques for Heavy Metals in Biological Fluids: Elsevier; 1983: 209-232.
236. Nurnberg, H. W,. The voltammetric appproach in trace metal chemistry of natural waters and atmospheric precipitation. **Anal. Chim. Acta;** 1984; **164:** 1-21.
237. Oddo, N., Magistrelli, C., Galli, P. and Spiga, G. Direct analysis for metallic

elements on membranes by flameless atomic absorption. **Chim. Ind. (Milan)**; 1978; **60:** 104-110.

238. Ogawa. K. A study of trace elements in human dental tissues, on concentrations of copper and lead. **Shika Izaku**; 1983; **46:** 66-78.

239. Ohata, K. and Suzuki, M. Atomic absorption spectrometry of tellurium with electrothermal atomisation in a molybdenum micro-tube. **Anal. Chim. Acta**; 1979; **110:** 49-54.

240. Omenetto, N. Review of spectrochemical techniques. in: Facchetti, S., Ed. Analytical Techniques for Heavy Metals in Biological Fluids: Elsevier; 1983: 155-169.

241. Onishi, H., Koshima, H. and Nagai, F. Collection of mercury from dilute aqueous solutions with activated carbon. **Bunseki Kagaku**; 1979; **28:** 451-454.

242. Orvini, E. Caramella-Crespi, V. and Genova, N. Activation analysis of As, Hg and Se in some marine organisms. in: Albaiges, J., Ed. Analytical Techniques in Environmental Chemistry: Pergamon; 1980: 441-448.

243. Ottaway, J. M. Heavy metals determination by atomic absorption and emission spectrometry. in: Facchetti, S., Ed. Analytical Techniques for Heavy Metals in Biological Fluids: Elsevier; 1983: 171-208.

244. Parr, R. M. On the need for improved quality assurance in biomedical neutron activattion analysis as revealed by the results of some recent IAEA intercomparisons. Quality assurance in biomedical neutron activation analysis, IAEA-TCHDOC-**323.**: IAEA; 1984: 53-69.

245. Paulhamus, J. A. Airborne contamination. in: Zief M. and Speights, R., Eds. Ultrapurity Methods and Techniques: Dekker; 1972: 255-286.

246. Paulson, A. J. Effects of flow rate and pretreatment on the extraction of trace metals from esturaine and coastal sea water by Chelex-100. **Anal Chem**; 1986; **58:** 183-187.

247. Peterson, P. J. Girling, C. A., Klumpp, D. W. and Minski, M. J. An appraisal of neutron activation analysis and other analytical techniques for the determination of arsenic, selenium and tin in environmental samples. Nuclear Activation Techniques in the Life Sciences 1978: IAEA; 1979: 103-114.

248. Pilipenko, A. T. and Samchuk, A. I. Extraction atomic absorption determination of indium in minerals and rocks. **Zh. Anal. Khim.**; 1979; **34:** 2128-2133.

249. Pinchin M.J., Newham J. & Thompson R.J.P. Lead, copper, and cadmium in teeth of normal and mentally retarded children. **Clin. Chim. Acta**; 1978; **85:** 89-94.

250. Pinta, M. Modern Methods for Trace Element Analysis: Ann Arbor Science; 1978.

251. Pleban, P. A.. and pearson, K. H. Determination of lead in whole blood and urine using Zeeman effect flameless atomic absorption spectroscopy. **Anal. Lett. Part B**; 1979; **12:** 935-950.

252. Potts, P. J. A Handbook of Silicate Rock Analysis: Blackie; 1987.
253. Pruszkowska, E., Carnrick, G. R. and Slavin, W. Direct determination of cadmium in costal sea water by atomic absorption spectrometry with the stabelized temperature platform furnace and Zeeman background correction. **Anal. Chem.;** 1983; **55:** 182-186.
254. Purchase, N. G. and Fergusson, J. E. Lead in teeth: the influence of teeth type and sample witin a tooth on lead level. **The Sci. Total Environ;** 1986; **52:** 239-250
255. Quality assurance in biomedical neutron activation analysis: IAEA; 1984; IAEA-TECDOC-323.
256. Rabberecht, H. Van Grieken, R. and Van der Sloot, H. A. Optimized selenite determination in environmental waters by X-ray fluoresence. in: Albaiges, J., Ed. Analytical Techniques in Environmental Chemistry: Pergamon; 1980: 463-468.
257. Radziuk, B., Thomassen, Y., Van Loon, J. C. and Chau, Y. K. Determination of alkyl lead compounds in air by gas chromatography and atomic absorption spectrometry. **Anal. Chim. Acta;** 1979; **105:** 255-262.
258. Radzuk, B., Thomassen, Y., Butler, L. R. P. Van Loon, J. C. and Chau, Y. K. Study of atomic absorption and atomic fluoresence systems as detectors in gas-chromatographic determination of lead. **Anal. Chim. Acta;** 1979; **108:** 31-38.
259. Raptis, S. and Mueller, K. New methods for determination of cadmium in blood serum. **Clin. Chim. Acta;** 1978; **88:** 393-402.
260. Raptis, S. E. Wegscheider, W., Knapp, G. and Tolg, G. X-ray fluoresence determination of trace selenium in organic and biological matrices. **Anal. Chem.;** 1980; **52:** 1292-1296.
261. Raspor, B. Distribution and speciation of cadmium in natural waters. In: Nriagu, J. O., Ed. Cadmium in the Environment Part I Ecological Cycling: Wiley; 1980: 147-236.
262. Reeves, R. D. and Brooks, R. R. Trace Element Analysis of Geological Materials: Wiley; 1978.
263. Rethfeldm H. Determination of thallium in plant material by X-ray fluoresence analysis. **Fresenius' Z. Anal. Chem.;** 1980; **301:** 308.
264. Richardson, J. H. and Peterson, R. V. Introduction to analytical methods. in: Richardson, J. H. and Peterson, R. V., Eds. Systematic Materials Analysis; 1974; 1: 1-37.
265. Richardson, J. H. and Peterson, R. V. (Eds.). Systematic Materials Analysis: Academic Press; 1974; **I, II, III.**
266. Rigin, V. P. Atomic flurosence analysis with atomisation by hot inert gas. **Zh. Anal. Khim.;** 1980; **35:** 863-869.
267. Robberecht, H. J. Van Grieben, R. E., Van den Bosch, P. A., Dielstra, H. and Van den Berghe, D. Losses of metabolically incorporated selenium in common digestion procedures for biological material. **Talanta;** 1982; **29:** 1025-1028.

268. Robbins, W. B. and Caruso, J. A. Development of hydride generation methods for atomic absorption analysis. **Anal. Chem.;** 1979; **51:** 889A-899A.

269. Robertson, D. E. The absorption of trace elements in sea water on various container walls. **Anal. Chim. Acta;** 1968; **42:** 533-536.

270. Robertson, D. E. Contamination problems in trace analysis and ultrapurification. in: Zief M. and Speights, R., Eds. Ultrapurity methods and Techniques: Dekker; 1972: 207-253.

271. Robinson, J. W. Undergraduate Instrumental Analysis. 3rd. Ed.: Dekker; 1982.

272. Rohbock, E. and Mueller, J. Measurement of gaseous alkyl-lead compounds in external air. **Mikrochim. Acta;** 1979; **1:** 423-.

273. Rothery, E. (Ed.). Analytical Methods for Graphite Tube Atomizer: Varian Techtron; 1982.

274. Sachs W.H. The elemental composition of human teeth with emphasis on trace constituents. Report BNL 50346, UC-48, USEPA and US Dept. Energy.; (1978).

275. Saeed, K., Thomassen, Y. and Langmyhr, F. J. Direct electrothermal atomic absorption spectrometric determination of selenium in serum. **Anal. Chim. Acta;** 1979; **110:** 285-289.

276. Salbe, I. D., Chaudhri, M. A. and Traxel, K. Surface profiling of trace elements across pre-carious lesions in teeth. **Nucl. Instr. Methods Phys. Rev.;** 1984; **32B:** 651-653.

277. Sansoni, B. and Panday, V. K. Ashing in trace analysis of biological materials. in: Facchetti, S., Ed. Analytical Techniques for Heavy Metals in Biological Fluids: Elsevier; 1983: 91-131.

278. Schafer, M. L., James, U. R., Peeler, J. T., Hamilton, C. H. and Campbell, J. E. A method for the determination of methylmercuric compounds in fish. **J. Agric. Food Chem.;** 1975; **23:** 1079-1083.

279. Schaule, B. and Patterson, C. C. The occurrence of lead in the Northeast Pacific and the effects of anthropogenic inputs. In: Branica, M. and Konrad, Z, Eds. Lead in the Marine Environment: Pergamon; 1980: 31-43.

280. Schöniger, W. A rapid microanalytical determination of halogen in organic substances. **Mikrochim. Acta;** 1955: 123-129.

281. Schreiker, B. E. and Frei, R. W. Determination of trace amounts of antimony by flameless atomic absorption spectroscopy. in: Frei, R. W. and Hutzinger, O., Eds. Analytical Aspects of Mercury and Other Heavy Metals in the Environment: Gordon Breach; 1975: 123-130.

282. Schrenk, W. G. Analytical Atomic Spectroscopy: Plenum; 1975.

283. Schwar, M. J. R. Sampling and measurement of environmental lead present in air, surface dust and paint. **London Environ. Supple. No.1;** 1983.

284. Scott, R. O. and Ure, A. M. Some sources of contamination in trace analysis. **Proc. Soc. Analyt. Chem.;** 1972: 288-293.
285. Scullman, J. and Widmark, G. Collection and determination of mercury in air. in: Frei, R. W. and Hutzinger, O., Eds. Analytical Aspects of Mercury and Other Heavy Metals in the Environment: Gordon Breach; 1975: 69-76.
286. Shapiro I.M., Burke A., Mitchell G. and Bloch, P. X-ray fluoresence analysis of lead in teeth of urban children in situ: correlation between the tooth lead level and the concentration of blood lead and free erythroporphyrins. **Environ. Res.;** 1978; **17:** 46-52.
287. Shapiro, I. M. and Marecek, J. Dentine lead comcentration as a perdictor of neuropsychological functioning of inner city children. **Biol. Trace Element Res.;** 1984; **6:** 69-78.
288. Shapiro I.M., Mitchell G., Davidson I. and Katz S.H. The lead content of teeth: evidence establishing new minimal levels of exposure. **Arch. Environ. Health;** 1975; **30:** 483-486.
289. Shrivastavant, A. K. and Tandon, S. G. Studies on lead pollution; atomic absorption spectrophotometric determination of lead in hair and teeth samples. **Int. J. Environ. Anal. Chem.;** 1984; **17:** 293-298.
290. Sibley, T. H. and Morgan, J. J. Equilibrium speciation of trace metals in fresh water sea water mixtures. Int. Conf. on Heavy metals in the Environment, Toronto; 1975: 319-338.
291. Sinex, S. A., Cantillo, A. Y. and Heiz, G. R. Accuracy of acid extraction methods for trace metals in sediments. **Anal. Chem.;** 1980; **52:** 2342-2346.
292. Sinko, I. and Kosta, L. Determination of lead, cadmium, copper and zinc in biological materials by anodic stripping polarography. in: Frei, R. W. and Hutzinger, O., Eds. Analytical Aspects of Mercury and Other Heavy Metals in the Environment: Gordon Breach; 1975: 111-122.
293. Skoog, D. A. and West, D. M. Principles of Instrumental Analysis. 2nd. Ed.: Holt Saunders; 1980.
294. Slavin, N. The determination of trace metals in sea water. **Atom. Spectros.;** 1980; **1:** 66-71.
295. Slovak, Z. Direct sampling of ion-exchanger suspensions for atomic absorption spectrometry with electrothermal atomisation. **Anal. Chim. Acta;** 1979; **110:** 301-306.
296. Slovak, Z. and Docekal, B. Sorption of arsenic, antimony and bismuth on glycolmethacrylate gels with bound thiol groups for direct sampling in electrothermal atomic absorption spectrometry. **Anal. Chim. Acta;** 1980; **117:** 293-300.
297. Smiley, I. E. and Kessler, W. V. Analytical methods for cadmium detection. In: Mennear, J. H., Ed. Cadmium Toxicity: Dekker; 1979: 1-27.
298. Smith, M., Delves, T. Lansdown, R. Clayton, B. and Graham, P. The effects of lead exposure on urban children : the Institute of Child Health/

Southampton study. **Dev. Med. Child Neurol., Suppl. 47**; 1983; **25:** 1-54.

299. Smythe, L. E. Analytical chemistry of pollutants. in: O'M Bockris, J., Ed. Environmental Chemistry: Plenum press; 1977: 677-747.
300. Smythe, L. E. and Finlayson, R. J. Rapid spectrometric methods for trace element investigations in natural waters. Research Project, Tech. Paper 32: Australian Waters Research Council; 1978: 1-31.
301. Sneddon, J. Collection and atomic spectroscopic measurement of metal compounds in the atmosphere: a review. **Talanta**; 1983; **30:** 631-648.
302. Sokal, R. R. and Rohlf, F. J. Biometry. 2nd. Ed.: Freeman; 1981.
303. Sokal, R. R. and Rohlf, F. J. Introduction to Biostatistics: Freeman; 1973.
304. Solera, J. J., Cristiano, L. C. Conley, M. K. and Kahn, H. L. Reduction of matrix interferences in furnace atomic absorption spectrometry. **Anal. Chem.**; 1983; **55:** 204-208.
305. Sperling, K-R. Determination of heavy metals in sea-water and in marine organisms by flameless atomic absorption spectrophotometry. IX Determination of cadmium traces in biological materials by a simple extraction method. **Fresenius' Z. Anal. Chem.**; 1979; **299:** 103-107.
306. Sperling, K-R. Determination of heavy metals in sea water and in marine organisms by flameless atomic absorption spectrophotometry XI Quality criteria for graphite tubes - a warning. **Fresenius' Z. Anal. Chem.**; 1979; **299:** 206-207.
307. Sprenger, F. J. Samplimg with regard to preservation. **Gewasserschutz Wasser Abwasser**; 1978; **26:** 15-37.
308. Stack, M. V. and Delves, H. T. Tooth lead analysis - an interlaboratory survey. Int. Symp. harmonization of Collaborative Anal. Studies, Finland.; 1981: 115-118.
309. Stainton, M. R., Capel, M. J. and Armstrong, F. A. J. The chemical analysis of fresh water. **Miscel. Spec. Pub. No. 25.** 2nd. ed.: Dept. Fisheries and Environment; 1977.
310. Stevenson, C. D. Analytical advances and changing perceptions of environmental heavy metals. **J. Roy. Soc. N. Z.**; 1985; **15:** 355-362.
311. Stewart D.J. Teeth as indicators of exposure of children to lead. **Arch. Dis. Child.**; 1974; **49:** 895-897.
312. Stoeppler. M. General analytical aspects of the determination of lead, cadmium and nickel in biological fluids. in: Facchetti, S., Ed. Analytical Techniques for Heavy Metals in Biological Fluids: Elsevier; 1983.
313. Stoeppler, M. Atomic absorption spectrometry - a valuable tool for trace and ultratrace determinations of metals and metalloids in biological material. **Spectrochim. Acta**; 1983; **38B:** 1559-1568.
314. Stoeppler, M., Dürbeck, H. W. and Nürnberg, H. W. Environmental specimen banking: a challenge in trace analysis. **Talanta**; 1982; **29:** 963-972.

315. Stoeppler, M. Present potentialities and limitations of atomic absorption spectroscopy in environmental and biological materials with particular reference to lead determinations. in: Branica, M. and Konrad, Z., Eds. Lead in the Marine Environment: Pergamon; 1980: 207-224.
316. Stoeppler. M., Valenta, P. and Nurnberg, H. W. Application of independent methods and standard materials an effective approach to reliable trace and ultratrace analysis of metals and metalloids in environmental and biological matrices. **Fresenius' Z. Anal. Chem.**; 1979; **297:** 22-34.
317. Struempler, A. W,. Adsorption characteristics of silver, lead, cadmium, zinc and nickel in borosilicate glass, polyethylene and polypropylene container surface. **Anal. Chem.**; 1973; **45:** 2251-2254.
318. Stuart, D. C. Factors affecting peak shape in cold vapour atomic absorption spectrometry for mercury. **Anal. Chim. Acta**; 1979; **106:** 411-415.
319. Sturgeon, R. E., Berman, S. S., Desaulniers, A. and Russell, D. S. Preconcentration of trace metals from sea-water for determination by graphite-furnace atomic absorption spectrometry. **Talanta**; 1979; **27:** 85-94.
320. Sukhoveeva, L. N., Butrimenko, G. G. and Spivakov, B. Ya. Atomic absorption determination of gallium and indium by evaporation from a platform in a graphite furnace. **Zh. Anal. Khim.**; 1980; **35:** 649-655.
321. Takeuchi, T., Shinogi, M. and Mori, I. Volatilization losses of mercury in neutron activation analysis. **J. Radioanal. Chem.**; 1979; **53:** 81-88.
322. Tatton, J. O'G. Identification of mercurial compounds. Mercury Contamination in Man and his Environment, (Tech. Report): IAEA; 1972; No. 137: 131-135.
323. Tertian, R. and Claisse, F. Principles of Quantitative X-Ray Fluoresence Analysis: Heyden; 1982.
324. Tessier, A., Campbell, P. G. C. and Bisson, M. Evaluation of the APDC-MIBK (ammonium pyrrolidine-1-carbodithioate - isobutylmethylketone) extraction method for atomic absorption analysis of trace metals in river water. **Int. J. Environ. Anal. Chem.**; 1979; **7:** 41-54.
325. Thompson, K. C. and Wagstaff, K. Simlified method for the determination of cadmium. chromium, copper, nickel, lead and zinc in sewage sludge using atomic absorption spectrophotometry. **Analyst (London)**; 1980; **105:** 883-896.
326. Till, R. Statistical Methods for the Earth Scientist: Macmillan; 1974.
327. Tolg, G. Extreme trace analysis of the elements I Methods and problems of sample treatment, separation and enrichment. **Talanta**; 1972; **19:** 1489-1521.
328. Tolg, G. Recent problems and limitations in the analytical characterisation of high purity materials. **Talanta**; 1974; **21:** 327-345.
329. Tomek, M. and Sefflova, A. Lead content in the hard dental tissues of the first dentitions. **Csek. Hay.**; 1984; **29:** 153-158.

330. Torsi, G. and Desimoni, E. Electrostatic accumulation furnace for electrothermal atomic spectrometry. **Anal. Lett. Part A;** 1979; **12:** 1361-1366.

331. Trincherini, P. R. and Facchetti. S. Isotope dilution mass spectrometry applied to lead determination. in: Facchetti, S., Ed. Analytical Techniques for Heavy Metals in Biological Fluids: Elsevier; 1983: 255-272.

332. Tsalev, D. and Petrov, I. Pulse nebulisation of chloroform and carbon tetrachloride extracts in flame atomic absorption spectrometry. **Anal. Chim. Acta;** 1979; **111:** 155-162.

333. Tsu Kai Jan and Young, D. R. Determination of microgram amounts of some transition metals in sea water by methylisobutylketone - nitric acid successive extraction and flameless atomic absorption spectrophotometry. **Anal. Chem.;** 1978; **50:** 1250-1253.

334. Tsujii, K., Kuga, K., Murayama, S. and Yasuda, M. Evaluation of new high frequency discharge lamps for atomic absorption and atomic fluopresence spectrometry of cadmium, lead and zinc. **Anal. Chim. Acta;** 1979; **111:** 103-109.

335. Uthe, J. F. and Armstrong, F. A. J. The micro-determination of mercury and organomercury compounds in environmental materials. in: Frei, R. W. and Hutzinger, O., Eds. Analytical Aspects of Mercury and Other Heavy Metals in the Environment: Gorden Breach; 1975: 21-53.

336. Vahter, M., Ed. Assessment of human exposure to lead and cadmium through biological monitoring: UNEP/WHO; 1982.

337. van der Linden, W. E. and Oostervink, R. The formation and properties of mixed cadmium sulfide - silver sulfide membranes for electrodes selective to cadmium(II) and mercury(II). **Anal. Chim. Acta;** 1979; **108:** 169-178.

338. Van Loon, J. C. How useful are environmental chemical data. Int. Conf. Heavy Metals in the Environment, Toronto; 1975; 1: 349-355.

339. Van Loon, J. C. Trace metal. in: Van Loon, J. C., Ed. Chemical Analysis of Inorganic Constituents of Water: CRC Press; 1982: 212-237.

340. Van Montfort, P. F. E., Agterdenbos, J. and Jatte, B. A. H. G. Determination of antimony and tellurium in human blood by microwave induced emission spectrometry. **Anal. Chem.;** 1979; **51:** 1553-1557.

341. Van Wyk C.W. and Grobler S.R. Lead levels in the teeth of children from selected areas in the Cape Peninsular. **Sth. African Med. J.;** 1982; **62:** 230.

342. Van Wyk, C. W. and Grobler, S. R. Lead in deciduous teeth from selected areas. **J. Dent. Res.;** 1983; **62:** 508.

343. Varian Analytical Methods for Flame Spectroscopy (Rev. Ed.): Varian; 1979.

344. Versheck, J., Barbier, F., Cornelis, R. and Hoste, J. Sample contamination as a source of error in trace-element analysis of biological samples. **Talanta;** 1982; **29:** 973-984.

345. Vijan, P. N., Pimenta, J. A. and Raynor, A. C. Determination of metals in air particulate matter. Int. Conf. on Heavy metals in the Environment, Toronto; 1975: 339-348.
346. Vijan, P. N. and Leung, D. Reduction of chemical interference and speciation studies in the hydride generation atomic absorption method for selenium. **Anal. Chim. Acta;** 1980; **120:** 141-146.
347. Vuan, P. N. and Sadana, R. S. Determination of lead in drinking waters by hydride generation and atomic absorption spectroscopy and three other methods. **Talanta;** 1980; **27:** 321-326.
348. Waidmann, E., Hilpert, K., Schladot, J. D. and Stoeppler, M. Determination of cadmium, lead, thallium in materials of the environmental specimen bank using mass spectrometric isotope dilution analysis (MS-IDA). **Fresenius' Z. Anal. Chem.;** 1984; **317:** 273-277.
349. Walton A., Ed. Methods of sampling and analysis of marine sediments and dredged materials. Ocean Dumping Report.: Dept. Fisheries and Environment, Ottawa, Canada; 1978.
350. Wang, J. and Hutchens-Kumar, L. D. Cellulose acetate coated mercury film electrodes for anodic stripping voltammetry. **Anal. chem.;** 1986; **58:** 402-407.
351. Ward, N. I., Stephens, R. and Ryan, D. E. Comparison of three analytical methods for the determination of trace elements in whole blood. **Anal. Chim. Acta;** 1979; **110:** 9-19.
352. Ward, N. I., Brooks, R. R. and Roberts, E. Contamination of a pasture by a New Zealand base metal mine. **N. Z. J. Sci.;** 1977; **20:** 413-418.
353. Waughman, G. J. and Brett, T. Interferences due to major elements during the estimation of trace heavy metals in natural materials by atomic absorption spectrophotometry. **Environ, Res.;** 1980; **21:** 385-393.
354. Welz, B. and Melcher, M. Decomposition of marine biological tissues for determination of arsenic, selenium and mercury using hydride generation and cold vapour atomic absorption spectrometries. **Anal. Chem.;** 1985; **57:** 427-431.
355. Welz, B. and Melcher, M. Use of a new anti-foaming agent for determination of arsenic in urine with the hydride atomic absorption technique. **At. Absorpt. Newsl.;** 1979; **18:** 121-122.
356. Westermark, T. and Ljunggren, K. The determination of mercuty and its compounds by destructive neutron activation analysis. Mercury Contamination in Man and his Environment, (Tech. Report): IAEA; 1972; No 137: 99-110.
357. Westoo, G. Determination of methylmercury compounds in foodstuffs. 1 methylmercury in fish, identification and determination. **Acta Chem. Scand.;** 1966; **20:** 2131-2137.
358. Westoo, G. Determination of methylmercury compounds in foodstuffs. 2 Determination of methylmercury in fish, egg, meat and liver. **Acta, Chem. Scand.;** 1967; **21:** 1790-1800.

359. WHO. Environmental Health Criteria 1 Mercury: WHO; 1976.

360. Willis, J. B. Atomic spectroscopy in environmental studies fact and artefact. in: Int. Conf. Heavy Metals in the Environment, Toronto; 1975; 1: 69-91.

361. Winefordner, J. D. and Ward, J. L. The reliability of detection limits in analytical chemistry. **Anal. Lett.;** 1980; **13:** 1293-1297.

362. Witters, L. E. Alich, A. and Aufderheide, A. C. Lead in bone I. Direct analysis for lead in milligram quantities of bone ash by graphite furnace atomic absorption spectroscopy. **Amer. J. Clin. Pathology;** 1981; **75:** 80-85.

363. Working party Bureau Int. Tech. du Chlore. Standardization of methods for the determination of traces of mercury Part 5 Determination of total mercury in water. **Anal. Chim. Acta;** 1979; **109:** 209-228.

364. Xi Feng and Ryan, D. E. Combination collectors in adsorption colloid flotation for multielement determination in waters by neutron activation. **Anal. Chim. Acta;** 1984; **162:** 47-55.

365. Yamaguchi, T., Bando, M., Nakajima, A., Teral, M. and Suzuki-yasumoto, M. An application of neutron activation analysis to biological materials IV approach to simultaneous determination of trace elements in human eye tissue with non-destructive neutron activation analysis. **J.Radioanal. Chem.;** 1980; **57:** 169-183.

366. Yeats, P. A. and Campbell, J. A. Nickel, copper, cadmium and zinc in the Northwest Atlantic ocean. **Marine Chem.;** 1983; **12:** 43-58.

367. Yeoman, W. B. Internal and external quality control with special reference to lead and cadmium. in: Facchetti, S., Ed. Analytical Techniques for Heavy Metals in Biological Fluids: Elsevier; 1983: 273-284.

368. Yoshida, Z. and Motojima, K. Rapid determination of mercury in air with gold-coated quartz wool as collector. **Anal. Chim. Acta;** 1979; **106:** 405-410.

369. Zdzojewski, A., Quickert, N., Dubois, L. and Monkman, J. L. The accurate measurement of lead in airborne particulates. in: Frei, R. W. and Hutzinger, O., Eds. Analytical Aspects of Mercury and Other Heavy Metals in the Environment: Gordon Breach; 1975: 89-103.

370. Zief, M. and Nesher, A. G. Clean environment for ultra-trace analysis. **Environ. Sci. Tech.;** 1978; **8:** 677-678.

371. Zief, M. and Mitchell, J. W. Contamination Control in Trace Elemental Analysis: Wiley; 1976.

PART III

HEAVY ELEMENTS

IN THE

ENVIRONMENT

CHAPTER 5

BIOGEOCHEMICAL CYCLES

In this chapter we will consider the ways which the heavy elements move throughout the world - called biogeochemical cycling. The broad features of the cycling is much the same for all the heavy metals, because the processes involved are controlled, in the main, by the interactions between the various outer sections of the earth, their energy content and the energy input to the earth.

SECTIONS OF THE EARTH

There are four main sections or spheres of the earth which we are in contact with - the atmosphere, the hydrosphere, the lithosphere (or the earth's crust) and the biosphere. The mass of each sphere is given in Table 5.1. The biosphere is minute compared with the others, and the sum of the mass of the four spheres is around 0.4% of the total mass of the earth. The compositions of the four spheres are quite different, as are the main physical state of each sphere, viz. gas, liquid, solid and mainly organic respectively. The major constituents of each sphere are listed in Tables 5.2. The values given for the lithosphere and biosphere are averages, as wide variations occur depending on the sample selected [6,15,19,25,27,36,51,54].

The concentrations of the ten heavy metals in the various spheres are mostly in the ppm range, and some typical figures are given for the hydrosphere (marine and fresh water), in Tables 5.3 and 5.4, the lithosphere (crust) in Table 5.5, the atmosphere in Table 5.6 and human tissue in Table 5.7 [6,15,19,27,36,51,54]. Cycling of the elements within and between these spheres is a result of various material and energy interactions. The energy sources are solar radiation (ultra-violet, visible and infra-red), mechanical (kinetic and potential), chemical and the earth's thermal energy, some of which comes from nuclear energy, such as the decay of ^{40}K. The elements may be released, transported and re-combined in different ways. One scheme which illustrates

TABLE 5.1 Mass of the Component Sections of the Earth's Surface

Sections of the earth	Mass x 10^{12} Tonnes	% of spheres
Biosphere (including water)	8	3.1×10^{-5}
Hydrosphere		
Freshwater (liquid)	126	4.9×10^{-4}
Ice	30,000	0.12
Oceans	1,420,000	5.57
Salts in oceans	49,000	0.19
Atmosphere	5140	2.0×10^{-2}
Crust (down to 17 km)	24,000,000	94.1
Total earth	6,000,000,000	

Sources of data; references 15, 27.

TABLE 5.2 Composition of the Sections of the Earth's Surface (% by weight)

Crust		Salts in oceans		Atmosphere		Biosphere	
Element	Wt. %	Element	Wt. %	Element	Wt. %	Element	Wt. %
O	46.6	Cl	55.1	N	78	H	49.8
Si	27.7	Na	30.6	O	21	C	24.9
Al	8.1	Mg	3.7	Ar	0.9	O	24.9
Fe	5.0	S	2.7	CO_2	0.03	N	0.27
Ca	3.6	Ca	1.2	Ne	0.002	Ca	0.07
Na	2.8	K	1.1	Others	<0.01	K	0.04
K	2.6	Br	0.2			Mg	0.03
Mg	2.1	C	0.08			Si	0.03
		Sr	0.02			P	0.03
		B	0.01			S	0.02

Sources of data; references 15,19,27,35,51,54.

TABLE 5.3 Levels of the Heavy Metals in the Oceans

Element	Concentration μg l^{-1}	Range μg l^{-1}	Species
As	3.7	0.5-3.7	$HAsO_4^{2-}$, $H_2AsO_4^-$
Bi	0.02	0.015-0.02	BiO^+, $Bi(OH)_2^+$
Cd	0.11	<0.01-9.4	Cd^{2+}, $CdCl_2$
Hg	0.03	0.01-0.22	$HgCl_2$, $HgCl_4^{2-}$
In	1×10^{-4}		$In(OH)_2^+$
Pb	0.03	0.03-13	$PbCl^+$, $PbCl_3^-$, $PbCO_3$, $Pb(CO_3)_2^{2-}$
Sb	0.3	0.18-5.6	$Sb(OH)_6^-$
Se	0.2	0.052-0.2	SeO_3^{2-}
Te	?		$HTeO_3^-$
Tl	0.02		Tl^+

Sources of data; references 6,27,35.

the variety of transport mechanisms, is given in Fig. 5.1 [19]. The scheme does not include the biosphere, which would add the cycling of elements between living material and the inanimate spheres. An interaction of the biosphere with the other spheres is shown in Fig. 5.2 [15,19]. However, sediments are included as they are major pools of trace elements, spaning the litho- and hydrospheres. There are a number of gaps in our knowledge of trace element movements, including volcanic emissions and deposition in marine sediments [25].

Weathering

One major process in the cycling of materials is weathering, both physical and chemical. The weathering step in Fig. 5.2 releases significant amounts of elements bound up in rocks.

The important chemical factors in weathering are water, its chemical composition, pH and temperature, reactivity of species with CO_2/H_2O (called carbonation), hydrolysis, solubility and redox properties of species. Some typical weathering reactions are;

$$\underset{\text{orthoclase}}{4KAlSi_3O_8(s)} + 22H_2O \rightarrow \underset{\text{kaolinite}}{Al_4Si_4O_{10}(OH)_8(s)} + 8H_4SiO_4(aq) + 4K^+(aq) + 4OH^-(aq),$$

$$4KAlSi_3O_8(s) + 18H_2O + 4H^+ \rightarrow Al_4Si_4O_{10}(OH)_8(s) + 4K^+(aq) + 8H_4SiO_4(aq),$$

$$4KAlSi_3O_8(s) + 22H_2O + 4CO_2 \rightarrow Al_4Si_4O_{10}(OH)_8(s) + 4K^+ + 8H_4SiO_4(aq) + 4HCO_3^-(aq).$$

The potassium ions and some of the silicon are leached out, and there is a structural change from the three dimensional feldspar to the layer structure of the clay. Heavy elements in the original minerals are released, but not necessarily leached out because of solubility factors.

TABLE 5.4 Levels of the Heavy Metals in Fresh Water

Element	Concentration $\mu g\ l^{-1}$	Range $\mu g\ l^{-1}$	Species
As	0.5	0.2-230	anionic
Bi	0.02?		
Cd	0.1	0.01-3	organic
Hg	0.1	0.0001-2.8	organic
Pb	3	0.06-120	
Sb	0.2	0.01-5	Sb(V)
Se	0.2	0.02-1	SeO_3^{2-}

Source of data; Bowen, 1979 [6].

TABLE 5.5 Levels of the Heavy Metals in the Crust

Element	Concentration $\mu g\ g^{-1}$	Element	Concentration $\mu g\ g^{-1}$
As	1.8	Tl	0.5
Bi	0.2	Cd	0.11
Hg	0.05	In	0.05
Pb	13	Sb	0.2
Se	0.05	Te	0.005

Sources; references 6,27,35.

TABLE 5.6 Levels of the Heavy Elements in the Atmosphere

Element	Shetlands ng m^{-3}	Sth. Pole ng m^{-3}	Enewetok ng m^{-3}
As	0.6	0.007	
Cd	<0.8	<0.015	0.0025
Hg	<0.04		
In	<0.02	0.00005	
Pb	21	0.63	0.01
Sb	0.4	0.0008	0.0024
Se	0.05	0.006	0.11

Source; Bowen, 1979 [6], Wiersma and Davidson, 1986 [57].

TABLE 5.7 Levels of the Heavy Elements in Human Tissue

Element	Ref. person mg /70 kg	Muscle mg kg^{-1}
As	18	0.009-0.65
Bi		0.032
Cd	50	0.14-3.2
Hg	13	0.02-0.7
In		0.004
Pb	40-120	0.23-3.3
Sb	8	0.042-0.19
Se	13	0.42-1.9
Te		0.017?
Tl	8	0.07

Source; Bowen, 1979 [6] and Table 13.21.

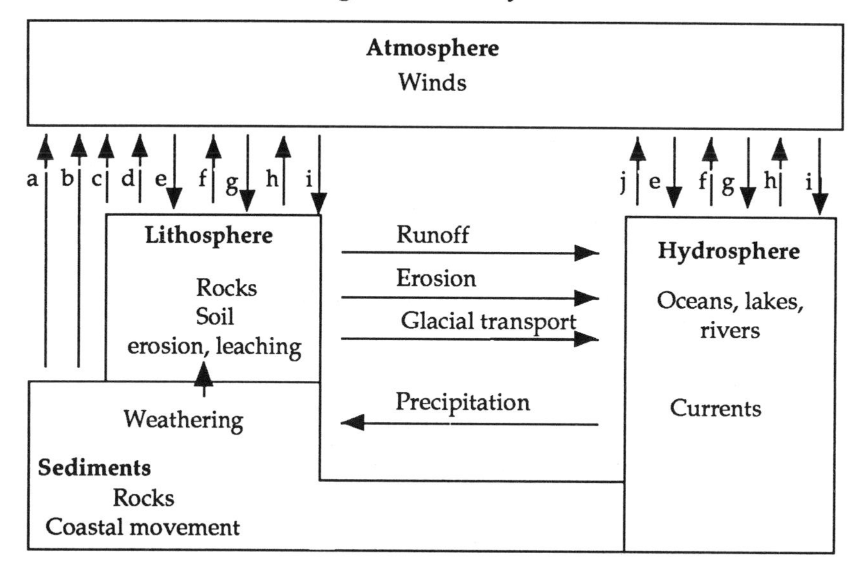

Fig. 5.1 Transport of materials within the physical world. a: volcanic activity, b: weathering, c: weathering. d: aerosol, e: fallout (solid), f: outgassing, g: gas absorption, h: evaporation, i: precipitation, j: spray.

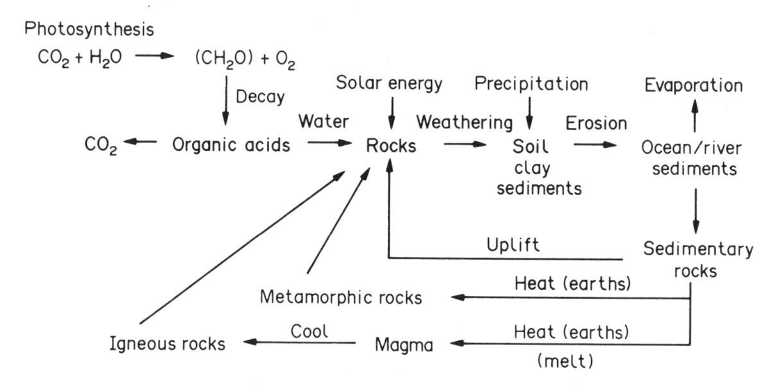

Fig. 5.2 Section of a biogeochemical cycle. Sources; references 15,59

The relative thermodynamic stabilities of the initial and weathered materials, can be determined from thermodynamic data and the results displayed as a stability field diagram [19]. For the two reactions;

$$\underset{\text{K-feldspar}}{3KAlSi_3O_8(s)} + 12H_2O + 2H^+ \rightarrow \underset{\text{mica}}{KAl_3Si_3O_{10}(OH)_2(s)} + 2K^+(aq) + 6H_4SiO_4(aq),$$

$$\underset{\text{mica}}{KAl_3Si_3O_{10}(OH)_2(s)} + 18H_2O + 2H^+ \rightarrow \underset{\text{gibbsite}}{3Al_2O_33H_2O} + 2K^+(aq) + 6H_4SiO_4(aq),$$

there is a loss of cation, and breakdown of the silicate structure, as indicated by the following changes in atom ratios.

Atom ratio	K-feldspar	Mica	Gibbsite
K/Al	1:1	1:3	0
Si/Al	3:1	1:1	0

The free energy of the first reaction is 73 kJ, calculated from free energy of formation data for the various species [19, 26]. The equilibrium constant K is;

$$K = \frac{[K^+]^2[H_4SiO_4]^6}{[H^+]^2},$$

where the value of K may be obtained from;

$$\log K = \frac{\Delta G}{2.303RT} = -12.8.$$

Hence

$$-12.8 = 2\log\frac{[K^+]}{[H^+]} + 6\log[H_4SiO_4]$$

For the second reaction the free energy is 106 kJ and logK = -18.6, hence;

$$-18.6 = 2\log\frac{[K^+]}{[H^+]} + 6\log[H_4SiO_4].$$

These two equations may be plotted as $\log([K^+]/[H^+])$ against $\log[H_4SiO_4]$ which gives the boundary lines between the three species (Fig. 5.3). The results indicate that if the silicic acid and potassium ions are removed from the site (i.e. their concentrations are reduced) and the acidity of the water increases the tendency is towards increased weathering, i.e. the system moves towards the bottom left hand corner of the diagram.

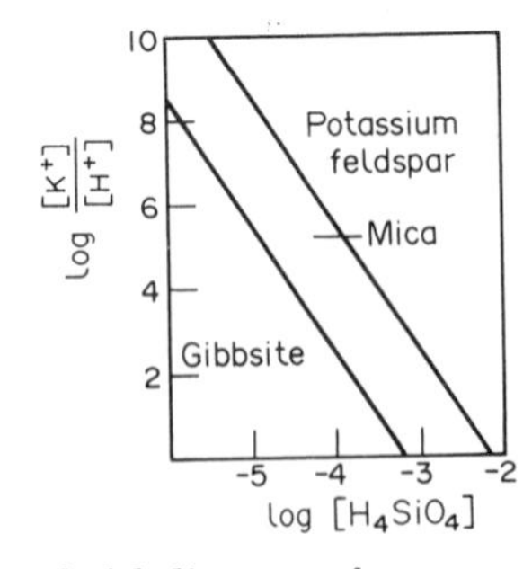

Fig 5.3 Stability field diagram for potassium feldspar, mica and gibbsite. Source of data; Fergusson, 1982 [19].

Redox processes in weathering are best considered with the use of E_h-pH diagrams, which were discussed in Chapter 3. Arsenic, for example is released into the environment when arsenopyrite FeAsS is exposed to dioxygen and water. The mineral is commonly associated with gold bearing ores, and when exposed, as in mine tailings, redox processes occur [8], namely oxidation of S^{2-} to SO_4^{2-} and As(III) to AsO_4^{3-}, both of which can be achieved through the reduction of dioxygen.

$$4FeAsS + 13O_2 + 6H_2O \rightarrow 4SO_4^{2-} + 4AsO_4^{3-} + 4Fe^{2+} + 12H^+.$$

Some of the ion-electron half reactions possible are

$$O_2 + 4H^+ + 4e \rightarrow 2H_2O, \quad E° = 1.23V,$$

$$S^{2-} + 4H_2O - 8e \rightarrow SO_4^{2-} + 8H^+, \quad -E° = -0.76V,$$

$$AsO_2^- + 2H_2O - 2e \rightarrow AsO_4^{3-} + 4H^+, \quad -E° = -0.56V.$$

The release of the arsenic from the mineral probably involves hydrolysis and oxidation. Once released from rocks and minerals the elements can be transported by various physical processes [25].

BIOGEOCHEMICAL CYCLES

There are some general features of the biogeochemical cycles of the elements that apply to all them. Each sphere contains a pool of the trace element, which is calculated from the product of the mass of material in the pool and the average concentration of the element in the pool. Large errors (such as ±20%) occur in both these values [25]. The pool consists of two parts, one that interacts with the biosphere (active pool) and one that does not interact with the biosphere (inactive pool).

The movement of material between pools is called the flux, and for most natural situations the systems are in a steady state. Anthropogenic inputs disturb the equilibrium. If a steady state is assumed the fluxes can be estimated, however, large errors also exist in these numbers [7,25,42].

From the pool and the flux values the residence times of the elements in the different sections of the earth may be determined i.e.,

$$t = m/\left(\frac{dm}{dt}\right) = m\left(\frac{dt}{dm}\right),$$

where t = residence time, m = mass of element in the pool, and dm/dt = the rate of input (or output) to the pool. Because of the large uncertainties in the various quantities the residence times are not precise [7, 38]. Residence times may also be determined experimentally using stable isotope or radioactive tracers. The residence times for some of the heavy elements in the different spheres are listed in Table 5.8. In general the times in the atmosphere are of the order of days to weeks, in fresh water from months to years, in the oceans thousands of years, in marine sediments around 10^8 years, in soils hundreds of years. Residence times for the heavy elements in the human body are from days to years, depending on the site of sorption in the body.

TABLE 5.8 Residence Times of the Heavy Elements

Elements	Atmos-phere (days)	Hydrosphere Oceans[a] (10^3 yrs.)	Lakes (days)	Litho-sphere[b] Soil (yrs.)	Biosphere Humans (days)
As	Short	550[c]-930[d]	415	2000	8-18[e], 360[f]
Sb	lived	40-210			160-230, 200
Se	volatile			<2500	65-90, 260
Hg	11-40-90	12-40	340	920	870, 2200
Cd	7	60-250	360-720	280	330-780, 500
Bi					
In	24-30				
Pb	5 nm -	0.29-0.85	25	400-3000	270-370, 2700
Te	20 μm	100			
Tl					

[a] In surface ocean waters residence times can be quite short, e.g. 2 years for lead. [b] Elements in marine sediments have long residence times (i.e. 2-5 $\times 10^8$ yrs.) as buried sediments eventually become rock [c] Estimated from inputs via rivers. [d] Estimated from outputs via sedimentation. [e] Estimated via diet. [f] Estimated via urine output. Sources; references 7,18,38,42.

Cadmium

The pools, concentrations and fluxes of cadmium in the earth's surface are given in Fig. 5.4. A number of units (see table below) are used in this and in following diagrams and tables.

Sphere	Concentrations	Pools	Fluxes
Atmosphere	ng m^{-3}, μg m^{-3}	kg, tonne (t)	kg y^{-1}, t y^{-1}
Lithosphere	ng g^{-1}, μg g^{-1}	kg, t	kg y^{-1}, t y^{-1}
Hydrosphere	ng l^{-1}, μg l^{-1}	kg, t	kg y^{-1}, t y -

where 1kg = 10^3 g, 1 tonne (t) = 10^3 kg = 10^6 g, and 1μg = 10^{-6} g and 1ng = 10^{-3}μg = 10^{-9} g.

The data in Fig. 5.4 refers mainly to the natural system, however, in some situations two concentration ranges are given. The first is for "natural" levels and the second for contaminated levels. At best the figures are a guide to levels that are found [1, 6, 7, 14, 21, 22, 34, 42, 46, 50]. Some of the natural sources and fluxes of cadmium (in 10^6 kg y^{-1}) are; windblown material 0.001- 0.22-0.85,

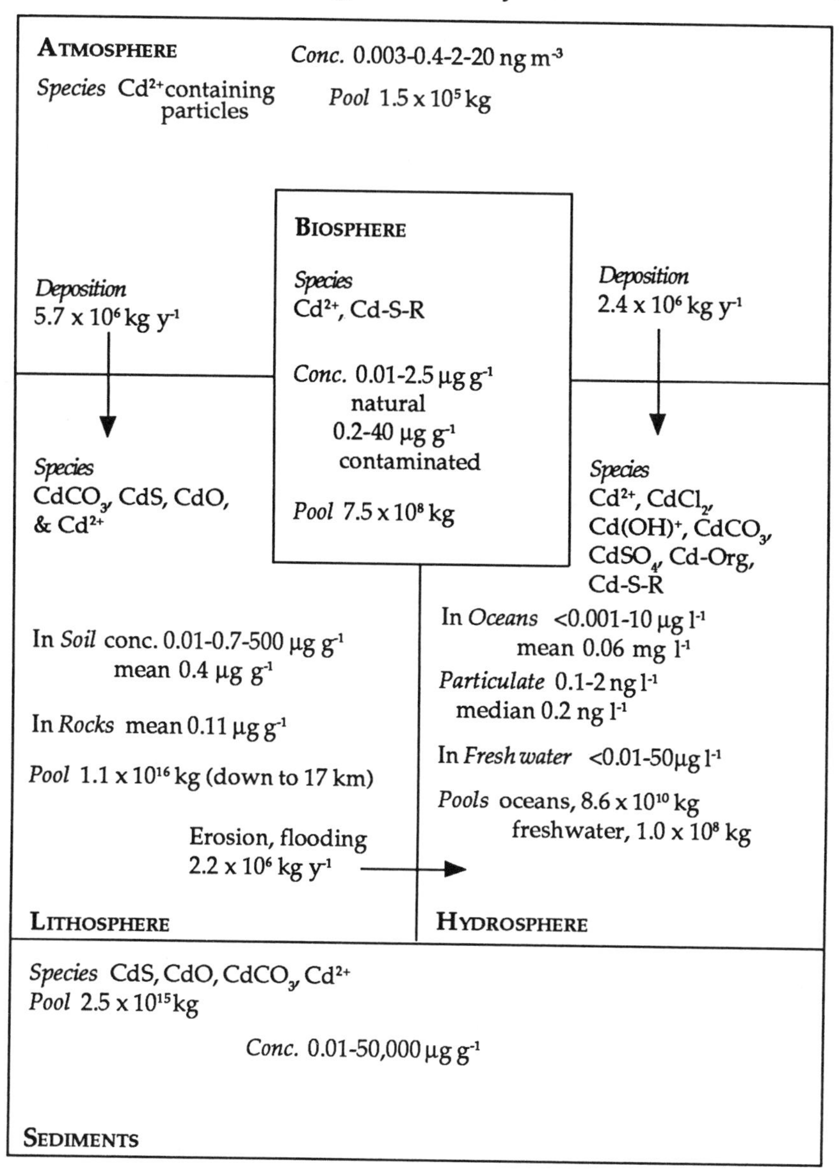

Fig 5.4 Biogeochemical cycling of cadmium. External inputs to atmosphere; coal and oil combustion, cement production, industrial e.g. cadmium production. To lithosphere/hydrosphere; electroplating, phosphatic fertilizers, pigments, sewage, stormwater. Sources; references 1,6,14,22,34,37,42,46,50

volcanic emissions 0.3-7.8, forest fires 0.001-0.07-1.5, vegetation emissions 0.05-0.2-2.7, and sea spray 0.001-0.4. This gives a total mean of around 0.83-0.96 x 10^{6} kg y^{-1} [28, 41, 46]. But in localized areas levels can be quite high, and a level of

3000 ng m^{-3} has been found above a hot vent in Mt. Etna [42]. Anthropogenic emissions from many sources, are listed in Fig. 5.4. These can produce an inbalance in fluxes and cadmium can build up in some of the pools, depending on the intensity of the the pollution. Natural emissions have been estimated at 0.3×10^6 kg y^{-1}, whereas anthropogenic emissions are 18 times higher at 5.5×10^6 kg y^{-1} [23, 37, 41]. Other studies suggest the natural and anthropogenic emissions are comparable [28]. The wide range of values in the literature arise from the different ways of calculating the figures, and because of the large errors in the data used in the calculations. Cadmium is a relatively mobile element in the environment (see Chapter 3), and the cation Cd^{2+} persists over a wide range of of pH values. Whereas cadmium tends to follow zinc in geological materials, with Cd/Zn ratios 1/100 to 1/1000, these are different in other materials because of selective vaporization of cadmium relative to zinc [45].

Lead

The biogeochemical cycle for lead is similar to that of cadmium, but with greater quantities (Fig. 5.5) [1,6,10,19,29,40,46,50]. The important pool, as regards the transport of lead, is the atmosphere, even though the concentration of lead is low compared with the lithosphere [40]. Anthropogenic emissions into the atmosphere are much greater than natural, being 400-450 x 10^6 kg y^{-1} compared with 2-6 x 10^6 kg y^{-1} [1,10,20,23,28,41,46,49,53].

Because of their toxicity the heavy elements are often discriminated against when taken in by biological systems, this is called biopurification [16, 47, 48]. The biopurification factor (BPF) for element X with respect to a nutrient element, such as calcium, is :

$$BPF = \left(\frac{[X]/[Ca](in\,food)}{[X]/[Ca](in\,consumer)} \right) = \left(\frac{[X](in\,food) \times [Ca](in\,consumer)}{[X](in\,consumer) \times [Ca](in\,food)} \right).$$

The factor should decrease, up the food chain, as the nutrient element is favoured over the toxic element. This is observed to be the case for lead up to the point where anthropogenic emissions significantly add to the body burden (Fig. 5.6). The BPF Pb/Ca ratios for rock to sedge leaves to vole to pine marten are 1/100, 1/16 and 1/1.1 respectively, and for Ba/Ca, are 1/16, 1/8 and 1/7 respectively. It is apparent from comparing the two sets of data, that an additional source of lead has been introduced at the second and third stages [16,17,48].

An estimation suggests that around 80% of environmental Me_4Pb is produced by chemical means, and 20% by biomethylation [11,12,37]. There is, however, debate over the biomethylation of lead within the biogeochemical cycle [11,12,60]. A modified scheme for the biogeochemical cycling of tetraalkyllead compounds in the environment is given in Fig. 5.7.

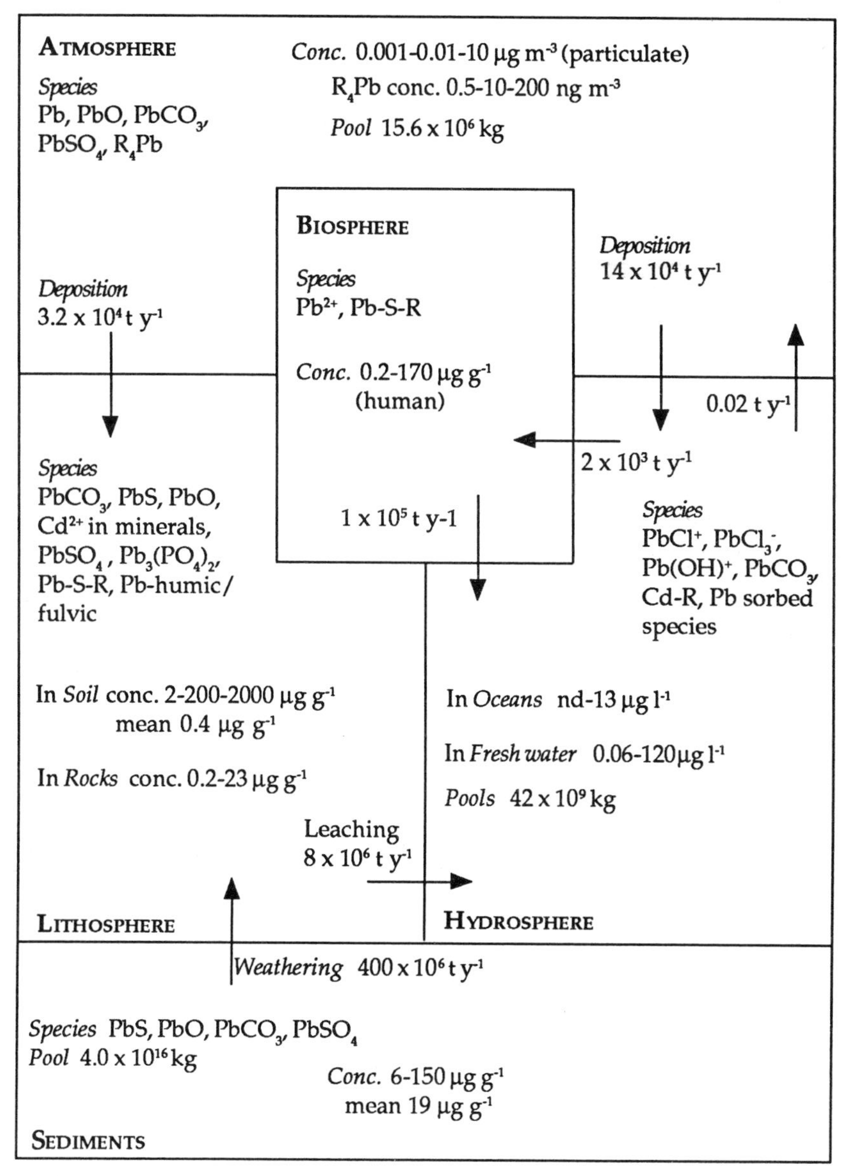

Fig 5.5 Biogeochemical cycling of lead. External inputs to atmosphere; coal and petrol burning, weathered paint, lead smelting, non-ferrous smelting. To lithosphere and hydrosphere; aerosol deposit, paint. Sources; references 1,6,10, 19,29, 37,40,44,46,50.

Mercury

One significant aspect of the global biogeochemical cycling of mercury, that is different to other metals, is the volatility of the metal. Therefore a significant

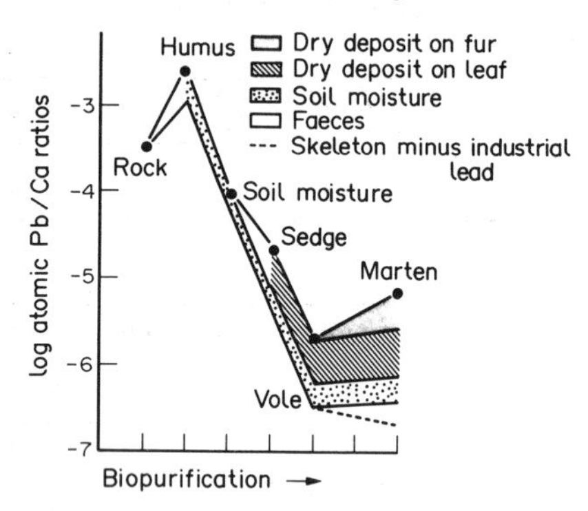

Fig 5.6 Biopurification of calcium with respect to lead from rock to carnivore. Sources of data; Elias et al., 1982 [16], Elias and Patterson, 1979 [17], Patterson, 1983 [48].

proportion of the mercury in the atmosphere is gaseous, generally 0.005-50 ng m^{-3} compared with aerosol mercury 1-10 ng m^{-3}. The cycle, levels, pools and fluxes of mercury are listed in Fig. 5.8 [1,46,13,18,19,24,30,31,33,36,50,56].

The natural sources of mercury emissions are, weathering of rocks, wind blown dust (0.005-0.035 x 10^6 kg y^{-1}), volcanic activity (0.01-0.24 x 10^6 kg y^{-1}), geysers, thermal fluids, degassing of the earth's mantle, emanations from the oceans (e.g. salt spray 0.001-0.003 x 10^6kg y^{-1}), transpiration and decay of vegetation (0.001-0.006 x 10^6 kg y^{-1}) and forest fires (0.04-0.16 x 10^6 kg y^{-1}) [28, 31, 46]. The emissions from volcanos fluctuate considerably, e.g. from Mt. St. Helens (USA) 0.9 x 10^6 kg y^{-1} and Arenal (Costa Rica) 3.0 x 10^6 kg y^{-1} [46]. The volcanic and thermal emissions of mercury are located around the main deposits of mercury, which are in areas of instability in the earth, such as along plate edges [30].

The main transport pathway for mercury is through the atmosphere [56], and various models have been developed to estimate the different fluxes [5, 28, 36]. For example it is estimated that around 13-18 x 10^6 kg y^{-1} of mercury is transported from the land into the atmosphere, (from industry, fuel burning, vapour and dust) and about the same amount returned by wet and dry deposition. Approximately 9-20 x 10^6kg y^{-1} of mercury from the oceans moves into the atmosphere, and 9-27 x 10^6 kg y^{-1} transported back via wet and dry precipitation [18, 28]. However, the actual amounts quoted differ considerably from study to study.

The volatilization of mercury may occur at any stage in the transport process. For example:

		Atmosphere		Atmosphere
	weathered	↑	bacteria	↑
Rocks	→	Hg/Hg^{2+}	→	RHg^+ & R_2Hg
(HgS)	chemical or bacterial			

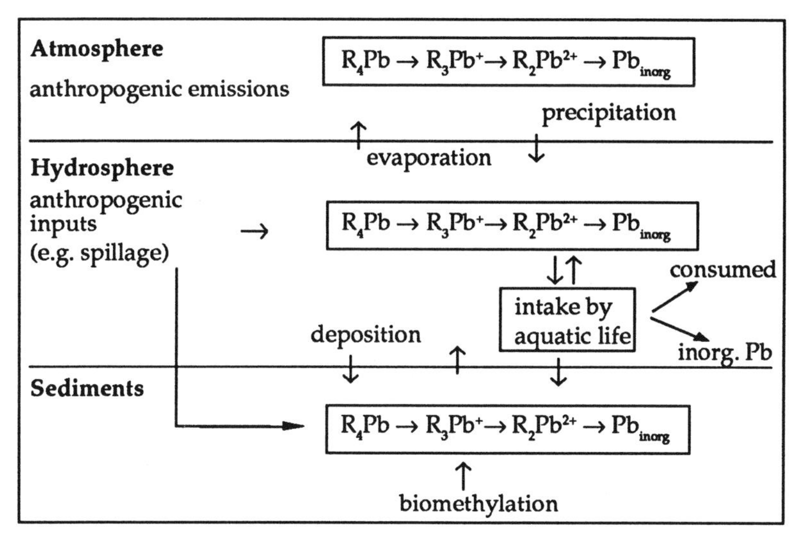

Fig 5.7 Biogeochemical cycling of tetraalkyllead compounds, R = CH_3 and C_2H_5. Sources; DeJonge and Adams, 1986 [12], Wood, 1980 [60].

Volatile species are produced by chemical or biochemical reduction of Hg^{2+} to Hg, and the biomethylation of mercury to give methylmercury compounds [31,55].

The main forms of mercury in the hydrosphere are, $HgCl_2$, $HgCl_4^{2-}$, Hg^{2+}, Hg, Hg-species sorbed onto clays or Fe/Mn oxides, and organomercury species [31]. In soils, mercury is probably present mainly chelated to S–containing aminoacids or proteins and humic acid like substances [31].

An important part of the biogeochemical cycling of mercury is the bioalkylation of mercury, especially in the aqueous environment, and its subsequent distribution throughout the earth [1,11,18,19,30,31,52,59,61,62]. The biogeochemical cycling that may occur in the hydrosphere is presented in Fig. 5.9 [19]. Bacteria feature prominently in the processes. We will consider the chemical and biochemical processes of alkylation in more detail in Chapter 12.

Arsenic

Of the four elements Cd, Pb, Hg and As, arsenic is the one where two oxidation states (As(V) and As(III)) feature prominently in the environment, as organic and inorganic species [39, 52]. It is clear from the E_h-pH diagram for arsenic (Fig. 3.28a) that both As(V) and As(III) species can exist under environmental pH and E_h conditions. The pools, concentrations and fluxes of arsenic species in biogeochemical cycling are listed in Fig. 5.10.

Natural emissions to the atmosphere total around 2.8×10^6 kg y^{-1} (dust) and 21×10^6 kg y^{-1} (vapour), whereas the anthropogenic emissions are greater at

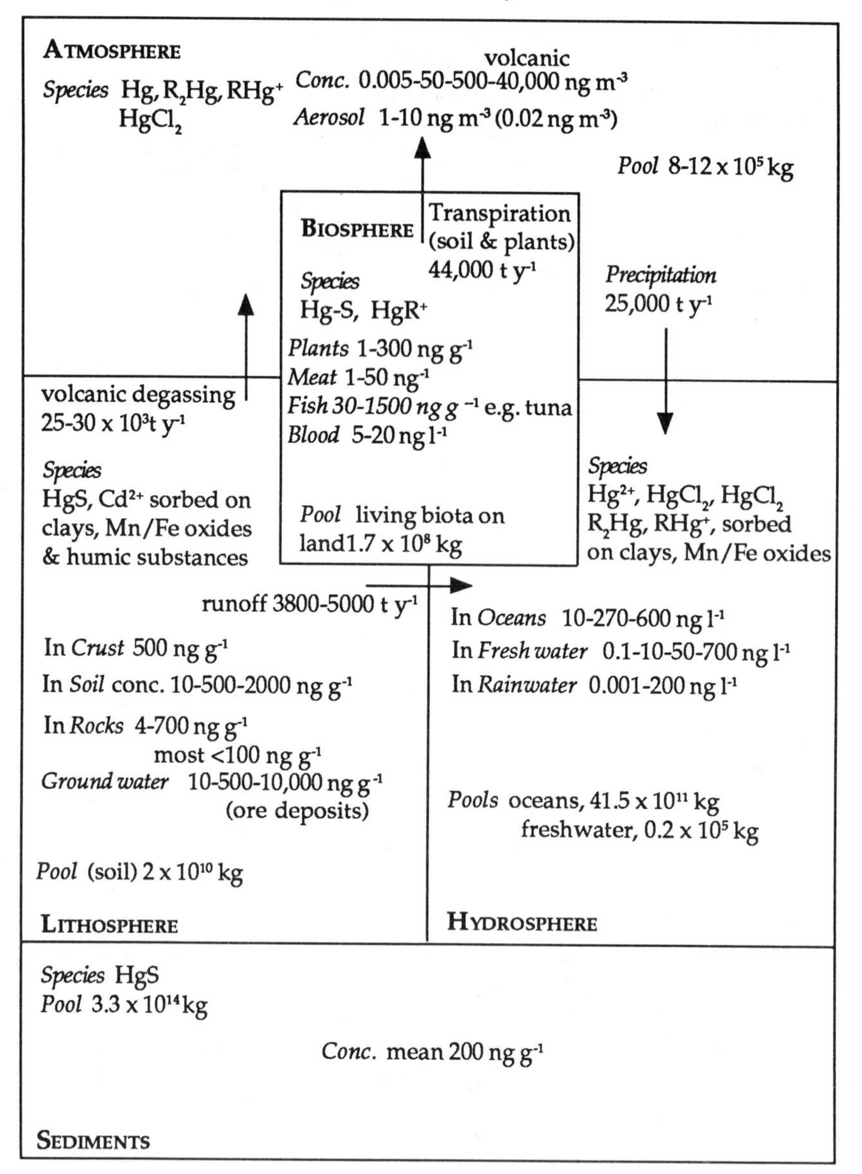

Fig 5.8 Biogeochemical cycling of mercury. Sources; references 1,4,5,6,13,18,19,24,28,30-33,36,37,50,56

around 78×10^6 kg y^{-1} [9, 23]. Some of the natural sources are windblown dust from crustal weathering, forest fires, vegetation emissions, volcanos and sea spray [9, 39, 46]. The amounts from individual volcanos vary considerably, for example 8.9×10^6 kg y^{-1} from Mt. St. Helens (USA) to 0.04×10^6 kg y^{-1} from Poas (Costa Rica) [46]. Most of the volcanic arsenic emissions are dust (e.g. 0.3×10^6

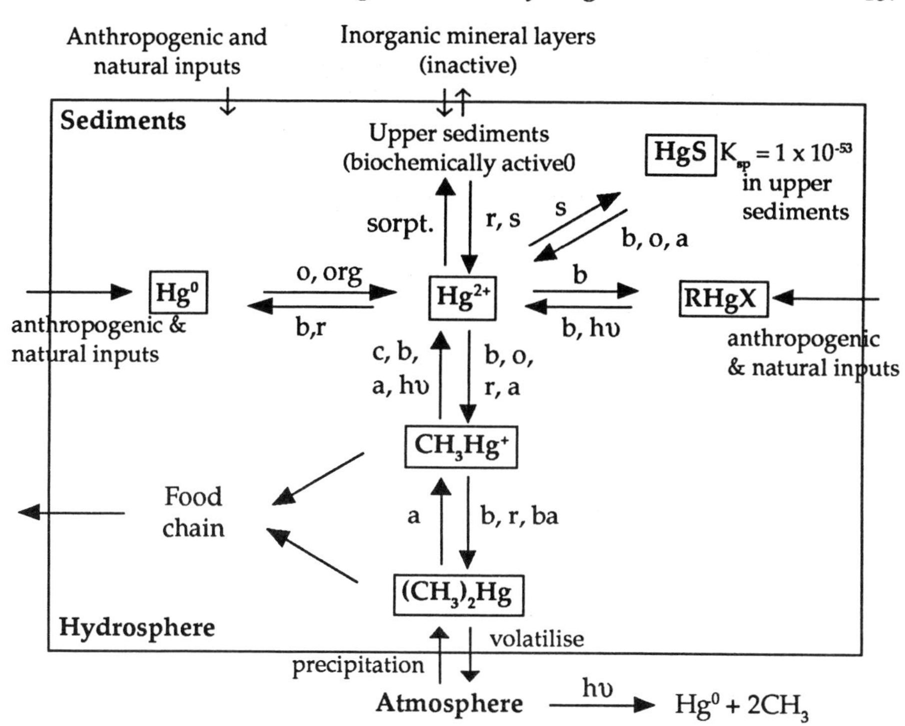

Fig 5.9 Chemical and biochemical pathways of mercury in the hydrosphere a = low pH, b = bacterial, ba = high pH, c = chemical, hυ = radiation, o = oxidation, aerobic, r = reduction, anaerobic, s = H_2S. Source; Fergusson, 1982 [19].

kg y^{-1}), compared with gaseous (0.01 x 10^6 kg y^{-1}) [32]. Man made emissions come from fuel combustion, smelting of ores and the use of arsenic in fertilizers and pesticides [39].

As for mercury and lead, bioalkylation of arsenic is an important feature of its biogeochemical cycle, producing organoarsenic compounds in the environment. A simplified cycle for some of the processes is given in Fig. 5.11 [11,37,39,52,58]. These involve both redox and methylation processes, the actual chemical form of the species present will also depend on the pH of the medium, either H_3AsO_4 or $H_2AsO_4^-$ or $HAsO_4^{2-}$ or AsO_4^{3-} see Fig. 3.28a.

Selenium

A reasonable amount is known about the concentrations, pools and fluxes of selenium in the environment, because it is an essential element for animals and because of the need to carefully control levels of intake by animals, due to the narrow range between essential levels, 0.1-2 ppm, and toxic levels, ≥4 ppm [2,3]. The biogeochemical data are given in Fig. 5.12. Primary sources of natural selenium are volcanic emissions and weathering of metal sulphides that contain selenium [1]. Volcanic emissions range from 0.1 x 10^6 kg y^{-1} (Poas, Costa

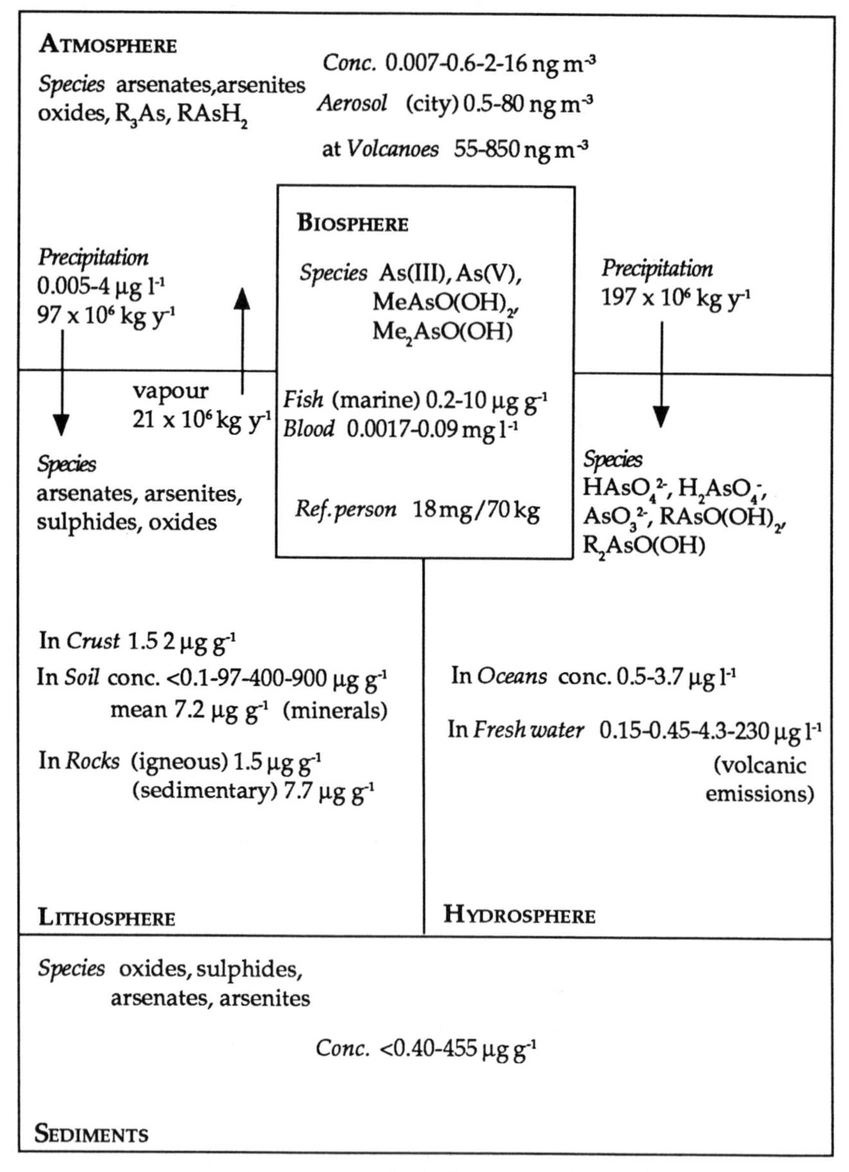

Fig 5.10 Biogeochemical cycling of arsenic.
Sources; reference 1,6,9,23,37,39,57

Rica) to 1.11 x 10^6 kg y^{-1} (Mt. St. Helens, USA) [46]. Overall, the mean volcanic emissions are about 1/3 of the flux of windblown dust, giving a total of approximately 0.4 x 10^6 kg y^{-1} [23,32,46,47]. Gaseous emissions are higher at around 3.0 x 10^6 kg y^{-1}, whereas anthropogenic are still higher at around 14 x 10^6 kg y^{-1} [23,47].

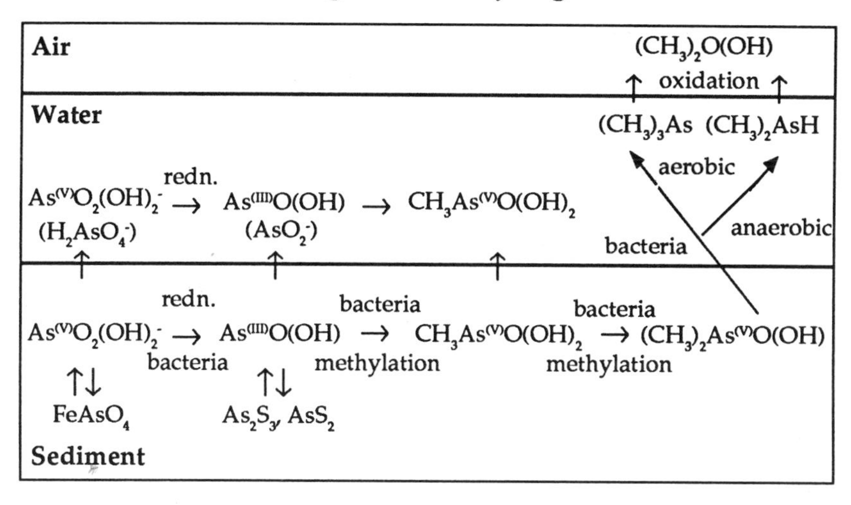

Fig 5.11 The chemical and biochemical processes for arsenic. Sources; references 11,37,39,52,58.

Both Se(IV) and Se(VI) may exist in the environment, as both are stable under the conditions of the natural environment. In addition selenides and elemental selenium may exist [1-3]. The change in both oxidation state and species with pH tends to follow the sequence;

low pH (& reducing)	→	→	→	high pH (& oxidising)
	Se^{2-}	Se^{0}	SeO_3^{2-}	SeO_4^{2-}
	Se(-II)	Se(0)	Se(IV)	Se(VI)

This trend has an influence on the mobility of selenium as selenate (SeO_4^{2-}) is the mobile form. Selenides and selenium have low solubilities and selenite forms insoluble compounds with iron, $Fe_2(SeO_3)_3$, $K_{sp} = 10^{-33}$, and $Fe_2(OH)_4SeO_3$, $K_{sp} = 10^{-63}$. The reduction of selenate to selenite,

$$2SeO_4^{2-} + 6H^+ + 4e \rightarrow 2HSeO_3^- + H_2O, \quad E° = 1.15V,$$

occurs more readily at lower pH (see the E_h-pH diagram Fig. 3.32a) even under oxidizing conditions.

Remaining Elements

Information on the remaining elements is sketchy, and some concentration data are listed in Table 5.9 [1,6,52,57]. Pools of the elements may be determined from the product of the pool mass and the average concentration in each pool.

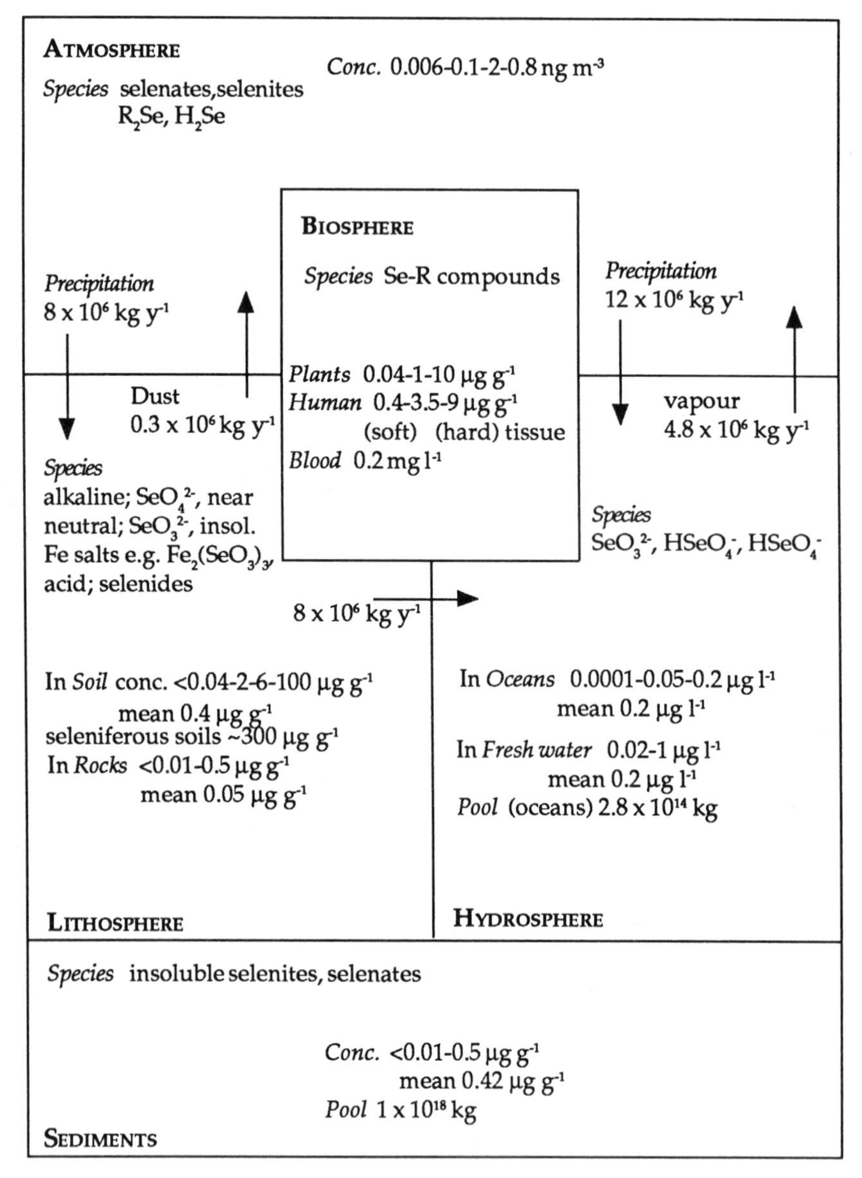

Fig 5.12 Biogeochemical cycling of selenium.
Sources; references 1-3,6,32,39,46

The biogeochemical cycling between the pools will be similar, in general terms, to those described above with some differences arising from dissimilar chemistries. There is little information on fluxes, but they may be approximately calculated by multiplying the average concentration in a material with the

TABLE 5.9 Levels of Indium, Thallium, Antimony, Bismuth and Tellurium on the Earth

Elements	Atmos-phere	Hydrosphere		Lithosphere		Sediments	Biosphere
		Ocean	Fresh	Rock	Soil		Blood
	ng m^{-3}	ng l^{-1}	ng l^{-1}	μg g^{-1}	μg g^{-1}	μg g^{-1}	mg l^{-1}
In	$5.4x10^{-5}$ -0.0002- -0.7	0.0001		0.049	0.7-3 1	0.009-0.07 0.04	
Tl	0.06-.22*	0.02		0.0005	0.1-0.8-5 0.2	0.1-0.9 0.95	0.0005
Sb	0.0004- -0.9-7.0	0.18-5.6 0.24	0.01-5 0.2	0.46	0.2-10 1.7	0.05-1.5 1.2	0.003
Bi	<1-3*	0.015-.02 0.02	0.02?	0.048	0.1-13 0.2	0.2-0.5 0.4	0.016
Te	0.18*			0.003		<0.1	0.006

Range of values, from natural to polluted levels or near ore bodies and mean or median levels. *For industrial countries. Sources: references. 1, 6, 52, 57.

amount of material that is transported. The estimated natural emissions for antimony into the atmosphere is 1 x 10^6 kg y^{-1} compared with 38 x 10^6 kg y^{-1} from anthropogenic sources [23,46,52]. Wet deposition for antimony has been estimated to have a concentration of 0.034 μg l^{-1} [23].

The chemical species and the transport of the elements within and between the various pools can be gauged from a consideration of their E_h-pH diagrams and the solubility data on their compounds (Chapter 3).

Anthropogenic Emissions

As indicated above, in addition to natural emissions of the heavy elements to the atmosphere, there are also emissions resulting from the actions of people. These are often the greater, and are localized in certain areas, especially urban and industrial areas. Therefore localized concentrations can become quite high. The subject of anthropogenic emissions will be covered in later chapters, but estimates of the relative magnitude of natural and anthropogenic emissions are given in Table 5.10. Data comes from different sources, and there are some divergent results, but generally they are of the same order of magnitude. Emissions to the atmosphere are normally considered the most important as the atmosphere is probably the main avenue for the transport of the heavy elements, the next being the hydrosphere.

The relative measure of the mobilization of an element in the environment can be expressed in diverse ways. One measure is the mobilization factor (MF) (Table 5.10) where

$$MF = \frac{\text{anthropogenic emission rate}}{\text{natural emission rate}}.$$

TABLE 5.10 Atmospheric Emissions (x 10^6 kg y^{-1})

Element	Emissions		Mobilization
	Natural	Anthropog.	Factor
As	2.8 + 21.0*	78.0	3.3
	7.8		
Cd	0.29	5.5	19
	0.83	7.3	8.8
	0.96		
	0.8	8.1	10
	1.5	0.7	0.5
Hg	0.04 + 25.0*	11.0	0.4
	0.16		
	0.07	11.0	157
Pb	5.9	2000	340
	18.6		
	54.0	358	6.6
	24.5	449	18.3
Sb	0.98	38	39
Se	0.4 + 3.0*	14	4.1
	0.4		

* Dust + vapour. Sources; references 23,28,32,41,42,57.

For the elements listed in Table 5.10, the factor decreases in the order Pb > Sb > Cd > Se,As > Hg [23, 32]. Alternatively, an enrichment factor (EF) may be considered,

$$EF = \left(\frac{[M]_{air}}{[M]_{crust}}\right)\left(\frac{[Al]_{crust}}{[Al]_{air}}\right).$$

When EF > 1 the element is enriched in the atmosphere relative to the crustal levels, assuming the levels of aluminium are not enriched [24]. For the heavy elements, with EF > 1, which have been measured at the South Pole, or over the North Atlantic, the order is Se > Pb > Sb > Cd [23].

REFERENCES

1. Adriano, D. C. Trace Elements in the Terrestial Environment.: Springer Verlag; 1986.
2. Allaway, W. H. Control of the environmental levels of selenium. **Trace Subst. in Environ. Health**; 1968; **2:** 181-206.
3. Allaway, W. H. Soil and plant aspects of the cycling of chromium, molybdenum and selenium. in: Int. Conf. Heavy Metals in the Environment, Toronto; 1975; **1:** 35-47.

4. Andren, A. W. and Nriagu, J. O. The global cycle of mercury. in: Nriagu, J. O., Ed. The Biogeochemistry of Mercury in the Environment: Elsevier/ North Holland; 1979: 1-21.
5. Berry, P. J. An introduction to the exposure commitment concept with reference to environmental mercury. MARC Tech. Rept.: MARC; 1979; **No 12.**
6. Bowen, H. J. M. Environmental Chemistry of the Elements.: Academic Press; 1979.
7. Bowen, H. J. M. Residence time of heavy metals in the environment. in: Int. Conf. Heavy Metals in the Environment, Toronto; 1975; **1:** 1-19.
8. Brooks, R. R., Fergusson, J. E., Holzbecher, J. Ryan, D. E., Zhang, H. F., Dale, J. M. and Freedman, B. Pollution by arsenic in a gold mining district of Nova Scotia. **Environ. Pollut.;** 1982; **4B:** 109-117.
9. Chilvers, D. C. and Peterson, P. J. Global cycling of arsenic. in: Hutchinson, T. C. and Meena, K. M., Eds. Lead, Mercury, Cadmium and Arsenic in the Environment, SCOPE: Wiley; 1987: 279-301.
10. Committee on Biological Effects of Atmospheric Pollution. Lead: Airborne Lead in Perspective.: Nat. Acad. Sci.; 1972.
11. Craig, P. J. Metal cycles and biological methylation. in: Huntzinger, O., Ed. **Handbook of Environmental Chemistry:** Springer Verlag; 1980; **1A:** 169-227.
12. De Jonghe, W. R. A. and Adams, F. C. Biogeochemical cycling of organic lead compounds. in: Nriagu, J. O. and Davidson, C. I., Eds. Toxic Metals in the Atmosphere: Wiley; 1986: 561-594.
13. D'Itri, F. M. The Environmental Mercury Problem: CRC Press; 1972.
14. Eaton, A. Marine geochemistry of cadmium. **Marine Chem.;** 1976; **4:** 141-154.
15. Ehrlich, P. R., Ehrlich, A. H. and Holdren, J. P. Ecoscience, Population, Resources, Environment: Freeman; 1977.
16. Elias, R. W., Hirao, Y. and Patterson, C. C. The circumvention of the natural biopurification of calcium along nutrient pathways by atmospheric inputs of industrial lead. **Geochim. et Cosmochim. Acta;** 1982; **46:** 2561-2580.
17. Elias, R. W. and Patterson, C. C. The toxicological implications of biogeochemical studies of atmospheric lead. (unpublished) 1979.
18. Environmental Studies Board Panel on Mercury. An assessment of mercury in the environment.: Nat. Acad. Sci.; 1978: 139-169.
19. Fergusson, J. E. Inorganic Chemistry and the Earth: Pergamon Press; 1982.
20. Fergusson, J. E. Lead: petrol lead in the environment and its contributions to human blood lead levels. **The Sci. Total Environ.;** 1986; **50:** 1-54.
21. Fleisher, M., Sarofim, A. F., Fassett, D. W., Hammond, P., Shacklette, H. T., Nisbet, I. C. T. and Epstein, S. Environmental impact of cadmium: a review by the panel on hazardous trace substances. **Environ. Health Perspec.;** 1974: 253-323.

22. Forstner, U. Cadmium. in: Huntzinger, O, Ed. Handbook of Environmental Chemistry: Springer Verlag; 1980; **3A:** 59-107.
23. Galloway, J. N., Thornton, J. D., Norton, S. A., Volchok, H. L. and McLean, R. A. N. Trace metals in atmospheric deposition: a review and assessment. **Atmos. Environ.;** 1982; **16:** 1677-1700.
24. Garrels, R. M., Mackenzie, F. T. and MacKenzie,. Chemical Cycles and the Global Environment: W. Kaufmann Inc.; 1975: 112-.
25. Garrels, R. M. and Lerman, A. The exogenic cycle: reservoirs, fluxes and problems. in: Global Chemical Cycles and Their Alteration by Man.: Dahlem Workshop; 1977: 23-31.
26. Garrels, R. M. and Christ, C. L. Solutions, Minerals and Equilibria: Harper Row; 1965.
27. Henderson, P. Inorganic Geochemistry: Pergamon; 1982.
28. Jawarowski, Z., Bysiek, M. and Kownacka, L. Flow of metals into the global atmosphere. **Geochim. et Cosmochim. Acta;** 1981; **45:** 2185-2199.
29. Jaworski, J. F. Group report: Lead. in: Hutchinson, T. C. and Meena, K. M., Eds. Lead, Mercury, Cadmium and Arsenic in the Environment, SCOPE: Wiley; 1987: 3-16.
30. Jonasson, I. R. and Boyle R. W. Geochemistry of mercury and origins of natural contamination of the environment. **Canad. Mining Met. Bull.;** 1972; **65:** 32-39.
31. Kaiser, G. and Tolg, G. Mercury. in: Hutzinger, O., Ed. The Handbook of Environmental Chemistry: Springer Verlag; 1980; **3A:** 1-58.
32. Lantz, R. J. and Mackenzie, F. T. Atmospheric trace metals: global cycles and assessment of man's impact. **Geochim. et Cosmochim. Acta;** 1979; **43:** 511-523.
33. Lindberg, S. Group report: Mercury. in: Hutchinson, T. C. and Meena, K. M., Eds. Lead, Mercury, Cadmium and Arsenic in the Environment, SCOPE: Wiley; 1987: 1-21.
34. Martin, J. H., Knauer, G. A. and Flegal, A. R. Cadmium in natural waters. in: Nriagu, J. O., Ed. Cadmium in the Environment Part 1. Ecological Cycles: Wiley; 1980: 141-145.
35. Mason, B and Moore, C. B. Principles of Geochemistry, 4th Ed.: J. Wiley; 1982.
36. Miller, D. R. and Buchanan, J. M. Atmospheric transport of mercury: exposure commitment and uncertainty calculations. MARC Tech. Rept.: MARC; 1979: No. 14.
37. Moore, J. W. and Ramamoorthy, S. Heavy metals in Natural Waters: Applied Monitoring and Impact Assessment.: Springer-Verlag; 1984.
38. Morel, F. M. M. and Hudson, R. J. M. The geobiological cycle of trace elements in aquatic systems: Redfield revisted. in: Stumm, W., Ed. Chemical Processes in Lakes: Wiley; 1985: 251-281.

39. Newland, L. W. Arsenic, beryllium, selenium and vanadium. in: Hutzinger, O, Ed. Handbook of Environmental Chemistry: Springer Verlag; 1982; **3B:** 27-68.
40. Newland, L. W. and Dawn, K. A. Lead. in: Hutzinger, O., Ed. Handbook of Environmental Chemistry: Springer Verlag; 1982; **3B:** 1-26.
41. Nriagu, J. O. Global inventory of natural and anthropogenic emissions of trace metals. **Nature;** 1979; **279:** 409-411.
42. Nriagu, J. O. Global cadmium cycle. in: Nriagu, J. O., Ed. Cadmium in the Environment Part 1. Ecological Cycles: Wiley; 1980: 1-12.
43. Nurnberg, H. W,. The voltammetric appproach in trace metal chemistry of natural waters and atmospheric precipitation. **Anal. Chim. Acta;** 1984; **164:** 1-21.
44. O'Brien, B. J., Smith, S. and Coleman, D. O. Lead pollution of the global environment. in: Progress Reports in Environmental Monitoring and Assessment. 1 Lead. Tech. Rept.: MARC; 1980: No. 16-18.
45. Oddo, N., Magistrelli, C., Galli, P. and Spiga, G. Direct analysis for metallic elements on membranes by flameless atomic absorption. **Chim. Ind. (Milan);** 1978; **60:** 104-110.
46. Pacyna, J. M. Atmospheric traces elements from natural and anthropogenic sources. in: Nriagu, J. O. and Davidson, C. I., Eds. Toxic Metals in the Atmosphere: Wiley; 1986: 33-52.
47. Patterson, C. C. An alternative perspective - lead pollution in the human environment: origin extent and significance. in: Lead in the Human Environment: NAS; 1980: 265-349.
48. Patterson, C. C. British mega exposures to industrial lead. in: Rutter, M. and Russel Jones, R, Eds. Lead versus Health: Wiley; 1983: 17-32.
49. Patterson, C. C. Natural levels of lead in humans. Address to Inst. Environ. Studies, Univ. Nth. Carolina, Chapel Hill.; 1982.
50. Piotrowski, J. K. and Coleman, D. O. Environmental hazards of heavy metals: summary evaluation of lead, cadmium and mercury. MARC Report No 20.; 1980.
51. Raiswell, R. W., Brimblecombe, P., Dent, D. L. and Liss, P. S. Environmental Chemistry.: Arnold; 1980.
52. Salomans, W. and Forstner, U. Metals in the Hydrocycle: Springer-Verlag; 1984.
53. Settle, D. M. and Patterson, C. C. Lead in Albacore: guide to lead pollution in Americans. **Science;** 1980; **207:** 1167-1176.
54. Skinner, B. J. Earth Resources. 2nd. Ed.: Prentice Hall; 1976.
55. Vostal, J. Transport and transformation of mercury in nature and possible routes of exposure. in: Friberg, L. and Vostal, J., Eds. Mercury in the Environment: CRC Press; 1972.
56. WHO. Environmental Health Criteria 1 Mercury: WHO; 1976.

57. Wiersma, G. B. and Davidson, C. I. Trace metals in the atmosphere of remote areas. in: Nriagu, J. O. and Davidson, C. I., Eds. Toxic Metals in the Atmosphere: Wiley; 1986: 201-266.
58. Wood, J. M. Biological cycles for toxic elements in the environment. **Science;** 1974; **183:** 1049-1052.
59. Wood, J. M. and Goldberg, E. D. Impact of metals on the biosphere. in: Global Chemical Cycles and Their Alteration by Man.: Dahlem Workshop; 1977: 137-153.
60. Wood, J. M. Lead in the marine environment some biochemical considerations. in: Branica, M. and Konrad, Z., Eds. Lead in the Marine Environment: Pergamon; 1980: 299-303
61. Wood, J. M., Segall, H. J., Ridley, W. P., Cheh, A., Cdudzk, W. and Thayer, J. S. Metabolic cycles for toxic metals in the environment. in: Int. Conf. Heavy Metals in the Environment, Toronto; 1975; **1:** 49-68.
62. Wood, J. M. and Wang, H. K. Strategies for microbial resistance to heavy metals. in: Stumm, W., Ed. Chemical Processes in Lakes: Wiley; 1985: 81-98.

CHAPTER 6

NATURAL CONCENTRATIONS OF THE HEAVY ELEMENTS

In this chapter we will consider the concentrations of the elements which may be regarded as natural i.e. concentrations existing before the advent of industrial activity. For some of the elements, these levels are not obtainable from direct measurement, because of global contamination of the earth. Hence indirect estimates may be used, such as concentrations in remote areas, background concentrations and concentrations in ancient or historical materials.

INTRODUCTION

An important reason for knowing the natural concentrations of the heavy elements, is that they provide a true reference point for estimating the extent of pollution from the elements. This is of particular importance when assessing the toxic effects of the elements. Natural levels allow contemporary levels to be seen in perspective, i.e. whether they are excessive or not. However, the concentrations are difficult to obtain because, either natural levels no longer exist, and this is probably true for lead, mercury, cadmium and arsenic, or substantial contamination of samples occur, during collection and/or analysis.

It is therefore necessary to obtain estimates of natural concentrations by other methods, but these are not to be identified or confused with natural levels. One approach is to measure levels in remote areas, such as in the Antarctica and Arctic. A second approach is to measure background levels, which are, in most cases, similar to remote levels. Background levels may be regarded as the levels that exist on top of which anthropogenic materials have been added. Concentrations of heavy elements deep down a sediment profile could be considered as background levels. A third approach is to investigate ancient levels of the heavy elements, especially in materials which existed before industrialization. Sediments, low down in a sediment profile, may be dated, and regarded as historical samples. Hence, whereas there are three methods of estimating natural concentrations, the distinction between the methods is arbitrary.

Terminology

There is confusion in the literature over the use of the term 'natural' concentration. The word 'normal' is often misinterpreted as 'natural. Levels found in the environment today, are often called normal, and mistakenly said to be natural. The false picture this gives of natural levels has been pointed out a number of times [26,34,58,59,69]. It is probably best to use the word 'typical' for contemporary trace element levels rather than 'normal' as it is less likely to be confused with 'natural'.

In Tables A6.1 toA6.11 are listed some of the reported values of remote (Tables A6.1-A6.3), background (Tables A6.4-A6.7), ancient (historical) (Tables A6.8-A6.9) and natural (Tables A6.10-A6.11) concentrations of the heavy elements in different materials, and locations around the earth. The Tables A6.1 to A6.11 are presented at the end of the chapter. A decision as to which table to put the data in is rather subjective and could be debated. In most cases the decision is based on the authors' description of their work. The units used for the concentrations reported differ considerably and in order for comparisons to be made the units have been standardized, as listed in Table 6.1. Before discussing the data some factors will be mentioned which will help in making a critical assessment of the information.

TABLE 6.1 Units Used in Tables

Material	Units used*	Equivalent units
Air, Aerosol	$\mu g\ m^{-3}$**	
	$ng\ m^{-3}$	
	$pg\ m^{-3}$	
Solids	$\mu g\ g^{-1}$ (ppm)	$mg\ kg^{-1}$
	$ng\ g^{-1}$ (ppb)	$\mu g\ kg^{-1}$
	$pg\ g^{-1}$ (ppt)	$ng\ kg^{-1}$
Water	$mg\ l^{-1}$ (ppm)	$\mu g\ ml^{-1}$, $mg\ kg^{-1}$***
	$\mu g\ l^{-1}$ (ppb)	$ng\ ml^{-1}$, $\mu g\ kg^{-1}$
	$ng\ l^{-1}$ (ppt)	$pg\ ml^{-1}$, $ng\ kg^{-1}$
Blood	$\mu g\ dl^{-1}$	
Urine	$\mu g\ l^{-1}$	

* Amounts are expressed on a mass basis, if moles are used (e.g. $\mu mol\ l^{-1}$) this has been converted to mass by multiplying by the the relative atomic weight (RAW) (e.g. $\mu g\ l^{-1} = \mu mol\ l^{-1} \times RAW$). **prefixes; m = milli = 10^{-3}, μ = micro = 10^{-6}, n = nano = 10^{-9}, p = pico = 10^{-12}. *** Take the density of water as 1. The factor 1/1.205 is used to convert an amount of air from kg to cubic meter (m^3).

FACTORS AFFECTING NATURAL LEVELS OF THE TRACE ELEMENTS

Contamination

The most serious problem, when analysing for low levels of trace elements, is contamination of the sample during collection and analysis. Some of the results given in the tables are suspect for this reason. The problem of obtaining reliable results increases the more an element is used and distributed. The problem is most critical for lead, the most widely used of the ten elements. Even though reported levels of lead are often elevated, and are difficult to obtain reliably, some of the best results have been obtained on lead, because of the endeavour put into sampling and analysis by some workers. The credit for this is due to Patterson and his coworkers, who have carried out very good and painstaking work [22,51,57,61,69]. Patterson has been critical of early work, and suggests that data obtained prior to 1960, and much since are not reliable [51,57,61]. It is very difficult to maintain the integrity of a sample, for example when a sediment section from low down in a sediment profile has been exposed to our environment.

It is essential to publish detailed information on the methods of sampling, handling and analysis so others may judge the reliability of the analytical data. This is frequently not done, but Patterson does give detailed information [6,22,48,51,57,61]. For example, in attempting to obtain natural levels of lead in ice cores, held in museums, a method was developed for producing ice veneers of a few millimeters thick. This made it possible to gradually move into the centre of the core, where natural levels would be less contaminated. For the surface veneer of one core the lead level was reported as 2.7×10^5 ng g^{-1}, and this fell almost regularly to 4.9 ng g^{-1} by the 9th ice veneer [51]. Plastic augers have been used to obtain ice cores in the field [5,6]. In carrying out the analysis it is necessary to have an authentic blank. This was achieved, in this case, by preparing an ice sample from ultra-pure water with a known lead content and treating the ice in exactly the same way as the core. It was found that, the sampling and analysis, added about 30% more lead to the lead level in the inner core [51]. Collecting soil moisture, and maintaining its integrity, is not an easy task when the metal levels are low. One method has been to allow the moisture to soak into an ultra-clean filterpaper, separated from a freshly exposed moist surface with a 0.2 μm Nucleopore filter [22].

The problems encountered for lead, are also problems for other heavy elements, and the reliability of data on mercury, antimony, selenium and arsenic, for remote and background samples, has been questioned [6]. The situation may be less of a problem for cadmium [6,46], but that has yet to be verified.

The importance of reliable background and natural levels has been demonstrated for the lead levels in albacore (tuna) [11,69]. The relevant data are listed in Table 6.2. Based on the results of the National laboratory (3rd. column) it would appear that putting the fish in lead soldered cans approximately

Table 6.2 Levels of Lead in Albacore Muscle

Sample	Lead levels ng g^{-1}	
	Caltech lab.	NMFS lab
Surface seawater (est. prehistoric)	0.0005	
Surface seawater (modern)	0.005	
Albacore muscle (est. prehistoric)	0.03	
Albacore muscle (fresh)	0.3	400*, 20**
Albacore muscle (in unsoldered can)	7	
Albacore muscle (in lead soldered can)	1400	700

* Cut at NMFS lab., **Cut at Caltech lab. Source; Settle and Patterson, 1980 [69].

doubled the lead level, from 400 to 700 ng g^{-1}, when in fact it had increased more than 2000 times, from 0.3 to 700 ng g^{-1} (2nd. column). The difference was due to contamination added, during the preparation stage and analysis. In addition there had been a 10 fold increase from prehistoric levels to current typical levels.

Alteration of Sample

A major requirement for samples, particularly those of historical interest, is that they represent the material of the time and have not altered in any way. Ancient samples, such as bones, may have absorbed trace elements from the soil moisture in contact with them, as well as have material leached from them [12,60,61]. Material that has altered needs to be recognized and removed, or avoided, or allowed for. The trace element environment of the ancient samples should be investigated, for example, what contact did the people have with the elements [61].

Atmospheric deposition over the years can be estimated from ice/snow cores, though often levels are too low to measure accurately [31]. Peat has been used, but the mobility of metal ions in a peat core is a problem [14,31]. Lake sediments may be one of the best materials to study for historical trends [31], however, there are problems, such as being sure that all the deposition is represented in the sediment, and there has been no bioturbation. The mobility of metal ions depend on the pH and the redox conditions of the system. Ideally the ions of the heavy elements must have an affinity for the sediments, e.g. by sorption, and remain fixed [2,31].

Materials such as human hair and teeth, as well as having some of the problems listed above, are also difficult to sample reliably, e.g. what position along the hair fibre, or from what part of the head, and what material to sample from the teeth, enamel, or dentine or the whole tooth? Direct comparisons between ancient and contemporary samples of hair is difficult because of variation in hygiene standards, diets, culture and environmental conditions [12].

Some materials used in historical monitoring have come from museums [13,24,32,51], and it is necessary in these cases to be particularly careful over sample selection. Ice cores samples, held at -30°C, need to be free of cracks and have no sign of surface melting [51]. Museum samples in general were not collected for analytical purposes, and therefore care was not taken to avoid contamination. During storage and preservation, loss or gain of trace elements may have occurred, and so it is necessary to be cautious over interpreting analytical results. It is not always possible to compare ancient animal samples with modern day ones because of differences, such as the age of the animal, the season it was caught and location. Many of these details are unknown for the ancient samples [13,32].

Enrichment

A method of determining if the concentration of a trace element is close to natural levels, is to calculate either the enrichment factor (EF) or mobilization factor (MF) for the element (see Chapter 5 for the equations) [31,36,38,74,75]. The enrichment factor, must be calculated using an element that is not a significant pollutant, such as Al, Sc, V, Fe, Si, Ti, Mn. If $EF_{crust} \leq 1$ the levels may be considered as close to natural or background. The mobilization factor tends to zero as human generated sources decrease [31].

The biopurification factor (BPF) (see Chapter 5) has been used as a measure of pollution [22,23,25,61], but it may also be used to estimate natural levels. It is based on the concept that toxic metals in rocks and soil are discriminated against relative to essential elements in living materials.

Dating

For many materials, including sediments, ice, peat and remains of once living things, dating methods can be used [1,2,4,33,37,48,50,71,73]. In the case of sediment cores, dating has been carried out by locating markers in the horizons, which can be identified with historical events, such as floods and winds. Radiochemical dating can be used to date post-1940 samples, by the measurement of fission products (e.g ^{137}Cs), that have been produced in nuclear explosions. Dating of earlier periods is done using radioisotopes such as ^{13}C, ^{210}Pb, and ^{228}Th. The presence of flora and fauna in sediments is a guide to their age, and pollen in samples can also be used. Tree rings may be used for wood dating, and thickness of the rings can be related to known seasonal variations. Estimation of present day fluxes can help determine the age of sediments, by utilizing a certain depth per flux rate. Remnant magnetization and varve stratigraphy in sediments has also been used. The velocity of ice movement is useful information for dating. For best results, at least two methods should be used, to ensure the dates are reliable.

Extent of Work Carried Out

The majority of the work on obtaining natural levels has been carried out on lead, as can be seen from the tables. The next most studied elements are cadmium and mercury, where global contamination is probably less of a problem, but cannot be ruled out. It is less easy to decide if the natural levels reported for the other elements are reliable as there are less data to compare. The data in Tables A6.1-A6.11 are a selection of some of the results for the elements arsenic, cadmium, mercury, indium, lead, antimony and selenium.

LEVELS IN REMOTE PLACES

Remote places include the Arctic, Antarctic, Greenland, mid-oceans and mountain areas. The most remote is the Antarctic, which until recently was uninhabited by humans (Tables A6.1-A6.3). An air lead level of 0.027 ng m^{-3} at the Antarctic [74] contrasts with levels of 0.1-0.2 ng m^{-3} in the Arctic [18,36], 0.86 ng m^{-3} in the Himalayas [9] and 3.6 ng m^{-3} at Jungfrau [67]. A similar trend occurs for most of the other elements, and a correlation coefficient between Cd and Pb of 0.999 in the different areas, suggests that the Antarctic is the most remote place. The higher levels of the elements in other remote places corresponds to some global contamination. Levels of arsenic, cadmium, mercury and lead in snow or ice are in the few pg g^{-1} range in Greenland and the Antarctic [5,6,17,18], whereas in the Himalayas lead levels are one to two orders of magnitude greater [19]. However, it is necessary to bear in mind that contamination, introduced during and after sampling, can also increase concentrations by similar amounts.

The mean blood lead concentrations of people in remote areas span 0.83 to 5.2 μg dl^{-1} [15,35,62,64]. The latter figure overlaps with the lower end of the scale found for urban/rural people. There has been a decline in blood lead levels over recent years, due to reduced lead intakes and better analytical data (see Chapter 13), therefore when making comparisons it is important to know the date when the work was done.

The careful work done in the collection and analysis of samples from Thompson Canyon in the High Sierra [22] (called a remote sub-alpine area in Table A6.3), is reflected in the very low levels found for a number of samples in the eco-system, such as 0.7-8.9 μg l^{-1} and 16.6 ng l^{-1} of lead in soil moisture and stream water respectively, and 5 ng g^{-1} lead in plants. The study demonstrates that, whereas the levels are low in the remote area, they are still not natural levels because of dispersed anthropogenic pollution.

BACKGROUND LEVELS OF HEAVY ELEMENTS

Background and remote levels are not easy to distinguish from each other, though background levels can be obtained close to urban/rural areas (Tables A6.4-A6.7). Background air lead concentrations, from different locations, extend from 3 to 100 ng m^{-3} [15,26,65]. It is necessary to be alert to the description of the background, for example in many studies a background for urban levels is taken as the levels found in rural areas. Deep sediments provide a source of background material laid down before major industrialization. The levels of lead in deep sediment profiles, are around 10-12 μg g^{-1} [50], and cadmium and mercury levels are one to two orders of magnitude lower [2,24,30,37]. Background levels of cadmium, mercury and lead in ice or snow, taken from different parts of the world, tend to be higher (a few ng g^{-1}) than for remote areas (a few pg g^{-1}) [1,39]. Background levels of lead in the oceans are found in deep water, and are around 1-6 ng l^{-1} [45,58,68,69]. In the case of cadmium, however, the levels deep in the ocean (< 125 ng l^{-1}) are usually greater than found in the surface water (4-5 ng l^{-1}) because of processes that deplete the concentrations in the surface waters (see Chapter 8) [46].

ANCIENT LEVELS OF HEAVY ELEMENTS

Ancient concentrations of the heavy elements have been determined on materials such as bone, teeth, hair, ice, sediments and peat (Tables A6.8, A6.9). If ancient levels are to be a guide to natural levels, it is necessary to know what contact the material has had with its environment. For example, the increase in lead levels in the bones of Nubians (0.5, 1.0, 2.0, 1.2, μg g^{-1}) and a similar increase of lead in their teeth (0.9, 2.1, 5.0, 3.2 μg $g^{-1)}$ with decreasing age (3300 - 2900 BC, 2000 - 1600 BC, 1650 - 1350 BC, 1 - 750 AD) can be related to an increase in the use of lead in the culture. This suggests that a natural level of lead in bone is at least <0.5 μg g^{-1} and lead in teeth <0.9 μg g^{-1}. Work on Peruvian remains suggest the natural lead levels in bone and teeth may be less than 2 ng g^{-1} (i.e. 0.002 μg g^{-1}) [61]. Overall there is a weak trend of increasing levels of lead in bone with proximity to the present day, from <0.1 μg g^{-1} (6000 bp) to around 2-3 μg g^{-1} 700-300 bp.

The heavy metal content in the hair of a group of 15th Century mummified bodies found in Greenland was Cd (0.31 μg g^{-1}), Hg (3.8 μg g^{-1}), Se (0.29 μg g^{-1}) and Pb (1.2 μg g^{-1}) [22]. Typical levels found in contemporary hair are 0.1-2.0, 3-5, 0.1-1.0 and 10-15 μg g^{-1} respectively. Only in the case of lead, has there been a significant rise from the 15th. Century to the present day.

The data in Table 6.3 are a summary of values, or range of values, for the elements As, Cd, Hg, In, Pb, Sb and Se estimated from the lowest background, remote and ancient levels. Where available estimated natural levels are given in parenthesis. The data are sparse except for lead, cadmium, and mercury, and more study is required for the other elements. The combined results are a guide to natural levels but cannot be considered as natural. For many of the materials the levels of the elements are quite low, parts per trillion for air, ice and water, parts per billion for fish, and parts per million for the rest of the materials i.e. rock, soil, sediment, animal muscle, bone, hair, blood, plant, and teeth.

NATURAL LEVELS OF THE HEAVY ELEMENTS

Only lead has been studied in detail as regards estimating its natural concentrations. This is clear by comparing Tables A6.10 and A6.11. However, because contamination is considered less of a problem for the other elements,

TABLE 6.3 Probable Background-Remote-Ancient Concentrations of the Heavy Elements

Material	Pb	Cd	Hg	As	Se	Sb	In
Air	6-100	0.4	25	5	5-7	0.5-20	0.02-1
pg m^{-3}	(40)*						
Ice/snow	1-2	1-25	0.1-3	1-30		18	
pg g^{-1}	(1)		(< 0.1)				
Rock	12-40	0.001-2					
μg g^{-1}	(12)						
Soil	12-20	0.01-0.2	0.01				
μg g^{-1}	(5)						
Sediment(f)	4-50	< 0.1-0.5	0.01-0.2	2-5			
μg g^{-1}	(10)		(0.004)				
Sediment(o)	10-12	0.04-0.2	0.01-0.2				
μg g^{-1}	(4)		(0.02)				
Water (f)	5-50	10-100	2-100				
ng l^{-1}	(20)						
Water (o)	1-5	1-5	2-10				
ng l^{-1}	(0.5)		(< 10)				
Animal (m)	1-12						
ng g^{-1}	(2)						
Blood	0.83-5.2	0.57	1.43				
μg dl^{-1}	(0.2)						
Bone	0.5-2						
μg g^{-1}	(0.4)						
Fish	0.3-0.4						
ng g^{-1}	(0.03)						
Hair	1.2	0.31	1.0-3.8		0.29		
μg g^{-1}							
Plant	0.005						
μg g^{-1}	(0.002)						
Teeth	< 1-5	< 0.1-1	4.0				
μg g^{-1}	(< 1)						

* Figures in parenthesis are estimated natural levels, f = fresh water, o = ocean.

it is likely that background and remote levels are closer to natural levels than they are for lead (see Table 6.3). Concentrations claimed as being natural levels represent, at best, the upper levels. A good summary of the situation for lead is given by Patterson in *Lead in the Human Environment* [51]. This review, and other sources, [1,12,58,61,62,69] lists natural (natural = prehistoric) levels as follows; air 0.04 ng m^{-3}, marine water 0.5 ng l^{-1}, fresh water 20 ng l^{-1}, soil moisture <2μg g^{-1}, ice ~1 pg g^{-1}, earth's biomass (trees) ~4 ng g^{-1}, humans 30 μg per 70 kg person, human food < 2 ng g^{-1}, fish ~0.03 ng g^{-1}, human blood 0.2 μg dl^{-1}, and human bone 0.36 μg g^{-1}. The significant aspect of these concentrations is how low they are, and therefore how polluted our environment is with lead. These values alter the significance of the terms normal or typical i.e. contemporary typical levels are highly elevated, and not just slightly elevated. A pictorial representation of these figures is given in Fig 6.1.

Two separate estimates have been made of the uptake of lead by human beings based on the natural levels [15, 57, 69]. The estimates differ by a factor of ten, one suggests a natural intake of lead <210 ng day^{-1} [57,69], whereas the other calculation indicates 2360 ng day^{-1} [15]. If the two studies are put on the same basis as regards the amount of air, food and water consumed, and the same absorption factor into the body the values are now <210 ng day^{-1} and 1370 ng day^{-1} respectively. The discrepancy lies in the values used for the natural levels of lead in the food water and air, the latter calculation includes 150 ng day^{-1} from soil. This situation highlights the problem of the true baseline levels, and the effect of using different estimates, particularly when comparing with present day intakes (see Chapter 14).

Estimates have been made of the contributions to the natural levels in materials such as air and ice [5,7,41,51,57]. The natural air lead concentration of 0.04 ng m^{-3} arises from wind blown soil (silicate dust), plants, sea salt spray, forest fire smoke and volcanic emissions [57]. For the natural ice lead concentration of 1 to 1.2 pg g^{-1}, 0.4-0.5 pg g^{-1} has been attributed to natural silicate dust, 0.4 pg g^{-1} from volcanic emissions and sea salt spray, leaving the remainder 0.2-0.4 pg g^{-1} from other sources [7,51]. Volcanic emissions have been given a prominent role as a natural source in many studies, on the other hand, it appears that the peaks in the lead and cadmium concentrations in dated ice do not relate to volcanic activity [5]. For the elements As, Cd, Hg, Pb, Sb, Se estimates of the natural sources gives continental dust as 70-90%, volcanic dust as 3-24% and volcanic gas as 0.02-3%. The volcanic contribution is greater for the volatile elements selenium and mercury [41].

The data summarized in the diagrams in Fig 6.1 [35,57,59] highlight the problem raised at the beginning of this chapter, viz. the relation between natural levels and contemporary typical levels for the element lead. The diagrams for blood lead and total body lead indicate clearly that contemporary levels are much closer to toxic than natural levels. The information in Fig. 6.2 [57] clarifies the main reason for incorrect results, viz. contamination during collection and analysis of the samples in the laboratory. The contamination is a serious effect in the measurement of natural levels (99% of the total lead measured), but not so important for the measurement of present day concentrations (13% of the total lead measured) because of the high contamination of our environment by lead.

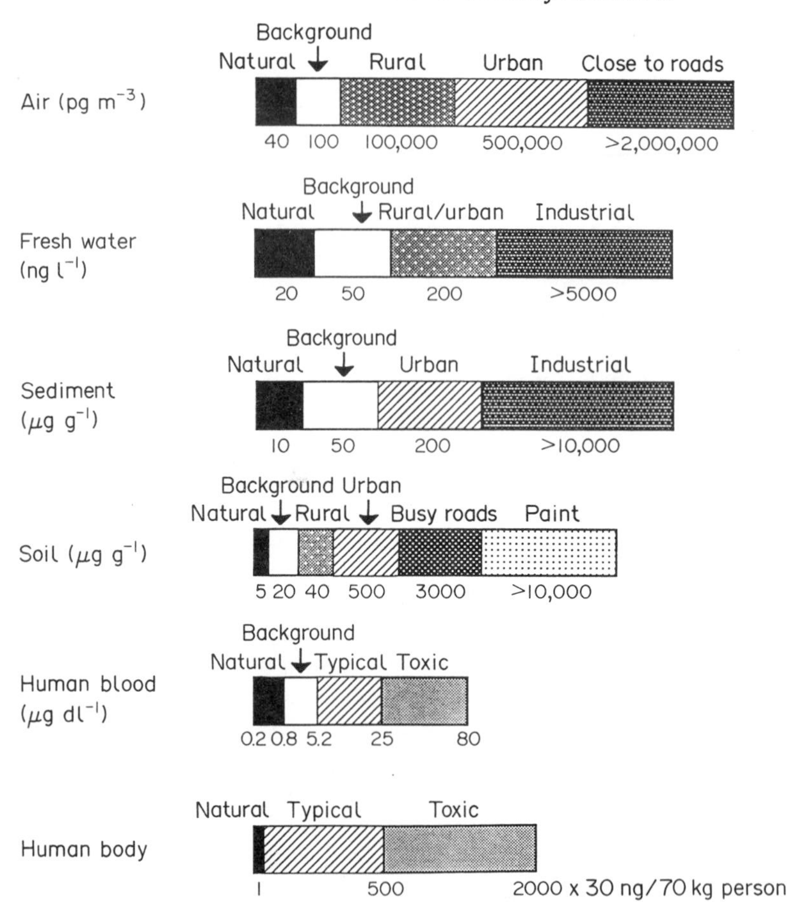

Fig 6.1 Natural and present day concentrations of lead. Source of data; Patterson, 1980 [57]. Note the log scale.

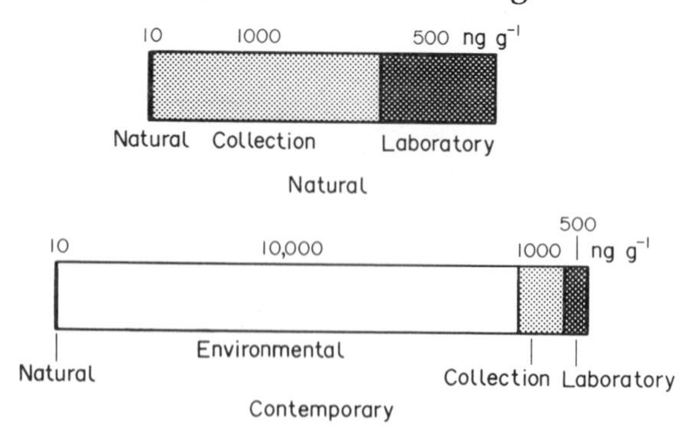

Fig 6.2 Natural concentrations of lead on a leaf comparedwith contemporary and contaminated concentrations. Source of data; Patterson, 1980 [57].

TABLE A6.1 Concentrations of Arsenic and Cadmium in Remote Places

Metal	Material	Location	Remote levels	Ref.
As	Aerosol	Antarctic	17(wint) 8.4(sum) pg m^{-3}	16
	Air	Greenland	42, 18-< 88 pg m^{-3}	17
		Hawaii	0.02 ng m^{-3}	74
		Jungfrau	0.16 ng m^{-3}	67
		Sth Pole	0.005 ng m^{-3}	67
		Sth Pole	14-41 pg m^{-3}	44
	Atmospheric deposition		3.1×10^{-4} kg ha^{-1} y^{-1}	31
	Snow (fresh)	Greenland	0.94-< 37 pg g^{-1} (old)	18
	Snow (old)	Greenland	19-< 22 pg g^{-1}	18
	Snow	Greenland	< 10-28 pg g^{-1}	17
	Water (rain, ice)	Arctic/Antarctic	0.019 μg l^{-1}	31
Cd	Aerosol	Antarctic	200(wint) 49(sum)pg m^{-3}	16
		Antarctic	0.049 ng m^{-3}	74
		Bolivia	0.006 ng m^{-3}	52
		Enewetok	0.0046 ng m^{-3}	74
		Greenland	15, <dl-33 pg m^{-3}	17
		Jungfrau	0.4 ng kg^{-1}	67
		Sth Pole	< 0.015 ng m^{-3}	52
		Sth Pole	3-36 pg m^{-3}	44
			Est. 0.04 (0.003-0.39) ng m^{-3}	52
	Atmospheric deposition		2.3×10^{-6} to 5×10^{-5} kg ha^{-1} y^{-1}	31
	Blood	Yanomamo (Venezuela)	0.57, 0.07-3.72 μg dl^{-1}	35
	Fluxes	Greenland	0(dry) 0.61(wet) ng cm^{-2} y^{-1}	18
	Rocks		0.001-1.6 0.03-2 μg g^{-1}	27
	Sediments (fresh)		5-19 μg g^{-1}	27
	Sediments (oceanic)	Nth. Atlantic	0.13-0.21 μg g^{-1}	21
			0.04-1.88 μg g^{-1}	27

Table A6.1 continued

Metal	Material	Location	Remote levels	Ref.
Cd	Snow (fresh)	Greenland	7.8-17 pg g^{-1}	18
	Snow (old)	Greenland	11-< 19 pg g^{-1}	18
	Snow	Greenland	< dl-24 pg g^{-1}	17
		Sth Pole	2-4 pg g^{-1}	5
	Soil		0.2 μg g-1	54
	Urine	Yanomamo (Venezuela)	1.2, 0.0-4.5 μg l^{-1}	35
	Water (fresh)	California	10-100 ng l^{-1}	47
	Water (ground)		0.1 μg l^{-1}	54
	Water (ocean)	Nth. Atlantic	60, 20-150 ng l^{-1}	21
			~ 20 ng l^{-1}	66
		Indian	70 ng l^{-1} (surface)	47
		Nth Atlantic	20-150 ng l^{-1} (50-100m)	47
		Pacific	0.8-12 ng l^{-1} (3-5000m)	47
	Water (ocean, solids)		1.0 mg l^{-1}	54
	Water (solids)	Nth. Atlantic	0.2 ng l^{-1}	21
	Water (rain)		< 0.1 μg l^{-1}	52
	Water (rain, ice)	Arctic/Antarctic	0.008, 0.004-0.639 μg l^{-1}	31

TABLE A6.2 Concentrations of Mercury, Indium, Antimony and Selenium in Remote Places

Metal	Material	Location	Remote levels	Ref.
Hg	Air	Jungfrau	0.039 ng m^{-3}	67
		Continental	4.0 ng m^{-3}	24
		NY	0.60 ng m^{-3}	74
		Oceanic & polar	0.7 ng m^{-3}	24
	Air (particulate)	Oceanic	<0.15, 0.005-0.06 ng m^{-3}	24

Table A6.2 continued

Metal	Material	Location	Remote levels	Ref.
Hg		Terrestial (non-mineral)	0.15, < 0.005-1.9 ng m^{-3}	24
	Air (vapour)	Oceanic	0.7, 0.6-0.7 ng m^{-3}	24
		Terrestial (non-mineral)	4.0, 1-10 ng m^{-3}	24
	Atmospheric deposition		< 2.0 x 10^{-2} to 2 x 10^{-4} kg ha^{-1} y^{-1}	31
	Blood	Yanomamo (Venezuela)	1.43, 0.3-5.7 μg dl^{-1}	35
	Fluxes (air)	Polar	22 x 10^{-6} g m^{-2} y^{-1}	24
	Hair	Yanomamo (Venezuela)	1.0, 0.3-1.4 μg g^{-1}	35
	Ice	Antarctic	0.1-3 pg g^{-1}	6
	Urine	Yanomamo (Venezuela)	7.2, 2.0-26.0 μg l^{-1}	35
	Water (fresh)	Lakes & ice caps	0.002-0.1,0.02-0.1 μg l^{-1}	47
	Water (ocean)		0.002-0.01-0.03 μg l^{-1}	47
	Water (rain, ice)	Arctic/Antarctic	0.079, 0.011-0.428 μg l^{-1}	31
In	Aerosol	Antarctic	0.19(wint),0.054(sum) pg m^{-3}	16
	Air	Bolivia	0.0078 ng m^{-3}	74
		Jungfrau	0.0007 ng m^{-3}	67
		Sth Pole	0.020-0.082 pg m^{-3}	44
Sb	Aerosol	Antarctic	2.1(wint), 0.45(sum) pg m^{-3}	16
	Air	Enenetok	0.0052 ng m^{-3}	74
		Jungfrau	0.13 ng m^{-3}	67
		Sth Pole	< 0.5-1.37 pg m^{-3}	44
	Atmospheric deposition		1.6 x 10-4 kg ha^{-1} y^{-1}	31
	Water (rain, ice)	Arctic/Antarctic	0.034 μg l^{-1}	31
Se	Aerosol	Antarctic	6.9(wint), 6.3(sum) pg m^{-3}	16
	Air	Hawaii	0.028 ng m^{-3}	74
		Jungfrau	0.028 ng $^{-3}$	67
		Sth Pole	0.004 ng m^{-3}	67
		Sth Pole	4.8-7.1 pg m^{-3}	44

TABLE A6.3 Concentrations of Lead in Remote Places

Material	Location	Remote levels	Ref.
Air	Antarctic	Est. 0.006 ng m^{-3}	61
	Antarctic	0.027 ng m^{-3}	74
	Arctic (late winter)	1.0-< 4.9 ng m^{-3}	36
	Arctic (summer)	0.1-0.2 ng m^{-3}	36
	Enewetok	0.13 ng m^{-3}	74
	Greenland	0.11-0.21 ng m^{-3}	18
	Greenland	490, 170-960 pg m^{-3}	17
	Himalayas	0.86 ng m^{-3} (out doors)	19
	Jungfrau	3.0 ng m^{-3}	67
	Nepal	0.22 µg m^{-3}	15
	Over remote land	1-10 ng m^{-3}	61
	Over remote waters	0.2-3 ng m^{-3}	61
	Sth Pole	45-120 pg m^{-3}	44
	Sth. Hemisphere	Est. 0.03 ng m^{-3}	61
	Variety	0.049-8.7 ng m^{-3}	15
		8-10 ng m^{-3} (contemp.)	57
		2-15 ng m^{-3}	61
	South Pole	0.63 ng m^{-3}	55
Animal (pine marten) skeleton	Remote sub-alpine area	0.64-1.41 µg g^{-1}	22
Animal (pine marten) muscle	Remote sub-alpine area	9.9, 12.0 ng g^{-1}	22
Animal (pine marten) hair	Remote sub-alpine area	6.4(unwash), 0.7(wash) µg g^{-1}	22
Animal (vole) skeleton	Remote sub-alpine area	1.1, 1.3 µg g^{-1}	22
Animal (vole) muscle	Remote sub-alpine area	1.2, 1.5 ng g^{-1}	22
Animal (vole) hair	Remote sub-alpine area	1.2, 2.2 nmol g^{-1}	22
Animals	Remote sub-alpine area	4 ng g^{-1} (fw)	61
Atmospheric deposition	Variety	0.01-4.2 mg m^{-2} y^{-1}	15
		1.2×10^{-5} to 9.3×10^{-3} kg ha^{-1} y^{-1}	31

Table A6.3 continued

Material	Location	Remote levels	Ref.
Blood	Nepal (children, adults)	3.5, 3.4 $\mu g\ dl^{-1}$	62
	Orinoco Indians	0.83 $\mu g\ dl^{-1}$	15
	Papua New Guinea	5.2 , 1.0-13.0 $\mu g\ dl^{-1}$	64
	Yanomamo (Venezuela)	0.83, 0-3.87 $\mu g\ dl^{-1}$	35
Blood (adults, children)		3.4, 3.5 $\mu g\ dl^{-1}$	62
Earth's biomass (trees)		~40 $ng\ g^{-1}$ (contemp.)	57
Fluxes	Greenland	1.5 (dry), 4.5 (wet) $ng\ cm^{-2}\ y^{-1}$	18
Human diet		~200 $ng\ g^{-1}$ (contemp.)	57
Plant (Sedge leaves)	Remote sub-alpine area	52-394(unwash) 14.5-70(wash) $ng\ g^{-1}$	22
Plants	Sub-alpine pond area	5 $ng\ g^{-1}$ (fw)	61
Rock	Remote sub-alpine area	12.4-41 $\mu g\ g^{-1}$	22
Snow	Greenland	46-390 $pg\ g^{-1}$	17
	South Pole	5, 5-9 $pg\ g^{-1}$	5
Snow (fresh)	Greenland	42-150 $pg\ g^{-1}$	18
	Himalayas	0.16 $ng\ g^{-1}$	19
Snow (old)	Greenland	120-140 $pg\ g^{-1}$	18
	Himalayas	0.18 $ng\ g^{-1}$	19
Soil	Global	12-20 $\mu g\ g^{-1}$	15
	Papua New Guinea	10 $\mu g\ g^{-1}$	64
	Remote sub-alpine area	14.5-20.7 $\mu g\ g^{-1}$	22
Soil moisture	Remote sub-alpine area	0.70-8.9 $ng\ g^{-1}$	22
Teeth (sec. dentine)	Eskimo	56.0, 9.6-97.8 $\mu g\ g^{-1}$	70
	Mexican Indians	4.3 $\mu g\ g^{-1}$	12
Urine	Yanomamo (Venezuela)	7.9, 0.7-21.6 $mg\ l^{-1}$	35
Water (fresh)		0.005-0.05 $ng\ g^{-1}$	57
		$\leq 3\ \mu g\ l^{-1}$	47
Water (rain)	American Samoa	< 0.2-1.8 $ng\ l^{-1}$	49

Table A6.3 continued

Material	Location	Remote levels	Ref.
Water (rain, ice)	Arctic/Antarctic	0.09, 0.02-0.41 $\mu g\ l^{-1}$	31
Water (stream)	Remote sub-alpine area	16.6 $pg\ g^{-1}$	22

TABLE A6.4 Background Concentrations of Arsenic, Antimony and Selenium

Metal	Material	Location	Background level	Ref.
As	Air	Belgium (rural)	Obs. 2.3 $ng\ m^{-3}$, Est. 0.7 $ng\ m^{-3}$	65
	Fluxes (air)		28.1 x 10^8 $g\ y^{-1}$	41
	Ice	Greenland	31 $pg\ g^{-1}$	1
	Sediments (fresh)	Adirondack lake area	1.2, 1.4 $\mu g\ g^{-1}$	37
		Gt. Lakes	< 2-5 $\mu g\ g^{-1}$	2
		Adirondack lake area	6.16 $\mu g\ g^{-1}$ (30 cm deep)	30
Sb	Air	Belgium (rural)	Obs. 1.7 $ng\ m^{-3}$, Est. 1.05 $ng\ m^{-3}$	65
	Fluxes (air)		9.813 x 10^8 $g\ y^{-1}$	41
	Ice	Greenland	18 $pg\ g^{-1}$	1
	Sediments (fresh)	Adirondack lake area	0.2 $\mu g\ g^{-1}$ (30 cm deep)	30
Se	Air	Belgium (rural)	Obs 0.62 $ng\ m^{-3}$, Est 0.17 $ng\ m^{-3}$	65
	Fluxes (air)		4.13 x 10^8 $g\ y^{-1}$	41
	Sediments (fresh)	Adirondack lake area	0.8, 1.6 $\mu g\ g^{-1}$	37

TABLE A6.5 Background Concentrations of Cadmium

Material	Location	Background levels	Ref.
Air	Belgium (rural)	Obs 0.78 $ng\ m^{-3}$, Est 0.37 $ng\ m^{-3}$	65
Fluxes (air)		2.901 x 10^8 $g\ y^{-1}$	41
Ice	Alaska	0.14, 0.13-0.15 $ng\ g^{-1}$	39

Table A6.5 continued

Material	Location	Background levels	Ref.
Ice	Austria	1.06 ng g^{-1}	39
	Global	0.76,0.13-8.6 ng g^{-1}	39
	Greenland	11 pg g^{-1}	1
	Greenland	3-14 pg g^{-1}	1
	Nepal	0.60 mg g^{-1}	39
	Nth Norway	0.33, 0.24-0.4 ng g^{-1}	39
	Peru	0.56 ng g^{-1}	39
	Spitzbergen	1.32, 0.18-2.55 ng g^{-1}	39
	Sth Norway	0.77, 0.36, 2.29 ng g^{-1}	39
	Uganda	1.0 ng g^{-1}	39
Sediments (fresh)	Adirondack lake area	0.6, 0.5 μg g^{-1}	37
	European	≤ 0.1 μg g^{-1}	2
	Gt. Lakes	< 2 μg g^{-1}	2
	Adirondack lake area	< 0.5 μg g^{-1} (30 cm deep)	30
Snow	Antarctic	1-25 pg g^{-1}	8
Soils		0.01-0.07 μg g^{-1}	63
Water (fresh)		0.01-0.1 μg l^{-1}	46
		< 1 μg l^{-1}	63
Water (ocean profile)	Pacific	11.2-12.3 ng l-1	9
		4.5-125 ng l-1	10
Water (ocean)	Calif. coast (100 km)	101 ng l^{-1} (1000 m deep)	45
	Mediterranen	0.02-1.9 μg l^{-1}	63
		0.1 μg l^{-1}	63
		60 ng l^{-1}	66
		4-5 ng l^{-1} (surface)	46
		< 125 ng l^{-1} (1-2 km deep)	46

TABLE A6.6 Background Concentrations of Mercury

Material	Location	Background levels	Ref.
Air	Belgium (rural)	Obs. 1.2 ng m^{-3}	65
		0.025, 0.005-0.06 ng m^{-3}	63
Fluxes (air)		0.3, 0.1, 0.001 x 10^8 g y^{-1}	41
Groundwater		0.05 ng g^{-1}	24
Ice	Alaska	0.18, 0.14-0.26 ng g^{-1}	39
	Antarctic	0.17 ng g^{-1}	39
	Austria	0.17 ng g^{-1}	39
	Global	0.25, 0.08-0.8 ng g^{-1}	39
	Nepal	0.23 ng g^{-1}	39
	Nth Norway	0.18, 0.08-0.42 ng g^{-1}	39
	Peru	0.38 ng g^{-1}	39
	Spitzbergen	0.27, 0.16-0.46 ng g^{-1}	39
	Sth Norway	0.37, 0.18-0.7 ng g^{-1}	39
	Uganda	0.16 ng g^{-1}	39
Sediments (fresh & oceanic)		300 ng g^{-1}	24
Sediments (fresh)	Adirondack lake area	0.28, 0.3 µg g^{-1}	37
	European	~ 10 ng g^{-1}	2
	European	100 ng g^{-1}	2
	Gt. Lakes	< 0.1 µg g^{-1}	2
Sediments (oceanic)	W. Mexico coast	0.012-0.173 µg g^{-1}	24
Snow	Canada rural	< 0.01-0.52 µg l^{-1}	24
	Greenland	0.013-0.169 µg l^{-1}	24
Soil		71 ng g^{-1}	24
Water (fresh)		0.02-0.06 µg l^{-1}	24
		< 0.1, 0.01-10 µg l^{-1}	63

Table A6.6 continued

Material	Location	Background levels	Ref.
Water (ocean)		0.05 µg l^{-1}	63
		0.01-0.03 µg l-1	24
Water (rain)	Rural UK	0.2 µg l^{-1}	24

TABLE A6.7 Background Concentrations of Lead

Material	Location	Background levels	Ref.
Air	Belgium (rural)	Obs. 125 ng m^{-3}, Est. 93 ng m^{-3}	65
	Jungfrau	8.7 ng m^{-3}	26
	Norway	5.6 ng m^{-3}	26
	Nth Hemisphere	~2.4 ng m^{-3}	15
	Tasmania	3.3 ng m^{-3}	26
Animal (vole)	High Sierra (Calif.)	0.06 µg g^{-1}	38
Blood plasma		0.1 ng g^{-1}	58
Fish (marine, muscle)		0.3-0.4 ng g^{-1}	58
Fish (tuna)		0.3 ng g^{-1}	69
Fluxes (air)		58.712 x 10^8 g y^{-1}	41
Ice	Antarctic	5 pg g^{-1}	1
	Austria	13.0 ng g^{-1}	39
	Global	7.0,1.03-54.6 ng g^{-1}	39
	Nepal	1.75 ng g^{-1}	39
	Norway (glaciers)	2.13 ng g^{-1}	1
	Peru	22.8 ng g^{-1}	39
	Spitzbergen	2.01, 1.82-2.22 ng g^{-1}	39
	Sth Norway	3.74, 1.03-9.02 ng g^{-1}	39
	Uganda	21.6 ng g^{-1}	39

Table A6.7 continued

Material	Location	Background levels	Ref.
Ice (old glacier)		1 pg g^{-1}	58
Leaves (sedge)	High Sierra (Calif.)	0.33 μg g^{-1}	38
Pine needles		10 ng g^{-1}	26
Sediments (fresh)	Adirondack lake area	3.9, 5.4 μg g^{-1}	37
	European	< 10 μg g^{-1}	2
	Gt. Lakes	< 50 μg g^{-1}	2
	Remote Scottish lakes	~50 μg g^{-1}	2
	Remote sub-alpine pond	9.6-12.8 μg g^{-1}	71
	Adirondack lake area	~10 μg g^{-1} (30 cm deep)	30
Sediments (oceanic)	Californian coast	~10 μg g^{-1}	2
	San Pedro (Calif.)	11.6 μg g^{-1} (at19 -20 cm)	50
	Santa Barbara (Calif.)	12.8 μg g^{-1} (at 71-72 cm)	50
	Santa Monica (Calif.)	12.4 μg g^{-1} (at 19-19.2 cm)	50
Snow	Poland	5-10 μg l^{-1}	1
	Antarctic	15-41 pg g^{-1}	8
Soil moisture	High Sierra (Calif.)	0.0007 mg l^{-1}	38
Tree trunks		1-3 ng g^{-1}	58
Tree rings	Appalachin	0.4 μg g^{-1} (1840-1870)	4
Water (fresh)		5-50 ng l^{-1}	58
Water (ocean profiles)		≤ 5 mg l^{-1} (20-25 cm deep)	20
Water (ocean)	Calif. coast (100 km)	2.9 ng l^{-1} (1000 m deep)	45
	NE Pacific	1 ng l^{-1} (2000-5000 m)	68
	Nth. Pacific	0.005 μg l^{-1}	69
Water (ocean, deep)		1-6 ng l^{-1}	58

TABLE A6.8 Concentrations of Arsenic, Cadmium, Mercury, Antimony, and Selenium in Ancient Samples

Metal	Material	Location	Ancient levels	Date	Ref.
As	Ice	Antarctic	≤ 4 pg g^{-1}	> 12,000 bp	6
Cd	Hair (mummified)	Greenland	0.31 μg g^{-1}	15th Century	12
	Ice	Antarctic	2.6 pg g^{-1}	> 12,000 bp	6
		Greenland	5-14 pg g^{-1}	800 bp	6
	Mosses	Sweden	0.23 μg g^{-1}	1916-42 AD	13
	Peat (annual storage)	Scandanavia	0.04-0.05 mg m^{-2} y^{-1}	1300-1400AD	14
	Teeth	Indian	< detection limit	200-600 AD	75
		Norway	0.2-0.7 μg g^{-1}	2-3rd Century	3
		Norway	< 0.1-0.4 μg g^{-1}	1500-1830AD	3
		Norway	1128, 312 ng g^{-1}	12th Century	28
Hg	Animals (Bison & moose muscle	Museum specimens	0.70, 0.95 mg g^{-1}	Pleistocene	24
	Animals (Horse bone)	Museum specimens	< detection limit	Pleistocene	24
	Animal (Mammoth foot)	Museum specimens	0.06 μg g^{-1}	Pleistocene	24
	Bats (hair)	Japan	2.63 μg g^{-1}	1890 AD	13
	Bats (kidney)	Japan	0.11 μg g^{-1}	1890 AD	13
	Bird feathers				
	(goshawk)	Sweden	2.2 μg g^{-1}	Pre 1940 AD	13
	(guillemots)	Baltic	0.9-4.4 μg g^{-1}	Pre 1940 AD	13
	(peregrine)	Sweden	~ 2000 ng g^{-1}	1834-1940 AD	13
	Hair (mummified)	Greenland	3.8 μg g^{-1}	15th Century	12
	Ice	Arctic	60, 30-75 pg g^{-1}	800 BC	1
		Arctic	2-19 pg g^{-1}	From 1727AD	1
		Norway, Alaska	3-400 pg g^{-1}	Pre industrial	1
		Greenland	2-19 pg^{-1}	1-3 Century	6
	Sediment (fresh)	L. Windemere	120 ng g^{-1}	Pre 1260AD	2

Table A6.8 continued

Metal	Material	Location	Ancient levels	Date	Ref.
Hg	Sediment (ocean)	Californian coast	< 0.04 μg g^{-1}	1400-1500bp	2
	Teeth	Hopi Indians	4.0 μg g^{-1}	17th Century	40
Sb	Ice	Antarctic	≤ 21 pg g^{-1}	> 12,000 bp	6
Se	Hair (mummified)	Greenland	0.29 μg g^{-1}	15th Century	12
	Ice	Antarctic	≤ 8-30 pg g^{-1}	> 12,000 bp	6

TABLE A6.9 Concentrations of Lead in Ancient Samples

Material	Location	Ancient levels	Date	Ref.
Bone	Nubian	0.6 μg g^{-1}	3300-2900 BC	34
	Bavaria	1.9 μg g^{-1}	1800-400 BC	12
	Denmark	0.2-6.8 μg g^{-1}	> 3000-250 bp	15
	Denmark	< 0.2 μg g^{-1}	4000-1000 BC	55
	Denmark	0.2-6.8 μg g^{-1}	1-1700 AD	55
	Egyptians	1.1 μg g^{-1}	200 BC	12
	Italy	< 0.1 μg g^{-1}	4000-1000 BC	55
	Italy	5 μg g^{-1}	1-1700 AD	55
	Nubian	0.6, 0.4-1.5 μg g^{-1}	3300-2900 BC	34
	Nubian	1.0, 0.5-1.6 μg g^{-1}	2000-1600 BC	34
	Nubian	2.0, 1.4-3.0 μg g^{-1}	1650-1350 BC	34
	Nubian	1.2, 0.9-1.6 μg g^{-1}	1-750 AD	34
	Peru	2 μg g^{-1}	1-1700 AD	55
	Peru	< 0.3 μg g^{-1}	1300 AD	12
	Peruvians	0.11, 0.16, 0.71, 1.4, 2.7 μg g^{-1}	2500 BC-600 AD	12
	Peruvians	0.56, 0.06-1.9 μg g^{-1}	500-1000 AD	12
	Poland	2.4 μg g^{-1}	300 AD	55

Table A6.9 continued

Material	Location	Ancient levels	Date	Ref.
Bone	Poland	5.8-199 $\mu g\ g^{-1}$	1100-1800 AD	55
	USA Indians	< detection limit	1400 AD	12
		0.3-0.5 $\mu g\ g^{-1}$	4000 BC-1000 AD	12
Bone & enamel	Peruvian	1, 0.11 $\mu g\ g^{-1}$	2000-1400 bp	15
Bone & dentine	Nubian	0.6, 0.9 $\mu g\ g^{-1}$	5300-4800 bp	15
Bone & teeth	Peruvians	$Pb/Ca = 3 \times 10^{-8}$	1600 bp	25
Hair (locket samples)	USA (adults, children)	93.4, 164 $\mu g\ g^{-1}$	1871-1923 AD	12
Hair (mummified bodies)	Greenland	1.2 $\mu g\ g^{-1}$	15th Century	12
Ice	Antarctic	1.4-2.2, < 1.2 $pg\ g^{-1}$	1490-2020 bp	51
	Antarctic	1.6 $pg\ g^{-1}$	1801-1797 AD	7
	Antarctic	< detection limit	pre 1940	48
	Antarctic	Calc. 0.03 $pg\ g^{-1}$	Old ice	48
	Antarctic	1.2 $pg\ g^{-1}$	2000 bp	73
	Antarctic	1-2 $pg\ g^{-1}$	> 12,000 bp	6
	Arctic	1.6-4.9, < 1.4 $pg\ g^{-1}$	2700-5500 bp	51
	Arctic	< 0.001 $ng\ g^{-1}$	800 BC	1
	Arctic	1.4 $pg\ g^{-1}$	5000 bp	73
	Greenland	< 0.0005 $ng\ g^{-1}$	800 BC	55
	Greenland	0.01, < 0.001 $ng\ g^{-1}$	1753 AD, 800 BC	48
Mosses	Sweden	61 $\mu g\ g^{-1}$	1916-42 AD	13
	Sweden	~20$\mu g\ g^{-1}$	1860 AD	13
Peat	UK	~4 $\mu g\ g^{-1}$	800 AD	14
	UK	3 $\mu g\ g^{-1}$	~850 AD	43
	UK, Derbyshire	8 $\mu g\ g^{-1}$	Pre 700 BC	42
	UK, Derbyshire	40-50 $\mu g\ g^{-1}$	Roman times	42
Peat (annual storage)	Scandanavia	0.8-1.1 $mg\ m^{-2}\ y^{-1}$	1300-1400 AD	14
Teeth	American Indian	1-5 $\mu g\ g^{-1}$	200-600 AD	12
	Hopi Indians	7.0 $\mu g\ g^{-1}$	17th Century	40

Table A6.9 continued

Material	Location	Ancient levels	Date	Ref
Teeth	Indian	1.3-5.3 $\mu g\ g^{-1}$	200-600 AD	75
	Norway	~ 2-3 $\mu g\ g^{-1}$	2-3rd Century	3
	Norway	2-140 $\mu g\ g^{-1}$	1500-1830 AD	3
	Norway	1.24 $\mu g\ g^{-1}$	Medieval	29
	Nubian	0.9 $\mu g\ g^{-1}$	3300-2900 BC	34
	Norway	1.22, 1.81 $\mu g\ g^{-1}$	12th Century	28
		Est. < 1 $\mu g\ g^{-1}$	Pre lead industry	12
Teeth & bones	Peruvian	2 ng g^{-1} (fw)	1600 bp	61
Teeth (enamel layers)	Indian Knoll	60 $\mu g\ g^{-1}$ (chips), 85 $\mu g\ g^{-1}$ (surface)	5000 bp	72
	Pueblo Indians	12 $\mu g\ g^{-1}$ (chips), 13-80 $\mu g\ g^{-1}$ (surface)	919-1130 AD	72
	Peruvians	0.04-0.23 $\mu g\ g^{-1}$	2500 BC-600AD	12
Teeth (sec. dentine)	Egyptian	9.7, 0.6-29.8 $\mu g\ g^{-1}$	1st, 2nd Millenia	70
	Machu Picchu (Peru)	13.6, 0.2-56.1 $\mu g\ g^{-1}$	12th Century	70
	Nubian	0.9, 0.4-1.4 $\mu g\ g^{-1}$	3300-2900BC	34
	Nubian	2.1, 0.9-3.1 $\mu g\ g^{-1}$	2000-1600BC	34
	Nubian	5.0, 2.4-7.7 $\mu g\ g^{-1}$	1650-1350BC	34
	Nubian	3.2, 2.4-4.4 $\mu g\ g^{-1}$	1-750 AD	34
	Peruvians	13.6 $\mu g\ g^{-1}$ med 3.7 $\mu g\ g^{-1}$	1100-1200AD	12

TABLE A6.10 Natural Concentrations of Arsenic Cadmium, Mercury, Antimony and Selenium

Metal	Material	Location	Natural Levels	Ref.
As	Fluxes	Atmosphere	7.8×10^{6} kg y^{-1}	56
	Sediments (fresh water)	USA, Canada	3.2-9.2, 2.7-13.2 $\mu g\ g^{-1}$	47
	Sediments (fresh, deposit)	Adirondack lake area	1900, 910 ng cm^{-2}	37

Table A6.10 continued

Metal	Material	Location	Natural Levels	Ref.
Cd	Emissions	Global	0.83×10^{-6} kg y^{-1}	53
	Fluxes	Adirondack lake area	0.0054 μg cm^{-2} y^{-1}	30
		Atmosphere	0.96×10^{6} kg y^{-1}	56
	Sediments (fresh, deposit)	Adirondack lake area	900, 350 ng cm^{-2}	37
Hg	Fluxes	Atmosphere	0.16×10^{6} kg y^{-1}	56
	Ice	Antarctic	< 0.1 pg g^{-1}	6
	Sediments (fresh)	USA, Canada lakes	0.004-0.07-0.3 μg g^{-1}	47
	Sediments (fresh, deposit)	Adirondack lake area	440, 200 ng cm^{-2}	37
	Sediments (ocean)	Hawaii	0.05, 0.02-0.24 μg g^{-1}	47
	Water (ocean)		< 0.01 μg l^{-1}	63
Sb	Fluxes	Adirondack lake area	0.0054 μg cm^{-2} y^{-1}	30
Se	Fluxes	Atmosphere	0.4×10^{6} kg y^{-1}	56
	Sediments (fresh, deposit.)	Adirondack lake area	1200, 1000 ng cm^{-2}	37

TABLE A6.11 Natural Concentrations of Lead

Material	Location	Natural levels	Ref.
Air	Nth Hemisphere	0.1, 0.02-0.5 ng m^{-3}	15
	Nth Hemisphere	0.025-1.0 ng m^{-3}	15
		0.01-0.1 ng m^{-3}	15
		0.04 ng m^{-3}	57
		0.005-0.04 ng m^{-3}	61
Air (overland)		0.1 ng m^{-3}	61
Air flux to oceans	NE Pacific	< 0.5 ng cm^{-2} y^{-1}	68
Air flux to remote places	Sub-alpine area	5 ng cm^{-2} y^{-1}	61
Animals terrestial		~ 2 ng g^{-1} (fw)	61

Table A6.11 continued

Material	Location	Natural levels	Ref.
Biomass (marine)		5 ng g^{-1} (fw)	61
Biomass (terrestial)		~2 ng g^{-1} (fw)	61
		~1 ng g^{-1} (fw)	61
Blood		Est. 0.25 μg dl^{-1}	59
		Est. 0.2 μg dl^{-1}	62
Bone & Teeth	Peruvians	Est. Pb/Ca = 2 x 10^{-8}	25
		< 1 μg g^{-1}	34
Bone	Nubian	0.6 (med) 0.4-1.5 μg g^{-1}	12
		< 1 & ~0.5 μg g^{-1}	12
		Est. 0.36 μg g^{-1}, 4 mg total	12
Earth's biomass (trees)		~4 ng g^{-1}	57
Emissions	Global	24.5 x 10^{-6} kg y^{-1}	53
Emissions to air		4000 t y^{-1}	61
Emissions to bio-systems		4000 t y^{-1}	61
Emissions to oceans/rivers		10,000 t y^{-1}	61
Eolian flux		0.03 ng cm^{-2} y^{-1}	61
Fish (tuna)		Est. ~0.03 ng g^{-1}	69
Fluxes	Off Californian coast	0.1-0.7 μg cm^{-2} y^{-1}	50
	Adirondack lake area	0.54 μg cm^{-2} y^{-1}	30
		2-200 mg m^{-2} y^{-1}	15
Fluxes to atmosphere	Cent. North Pacific	~3, 1.2-7.5 ng cm^{-2} y^{-1}	67
		18.6 x 10^{6} kg y^{-1}	56
Food (intake)		< 2ng g^{-1}	69
Food		0.0001-0.1 μg g^{-1}	15
Human		30 μg/70 kg person	58
Human diet		2 ng g^{-1} (wet weight)	57
Ice	Antarctic	~1 pg g^{-1}	1
	Antarctic	1.7, 1.8 pg g^{-1}	7

Table A6.11 continued

Material	Location	Natural levels	Ref.
Ice	Arctic	~1 pg g^{-1}	1
	Arctic & Antarctic	0.8 ng kg-1 (0.4 silicate dust, 0.4 volcanic/ sea spray)	51
	Greenland & Antarctic	~1 pg g^{-1}	61
Lead input		0.8 ng kg^{-1}	73
Lead intake (air)	Human	0.04 ng m^{-3}	69
	Human	0.0008 μg day^{-1} (0.0001 μg m^{-3})	15
	Human	0.3 ng day^{-1} (0.04 ng m^{-3})	57
Lead intake (food)	Human	2.0 μg day^{-1} (0.01 μg g^{-1})	15
	Human	< 210 ng day^{-1} (< 2.0 ng g^{-1})	57
Lead intake (soil)	Human	0.15 μg day^{-1} (15 μg g^{-1})	15
Lead intake (total)	Human	2.35 μg day^{-1}	15
	Human	< 210 ng day^{-1}	57
	Human	< 210 ng day^{-1}	69
Lead intake (water)	Human	0.2 μg day^{-1} (1 μg l^{-1})	15
	Human	< 2.0 ng day^{-1} (< 20 ng kg^{-1})	57
River input to oceans	NE Pacific	3 ng cm^{-2} y^{-1}	68
Rocks		12 μg g^{-1}	61
Sea spray to ice	Lead input	500 t y^{-1}	73
Sediments (fresh)	Arctic lakes	10-33 μg g^{-1}	47
Sediments (fresh, deposit.)	Adirondack lake area	6200,3500 ng cm^{-2}	37
Sediments (oceanic)		≤ 4 μg g^{-1}	47
	San Pedro (Calif.)	5.9 μg g^{-1} residue of acid leach	50
	Santa Barbara (Calif.)	5.6 μg g^{-1} residue of acid leach	50
	Santa Monica (Calif.)	5.5 μg g^{-1} residue of acid leach	50
Silicate dust to ice	Lead input	2000 t y^{-1}, 0.4 ng kg^{-1}	73
Soil	Global	Annual increase 0.07%	15
		5-25 μg g^{-1}	15

Table A6.11 continued

Material	Location	Natural levels	Ref.
Soil moisture		< 2 ng g^{-1}	57
		1 ng g^{-1}	61
Volcanic dust to ice	Lead input	1500 t y^{-1}	73
Water (fresh)		0.5, < 0.01-10 µg l^{-1}	15
		0.005-10 µg l^{-1}	15
		0.02 ng g^{-1}	57
		< 20 pg g^{-1}	61
Water (intake)		< 20 ng kg^{-1}	69
Water (ocean)	NE Pacific	0.6 ng kg^{-1} (0-100 m)	68
	NE Pacific	1 ng kg^{-1} (100-4300 m)	68
	North Pacific	0.0005 ng g^{-1}	69
		0.0006, 0.001 µg l-1 (deep)	15
		0.001 µg l^{-1}	15
		0.0005 ng g^{-1}	57
		~2 pg g^{-1}	61
Water (ocean, deep)	Atlantic	~4 pg g^{-1}	61
	Pacific	~ 0.8 pg g^{-1}	61

REFERENCES

1. Alderton, D. H. M. and Coleman, D. O. Ice cores and snow. in: Historical Monitoring: MARC Report No. 31; 1985: 97-143.
2. Alderton, D. H. M. Sediments. in: Historical Monitoring: MARC Report No. 31; 1985: 1-95.
3. Attramadal, A. and Jonsen, J. Heavy trace elements in ancient Norwegian teeth. **Acta Odontol. Scand.**; 1977; **36:** 97-101.
4. Bertine, K. K. Lead in the historical sedimentary record. in: Branica, M. and Konrad, Z., Eds. Lead in the Marine Environment: Pergamon; 1980: 319-324.
5. Boutron, C. Atmospheric trace metals in the snow layers deposited at the South Pole from 1928 to 1977. **Atmos. Environ.**; 1982; **16:** 2451-2459.
6. Boutron, C. F. Atmospheric toxic metals and metalloids in the snow and ice layers deposited in Greenland and Antarctica from prehistoric times to present. in: Nriagu, J. O. and Davidson, C. I., Eds. Toxic Metals in the Atmosphere: Wiley; 1986: 467-505.
7. Boutron, C. and Patterson, C. C. The occurrence of lead in Antarctic recent snow, firn deposited over the last two centuries and prehistoric ice. **Geochim. et Cosmochim. Acta**; 1982; **47:** 1355-1368.
8. Boutron, C. and Lorius, C. Trace metals in Antarctic snows since 1914. **Nature**; 1979; **277:** 551-554.
9. Boyle, E. A., Sclater, F. and Edmonds, J. M. On the marine geochemistry of cadmium. **Nature**; 1976; **263:** 42-44.
10. Bruland, K. W., Knauer, G. A. and Martin, J. H. Cadmium inNortheast Pacific waters. **Limnol. Oceanogr.**; 1978; **23:** 618-625.
11. Burnett, M. and Patterson, C. C. Analysis of natural and industrial lead in the marine ecosystems. in: Branica, M. and Konrad, Z., Eds. Lead in the Marine Environment: Pergamon; 1980: 15-30.
12. Coleman, D. O. Human remains. in: Historical Monitoring: MARC Report No. 31; 1985: 282-315.
13. Coleman, D. O. and Hutton, M. Museum specimens. in: Historical Monitoring: MARC Report No. 31; 1985: 203-268.
14. Coleman, D. O. Peat. in: Historical Monitoring: MARC Report No. 31; 1985: 155-173.
15. Committee on Lead in the Human Environment. Lead in the Human Environment: Nat. Acad. Sci.; 1980.
16. Cunningham, W. C. and Zoller, W. H. The chemical composition of remote area aerosols. **J. Aerosol. Sci.**; 1981; **12:** 367-384.
17. Davidson, C. I., Santhanam, S., Fortmann, R. C. and Olson, M. P. Atmospheric transport and deposition of trace elements onto the Greenland ice sheet. **Atmos. Environ.**; 1985; **19:** 2065-2081.
18. Davidson, C. I., Chu, L., Grimm, T. C., Nasta, M. A. and Qamoos, M. P. Wet and dry deposition of trace elements onto the Greenland ice sheet. **Atmos. Environ.**; 1981; **15:** 1429-1437.

19. Davidson. C. I., Grimm, T. C. and Nasta, M. A. Airborne lead and other elements derived from local fires in the Himalayas. **Science;** 1981; **214:** 1344-1346.
20. Davis, A. O., Galloway, J. N. and Nordstrom, D. K. Lake acidification: its effect on lead in the sediment of two Adirondack lakes. **Limnol. Oceanogr.;** 1982; **27:** 163-167.
21. Eaton, A. Marine geochemistry of cadmium. **Marine Chem.;** 1976; **4:** 141-154.
22. Elias, R. W., Hirao, Y. and Patterson, C. C. The circumvention of the natural biopurification of calcium along nutrient pathways by atmospheric inputs of industrial lead. **Geochim. et Cosmochim. Acta;** 1982; **46:** 2561-2580.
23. Elias, R. W. and Patterson, C. C. The toxicological implications of biogeochemical studies of atmospheric lead. (unpublished); 1979.
24. Environmental Studies Board Panel on Mercury. An assessment of mercury in the environment: Nat. Acad. Sci.; 1978: 139-169.
25. Ericson J.E., Shirahata, H. & Patterson C.C. Skeletal concentrations of lead in ancient Peruvians. **New Eng. J. Med.;** 1979; **300:** 946-951.
26. Fergusson, J. E. Lead: petrol lead in the environment and its contributions to human blood lead levels. **The Sci. Total Environ.;** 1986; **50:** 1-54.
27. Fleisher, M., Sarofim, A. F., Fassett, D. W., Hammond, P., Shacklette, H. T., Nisbet, I. C. T. and Epstein, S. Environmental impact of cadmium: a review by the panel on hazardous trace substances. **Environ. Health Perspec.;** 1974: 253-323.
28. Fosse, G. and Wesenberg, G. B. R. Lead, cadmium, zinc and copper in deciduous teeth of Norwegian children in the pre-industrial age. **Int. J. Environ. Studies;** 1981; **16:** 163-170.
29. Fosse, G. and Berg Justen, N-P. Lead in deciduous teeth of Norwegian children. **Arch. Environ. Health;** 1978; **33:** 166-175.
30. Galloway, J. N. and Likens, G. E. Atmospheric enhancement of metal deposition in Adirondack lake sediments. **Limnol. Oceanogr.;** 1979; **24:** 427-433.
31. Galloway, J. N., Thornton, J. D., Norton, S. A., Volchok, H. L. and McLean, R. A. N. Trace metals in atmospheric deposition: a review and assessment. **Atmos. Environ.;** 1982; **16:** 1677-1700.
32. Gibbs, R. H., Jarosewich, E. and Windom, H. L. Heavy metal concentrations in museum fish specimens: effects of preservation and time. **Science;** 1974; **184:** 475-477.
33. Goffer, Z. Archaeological Chemistry: Wiley; 198.
34. Grandjean, P., Nielson, O. V. and Shapiro, I. M. Lead retention in ancient Nubian and contemporary populations. **J. Environ. Path. Toxicol.;** 1979; **2:** 781-787.

35. Hecker, L. H., Allen, H. E., Dinman, B. D. and Neel, J. V. Heavy metal levels in acculturated and unacculturated populations. **Arch. Environ. Health**; 1974; **29:** 181-185.

36. Heidam, N. Z. Trace metals in the Arctic aerosol. in: Nriagu, J. O. and Davidson, C. I., Eds. Toxic Metals in the Atmosphere: Wiley; 1986: 267-293.

37. Heit, M., Tan, Y., Klusek, C. and Burke, J. C. Anthropogenic trace elements and polycyclic aromatic hydrocarbon levels in sediment cores from two lakes in the Adirondack acid lake region. **Water, Air Soil Pollution**; 1981; **15:** 441-464.

38. Hirao, Y. and Patterson, C. C. Lead aerosol pollution in the High Sierra overrides natural mechanisms which exclude lead from a food chain. **Science**; 1974; **184:** 989-992.

39. Jawarowski, Z., Bysiek, M. and Kownacka, L. Flow of metals into the global atmosphere. **Geochim. et Cosmochim. Acta**; 1981; **45:** 2185-2199.

40. Kuhnlein, H. V. and Calloway, D. H. Minerals in human teeth: differences between pre-industrial and contemporary Hopi Indians. **Amer. J. Clin. Nutrit.**; 1977; **30:** 883-886.

41. Lantz, R. J. and Mackenzie, F. T. Atmospheric trace metals: global cycles and assessment of man's impact. **Geochim. et Cosmochim. Acta**; 1979; **43:** 511-523.

42. Lee, J. A. and Tallis, J. H. Regional and historical aspects of lead pollution in Britain. **Nature**; 1973; **245:** 216-218.

43. Livett, E. A., Lee, J. A. and Tallis, J. H. Lead zinc and copper analysis of British blanket peats. **J. Ecol.**; 1979; **67:** 865-891.

44. Maenhaut, W., Zoller, W. H. Druce, R. A. and Hoffman, G. L. Concentration and size distribution of particulate trace elements in the South Polar atmosphere. **J. Geophys. Res.**; 1979; **84 (C5):** 2421-2431.

45. Mart, L., Nurnberg, H. W. and Valenta, P. Comparative base line studies on Pb-levels in European coastal waters. in: Branica, M. and Konrad, Z., Eds. Lead in the Marine Environment: Pergamon; 1980: 155-179.

46. Martin, J. H., Knauer, G. A. and Flegal, A. R. Cadmium in natural waters. in: Nriagu, J. O., Ed. Cadmium in the Environment Part 1. Ecological Cycles: Wiley; 1980: 141-145.

47. Moore, J. W. and Ramamoorthy, S. Heavy metals in Natural Waters: Applied Monitoring and Impact Assessment: Springer-Verlag; 1984.

48. Murozumi, M., Chow, T. J. and Patterson, C. C. Chemical concentrations of pollutant lead aerosols terrestrial dusts and sea salt in Greenland and Antarctic snow strata. **Geochim. et Cosmochim. Acta**; 1969; **33:** 1247-1294.

49. Nagourney, S. J. and Bogen, D. C. Determination of Mn and Pb in atmospheric deposition at a remote coastal site. **Water, Air and Soil Pollut.**; 1981; **15:** 425-432.

50. Ng, A. and Patterson, C. C. Changes of lead and barium with time in California off-shore basin sediments. **Geochim. et Cosmochim. Acta;** 1982; **46:** 2307-2321.
51. Ng, A. and Patterson, C. C. Natural concentrations of lead in ancient Arctic and Antarctic ice. **Geochim. et Cosmochim. Acta;** 1981; **45:** 2109-2121.
52. Nriagu, J. O. Cadmium in the atmosphere and precipitation. in: Nriagu, J. O., Ed. Cadmium in the Environment Part I Ecological Cycling: Wiley; 1980: 71-114.
53. Nriagu, J. O. Global inventory of natural and anthropogenic emissions of trace metals. **Nature;** 1979; **279:** 409-411.
54. Nriagu, J. O. Global cadmium cycle. in: Nriagu, J. O., Ed. Cadmium in the Environment Part 1. Ecological Cycles: Wiley; 1980: 1-12.
55. O'Brien, B. J., Smith, S. and Coleman, D. O. Lead pollution of the global environment. in: Progress Reports in Environmental Monitoring and Assessment. 1 Lead. Tech. Rept.: MARC; 1980: No. 16-18.
56. Pacyna, J. M. Atmospheric traces elements from natural and anthropogenic sources. in: Nriagu, J. O. and Davidson, C. I., Eds. Toxic Metals in the Atmosphere: Wiley; 1986: 33-52.
57. Patterson, C. C. An alternative perspective - lead pollution in the human environment: origin extent and significance. in: Lead in the Human Environment: NAS; 1980: 265-349.
58. Patterson, C. C. British mega exposures to industrial lead. in: Rutter, M. and Russel Jones, R., Eds. Lead versus Health: Wiley; 1983: 17-32.
59. Patterson, C. C. Contaminated and natural lead environments of man. **Arch. Environ. Health;** 1965; **11:** 344-360.
60. Patterson, C. C. Lead in ancient human bones and its relevance to historical developments of social problems with lead. **The Sci Total Environ.;** 1987; **61:** 167-200.
61. Patterson, C. C. Natural levels of lead in humans. Address to Inst. Environ. Studies, Univ. Nth. Carolina, Chapel Hill.; 1982.
62. Piomelli, S., Corash, L., Corash, M. B., Seaman, C., Mushak, P., Glover, B. and Padgett, R. Blood lead concentrations in a remote Himalayan population. **Science;** 1980; **210:** 1135-1137.
63. Piotrowski, J. K. and Coleman, D. O. Environmental hazards of heavy metals: summary evaluation of lead, cadmium and mercury. MARC Report No 20.; 1980.
64. Poole, C., Smythe, L. E. and Alpers, M. Blood lead levels in Papua New Guinea children living in a remote area. **The Sci. Total Environ.;** 1980; **15:** 17-24.
65. Priest, P., Navarre, J. L. and Ronneau, C. Elemental background concentration in the atmosphere of an industrialized country. **Atmos. Environ.;** 1981; **15:** 1325-1336.

66. Raspor, B. Distribution and speciation of cadmium in natural waters. in: Nriagu, J. O., Ed. Cadmium in the Environment Part I Ecological Cycling: Wiley; 1980: 147-236.
67. Salamons, W. Impact of atmospheric inputs on the hydrospheric trace metal cycle. in: Nriagu, J. O. and Davidson, C. I., Eds. Toxic Metals in the Atmosphere: Wiley; 1986: 409-466.
68. Schaule, B. and Patterson, C. C. The occurrence of lead in the Northeast Pacific and the effects of anthropogenic inputs. Branica, M. and Konrad, Z. Lead in the Marine Environment: Pergamon; 1980: 31-43.
69. Settle, D. M. and Patterson, C. C. Lead in Albacore: guide to lead pollution in Americans. **Science;** 1980; **207:** 1167-1176.
70. Shapiro I.M., Mitchell G., Davidson I. & Katz S.H. The lead content of teeth: evidence establishing new minimal levels of exposure. **Arch. Environ. Health;** 1975; **30:** 483-486.
71. Shirahata, H., Elias, R. W. and Patterson, C. C. Chronological variations in concentrations and isotopic compositions of anthropogenic atmospheric lead in sediments of a remote subalpine pond. **Geochim. et Cosmochim. Acta;** 1980; **44:** 149-162.
72. Steadman, L. T., Brudevold, F., Smith, F. A., Gardner, D. E. and Little, M. F. Trace elements in ancient Indian teeth. **J. Dent. Res.;** 1959; **38:** 285-292.
73. Tite, M. S. Methods of Physical Examination in Archaeology: Seminar press; 1972.
74. Wiersma, G. B. and Davidson, C. I. Trace metals in the atmosphere of remote areas. in: Nriagu, J. O. and Davidson, C. I., Eds. Toxic Metals in the Atmosphere: Wiley; 1986: 201-266.
75. Wilkinson, D. R. and Palmer, W. Lead in teeth as a function of age. **Amer. Lab.;** 1975: 67-70.

CHAPTER 7

THE HEAVY ELEMENTS IN THE ATMOSPHERE

In this chapter we will survey the heavy elements in the atmosphere, including the properties of aerosols, and the chemistry of the heavy elements in the air. In addition we will review the levels of the elements in the air, and efforts to determine their sources.

THE ATMOSPHERE

Composition

The earth's atmosphere consists of four principal zones the troposphere, the stratosphere, the mesosphere and the thermosphere. The zones are separated by regions of temperature inversion, called the tropopause, the stratopause and the mesopause respectively. The relationship between the height of the zones above the earth and the temperature changes are shown in Fig 7.1. The principal species in each sphere are also listed. We will be concerned in this chapter with the troposphere, the sphere closest to the earth. The composition of which near sea level, is given in Table 7.1 [91]. Around 50% of the material in the troposphere occurs up to 5 km and 90% up to 12 km. The majority of the heavy element species are in the particulate matter or aerosol. However, because of the volatility of mercury, and of some heavy element compounds, they may also occur in the vapour state.

Meteorology of the Troposphere

The movement of the air around the world is dominated by two processes, air currents giving vertical movement and winds giving horizontal movement.

Vertical movement The stability of the air determines its vertical movement. If a parcel of air in the atmosphere rises it will expand as the pressure of the surrounding air drops. If the expansion is adiabatic, i.e. if there is no transfer of

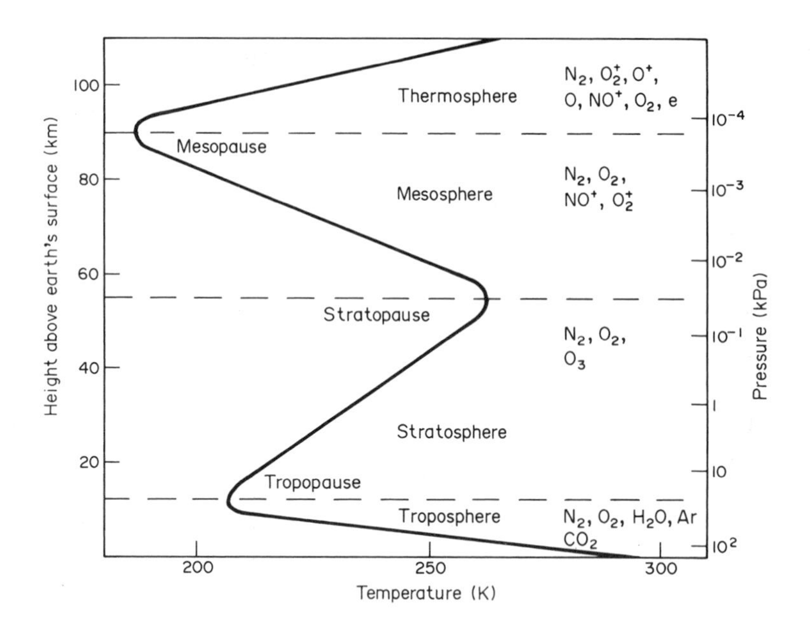

Fig. 7.1 Structure of the earth's atmosphere.

TABLE 7.1 The Composition of the Atmosphere Near Sea level

Component	Level (ppm)	Component	Level (ppm)
Major		**Trace**	
Dinitrogen, N_2	780,840	Neon, Ne	18.18
Dioxygen, O_2	209,460	Helium, He	5.24
		Methane, CH_4	1.4
		Krypton, Kr	1.14
		Dihydrogen, H_2	0.5
Minor		Nitrous oxide, N_2O	0.25
Argon, Ar	9340	Carbon monoxide, CO	0.08
Carbon dioxide, CO_2	325	Ozone, O_3	0.025
Water vapour, H_2O	variable	Ammonia, NH_3	6×10^{-3}
		Nitrogen dioxide, N_2O	4×10^{-3}
		Sulphur dioxide, SO_2	2×10^{-4}

heat between the parcel of air and its surroundings, the gas temperature will fall. When the temperature drop, called the adiabatic lapse rate, is greater than the temperature change of the surrounding air the parcel of air will begin to decend. Such a situation is called stable. Therefore for a parcel of gas (p), using the ideal gas law,

$$P_pV_p = nRT_p,$$

and replacing n/V_p by d the density of the air we get,

$$P_p = d_pRT_p.$$

Similarly for the surrounding air (a),

$$P_a = d_aRT_a.$$

Since the two pressures P_p and P_a will be the same, we have,

$$d_p = d_a\frac{T_a}{T_p}.$$

Initially the two temperatures will be the same, but as the parcel of air rises and its temperature falls, $T_p < T_a$, and therefore $d_p > d_a$ and the gas will decend, a stable situation. However, if $T_p > T_a$ then $d_p < d_a$ and the gas will continue to rise, an unstable situation, which allows for the vertical dispersion of material emitted into the atmosphere. When the temperature of the surrounding air increases with height a particularly stable situation results, called a temperature inversion.

Atmospheric temperature inversion Normally in the troposphere the temperature falls with increase in altitude, but the reverse, a temperature inversion can also occur. A subsidence inversion happens when the hot air, produced around the equator, rises until it reaches the tropopause. As the air rises it will expand, cool and lose its moisture in the humid tropics. The cooled air then moves towards the poles until around latitude 30°N or S when it begins to decend. The resulting increase in pressure raises the air temperature and cooler air is trapped below (Fig. 7.2). Seasonal variation causes the subsidence to move north or south with the zone of greatest heating.

A radiative inversion can occur on clear cold nights, when the earth's surface cools rapidly, radiating heat into the atmosphere. The air near the ground cools compared with the air higher up (Fig.7.3a). The following day the surface air is warmed giving a limited mixing area above which sits warm air trapping in the low level material (Fig. 7.3b). During the day as the air warms up the inversion is likely to disappear (Fig. 7.3c).

Cold air from the ocean, being heavier than warm air on the land, can slip under the the warm air. This air produces a sea breeze, but when the movement is slow a frontal inversion occurs on the land near to the coast (Fig 7.4).

A fourth temperature inversion called an advective inversion occurs when the sides of a valley cools at night and the adjacent cooled, and more dense air, falls into the valley. Warmer air may flow across the top of the valley trapping the cold air (Fig. 7.5).

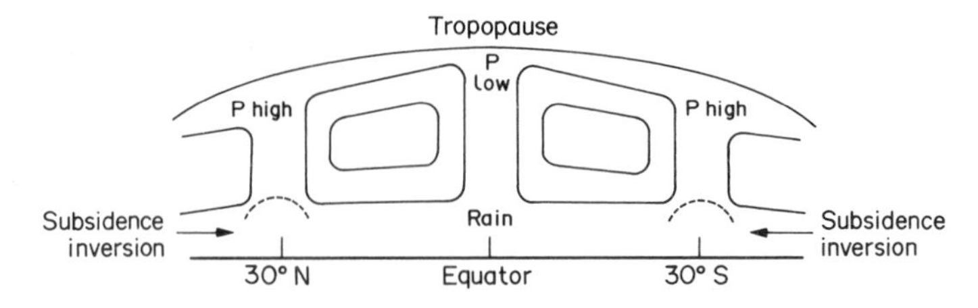

Fig. 7.2 The formation of a subsidence inversion.

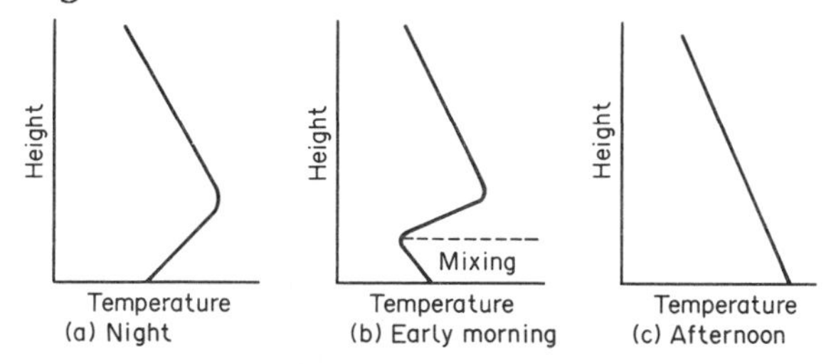

Fig. 7.3 Development of an inversion after a clear cold night.

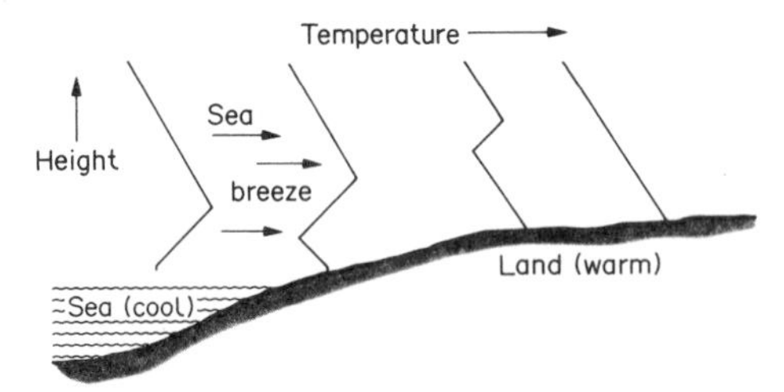

Fig. 7.4 Development of an inversion on land close to the sea.

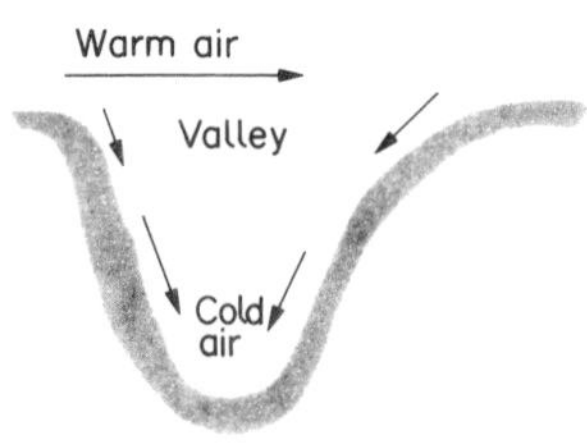

Fig. 7.5 Formation of an inversion in a valley.

Winds Wind, which is the horizontal movement of air, varies from light to very strong. It promotes the dilution and dispersal of air pollutants. The horizontal movement arises from the interplay of three factors; a pressure gradient (the air moving from high to low pressure), the Coriolis deflection and friction with the earth's surface. The coriolis deflection (Fig. 7.6) arises from the influence of the spin of the earth on the surrounding air. As a parcel of air moves north (in the northern hemisphere) its velocity is greater than at any point north of it, due to

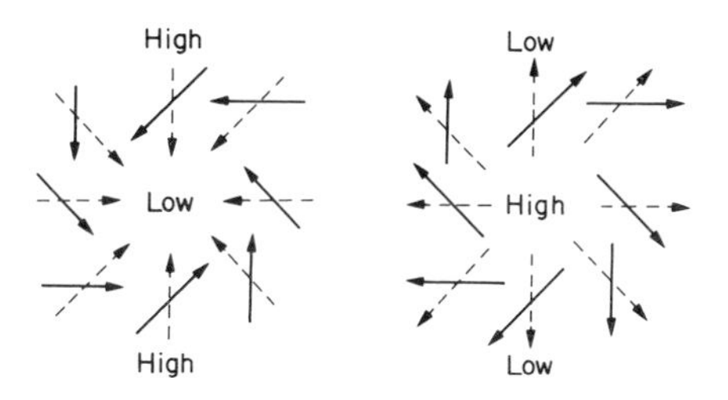

Fig. 7.6 Deflection of the wind by the Coriolis effect (the broken arrows indicate the direction of the wind if the earth did not rotate).

the fact that a point on the equator has a greater velocity than any other points on the earth's surface. Therefore to an observer the parcel of air appears to be deflected to the east (in the northern hemisphere). The result is that the wind moves along rather than across the isobars (lines of constant pressure) with some deviation because of the effect of friction. Wind strength is influenced by terrain, such as buildings in cities, and air turbulence is due to eddies, i.e. packets of air moving randomly, but with a circular motion.

Emission behavior An emission plume behaves in a variety of ways depending on the vertical temperature gradient, and the presence or absence of temperature inversions. Some behaviors are illustrated in Figs. 7.7a-f. The dashed lines represent the adiabatic lapse rate and the full lines the vertical temperature gradient of the air. Fastest dispersal of the emission occurs through looping (Fig. 7.7a), followed by coning (Fig.7.7b). In the last four examples (Fig. 7.7c-f) the emission is entrapped in inversions, the worst being fumigation (Fig. 7.7e).

The climate in cities The climate within cities can be different to surrounding rural areas, and is influenced by the terrain, i.e. buildings, high energy consumption and subsequent loss to the atmosphere, and reflecting surfaces. Some of the differences are highlighted in Table 7.2 [60]. The reduced wind speed, loss of heat to the atmosphere at night from surfaces, which are good heat conductors, provide the conditions that trap pollutants. In addition cities become heat islands and air circulates within them which helps to keep the material within the city. The build up of material within a city can be estimated using a box model. In a steady state, with good mixing, the concentration of material in a city's atmosphere is given by;

$$C(\text{mass/vol.}) = \frac{\text{Emission rate(E)(mass/time)}}{\text{Air flow rate (vol./time)}}.$$

The air flow rate into a box of width l and mixing height h and a wind speed v is vlh, hence;

$$C = \frac{E}{vlh}.$$

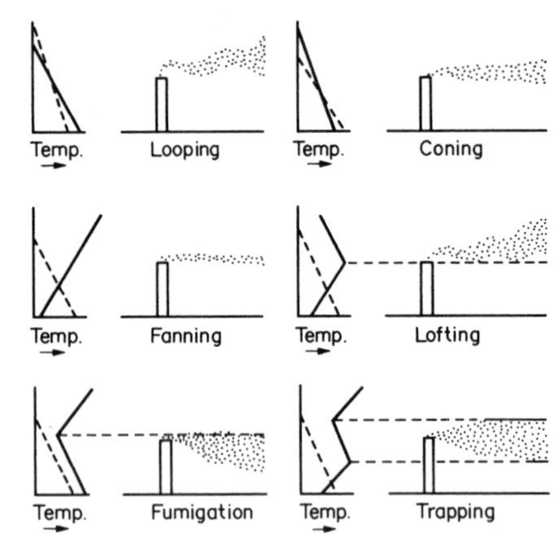

Fig. 7.7 Behaviour of emissions from smoke stacks in relation to air temperature (the full line is the temperature gradient, and the broken line the adiabatic lapse rate).

TABLE 7.2 Climate Differences Between Urban and Rural Areas

Element	Comparison with rural areas
Temperature	0.5-1.0 °C higher
Relative humidity	6% lower
Dust particles	10 times more
Cloudiness	5-10% more
Radiation	15-20% less
Wind speed	20-30% lower
Calm	5-20% more
Precipitation	5-10% more

Source of data; Landsberg, 1962 [60].

The build up of an emission in a box will increase when v and h are low. For a city with $l = 30$ km, $h = 200$ m and $v = 8000$ m h^{-1} and E = 50 x 10^9 mg h^{-1} of particulate matter.

$$C = \frac{50 \times 10^9 \, (\text{mg h}^{-1})}{8 \times 10^3 \, (\text{m h}^{-1}) \times 3 \times 10^4 \, (\text{m}) \times 2 \times 10^2 \, (\text{m})}$$

$$C \approx 1.0 \text{ mg m}^{-3}.$$

If the wind speed is halved the concentration would double, demonstrating the importance of wind.

Atmospheric Particulate Material (Aerosols)

Particle size The size of atmospheric particles range from molecular species to material that settles out rapidly, i.e. from 10^{-9} to 10^{-4} m (0.001 to 100 μm) in diameter [85]. The particles, whose size varies with the source, are either dispersed into the air, or formed from condensation of species already present in the atmosphere. The diameter of atmospheric particles has been defined in different ways, and it is not always clear what definition is used in some publications. The aerodynamic diameter, is the diameter of a spherical particle with unit density, but with the same properties of the real particle. It is similar to the Stokes' diameter, which is the diameter of a sphere with the same falling velocity and density as the real particle. The mass size factor describes the variation in mass with the size of the particles, and the mass median diameter, MMD (or mass median equivalent diameter, MMED) is the diameter below which 50% of the total mass of the particles, or the total mass of an element occurs [77].

The shape of aerosols range from spherical to quite irregular. The range of particle sizes for various aerosols are listed in Fig. 7.8 [6,16,85,94,102], the majority of particles lie within 0.01-100 μm. Particles with diameters in the range 0.38-0.76 μm have comparable dimensions to the wavelength of visible radiation, and will therefore affect its transmission producing haze.

The distribution of the sizes of aerosols is log-normal [29, 70], skewed at the higher end of the size range. Within this distribution most of the mass of aerosols is in the 0.01-10 μm range [106] with a mean around 1 μm [20]. However, the plot of mass distribution versus size, for a chemical element, is of the form;

$$\left(\frac{\Delta C}{C_T}\right) / \Delta \log d_t \text{ versus } dp_{\mu m}.$$

It is often bimodal with peaks at <1 and >5 μm; where ΔC is the airborne mass concentration of metal M, C_T is total airborne mass concentration, $dp_{\mu m}$ is the particle size and $\Delta \log d_t$ is the difference $dp_{max} - dp_{min}$, [21]. Other studies have found that the particle size distribution has two or three modes, around ~0.02 μm, ~0.4 μm and ~10 μm in mass or size distributions [89]. The Aitken nuclei range mode is characteristic of the freshly formed aerosol, the accumulation range mode is characteristic of the degree of aging and the coarse particle mode is characteristic of the amount of windblown dust, sea spray or mechanically produced dust such as fly ash [105].

Coagulation and sedimentation The small sized particules (i.e. <0.1 μm) in the atmosphere can combine into larger particles by coagulation, the movement and contact being controlled mainly by Brownian motion. If each contact produces coagulation the reduction in the number of particles is given by;

$$n = \frac{n_0}{1 + kn_0t},$$

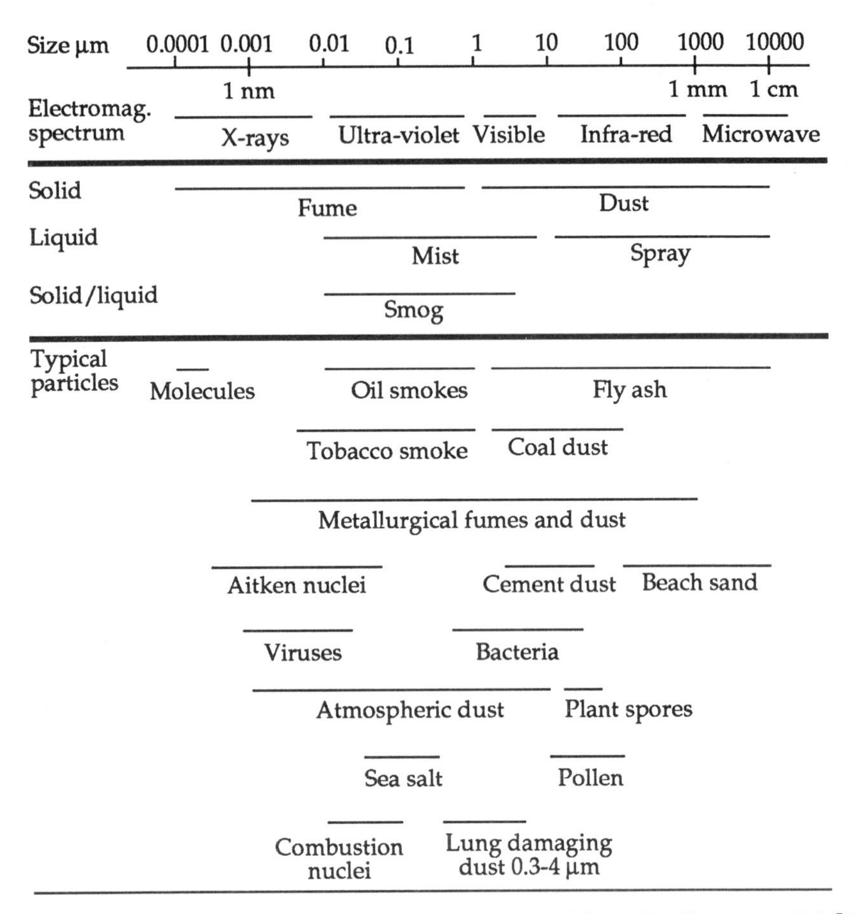

Fig. 7.8 The sizes of atmospheric particles. Source of data; Paulhamus, 1972 [85].

where n = final concentration, n_o = initial concentration, t = time and k is a constant 10^{-10} to 10^{-8} $cm^3 s^{-1}$. For particles < 0.01 µm the decrease is 50% in a hour, and for < 0.05 µm the decrease is 50% in a day [29].

The larger particles in the atmosphere fallout rapidly, and in still air the settling or sedimentation velocity is given by Stoke's law;

$$v_d = \frac{2(\rho - \rho_0)g}{9\eta}\left(\frac{d}{2}\right)^2,$$

where v_d = velocity, $\rho - \rho_o$ = density difference between the air and particle, η = viscosity of the air, g = gravitational acceleration constant, and d = diameter of the particle [106]. The velocity will range from 9×10^{-5} to 1.3×10^{-1} cm s^{-1} for particles of diameter 0.1 to 20 µm.

PARTICULATE MATTER AND THE HEAVY ELEMENTS

The Size of Heavy Element Particles

Particles <2 μm generally come from anthropogenic sources, whereas when they are above 2 μm the main source is wind blown and re-entrained dust [78]. For a number of cities, where anthropogenic sources dominate, the aerosol sizes mainly span 0.12-0.7 μm, of which 20-25% lies at the lower end of the range [78]. For example the elements Sb, As, Pb, coming from man made sources, are mostly associated with the smaller sized fraction of the aerosol, i.e. <1.1 μm [80,83]. Particles containing cadmium have been identified in the size range 0.6-10 μm with a mean diameter of 2.2 μm [77,83]. A range of MMD's 0.28-5.5 μm have also been reported for cadmium containing particles [21]. A good proportion of these particles are respirable [77]. Around 50-90% of the cadmium in fly ash aerosol is <5 μm size [98].

Lead The size of lead particles have been extensively studied, because of concern over automobile lead emissions. Three particle types are emitted by cars, the primary exhaust particles (0.01-0.1 μm), chain aggregates, mostly diesel smoke (0.3-1 μm), and large material (>1 μm) [14]. The MMD's of lead containing particles are reported to be approximately 0.25-0.3 μm [15,16,66], but for aerosol near to motorways, the MMD is around 0.02-0.05 μm [14,65,66]. A plot of lead concentration against particle size displays two concentration minima, around the 1.1-2.0, and 2.0-3.3 μm sizes, and the concentration of lead is higher in the smaller particles, i.e. <1.1 μm [98].

The size of particles from motor vehicles decreases with distance from the road, up to 10% of the particles are coarse within 3.7 m from the road, and around 50% of the particles are greater than 6.5 μm, 600 m away [16,78]. Lead containing particles are bigger in urban areas compared with rural areas or along motorways, [15,65]. For example around 30% of the particles in the urban situation are <0.3 μm, whereas along a motorway the proportion is 67% [15,65,66]. The lead containing particles are probably larger in urban areas because of the association between the aerosol and general urban aerosol [65].

Residence Times

The lifetime of aerosols in the air, which contain heavy elements, is a function of the particle size. The smallest particles, 0.001-0.08 μm, have a lifetime of <1 hour, because of coagulation into bigger particles, whereas in the accumulation range, 0.08-1.0 μm, the life time is 4-40 days, and the large particles >1.0 μm have a life time of minutes to days [6,39,77,78,102,105].

Because of the volatility of mercury it is recycled through the environment, including the atmosphere, which increases its residence time over land from 7 to 315 days. The oceanic residence time is about 14 days, because in this case, there is less recycling of the element [71].

Transport of Aerosols

Because of the long residence times, transport of particulate material in the atmosphere can extend over long distances e.g. 100 to 1000 km. In ice cores of the Arctic and Antarctic, pre-1940 lead levels are <0.08 ng kg^{-1}, whereas those found for 1965 are 0.15-0.42 ng kg^{-1}, a 2 to 5 fold increase, attributed to transport of lead from industrial areas [106]. The relationship between the aerosol levels of trace elements and their concentration in snow is expressed by;

$$C_a = kC_s,$$

where C_a is the concentration in the air, k is a transfer constant, and C_s is the concentration in the snow. Another expression is;

$$C_a = ke_n\eta/L,$$

where e_n is the mass fraction of aerosol used in condensation nucleation, η is a factor linked to evaporation below the cloud, and L is the liquid water content of the cloud. Taking h close to 1 and $0.1 \le e_n \le 1$, $1 \le L \le 3$ g m^{-3}, the estimated value of k is 1.0 to 6.0 g m^{-3} which is close to experimental values [7].

A model for the transport of mercury which includes a number of recycling pathways is [71];

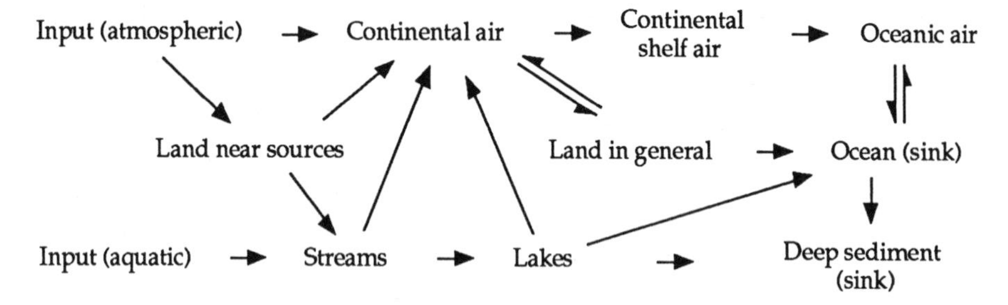

To a lesser extent the same processes will occur for selenium and arsenic. Cadmium has been observed in air, which has not originated from local sources [40].

HEAVY ELEMENT EMISSIONS AND FLUXES

Heavy element emissions into the atmosphere, globally or locally, are either natural or anthropogenic. Heavy element atmospheric fluxes are expressed as emissions or fluxes into the atmosphere or as fluxes from the atmosphere to the earth. The flux depends on the amount of aerosol and the concentration of the element in the aerosol. Natural fluxes of total particulate matter are estimated to be around 1000 Tg y^{-1} or 7.1×10^{12} g m^{-2} s^{-1}. The sources are desert material, grasslands, forests, oceans and point sources such as volcanoes, and forest fires.

The urban flux is estimated to be 100 Tg y^{-1} or 7.1×10^{11} g m^{-2} s^{-1}. On the basis of these fluxes the concentrations of total particulate matter are; troposphere 1-10 μg m^{-3}, over deserts 1-1000 μg m^{-3}, over steppes and grasslands 10-100 μg m^{-3}, over forests 10-500 μg m^{-3}, over marshlands 10-100 μg m^{-3}, over oceans 1-50 μg m^{-3}, and urban 1-1000 μg m^{-3} [39]. A separate estimate suggests higher levels i.e. the natural flux of particulates in the air as 773-2200 Tg y^{-1} and 185-415 Tg y^{-1} for man made aerosol [70].

Some estimates of natural emissions of the heavy elements into the air are listed in Table 7.3 [55, 61, 84] in units of ng kg^{-1} (or pg g^{-1}) of dust emitted. The highest concentrations occur in the volcanic emissions, however, compared with the amount of windblown material the total contribution of volcanic emissions is less. Around 60-85% of the total natural emissions of lead (18.6×10^6 kg y^{-1} [78] or 29.5×10^6 kg y^{-1} [88]) come from the windblown material, 5-10% from emissions from vegetation and the rest from sea-spray, volcanic emissions, forest fires and meteoritic material [78, 88]. The natural emissions of cadmium to the atmosphere are estimated to be (in 10^6 kg y^{-1}); windblown dust 0.1, forest fires 0.012, volcanic emissions 0.52, vegetation emissions 0.2 and sea spray 0.4, a total of 1.232×10^6 kg y^{-1} [88]. For mercury the natural degassing of the element is estimated to be 0.02-0.03 μg m^{-2} h^{-1}, whereas around mineralized areas the emissions are as high as 1.7 μg m^{-2} h^{-1} [63]. Global natural emissions of mercury vapour are estimated as 6×10^6 kg y^{-1} and anthropogenic as 2×10^6 kg y^{-1} [62].

Global estimates of anthropogenic emissions from a variety of sources and emissions from coal and oil combustion are listed in Table 7.4, in 10^9 g y^{-1} [4,62,69,71,82,83,88,101,107]. The data suggests that oil is a significant source of mercury, and coal is a significant source of selenium.

The emissions of lead to the atmosphere, from smelting and refining processes, have decreased from 5% in Roman times to 0.5% today. However, over the last 50-60 years the use of tetraalkyllead in petrol has increased emissions 3 to 4 fold. More than 1/2 of the lead produced today gets into the air in some form or other, around 60-68% from tetraalkyllead and 28% from metal production [28,95]. Emissions vary considerably with location, for example in New Zealand they are estimated to be 1.33×10^6 kg y^{-1}, over US cities it is 1.2-

TABLE 7.3 Natural Inputs of Heavy Elements into the Atmosphere (ng kg^{-1} of dust)

Element	Volcanic	Windblown	Forest fires	Vegetation	Sea spray
As	300-800	0.5-2.0	0.5-4.4	3.5	0.1-0.6
Cd	30-800	0.002-1.7	0.03-2	2.7-36	0.001-0.003
Pb	100-9600	0.4-70	1.1-78	21-280	0.001-0.09
Se	10-1700	0.6			
Sb	30				

Sources of data; references 55,61,78,84.

TABLE 7.4 Global emissions of Heavy Elements and the Contribution from the Combustion of Coal and Oil (10^9 g y^{-1})

Element	Total	Coal	% of total	Oil	% of total
As	23.6	0.7	3.0	0.002	0.008
Se	1.1	0.42	38.2	0.03	3.3
Cd	7.3	-	-	0.002	0.03
Hg	2.4	0.0017	0.07	1.6	66.7
Pb	449	3.5	0.8	0.05	0.01
Bi	-	0.75	-	-	-

Sources of data; references 4,62,69,71,82,83,88,101,107.

2.0 x 10^4 ng cm^{-2} y^{-1}, remote rural 2.5 x 10^3 ng cm^{-2} y^{-1}, remote mountain 4 x 10^2 ng $cm^{-2}y^{-1}$, and over the oceans the flux is 7 x 10^{-2} to 50 ng cm^{-2} y^{-1} [95]. At a site in Antarctica the lead concentration in the snow is ~4 pg g^{-1} and the deposition of snow is ~3 g $cm^{-2}y^{-1}$, which gives a flux of 12 pg cm^{-2} y^{-1}, for both dry and wet deposition. Extrapolation to the whole of the Antarctic, with a snow accumulation of 17 g H_2O cm^{-2} y^{-1} the total lead flux is ~70 pg cm^{-2} y^{-1}, approximately 9 tons y^{-1}. In the South Pacific easterly the lead flux is ~2 ng cm^{-2} y^{-1}, hence the flux ratio Antarctica/Sth Pacific ~1/30. Similar ratios occur for other land masses and ocean environments, for Greenland/Nth Pacific ~1/22 and Greenland/ Nth Atlantic ~1/34 [8].

The principal source of anthropogenic lead emissions is the combustion of tetraalkyllead, but as more countries reduce the amount of lead used this source will decrease. The quantity of lead consumed by a car engine is a function of the concentration of lead in the petrol, rate of petrol consumption, the size, efficiency and running speed of the engine, and the engine load. The proportion of lead exhausted depends on the age, temperature and physical state of exhaust system, and the manner in which car is operated. The amounts exhausted can range from 20-300% with an average around 75% [15]. Therefore, for 1000 cars h^{-1}, a petrol lead concentration of 0.49 g l^{-1}, fuel consumption of 0.24 l km^{-1}, and for a 75% emission of lead, the source strength is 10 μg s^{-1} m^{-1} [14]. In Los Angeles, for a lead in petrol concentration of 0.56 g l^{-1}, and a 75% emission factor, the total emissions added up to 17.9 t d^{-1}, (0.3 from evaporation, 16.7 as aerosol and 0.9 as organic vapour). The material dispersed by the wind was 5.6 t d^{-1} (5.3 aerosol, 0.3 organic vapour), whereas 11.5 t d^{-1} deposited on the ground, (9.5 nearby and 2.0 further away) [54].

Other aerosol lead sources are, workroom exhausts, e.g. where soldering is carried out, natural weathering and burning of paint, incineration of plastics and the recycling of lead batteries [16].

ATMOSPHERIC LEVELS OF THE HEAVY ELEMENTS

The levels of the heavy elements in the atmosphere are diverse over the earth's surface, depending on the particular environment and prevailing conditions. There are difficulties in comparing results from different studies, because of the factors that can influence the actual level measured. Details of the sampling time, the period over which the mean level was determined, and the year when the work was done are necessary for making any comparisons. The data given in Table 7.5, are for air lead levels for a number of environments, where some of the above details are available, and comparisons can be made.

Levels of Heavy Elements in Urban Aerosol

Much urban aerosol is generated from polluting sources within the urban area. There is a wealth of data on air lead levels, and typical urban air lead levels lie in the range 0.5-2 $\mu g\ m^{-3}$, but approach 10 $\mu g\ m^{-3}$ in heavy traffic areas (Table 7.5) [6,17,37,38,42,78,80,81,88,93,95]. The levels closely relate to the combustion of petrol lead, this being the main lead source (~90%) in urban areas. For example in Belgium on traffic free days the air lead levels in urban areas were around 0.6 $\mu g\ m^{-3}$ compared with the usual 5 $\mu g\ m^{-3}$. Other environments where alkyllead sources are clearly implicated are in tunnels with air lead concentrations in the range 20-100 $\mu g\ m^{-3}$, close to traffic, (8-40 $\mu g\ m^{-3}$) and close to highways (5-20 $\mu g\ m^{-3}$) [17,78,79,80]. An approximate order of air lead levels in different urban settings are: central business area > heavy density residential > shopping commercial > heavy industrial > central park areas > medium density residential > light traffic residential > open fields. Air lead levels also relate to population densities as shown by the following mean data for USA cities [78]; < 0.1 million people, 1.47 $\mu g\ m^{-3}$; 1-2 million people, ~2.0 $\mu g\ m^{-3}$; and >3 million people, >3 $\mu g\ m^{-3}$.

Other aerosol heavy metal levels have not been as well studied as for lead, cadmium being the next most investigated. Some data for arsenic, cadmium, mercury, indium, antimony and selenium are listed in Table 7.6. The concentrations are generally, 2 to 3 orders of magnitude, less than for lead.

Levels of Heavy Elements in Rural Aerosol

The levels of the heavy elements are significantly less in rural areas, because of the dilution of the urban materials as they are transported away. The heavy element concentrations in rural air in Belgium, given in Table 7.7, were measured at times of low pollution. The background levels were estimated by comparing the measured values with those of sulphur, using the equation,

$$[E] = a\,[S] + b,$$

where [E] and [S] are the concentrations of the metal and sulphur respectively, at low pollution times, a is the E/S ratio from the correlation coefficient calc-

TABLE 7.5 Some Selected Levels of Lead in the Atmosphere

City	Location	Years	Sampling interval	Level $\mu g\ m^{-3}$	Comments
Osaka	Main streets	1955-56	30 min	10.2	
Los Angeles	Freeway	1963-64		8.2-18.3	Mean daily range
Warwick	City centre	1965-66	weekly	2.80-4.46	Monthly means
Palo Alto	Freeway	1966	5 min	1-19	Range during day
Los Angeles	Downtown	1966-75	24 h	~5.1 ~2.8	Ann. mean 1971 & 76
San Diego	Mission Val.	1969	weekly	2.16-4.61	Quartely means
New York	45th. St.	1969	2 h	9.3	Two hourly means
London, UK	Fleet street	1972-73	24 h	3.3	Daily means
London, UK	Fleet street	1972-73	daytime	6.0	Daytime only
Melbourne	City centre	1974-75	hourly	1.4-3.9	Means 2-7 h
Birmingham	Various	1975	weekly	0.5-1.40	Not exposed to cars
Brisbane	Freeway	1975	24 h	5.17	Mean
London, UK	M4	1978	daytime	8.9	2 m from berm
London, UK	M40	1978	daytime	3.3	2 m from berm
Sydney	Urban	1979	2.5-5 h	2.44	Daytime mean
Los Angeles	Lennox	1979	24 h	0.62-3.91	Monthly means
Christchurch	Urban	1981	2 h	3.3	Daily mean

Source of data; Simmonds et al. 1983 [93].

ulation, and *b* is the estimated background level. In the rural areas the levels are high compared with remote places, indicating some pollution material was present [86].

On a mass basis the concentration of lead in particles in cities is around 1-10%, which drops to 0.1-1% in rural areas, and still further, to < 0.1%, in remote areas. The fall is due to dilution of the aerosol with other dusts [78]. The air lead levels in northern hemisphere rural areas lie in the range 0.05-0.2 $\mu g\ m^{-3}$, whereas in the southern hemisphere the levels are around 0.02 $\mu g\ m^{-3}$ [79]. The difference reflects the greater anthropogenic inputs in the north.

Levels of aerosol cadmium in rural areas are generally <1 $ng\ m^{-3}$ [30,38,77], and the air levels of mercury are reported as; continental air, 20 $ng\ m^{-3}$, continental shelf air 2.9 $ng\ m^{-3}$ and oceanic air 0.7 $ng\ m^{-3}$ [71]. Similar levels of atmospheric arsenic, antimony and selenium are reported in rural areas, i.e. 1.4-3.8 $ng\ m^{-3}$, 2.4 $ng\ m^{-3}$ and 1.1-2.6 $ng\ m^{-3}$ respectively [38].

Levels of the Heavy Elements in Remote Areas

Rural and remote areas merge into each other, and the aerosol levels of some of the heavy elements are given in Table 7.8. The concentration ranges vary considerably, but depend on the closeness of the area to urban, industrial or high emission natural sources.

TABLE 7.6 Levels of Arsenic, Cadmium, Mercury, Indium, Antimony and Selenium in Urban Aerosol

Element	Location	Concentration ng m^{-3}	Year Reported	Reference
As	Washington DC	3.2	1986	38
	Portland Or.	5.0	1986	38
	UK cities	6.4	1979	88
	Swansea	15	1974	88
	Toronto	2-20, mean 15	1976	80
	Toronto freeway	2-43, mean 10	1976	80
Cd	Hobart	1-9, mean 3	1977	6
	Urban	2-15, 1-50	1980	77
	UK cities	2.8	1979	88
	Urban	2-370	1974	30
	Bronx, NY	6-22	1974	30
	Cincinnati	80	1974	30
	Chicago	19	1974	30
	Polish towns	2-51	1974	30
	Ann Arbor	100-300	1971	42
	Washington DC	2.5	1986	38
Hg	Long Is USA	2.9	1985	95
	New Zealand	1.15 mean	1985	95
	Toronto	16	1976	80
	Toronto expressway	19	1976	80
	Urban	2-30	1982	62
	Large cities	5-50	1974,75	62
In	Swansea	<0.7	1974	88
Sb	UK cities	7.3	1979	88
	Swansea	2.9	1974	88
	Toronto	0.9-36, mean 8	1976	80
	Toronto expressway	1.6-24, mean 6	1976	80
	Washington DC	2.1	1986	38
	Boston	0.55-40, 0-58, 8.1	1976	53
Bi	L. Michigan	≤0.05-3, mean 6	1971	41
Se	Swansea	2.7	1974	88
	Boston	0.19-9.1, 0-3.8, 1.23	1976	53
	Washington DC	2.4	1986	38
	Portland Or.	3.0	1986	38

Other reference 13.

Lead Lead levels decrease with height into the troposphere, but are reported to increase again in the stratosphere [78]. Remote continental and oceanic levels in the northern hemisphere are reported as 0.5-1.5 ng m^{-3} and 0.2-6.0 ng m^{-3} respectively, in the southern hemisphere the levels are 0.5-5.0 ng m^{-3} and <10 ng m^{-3} respectively, the latter results are from limited data [79]. Lead levels in the Arctic region have been observed to vary with the season, and over different

TABLE 7.7 Levels of the Heavy Elements in Rural Belgium

Element	Average ng m^{-3}	Range ng m^{-3}	Estimated background ng m^{-3}
As	2.3	0.39-5.1	0.7
Se	0.62	0.18-1.21	0.17
Cd	0.78	0.29-1.9	0.37
Sb	1.7	0.42-3.0	1.05
Hg	1.2	0.07-4.1	
Pb	125	54-230	93

Source of data; Priest et al., 1981 [86].

TABLE 7.8 Levels of Heavy Elements in Remote Areas (pg m^{-3})

Metal	Remote (range)	UK Lake	Jungfrau District	Antartica Sum.	Antartica Wint.	Enewetok Dry	Enewetok Wet	Green-land
As	8.4-2300	1990	158	17	8.4			42
Cd	2.5-720	830	332	<200	49	4.6	2.5	15
Hg	600-3400	39	20					
In	0.054-78	22	7	0.19	0.054			
Pb	46-97,000	31,540	2988			130	96	490
Sb	0.45-930	655	133	2.1	0.45	5.2	2.4	
Se	6.3-1400	672	28	6.9	6.3	150	110	

Sources of data; references 6,7,10,13,18,19,22,26,38,62,77,78,88,95,98.

sites the range is 1.7-3.84 ng m^{-3}, with higher levels occurring in the winter months 3.49-6.38 ng m^{-3} [3,50].

At Enewetok in the North Pacific(11°N, 162°E) atmospheric lead levels are 90-260 pg m^{-3}. The lead rich particles are smaller in size than the silicate dust particles, and there is no temporal association between the two particles. This suggests that the lead derives from a source different from the silicate dust, and is probably a continental source [95]. The lead and other heavy elements display a seasonal change (Table 7.8) which is related to the amount of precipitation, the concentrations being less in the wet season [26].

At Te Paki (Nth tip of New Zealand) the rainfall contained 17 pg g^{-1} of lead and was associated with the air flow from the Indian Ocean. In the mid Tasman Sea, however, a level of 490 pg g^{-1} was associated with the air flow from Australia, which fell to 41 pg g^{-1} when the flow swung further south [95]. In remote areas of Tasmania levels of 1-160 ng m^{-3} have been reported [6]. Air lead concentrations in the Himalayas of 110-160 ng m^{-3} are enriched 20 fold, and may arise from open fires [23].

Other elements Over the South Polar regions the concentrations of the elements As, Se, Sb, Cd and In are in the <1 to few pg m $^{-3}$ range [67]. The levels of cadmium in remote areas are around 3-620 pg m^{-3}, but in areas of high natural sources, such as Mt. Etna, levels in the volcanic plume of 92 ng m^{-3}, and above a hot vent of 30,000 ng m^{-3} have been recorded [77]. Around natural mercury deposits, concentrations of the element in the air are around 30-1600 ng m^{-3}, and at geothermal areas concentrations of 10-40,000 ng m^{-3} are reported [63]. In Norway background levels of some heavy elements (measured at two sites during 1974) were as follows: Hg 20-30,15-31 ng m^{-3}, Cd 20-31, 3-29 ng m^{-3} and Pb 23-31, 3-29 ng m^{-3}. Most of the mercury comes from natural sources, whereas some of the lead and cadmium is anthropogenic. There was a direct relationship between the concentration of Cd and Pb and the wind direction, with the concentrations increasing when the flow was from pollution sources [98]. In remote areas the range of mean concentrations of mercury in aerosols is 0.015-0.4 ng m^{-3}. The total atmospheric levels of mercury fall rapidly with height, suggesting the flux comes from the soil [69]

Industrial Levels of the Heavy Elements

Concentrations of the heavy elements around industrial areas, such as smelters, can reach very high values and, unlike other areas it is not possible to give typical levels. Cadmium levels of around 500, 200, and 160-320 ng m^{-3}, have been reported around Japanese smelters at distances 100, 400 and 500 m away respectively. Around Swedish smelters weekly means of cadmium aerosol levels were 600, 300 ng m^{-3}, at distances 100, 500 m away. A maximum value of 54,000 ng m^{-3} has also been recorded [30].

Levels of mercury around chlor-alkali cell waste ponds were, 18, 64 and 90 ng m^{-3} at temperatures around 6° C, and 991 ng m^{-3} at 29° C. Approximately 2 km away the level had dropped to 3-9 ng m^{-3}. A similar fall off in concentrations was observed at different distances from a coal fired power station as shown by the data in Table 7.9 [63].

A study of the lead levels around a smelter in El Paso (during 1969-1971) indicated that at the smelter's boundary, and downwind, the air lead concen-

TABLE 7.9 Levels of Mercury Associated with a Coal Fired Power station

Source	Hg vapour in air, ng m-3	Hg aerosol ng m-3	Total particles µg m-3	Hg in solid µg g^{-1}	µg/g particle / µg/g coal
Coal				0.28	1
Ash				0.0037	0.01
Plume					
0.25 km	1700	150	3460	43	150
7 km	1000	30	740	40	140
22 km	200	2	95	20	70
Background	12	0.1	17	6	-

Sources of data; Lindberg, 1987 [62] and 1986 [63].

tration was 15-269 $\mu g\ m^{-3}$ (annual mean = 92 $\mu g\ m^{-3}$) and this had fallen to background levels 4-5 km away. Later, in1972-1973, the annual mean had fallen to 43 $\mu g\ m^{-3}$ after some changes had been made. The concentration of lead was highest in the particles < 1 μm, and 42% of the lead at 250 m from the smelter was < 2 μm. At the smelter the 10 monthly mean for the particulate matter was 204 mg m^{-2} $month^{-1}$ [59].

VARIATIONS IN ATMOSPHERIC LEVELS OF THE HEAVY ELEMENTS

One of the central problems in comparing the concentrations of the heavy elements in the air, determined in different studies, is the variety of factors that influence the levels. Some of the factors can be controlled by the experimenter e.g. position of the sampler, whereas some are outside the control of the experiment e.g. the weather.

Factors Influencing Heavy Element Levels

Some of the factors that have a bearing on the measured level of a heavy element are: height of sampler above ground, distance from the source, distance from buildings, type of sampler, wind speed, wind direction, air temperature, air stability, season, topography, vertical spread of the plume, and in the case of lead arising from petrol; traffic density, date of sampling, amount of lead in the petrol, driving mode, traffic turbulance, bouyancy of the exhaust gases, petrol consumption, age and condition of the car [1,14,15,66].

Models for estimating atmospheric levels take into consideration some of the above factors. One model for the atmospheric concentration of lead is [27];

$$\text{Concentration} = \frac{\text{emission rate}}{\text{mean wind speed x vertical mixing height}}$$

Another model, relating to traffic as the source of lead, is [93];

$$\text{Aerosol lead levels} = K\left(\frac{VE}{TSH}\right)D,$$

where K is a constant, V is traffic volume, E is the lead emitted, T is the temperature, S the wind speed, H the mixing height, and D the wind direction factor. It is not possible to produce a satisfactory quantitative model that encompasses all factors.

Seasonal or climatic effects Seasonal variations in heavy metal aerosol concentrations occur, but differ with site, e.g. in New York levels were highest during the fall and spring, whereas in Los Angeles (see Table 7.10) and Boston the highest levels were in the winter. Seasonal variations maybe due to variations in source intensity, climatic variations, scavenging efficiency of precipitation, and temperature inversions [16, 78]. The seasonal pattern of air lead levels in Denver during Jan 72-June 75, was peaks in the summer, with a gradual decrease in levels over the years. The seasonal pattern was related to the lead

TABLE 7.10 Atmospheric Lead Levels in Los Angles (μg m^{-3})

Place		Season		Diurnal	
Downtown	3.0	Summer	1.9	11pm-3am	1.0
Outlying	2.0	Fall	2.8	3am-7am	1.1
		Winter	3.1	7am-11am	1.2
		Spring	2.1	11am-3pm	0.7
				3pm-7pm	0.8
				7pm-11pm	1.1

Source of data; NAS, 1972 [16].

in the petrol, which was at a higher concentration in the summer, and the over all decrease to the introduction of catalysts to cars, which could not use leaded petrol. However, at six sites the plot of monthly mean air lead levels was at a maximum in the winter, when the input from petrol sales were at a minimum. This was due to the mixing heights being lower in the winter (800-1000 m) compared with the summer (2600 m). The dispersion factor, hv, where h is the mixing height, and v the speed of the wind, correlated well with the concentration of air lead ($p < 0.001$) [27].

Wind speed is a significant factor in determining the atmospheric levels of the heavy elements, whereas air temperature is relatively less important. These factors, and traffic density, are compared with the levels of aerosol lead in Fig 7.9. The dramatic effect of increase in wind velocity on the air lead concentrations is clear from the diagram [93]. However, low concentrations because of the wind, does not mean that the element is not being emitted from the source, but rather it is more rapidly dispersed and diluted [93]. Cadmium levels, on the other hand, 0.0024 μg m^{-3} (urban) and 0.0020 μg m^{-3} (rural), are reported to be less influenced by these factors [72].

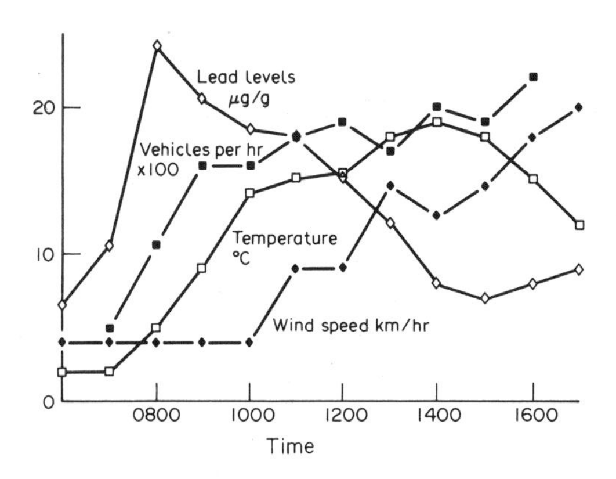

Fig. 7.9 Variation of lead levels with number of vehicles, wind speed, and temperature. Source of data; Simmonds et al., 1983 [93].

Seasonal affects have also been reported for arsenic antimony and cadmium, with higher levels in the winter. This is probably due to extra emissions in associated with combustion for heating and production of power [77,88]. However, the reverse seasonal situaton has also been observed, and this may relate to different dispersal modes in the summer and winter [77].

Diurnal variations Diurnal variations correlate with traffic density for lead (Table 7.10, Fig 7.9), and concentrations are also low in the weekends [16,78,93]. In urban and rural areas cadmium levels show diurnal and day to day variations. The reason is due to the variation in source emissions, wind and local ventilation characteristics.

Sampling factors A number of sampling factors influence the observed levels of the heavy elements. Levels of lead in streets vary with height above the ground, e.g. 10.4 $\mu g\ m^{-3}$ at 0.3 m, 8.3 $\mu g\ m^{-3}$ at 1.5 m and 5.1 $\mu g\ m^{-3}$ at 9.2 m [78]. The length of sampling time has a marked effect on what is observed. Short intervals may pick up peak or low emission times that would be missed using longer sampling times [6]. Other factors are; the direction of the wind with respect to the sampler, the distance from the source, the porosity of the filter, and the relation of the sampler to vertical or inclined surfaces that modify air movement [78,93].

The influence of the automobile The concentration of lead in air relates to traffic density and mode of operation. More lead is emitted along highways than in city driving, approximately 20% is emitted in suburban driving, and 200% at full throttle when accelerating to 60 mph. From an analysis of exhaust dust it appears that about 21-28% of the petrol lead is retained in the exhaust system [78, 93, 51]. Both the physical and chemical characteristics of the particulate matter is affected by fuel consumption, engine wear, nature of the engine oil, the exhaust system and air/fuel ratio [78].

In West Germany petrol lead was reduced from 0.6 to 0.4 $g\ l^{-1}$ in 1972 and then to 0.15 in 1976. The reduction in the lead content by 60% (1976) produced a reduction in air lead by 55-60% in streets carring heavy traffic. For example in Frankfurt the mean air lead was 2.56 $\mu g\ m^{-3}$ before reduction in the petrol lead, and 1.04 $\mu g\ m^{-3}$ after. However, in streets with lower traffic levels the reduction in air lead was less, around 30%. This may be because of an additional source of lead, but could also be due to the nature of the exponential fall off in lead levels with distance from the source. In the Ruhr area the drop in air lead was only 20%, whereas in the lead mining area, Stolberg, the reduction of lead in petrol had no effect on the air lead levels [56].

THE DEPOSITION OF HEAVY ELEMENT AEROSOLS

Aerosols eventually deposit on the earth, either as a dry deposit or as wet deposit (in rain or snow). The data in Table 7.11 indicates the magnitude of

TABLE 7.11 Wet Deposition for Some Heavy Elements (μg l^{-1})

Element	Marine	Rural	Urban
As	5.8	0.005-4 (0.286)*	0.019
Cd	0.48-2.3 (0.7)*	0.08-46 (0.5)	0.004-0.639 (0.008)
Hg	0.002-3.8 (0.745)	0.005-2.2 (0.09)	0.011-0.428 (0.079)
Pb	5.4-147 (44)	0.59-64 (12)	0.02-0.41 (0.09)
Sb	-	-	0.034

* Values in parenthesis are mean levels. Sources of data; Buat-Mènard and Duce, 1987 [10], Galloway et al., 1982 [33].

deposition levels (in μg l^{-1}) in rain found in different environments.

Deposition Levels

Most information available is on lead, and some data are given in Table 7.12. The deposition is expressed in mg m^{-2} y^{-1}, or μg l^{-1} (for wet deposition), though a number of other units have been used. The deposition levels of lead range from < 5 mg m^{-2} y^{-1} in remote areas, to 25-300 mg m^{-2} y^{-1} in urban areas, and much higher close to lead based industries, and roads busy with traffic. For deposition of cadmium, the bulk precipitate (wet and dry) has a mean concentration of the metal in the wet material of 0.6-37 μg l^{-1}, with most concentrations around 1-2 μg l^{-1}. In the case of dustfall (dry deposition), concentrations of 13 and 200-1000 μg g^{-1} have been reported [77].

A number of factors influence the level of deposition in any area, such as the locality i.e. remote, rural, urban or industrial. Wet precipitation depends on the existence of rain or snow, the amount, its duration and intensity. The concentration of lead in precipitation decreases with increasing intensity of the precipitation. The rates of deposition of cadmium containing aerosol have increased over the years. This relates to increasing population density, local variations and is a function of land use [77].

Deposition Models

Different models for both wet and dry deposition have been developed. For wet deposition the flux rate is given by,

$$\text{Flux} = WRA\left(\frac{C}{\rho}\right),$$

where W is the washout coefficient, R the annual rainfall (m), A the area (m^2), C is the metal concentration in air, and r is the density of the air = 1200 g m^{-3} [78]. Wet deposition is intermittent and the washout factor or scavenging ratio W is given by [10];

$$W = \frac{C_R \cdot \rho}{C_A},$$

where C_R is the concentration of the element in the rain μg kg^{-1}, ρ the density of air, and C_A the concentration of the element in the air, μg m^{-3}. The washout factor

TABLE 7.12 Levels of Deposited Lead Aerosol

Location	Deposition level	Reference
Northern hemisphere	0.8 mg m^{-2} y^{-1}	78
Southern hemisphere	0.4 mg m^{-2} y^{-1}	78
Remote	< 5 mg m^{-2} y^{-1}	78
France, mountains (wet)	9 μg l^{-1}	92
Remote terrestial		
winter snow	0.5 mg m^{-2} y^{-1}	28
summer snow & rain	0.4 mg m^{-2} y^{-1}	28
dry	2.0 mg m^{-2} y^{-1}	28
Remote	0.01-4 mg m^{-2} y^{-1}	17
Rural areas	5-20 mg m^{-2} y^{-1}	78
France, rural (wet)	16 μg l^{-1}	92
Rural	6-34 mg m^{-2} y^{-1}	17
Polluted areas	> 20 mg m^{-2} y^{-1}	78
Paris traffic (wet)	43 μg l^{-1}	92
Urban areas	10-30 mg m^{-2} y^{-1}	28
Urban	25-300 mg m^{-2} y^{-1}	17
France, industrial (wet)	5052 μg l^{-1}	92
Near lead sources	600-6000 mg m^{-2} y^{-1}	17
Near motorway, 0 to 33 m	13.4-0.5 mg m^{-2} d^{-1}	66
	7.1-1.6 mg m^{-2} d^{-1}	66

tends to be greatest for the larger particles [10,89], particularly over land masses, but over remote marine areas (e.g at Enewetok) there appears to be no relationship [10]. The efficiency of the washout of aerosols is given by;

$$\text{Efficiency} = \frac{[M]_{\text{(in unit volume of rain)}}}{[M]_{\text{(in corresponding volume of cloud)}}},$$

where M is the element [10,100].

The velocity of dry deposition (V_d) in rural and remote areas is < 1.0 cm s^{-1} (mean is 0.3 cm s^{-1}), but the velocity depends on the particle diameter and wind speed. The flux rate (F) (mg m^{-2} y^{-1}) for dry deposition is given by;

$$V_d(\text{cm s}^{-1}) = \frac{-\text{F(downward flux) mg m}^{-2}\text{ y}^{-1}}{\text{C(airborne conc.) mg m}^{-3}},$$

where C is the metal concentration in the air [10,20,78,89]. In polar regions V_d = 0.5 cm s^{-1}, and over the oceans 0.8 cm s^{-1}. The removal of large particles take place by sedimentation, i.e. gravitational settling, and small particles by diffusional transport and inertial impaction and interception [89]. The model is not adequate if a small fraction of the airborne mass is responsible for the bulk of the deposition. The deposition velocity is a function of the aerosol aerodynamic diameter (the diameter of spherical particles of unit density which behaves like the particle) the wind speed and surface conditions e.g. the roughness. The deposition velocity of cadmium varies by a factor of 20 because of different size distributions [20].

Dry Versus Wet Deposition

The ratio of dry to wet deposition is controlled by the amount of precipitation, and the dry fraction can vary from 0.2-0.6, or 0.2-0.9, of the total deposition [9,22,33,77,89]. It appears that wet deposition for trace elements is as important, or more so, than dry deposition.

Moss for Detecting Aerosol Fallout

Atmospheric deposition has been measured using the levels of the trace elements accumulated in moss, such as *sphagnum*. For reliable results the site needs to be treeless, ombrotropic (wet and rainy), contain hummocks with as little shrub cover as possible. The deposition rate is given by;

$$R = (CP) - L + U,$$

where R = deposition in mg m^{-2} y^{-1}, C = metal concentration in mg kg^{-1}, P = biomass production in kg m^{-2} y^{-1}, L = leaching of element from the moss, and U = uptake of the element from older organic material. In Europe the annual rate of metal accumulation in *sphagnam* mosses is 2.1-63 mg m^{-2} y^{-1} [35]. The reproducibility of the method for Pb (59.3 μg g^{-1} d^{-1}), and cadmium (2563 ng g^{-1} d^{-1}), is demonstrated by the coefficients of variation of 16.7% and 16.3% respectively. By using a network of bags it is possible to build up contour diagrams of equal concentrations of the elements in any area [64,96].

Fallout From Automobiles

Numerous studies have been carried out on the deposition of lead around roadways from automobiles. The following data (in mg m^{-2} d^{-1}) clearly shows a fall off in the deposited lead with distance from the road.

Central reserve	2	8	18	33 meters
13.40	5.20	2.91	1.00	0.52
7.05	7.36	5.58	2.82	1.61

A deposition rate of >3.3 mg m^{-2} d^{-1} is considered excessive for residential areas. This is not normally encountered at more than 15 m from the road [66]. Approximately 10% of the lead emitted from a car at cruise speeds is deposited within 100 m of the road, when there is no intercepting surfaces. The fall in atmospheric lead concentrations with distance from the road is in part due to depostition and in part due to the upward movement of the plume. The large particles deposit close to the road (> 90% within 1.5 m when the size >5 μm) [15].

HEAVY ELEMENT SPECIES AND REACTIONS IN THE ATMOSPHERE

The identification of the chemical forms of the trace elements in aerosols is a difficult problem, because, for most methods available for solids, relatively high concentrations of the species are required. The instrumental methods employed are the scanning electron microscope, electron microprobe and XRD. The first two methods are only useful for particles >0.5 μm, though the transmission electron microscope can be used on particles down to 0.1 μm. In the case of XRD only crystalline materials are observed and each species needs to be >5% in concentration [46].

Lead Species

Automobile lead Most work has been done on lead species, because of the high levels of lead in the particulate matter. The list of lead compounds reported in aerosols from a variety of sources is given in Table 7.13 [78]. The actual lead compounds in a particular aerosol will depend on other constituents in the atmosphere, and the age of the aerosol. The primary material from the exhausts of motor cars is mainly PbClBr (when both CH_2Cl_2 and CH_2Br_2 are used in the petrol). However, PbO, Pb(OH)X (X = Cl, Br), and some $PbSO_4$, $Pb_3(PO_4)_2$, $PbO{\cdot}PbSO_4$ and Pb may also occur in the larger particles, whereas in the smaller particles α and β $NH_4Cl{\cdot}2PbClBr$, $2NH_4Cl{\cdot}PbClBr$ are found. When phosphorus is present in petrol the compound $Pb_5(PO_4)_2(Cl,Br)$ may form. [5,43,46,47,51,78]. Some organolead compounds, such as R_4Pb and R_3PbCl are also emitted. A

TABLE 7.13 Lead Species in Aerosols

Source	Lead species in aerosols
Automotive	$PbCl_2$, $PbBr_2$, PbClBr, Pb(OH)Cl, $PbCl_2{\cdot}PbClBr$, $PbO{\cdot}PbBr_2$, $PbO{\cdot}PbClBr$, $PbO{\cdot}PbCl_2$, PbO_x, $PbSO_4$, $PbO{\cdot}PbSO_4$, PbP_2O_7, $Pb_3(PO_4)_2$, $Pb_3(PO_4)_2{\cdot}PbClBr$, $Pb_5(PO_4)_3(Cl,Br)$, $Pb_4O(PO_4)_2$, $2NH_4Cl{\cdot}PbClBr$, $\alpha NH_4Cl{\cdot}2PbClBr$, $PbCO_3$, $(NH_4)_2ClBr{\cdot}2PbClBr$, $\beta NH_4Cl{\cdot}2PbClBr$, $PbO{\cdot}PbCO_3$ $(PbO)_2{\cdot}PbCO_3$
Mining activities	PbS, $PbCO_3$, $PbSO_4$, $Pb_5(PO_4)_3Cl$, $PbS{\cdot}Bi_2S_3$, PbO_x,Pb-silicates
Base metal smelting and refining	Pb, PbO_x, $PbCO_3$, $PbSO_4$, $PbO{\cdot}PbSO_4$, $(PbO)_2{\cdot}PbCO_3$, Pb in metal oxides, Pb-silicates, PbS
Coal-fired power stations	PbO_x, $Pb(NO_3)_2$, $PbSO_4$, $PbO{\cdot}PbSO_4$, surface sorbed material $PbCl_2$, PbS, Pb
Cement manufacture	$PbCO_3$, $Pb_5(PO_4)_3Cl$
Fertilizer production	$PbCO_3$, PbO_x, $Pb_5(PO_4)_3Cl$
Ferroalloys	Pb, Pb-alloy particles
Lead products	Including lead arsenate, antimonate, chromate, cyanamide, iodide, fluorosilicate, molybdate, nitrate, selenide, silicates, titanate, vanadate

Sources of data; Harrison, 1986 [43], Nriagu, 1978 [78], Pacyna, 1987 [83].

study of the main crystalline lead compound in the primary lead aerosol was found to be the solid solution $PbBr_{1.4}Cl_{0.6,}$ based on the unit cell dimensions [93]. From the speciation results given in Table 7.14, it is clear, that the proportion of halogen containing compounds decrease with time and distance, while there is a corresponding increase in the oxy-compounds. There is a close similarity between the lead species in the aerosol collected directly from the exhaust, and collected near the road, also there is a similarity between aged aerosol and aerosol collected some distance from the roadside (Table 7.14) [57]. The principal components in aged vehicle aerosol are oxy-lead species, $PbCO_3$, $(PbO)_2$-$\cdot PbCO_3$, $PbSO_4 \cdot (NH_4)_2SO_4$ and PbO and $PbSO_4$ (a total of about 80%) [43,78,97].

Smelter lead The lead species from smelters are similar to aged automobile aerosol (cf. Table 7.13) [78,83]. Particulate materials from smelters contain PbS (from the primary ore), $PbSO_4$, and $PbO \cdot PbSO_4$ (both from the sintering of the PbS ore) in the stack and in the ambient and internal atmospheres, Lead oxide comes from oxidation of Pb during smelting, and the metal from reduction of compounds added to sawdust [43]. The lead species, in the stack emissions from a primary lead/zinc smelter, identified using XRD, differ in form depending on the position of the samples in the stack. For different positions the species were; (a) $PbSO_4$, $PbO \cdot PbSO_4$, (b) PbS, (c) PbO, Pb, and (d) CdO. Outside the smelter the aerosol lead species varied with the size, as follows; PbS, $PbSO_4$ in >7 μm particles, PbS, $PbSO_4$ in 3.3-7 μm particles, and $PbSO_4$ in 2-3.3 μm particles [45].

Reactions of lead species In a study [43,47] of the chemistry of PbClBr the following chemical processes were identified;

$$PbClBr \xrightarrow[\text{hv (natural)}]{} \text{no reaction,}$$

$$PbClBr + SO_2 \rightarrow \text{negligible reaction,}$$

$$PbClBr + NH_3 + H_2O \underset{\text{slow reaction}}{\rightarrow} Pb(OH)Br + \text{other products,}$$

$$2PbClBr + 2(NH_4)_2SO_4 \rightarrow \underset{\text{main product}}{PbSO_4 \cdot (NH_4)_2SO_4} + (NH_4)_2BrCl \cdot PbClBr,$$

The last reaction, proceeds in moist air, and is the principal one because ammonium sulphate is a major component in aerosols. In addition the reactions;

$$PbClBr + H_2SO_4 \rightarrow PbSO_4 + HCl + HBr$$

$$6PbClBr + 2NH_4HSO_4 \rightarrow 2PbSO_4 + \alpha NH_4Cl \cdot 2PbClBr + NH_4Br \cdot 2PbClBr + HCl + HBr$$

TABLE 7.14 Lead Compounds in Automobile Exhaust Material

	1	2	3	4	5	6	7	8	9	10	11	12	13	14
Exhaust gas														
0 h	10.4	5.5	32.0	7.7	2.2	5.2	1.1	31.4	1.2	-	2.2	1.0	-	0.1
18 h	8.3	0.5	12.0	7.2	0.1	5.6	0.1	1.6	13.8	-	21.2	29.6	0.1	-
Near a busy road														
Near	11.2	4.0	4.4	4.0	2.0	2.8	0.7	2.0	15.6	0.2	12.0	37.9	1.0	2.2
400 y	10.5	0.7	0.6	8.8	1.1	5.6	0.3	0.6	14.6	0.3	25.0	21.3	4.6	6.0
Rural	5.4	0.1	1.6	4.0	-	1.5	-	1.0	30.2	-	20.5	27.5	5.0	3.2

Where 1= $PbCl_2$, 2 = $PbBr_2$, 3 = PbClBr, 4 = Pb(OH)Cl, 5 = Pb(OH)Br, 6 = $(PbO)_2 \cdot PbCl_2$, 7 = $(PbO)_2 \cdot PbBr_2$, 8 = $(PbO)_2 \cdot PbClBr$, 9 = $PbCO_3$, 10 = $Pb_3(PO_4)_2$, 11 = PbO_x, 12 = $(PbO)_2 \cdot PbCO_3$, 13 = $PbO \cdot PbSO_4$, 14 = $PbSO_4$. h = hours, y = yards. Source of data; Servant, 1986 [92].

occur whereby bromine is lost from the aerosol [43,47,93]. The lead halides in the atmosphere are said (in contrast with above) to be photochemically decomposed liberating free dichlorine and dibromine [0519]. The lead halides (e.g. PbClBr), react with acid sulphates and nitrates also forming the dihalogens and HX. Many of these reactions reduce the Br/Pb ratio in the aerosol [44]. When heated in air to 250° C the PbClBr compound showed no change, but at 400° C PbO and Pb_3O_4 were formed [93].

Tetraalkyllead compounds react in the atmosphere with ozone and the radcals ·O and ·OH, and are photoreactive [24,43].

e.g. $$(CH_3)_3PbCH_3 + \cdot OH \rightarrow (CH_3)_3PbCH_2\cdot + H_2O.$$

The Pb-C bond can be broken [43], producing other members of the organolead species, as well as inorganic lead.

$$2R_3PbX \rightarrow R_2PbX_2 + R_4Pb,$$

$$2R_2PbX_2 \rightarrow R_3PbX + RX + PbX_2.$$

Species for Other Elements

Much less is known about the species of the other heavy elements in aerosols, mainly because of their low concentration, making it difficult to identify the compounds. The cadmium species in aerosols are probably Cd, CdS, CdO, $Cd(OH)_2$ and mixed oxides with copper and zinc [77,83]. From smelters there is evidence that the oxide CdO occurs in the stack, and CdO, Cd and $Cd(OH)_2$ have been identified in the internal atmosphere [43,55]. Cadmium dichloride may also form during the incineration of refuse [83].

Mercury in the vapour state, which may be >90% of the total atmospheric mercury, is most likely to be Hg, $HgCl_2$, CH_3HgCl, $(CH_3)_2Hg$ and other organomercury compounds. Elemental mercury may account for 50% of the volatile species [43,71,83].

The major chemical species for arsenic are As, As_2O_3, As_2S_3 and organoarsenic compounds from combustion and metallurgical processes. Also, because of the possibility of chlorides in refuse, incineration could produce volatile $AsCl_3$ [83].

Solubility of Aerosols

The solubility of the heavy elements in aerosols in water, such as in precipitation, will influence the reactivity of the elemets in the air. The solubility of lead aerosols is reported as being around 50-90% in water, whereas for deposited dust it is <1% [78,92]. Lead species are relatively soluble in wet samples, but are less so (<10%) in dry samples [34]. Cadmium aerosols are relatively soluble in water and acids, so that as much as 30 to 95% of the metal will be in solution in precipitation. Cadmium is reported to be more concentrated on the surface of particles produced by combustion. The ratio of surface to bulk levels of cadmium in aerosols is ~30, and this may be why its solubility is high [77]. In general, particulate cadmium is more soluble than particulate lead [34], and both their solubilities increase with decrease in particle size [88].

SOURCES OF THE HEAVY ELEMENTS IN AEROSOLS

A number of ways have been developed to determine the source of the elements in aerosols. The methods fall into three main categories, (a) calculation of enrichment factors and relating these to sources, (b) direct measurement, such as historical changes, and (c) determination of the composition of source receptors and the relation of these to the sources by use of techniques such as a chemical element balance.

Enrichment Factors

Relative enrichment of the elements The calculation of enrichment factors (EF) for some heavy elements in aerosols produced the order Se > Pb > Sb > Cd for the elements with the EF > 1. The factors are calculated with reference to elements, such as Al, Ce, Fe, Si and Ta [13]. Another measure is the mobilzation factor (MF), which is the ratio of emissions from human and natural sources. The order in this case is Pb > Sb > Cd > As, Se > Hg. A study of historical samples gives the orders Pb >> As > Cd > Se and Sb > Pb > Cd. A summary of these results, for low, moderate and high enrichments, is given below indicating a good level of consistency [33,61,88].

Factor	Low	Mod.	High	No data
Mobilization	Hg	As, Se	Cd, Pb, Sb, Se	Te, Tl
Enrichment			Cd, Pb, Sb, Se	Te, Tl, Hg,As
Historical		Se	Cd, Pb, Sb, As	Te, Tl, Hg

Another measure is the 'human activity' factor (similar to the mobilization factor) which is the ratio of the concentrations of a particular element in urban to remote areas, or rural to remote areas. The order is Pb (489, 128) > As (305, 15) > Cd (88, 66) > Hg (9.4, 1.1). The values for mercury are low because of inefficient collection in wet deposition, and there are insufficient data for Sb, Se, Te, and Tl [33].

In remote areas the order of enrichment factors, calculated with respect to aluminium, are Se > Cd > Pb > Sb > As, (EF's range 3500-49) [18,104,107]. The volatile elements appear to be under-estimated. The enrichment can not be entirely accounted for by crustal dust, sea spray and meteoritic dust. However, volcanic dust appears to be significant. In the Antartica the enrichment of As and Se (and Sb, In) increased in 1975, a few weeks after the eruption of Mt. Ngauruhoe in New Zealand [18]. The selenium and antimony levels in the air at Heimaey (Iceland) were elevated after the 1973 volcanic eruption. Both elements were enriched in the aerosol, lava ash, and fumerole deposits, and for two aerosol samples, the enrichments, relative to crustal aluminium, were Se 6200 and 21,600, and Sb 97 and 115. These factors are similar to what was found at Kilauea 1400-36,000, 17000, Sth pole 18,000, 1300, and Nth Atlantic 10,000, 2300 for Se and Sb respectively. This suggests that volcanoes may be a significant source of both metals [67,73]. In the Nth. Atlantic aerosol the enrichment factors for Cd 730, Pb 2200, Sb 2300 and Se 10,000 (GM) are greater than expected from crustal weathering and from the oceans. It is suggested, however, that the enrichment has a natural origin, and arises from the volatility of the elements and their compounds. This is based on the similarity of the enrichments in the Nth. Atlantic with those at the Sth Pole, as it was suggested that not all of the material from the north would get to the south, because of losses by precipitation [25]. However, it is not so much the amount that is transported that is important, but the concentration, and this could well increase with movement from the north to the south, because of the increase in the proportion of small particles during transport.

The EF's in urban air, such as in Toronto, are as follows: As 260, Hg < 6300, Sb 1300 and Pb 2500 [80]. The heavy metal enrichment factors in dry deposition, suspended particles and rainwater, generally increase in the orders portrayed by the data in Table 7.15 [38,107]. Greater enrichment occurs in the wet deposition, which may be because the finer particles tend to be removed by rain and these generally have the higher concentration of the heavy elements. Around a chemical factory using antimony, its EF is elevated compared with other elements: Sb 20,000, As 200, Se 4000, Hg 500, Cd 400, Pb 2000 and Bi 200 [75].

The estimated EF's for some heavy elements measured in the air at Enewetok in the North Pacific, with respect to crustal Al, and to Na (in sea water) are given in Table 7.16. The factors for selenium and lead have also been estimated with respect to the size of the particles, and the results are listed in Table 7.17. Lead is enriched relative to the crust and seawater but in the dry season in the size range 1.5-3.0 μm the EF tend towards the crustal level,

TABLE 7.15 Enrichment Factors in Different Portions of Urban Aerosols

Element	Dry Deposition	Suspended Particle	Rainwater*
In	3.4	16	10
As	24	100	800
Sb	380	580	970
Se	170	1400	4700

* also contains fine particles. Source; Gordon, 1986 [38].

TABLE 7.16 The Enrichment Factors for some Heavy Elements at Enewetok

Elements	Dry season	Wet season	All
Se	3700 (8300)*	48,000 (8000)	23,000 (8200)
Cd	75 (71,000)	180 (10,000)	130 (41,000)
Sb	29 (39)	180 (27)	93 (34)
Pb	11 (23,000)	110 (30,000)	40 (26,000)

* EF's in parenthesis are with respect to sodium (seawater)
Source of data; Duce et al., 1983 [26].

TABLE 7.17 Enrichment Factors for Se and Pb with Respect to the Size of the Aerosol

	Sizes					
Element	7.4	3.0	1.5	0.95	0.49	<0.49 μm
Se (D)*	1900§	2400	2800	7900	4700	4600
	640§§	5500	15000	30000	55000	130000
Se (W)**	32,000	32,000	57,000	75,000	66,000	37,000
	870	5600	32,000	48,000	99,000	18,0000
Pb (D)	23	4.9	4.6	15	40	-
	5500	10,000	28,000	14,0000	56,0000	-
Pb (W)	360	35	70	210	810	-
	6600	17,000	64,000	22,0000	19,00000	-

* dry season, ** wet season, § EF with respect to Al, §§ EF with respect to Na.
Source of data; Duce et al., 1983 [26].

indicating that some factor is masking the anthropogenic inputs. Selenium is very enriched, especially in the size range 1-3 μm. The size distribution is bimodal suggesting multiple sources [26], and it is very likely that volcanic activity is contributing to the high enrichment factors.

Enrichment of lead Enrichment factors are high for lead in remote places e.g. 130-3500 in polar and oceanic regions. They are generally lower over the oceans compared with land. The factors are around an order of magnitude greater than enrichments attributable to soil (~1), fuel combustion (~ 9) and volcanic sources (~ 60-100). Relative to titanium the EF of lead in Greenland aerosol (1979-1981) was 122-1330, and bromine was also highly enriched, 2000-7000, [48,49] suggesting that the high enrichments are from lead bromide emissions from cars. Approximately 10 times more lead has been found in precipitation compared with what was measured in the air. This may be because the small particles, which are lead rich, are missed in air sampling, i.e. they would go through the filter, whereas precipitation removes a greater proportion of the small particles. Levels of lead in snow have increased 300 times over 300 years, and concentrations today, are around 50 times more than would be expected from air lead levels, suggesting an efficient removal of the lead particles by the snow [76,90].

Another indicator for lead sources is the bromine to lead concentration ratio. The [bromine]/[lead] ratio may be used because of the addition of CH_2X_2 (X = Cl, Br) to petrol to scavenge the lead. Ratios are often <0.386 (the value expected from petrol lead) because of loss of volatile bromine compounds, loss of bromine by reaction in the atmosphere, other sources of lead, some lead emitted as organolead, aged particles containing re-entrained dust, over compensation for marine bromine, loss during storage and analytical errors, especially bromine loss during XRD analysis [44]. The bromine loss can be as much as 53%, with a mean of 36% [68]. On the other hand, the ratio may be high, because of the lead retained in the exhaust system relative to bromine, evaporation of ethylenedibromide, analytical errors, inadequate correction for other sources of bromine including agriculture, industrial and coal burning sources. Near industrial sources of lead the ratio can be less than 0.1 [44]. The [Br]/[Pb] ratio 200 m from a lead smelter at El Paso was 0.016, which rose to 0.385, 5-6 km away, the ratio expected for automobile lead [59]. In another example, sampling for lead and bromine two hour intervals revealed a concentration profile which was the same for each element, suggesting a common source, and probably the automobile [103]. Seasonal variations in the lead levels in the Arctic, with a peak around late Jan and Feb suggests an anthropogenic source, which was confirmed by the bromine levels having a similar seasonal profile [3]. A similar association is found for Greenland aerosol (1979-1981) [48,49].

Direct Measurement

Frequently direct measurement of the levels of the heavy elements in aerosols enables identification of the source of the heavy element. For example seasonal changes in levels point to a source as discussed above for lead [3,48,49]. Some of the air lead in New Zealand comes from Australia, and this is seen from the relatively high level of lead in the Tasman Sea (490 pg g^{-1}), when the air flow came from Australia [95].

Historical changes in emissions help in identifying sources of the heavy elements. A decrease in lead emissions from smelting has occurred over recent years even though the amount of lead smelted has increased. This has, however, been offset by petrol lead emissions. The data in Table 7.18, for estimated aerosol lead emissions and the lead levels in air and snow, show the relationship between the quantites of lead smelted and emitted and the environmental levels. The last two columns of the table may be compared directly, if it is assumed that the lead in the snow contains a constant proportion of the aerosol lead. It appears that the increase in the environmental lead since 1933 has been mainly due to petrol lead emissions [74]. Virtually all present day lead that is in excess over natural in the Arctic and Antarctic is due to anthropogenic sources, as shown from historical trends in concentrations, mass inventories and measurement of lead in volcanic plumes [76].

A sampling time of two hours allows for the resolution of changes in sources over time. For example, in a study it was found that on certain days the lead and bromine concentration profiles differed from each other, whereas the profiles for the elements Pb, Hg, Sb, As, In and Se were similar. This indicated a source of lead different from automobiles. The results were also found to relate to the direction from which the wind blew [103].

A comparison of the levels of lead in air, rain and rocks, has provided the suggestion that some part of the metal in the air has come from volatilization from the earth's surface [36]. Lead isotopes may also be used to identify the lead source because ^{210}Pb only forms in the air from ^{222}Rn. The isotope ^{222}Rn emanates from the earth and in the atmosphere decays to give ^{210}Pb (^{222}Rn ($t_{1/2} = 3.8$ d) $\rightarrow$ ^{210}Pb ($t_{1/2} = 22$ y)). The lead has a mean lifetime in the air of 5 days, not long enough for much of the lead isotope to have decayed to ^{210}Po. Therefore the activity of ^{210}Pb in aerosols is a guide to its source. At Enewetok a relationship exists between the ^{210}Pb activity and the aluminium concentration in the dust flux from Asia, indicating the same source [99].

TABLE 7.18 Historical Trend in Lead Emissions and Environmental Lead

Date	Pb smelted 10^5t/y	% in aerosol	Pb aerosol from smelting 10^3/y	R_4Pb burned 10^5/y	%in aerosol	Pb aerosol from R_4Pb 10^3/y	Total Pb aerosol 10^3/y	Pb pg g^{-1} in snow
1753	1	2	2	-	-	-	2	0.01
1815	2	2	4	-	-	-	4	0.03
1933	16	0.5	8	0.1	40	4	10	0.07
1966	31	0.06	2	3	40	100	100	0.2

Source of data; Murozumi et al., 1969 [74].

Source Receptor Models

The third method of locating the source of the heavy elements in an aerosol sample is comparing the elemental composition of the source and receptor. Some of the methods used are, pairwise comparisons, chemical element balances, cluster analysis, dendograms, principal component and factor analysis [11,57,58,75,84,87]. The methods of factor analysis and chemical element balances are the most commonly used, and are based on multi-element analytical results on aerosols and sources. The results reported here are just for the heavy elements, and mainly lead, but normally many other elements are considered in the analyses.

Factor analysis results By the use of factor analysis, the main sources of lead in Greenland aerosol were estimated to be combustion, engine exhaust, and to a lesser extent metal sources [48,49]. From a factor analysis of Boston aerosol it was found the main sources of selenium in urban areas were: oil > soil > road dust; and in suburban areas: soil > road dust. For antimony in urban areas the main sources were: refuse > soil = road dust, and in suburban areas: road dust > refuse > soil. There is reasonable agreement between the predicted and observed levels of the elements as shown by the data in Table 7.19 [2,53].

Chemical element balance results In determining the source of trace elements in aerosols, the chemical mass balance method assumes that the unique trace element content of a source can be used as a tracer for the presence of that source in the aerosol. The equation is:

$$x_{ij} = \Sigma \alpha_{ik} a_{ik} m_{kj}$$

where x_{ij} = concentration of ith element in the jth aerosol sample, α_{ik} = coefficient of fractionation (due to settling) between source and receptor (often neglected, or taken as 1, i.e. no fractionation, which is unlikely to be correct), a_{ik} = concentration of ith element in the kth source, and m_{kj} = fraction of particles from the kth source in the jth aerosol. If there are more sources than elements analysed, or if sources can not be identified on the basis of their chemical composition then the number may be reduced using principal component analysis [11,52,58].

The sources of lead in the fine aerosol collected in Houston Texas, were 8% from a tetraethyllead plant, 29% from a steel mill and 60% from mobile sources, such as cars [11]. In a different study in a different environment, the lead sources in aerosol at E. Helena are estimated to be: road and soil dust 18.5%, blast furnace 26%, fugative emissions 35.3%, copper kiln, 0.6%, ZnO materials 9.4%, slag pouring, 2% and motor vehicles 3.8% [52]. In the case of Los Angeles aerosol the sources considered were sea salt, soil and dust, auto exhaust, fuel oil, fly ash, portland cement and tyre dust. The lead in the aerosol was 3.3% by mass, and came from auto exhaust [31]. For the fine particle emissions 83% of lead was said to arise from car exhausts, and 9% from re-suspended road dust [12].

TABLE 7.19 Sources of Selenium and Antimony in Boston Aerosol (ng m^{-3})

Element	Soil	Oil combust.	Refuse inciner.	Marine	Motor vehicle	Road dust	Predicted	Observed
Suburban								
Se	0.5	0	0.02	0	0.09	0.4	1.1	1.3
Sb	0.4	0.2	1.6	0	0.3	2.6	5.1	6.3
Urban								
Se	0.3	0.6	0.08	0	0	0.2	1.1	1.2
Sb	2.6	0.7	4.7	1.7	1.1	2.6	13	15

Source of data; Alpert and Hopke, 1980 [2].

The aerosol in Washington DC was investigated using the method of chemical element balance. The range of levels of some elements in the aerosol were (in ng m^{-3}): Pb 82-5800, Se 1.6-10.1, As 0.82-14.7, Cd 0.81-7.0, Sb 1.75-26. The sources investigated were soil, marine, coal, oil, refuse, motor vehicles and the results are listed in Table 7.20. These indicate that the main sources are; Cd and Sb from refuse, As and Se from coal, and Pb from motor vehicles. There are limitations in the method, as Se, Cd and Sb are not completely accounted for, as

TABLE 7.20 Sources of the Heavy Elements in Aerosol from Washington DC (ng m^{-3})

Metal	Soil	Marine	Coal Comb.	Oil	Refuse Inciner.	Motor Vehicles	Prediced	Observed
Pb	0.23	<0.001	2.9	0.86	81	1300	1380	1400
As	0.09	<0.001	4.4	0.061	0.24	-	4.8	5.7
Se	0.001	<0.001	1.1	0.076	0.037	-	1.2	3.5
Cd	0.002	<0.001	0.19	0.006	1.5	-	1.7	3.5
Sb	0.012	<0.001	0.18	0.015	2.1	-	2.3	9.7

Source of data; Kowalczyk et al., 1978 [57].

TABLE 7.21 Sources of the Heavy Elements in Aerosols*

M	Soil	Limestone	Coal Comb.	Oil	Refuse Inciner.	Motor vehicle	Marine	Predicted	Obs.
As	0.061	0.002	3.1	0.028	0.10	-	0.0001	3.32	3.25
Se	0.0009	0.0002	0.78	0.035	0.016	0.035	0.0001	0.87	2.5
Cd	0.0011	0.0001	0.13	0.0028	0.064	1.03	<0.0001	1.80	2.4
In	0.7	0.1	2.3	<0.1	2.4	-	0.4	5.9	20
Sb	0.0081	0.0004	0.13	0.007	0.89	0.60	<0.0001	1.6	2.1
Ba	7.1	0.02	4.7	2.0	0.30	6.4	0.0006	21	19
Pb	0.15	0.019	2.1	0.39	34	428	<0.0001	465	440

* Concentrations ng m^{-3}, except for indium pg m^{-3}. Source; Kowalczyk et al., 1982 [58].

seen by comparing the last two columns of the table. It is possible, that because of the volatility of these elements, some fraction of them had been missed. Other factors that may influence the results are; that cement was not used as a source, there is a wide varibility in the trace element composition of oils, it is assumed that all the particles leaving the source reach the receptor, and it is possible that some fractionation of the elements that derive from soil had taken place [57,107]. The results of another analysis are given in Table 7.21. Again the predicted and observed levels of selenium are not in agreement [57,58,138]. The contribution of the sources to all sampling sites (i.e. average of each site) is for urban sites, in $\mu g\ m^{-3}$, soil 16.2, limestone 1.8, coal 4.3, oil 0.39, refuse 0.65, motor vehicles 4.8, marine 0.69 (total 28.9), and for rural sites soil 10.7, limestone 1.6, coal 2.5, oil 0.19, refuse 0.42, motor vehicles 2.7, marine 0.54 (total 18.6) [58]. From another study it was estimated that the aerosol arises from car emissions, 20%, from fuel and fly ash, 1-2%, and from soil and road dust 20-50%. Most cadmium comes from stationary sources and industrial processes [12].

Pattern recognition and construction of dendograms have been used to identify sources of lead and cadmium in desert aerosol. It appears that they have not come from the soil, but from some long range source such as combustion and transport emissions [32].

REFERENCES

1. Ali, E A., Nasralla, M. M. and Shakrur, A. A. Spatial and seasonal variation of lead in Cairo atmosphere. **Environ. Pollut.;** 1986; **11B:** 205-210.
2. Alpert, D. J. and Hopke, P. K. A quantitative determination of sources in the Boston urban aerosol. **Atmos. Environ.;** 1980; **14:** 1137-1146.
3. Barrie, L. A. and Hoff, R. M. Five years of air chemistry observations in the Canadian Arctic. **Atmos. Environ.;** 1985; **19:** 1995-2010.
4. Bertini, K. K. and Goldberg, E. D. Fossil fuel combustion and the major sedimentary cycle. **Science;** 1971; **173:** 233-234.
5. Biggins, P. D. E. and Harrison, R. M. Identification of lead compounds in urban air. **Nature;** 1978; **272:** 531-532.
6. Bloom, H. and Noller, B. N. Application of trace analysis techniques to the study of atmospheric metal particulates. **Clean Air Symp. (May);** 1977.
7. Boutron, C. F. Atmospheric toxic metals and metalloids in the snow and ice layers deposited in Greenland and Antarctica from prehistoric times to present. in: Nriagu, J. O. and Davidson, C. I., Eds. Toxic Metals in the Atmosphere: Wiley; 1986: 467-505.
8. Boutron, C. and Patterson, C. C. The occurrence of lead in Antarctic recent snow, firn deposited over the last two centuries and prehistoric ice. **Geochim. et Cosmochim. Acta;** 1982; **47:** 1355-1368.
9. Buat-Mènard, P. E. Fluxes of metals through the atmosphere and oceans. In: Nriagu, J. O., Ed. Changing metal Cycles and Human Health, Dahlem Konferenzen: Springer-Verlag; 1984: 43-69.

10. Buat-Mènard, P. and Duce, R. A. Metal transfer across the air-sea interface: myths and mysteries. In: Hutchinson, T. C. and Meema, K. M., Eds. Lead, Mercury, Cadmium and Arsenic in the Environment: Wiley, SCOPE; 1987: 147-173.
11. Cass, G. R. and McRae, G. J. Emissions and air quality relationships for atmospheric trace metals. in: Nriagu, J. O. and Davidson, C. I., Eds. Toxic metals in the Atmosphere: Wiley; 1986: 145-171.
12. Cass, G. R. and Mcrae, G. J. Source-receptor reconcillation of routine air monitoring data for trace metals: an emission inventory approach. **Environ. Sci. Techn.;** 1983; **17:** 129-139.
13. Cawse, P. A. Inorganic particulate matter in the atmosphere. In: Bowen, H. J. M., Rept. Environmental Chemistry: Roy. Soc. Chem.; 1982; **2:** 17-37.
14. Chamberlain, A. C., Heard, M. J., Little, P. and Wiffen, R. D. The dispersion of lead from motor exhausts. **Phil. Trans. Roy. Soc. London;** 1979; **290A:** 577-589.
15. Chamberlain, A. C., Heard, M. J., Little, P. Newton, D., Wells, A. C. and Wiffen, R. D. Investigations into lead from motor vehicles. Rept. AERE Harwell, AERE-R 9198; 1978.
16. Committee on Biological Effects of Atmospheric Pollution. Lead: Airborne Lead in Perspective: Nat. Acad. Sci.; 1972.
17. Committee on Lead in the Human Environment. Lead in the Human Environment: Nat. Acad. Sci.; 1980.
18. Cunningham, W. C. and Zoller, W. H. The chemical composition of remote area aerosols. **J. Aerosol. Sci.;** 1981; **12:** 367-384.
19. Davidson, C. I., Santhanam, S., Fortmann, R. C. and Olson, M. P. Atmospheric transport and deposition of trace elements onto the Greenland ice sheet. **Atmos. Environ.;** 1985; **19:** 2065-2081.
20. Davidson, C. I. Dry deposition of cadmium from the atmosphere. Nriagu, J. O., Ed. Cadmium in the Environment, Part I, Ecological Cycling: Wiley; 1980: 115-139.
21. Davidson, C. I. and Osborn, J. F. The size of airborne trace metal containing particles. in: Nriagu, J. O. and Davidson, C. I., Eds. Toxic metals in the Atmosphere: Wiley; 1986: 35-390.
22. Davidson, C. I., Chu, L., Grimm, T. C., Nasta, M. A. and Qamoos, M. P. Wet and dry deposition of trace elements onto the Greenland ice sheet. **Atmos. Environ.;** 1981; **15:** 1429-1437.
23. Davidson. C. I., Grimm, T. C. and Nasta, M. A. Airborne lead and other elements derived from local fires in the Himalayas. **Science;** 1981; **214:** 1344-1346.
24. De Jonghe, W. R. A. and Adams, F. C. Biogeochemical cycling of organic lead compounds. in: Nriagu, J. O. and Davidson, C. I., Eds. Toxic Metals in the Atmosphere: Wiley; 1986: 561-594.

25. Duce, R. A., Hoffman, G. L. and Zoller, W. H. Atmospheric trace metals at remote northern and southern hemisphere sites: pollution or natural. **Science**; 1975; **187**: 56-61.
26. Duce, R. A., Arimoto, R., Ray, B. J., Unni, C. K. and Hardner, P. J. Atmospheric trace elements at Enewetok atoll: 1 concentration, sources and temporal variability. **J. Geophys. Res.**; 1983; **88, C9**: 5321-5342.
27. Edwards, H. W. and Wheat, H. G. Seasonal trends in Denver atmospheric lead concentrations. **Environ. Sci. Techn.**; 1978; **12**: 687-692.
28. Elias, R. W. and Patterson, C. C. The toxicological implications of biogeochemical studies of atmospheric lead. Private Commum.; 1979
29. Fergusson, J. E. Inorganic Chemistry and the Earth: Pergamon Press; 1982.
30. Fleisher, M., Sarofim, A. F., Fassett, D. W., Hammond, P., Shacklette, H. T., Nisbet, I. C. T. and Epstein, S. Environmental impact of cadmium: a review by the panel on hazardous trace substances. **Environ. Health Perspec.**; 1974: 253-323.
31. Friedlander, S. K. Chemical element balances and identification of air pollution sources. **Environ. Sci. Techn.**; 1973; **7**: 235-240.
32. Gaarenstroom, P. D., Perone, S. P. and Moyers, J. L. Application of pattern recognition and factor analysis for characterisation of atmospheric particulate composition in southwest desert atmosphere. **Environ. Sci. Techn.**; 1977; **11**: 795-800.
33. Galloway, J. N., Thornton, J. D., Norton, S. A., Volchok, H. L. and McLean, R. A. N. Trace metals in atmospheric deposition: a review and assessment. **Atmos. Environ.**; 1982; **16**: 1677-1700.
34. Gatz, D. F. and Chu, L-C. Metal solubility in atmospheric deposition. in: Nriagu, J. O. and Davidson, C. I., Eds. Toxic metals in the Atmosphere: Wiley; 1986: 391-408.
35. Glooschenko, W. A. Monitoring the atmospheric deposition of metals by use of bog vegetation and peat profiles. in: Nriagu, J. O. and Davidson, C. I., Eds. Toxic metals in the Atmosphere: Wiley; 1986: 507-533.
36. Goldberg, E. D. Rock volatility and aerosol composition. **Nature**; 1976; **260**: 128-129.
37. Goodman, H. S., Noller, B. N., Pearman, G. I. and Bloom, H. The heavy metal composition of atmospheric particulates in Hobart Tasmania. **Clean Air**; 1976; **10**: 38-41.
38. Gordon, G. E. Sampling, analysis and interpretation of atmospheric particles in rural continental areas. in: Legge, A. H. and Krupa, S. V., Eds. Air Pollutants and their Effects on the Terrestrial Environment: Wiley; 1986: 137-158.
39. Graedel, T. E. Atmospheric photochemistry. in: Hutzinger, O., Ed. Handbook of Environmental Chemistry: Springer Verlag; 1980; **2A**: 108-143.
40. Harrison, R. M. and Williams, C. R. Atmospheric cadmium pollution at rural and urban sites in North-West England. In: Managemnet and Control of Heavy Metals in the Environment: CEP Consultants; 1979: 262-266.

41. Harrison, P. R. and Winchester, J. W. Area-wide distribution of lead, copper and cadmium in air particulate from Chicago and northwest Indiana. **Atmos. Environ.;** 1971; **5:** 863-880.
42. Harrison, P. R., Matson, W. R. and Winchester, J. W. Time variations of lead, copper and cadmium concentrations in aerosol in Ann Arbor, Michigan. **Atmos. Environ.;** 1971; **5:** 613-619.
43. Harrison, R. M. Chemical speciation and reaction pathways of metals in the atmosphere. in: Nriagu, J. O. and Davidson, C. I., Eds. Toxic metals in the Atmosphere: Wiley; 1986: 319-333.
44. Harrison, R. M. and Sturges, W. T. The measurement and interpretation Br/Pb ratios in airborne particles. **Atmos. Environ.;** 1983; **17:** 311-328.
45. Harrison, R. M. and Williams, C. R. Physico-chemical characterisation of atmospheric trace metal emissions from a primary zinc-lead smelter. **The Sci. Total Environ.;** 1983; **31:** 129-140.
46. Harrison, R. M. Physico-chemical speciation techniques for atmospheric particles. In: Harrison, R. M. and Perry, R., Eds. Handbook of Air Pollution Analysis. 2nd ed.: Chapman and Hall; 1986: 523-533.
47. Harrison, R. M. and Biggins, P. D. E. The speciation and atmospheric chemistry of automotive inorganic lead compounds in urban air. In: Management and Control of Heavy Metals in the Environment: CEP Consultants; 1979: 381-385.
48. Heidam, N. Z. The composition of the Arctic aerosol. **Atmos. Environ.;** 1984; **18:** 329-343.
49. Heidam, N. Z. Crustal enrichments in the Arctic aerosol. **Atmos. Environ.;** 1985; **19:** 2063-2097.
50. Heidam, N. Z. Trace metals in the Arctic aerosol. in: Nriagu, J. O. and Davidson, C. I., Eds. Toxic Metals in the Atmosphere: Wiley; 1986: 267-293.
51. Hirschler, D. A., Gilbert, L. F., Lamb, F. W. and Niebylski, L. M. Particulate lead compounds in automobile exhaust gas. **Ind. Eng. Chem.;** 1957; **49:** 1131-1142.
52. Hopke, P. K. Quantitative source attribution of metals in the air using receptor models. in: Nriagu, J. O. and Davidson, C. I., Eds. Toxic metals in the Atmosphere: Wiley; 1986: 173-200.
53. Hopke, P. K., Gladney, E. S., Gordon, G. E. and Zoller, W. H. The use of multivariate analysis to identify sources of selected elements in the Boston urban aerosol. **Atmos. Environ.;** 1976; **10:** 1015-1025.
54. Huntzicker, J. J., Friedlander, S. K. and Davidson, C. I. Material balance for automobile emitted lead in Los Angeles basin. **Environ. Sci. Techn.;** 1975; **9:** 448-457.
55. Jawarowski, Z., Bysiek, M. and Kownacka, L. Flow of metals into the global atmosphere. **Geochim. et Cosmochim. Acta;** 1981; **45:** 2185-2199.
56. Jost, D. and Sartorius, R. Improved ambient air quality due to lead in petrol regulation. **Atmos. Environ.;** 1979; **13:** 1463-1465.

57. Kowalczyk, G. S., Choquette, C. E. and Gordon, G. E. Element balances and identification od air pollution sources in Washington, DC. **Atmos. Environ.**; 1978; **12:** 1143-1153.

58. Kowalczyk, G. S. Gordon, G. E. and Rheingrover, S. W. Identification of atmospheric particulate sources in Washington DC, using chemical elemnt balances. **Environ. Sci. Techn.**; 1982; **16:** 79-90.

59. Landrigan, P. J., Stephen, H. Gehlbach, M. D., Rosenblum, B. F., Shoults, J. M., Candelaria, R. M., Barthel, W. F., Liddle, J. A., Smrek, A. L., Staehling, N. W. and Sanders, J. F. Epidemic lead absorption near an ore smelter: the role of particulate lead. **New Engl. J. Med.**; 1975; **292:** 123-129.

60. Landsberg, H. H. Air over cities, Symposium: US Dept. Health, Education and Welfare; 1962.

61. Lantzy, R. J. and Mackenzie, F. T. Atmospheric trace metals: global cycles and assessment of man's impact. **Geochim. et Cosmochim. Acta**; 1979; **43:** 511-523.

62. Lindberg, S. E. Emission and deposition of atmospheric mercury vapour. In: Hutchinson, T. C. and Meema, K. M., Eds. Lead, Mercury, Cadmium and Arsenic in the Environment: Wiley, SCOPE; 1987: 89-105.

63. Lindberg, S. E. Mercury vapour in the atmosphere: three case studies on emission, deposition and plant uptake. in: Nriagu, J. O. and Davidson, C. I., Eds. Toxic metals in the Atmosphere: Wiley; 1986: 535-560.

64. Little, P. and Martin, M. H. Biological monitoring of heavy metal pollution. **Environ. Pollut.**; 1974; **6:** 1-19.

65. Little, P. and Wiffen, R. D. Emissions and deposition of petrol engine exhaust lead. I Deposition of exhaust Pb to plant and soil surfaces. **Atmos. Environ.**; 1977; **11:** 437-447.

66. Little, P. and Wiffen, R. D. Emissions and deposition of petrol engine exhaust lead. II Airborne concentration, particle size and deposition of lead near motorways. **Atmos. Environ.**; 1978; **12:** 1331-1341.

67. Maenhaut, W., Zoller, W. H. Druce, R. A. and Hoffman, G. L. Concentration and size distribution of particulate trace elements in the South Polar atmosphere. **J. Geophys. Res.**; 1979; **84 (C5):** 2421-2431.

68. Martens, C. S., Wesolowski, J. J., Kaifer, R. and John, W. Lead and bromine particle size distributions in the San Francisco Bay area. **Atmos. Environ.**; 1973; **7:** 905-914.

69. Matheson, D. H. Mercury in the atmosphere and in precipitation. In: Niragu, J. O., Ed. The Biogeochemistry of Mercury in the Environment: Elsevier/Nth. Holland; 1979: 113-129.

70. Meszaros, E. Atmospheric Chemistry: Elsevier; 1981.

71. Miller, D. R. and Buchanan, J. M. Atmospheric transport of mercury: exposure commitment and uncertainty calculations. MARC Tech. Rept.: MARC; 1979: No. 14.

72. Moyers, J. L., Ranweiler, L. E., Hopf, S. B. and Korte, N. E. Evaluation of particulate trace species in southwest desert atmosphere. **Environ. Sci. Techn.**; 1977; **11:** 789-795.
73. Mroz, E. J. and Zoller, W. H. Composition of atmospheric particulate matter from the eruption at Heimaey Island. **Science**; 1975; **190:** 461-464.
74. Murozumi, M., Chow, T. J. and Patterson, C. C. Chemical concentrations of pollutant lead aerosols terrestrial dusts and sea salt in Greenland and Antarctic snow strata. **Geochim. et Cosmochim. Acta**; 1969; **33:** 1247-1294.
75. Neustadter, H. E., Fordyce, J. S. and King, R. B. Elemental composition of airborne particulates and source identification: data analysis techniques. **J. Air Pollut. Control Assoc.**; 1976; **26:** 1079-1084.
76. Ng, A. and Patterson, C. C. Natural concentrations of lead in ancient Arctic and Antarctic ice. **Geochim. et Cosmochim. Acta**; 1981; **45:** 2109-2121.
77. Nriagu, J. O. Cadmium in the atmosphere and precipitation. in: Nriagu, J. O., Ed. Cadmium in the Environment Part I Ecological Cycling: Wiley; 1980: 71-114.
78. Nriagu, J. O. Lead in the atmosphere. in: Nriagu, J. O., Ed. **The** Biogeochemistry of Lead in the Environment, Part A, Ecological Cycles: Elsevier/ Nth. Holland; 1978: 137-184.
79. O'Brien, B. J., Smith, S. and Coleman, D. O. Lead pollution of the global environment. Progress Reports in Environmental Monitoring and Assessment. 1 Lead. Tech. Rept.: MARC; 1980: No. 16-18.
80. Paciga, J. J. and Jervis, R. E. Multielement size characterisation of urban aerosols. **Environ. Sci. Techn.**; 1976; **10:** 1124-1128.
81. Paciga, J. J., Roberts, T. M. and Jervis, R. E. Particle size distributions of lead, bromine and chlorine in urban-industrial aerosol. **Environ. Sci. Techn.**; 1975; **9:** 1141-1144.
82. Pacyna, J. M. Atmospheric trace elements from natural and anthropogenic sources. in: Nriagu, J. O. and Davidson, C. I., Eds. Toxic Metals in the Atmosphere: Wiley; 1986: 33-52.
83. Pacyna, J. M. Atmospheric emissions of arsenic, cadmium, lead and mercury from high temperature processes in power generation and industry. In: Hutchinson, T. C. and Meema, K. M., Eds. Lead, Mercury, Cadmium and Arsenic in the Environment: Wiley, SCOPE; 1987: 69-87.
84. Pacyna, J. M. Emission factors of atmospheric elements. in: Nriagu, J. O. and Davidson, C. I., Eds. Toxic metals in the Atmosphere: Wiley; 1986: 1-32.
85. Paulhamus, J. A. Airborne contamination. in: Zief, M. and Speights, R., Eds. Ultrapurity Methods and Techniques: Dekker; 1972: 255-286.
86. Priest, P., Navarre, J. L. and Ronneau, C. Elemental background concentration in the atmosphere of an industralised country. **Atmos. Environ.**; 1981; **15:** 1325-1336.

87. Roscoe, B. A., Hopke, P. K. Dattner, S. L. and Jenks, J. M. The use of principal component factor analysis to interpret particulate compositional data sets. **J. Air Pollut. Control Assoc.**; 1982; **32:** 637-642637-642.

88. Salamons, W. Impact of atmospheric inputs on the hydrospheric trace metal cycle. in: Nriagu, J. O. and Davidson, C. I., Eds. Toxic Metals in the Atmosphere: Wiley; 1986: 409-466.

89. Salomans, W. and Förstner, U. Metals in the Hydrocycle: Springer-Verlag; 1984.

90. Schaule, B. and Patterson, C. C. The occurrence of lead in the Northeast Pacific and the effects of anthropogenic inputs. in: Branica, M. and Konrad, Z., Eds. Lead in the Marine Environment: Pergamon; 1980: 31-43.

91. Schialowski, M. The atmosphere. in: Hutzinger, O., Ed. Handbook of Environmental Chemistry: Springer Verlag; 1980; **1A:** 1-16.

92. Servant, S. Airborne lead in the environment in France. in: Nriagu, J. O. and Davidson, C. I., Eds. Toxic metals in the Atmosphere: Wiley; 1986: 595-619.

93. Simmonds, P. R., Tan, S. Y. and Fergusson, J. E. Heavy metal pollution at an intersection involving a busy urban road in Christchurch New Zealand 2 Aerosol lead levels. **N. Z. J. Sci.**; 1983; **26:** 229-242.

94. Sneddon, J. Collection and atomic spectroscopic measurement of metal compounds in the atmosphere: a review. **Talanta**; 1983; **30:** 631-648.

95. Steimer, J. T. and Clarkson, T. S. Heavy metals in the New Zealand atmosphere. **J. Roy. Soc. N. Z.**; 1985; **15:** 389-398.

96. Temple, P. J., McLaughlin, S. N., Linzon, S. N. and Wills, R. Moss bags as monitors of atmospheric deposition. **J. Air Pollut. Control Assoc.**; 1981; **31:** 668-670.

97. Ter Haar, G. L. and Bayard, M. A. Composition of airborne lead particles. **Nature**; 1971; **232:** 553-554.

98. Thrane, K. E. Background in air of lead, cadmium, mercury and some chlorinated hydrocarbons measured in south Norway. **Atmos. Environ.**; 1978; **12:** 1155-1161.

99. Turekian, K. K. and Cochran, J. K. 210-Pb in surface air at Enewetok and the Asian dust flux to the Pacific. **Nature**; 1981; **292:** 522-524.

100. Twomey, S. Atmospheric aerosols: Elsevier; 1977.

101. Walsh, P. R., Duce, R. A. and Fasching, J. L. Consideration of the enrichment, sources and flux of arsenic in the troposphere. **J. Geophys. Res.**; 1979; **C84:** 1719-1726.

102. Wayne, R. P. Chemistries of Atmospheres: Oxford UP; 1985.

103. Wesolowski, J. J., John, W. and Kaifer, B. Lead source identification by multielement analysis of diurnal samples of ambient air. in: Kothney, E. L., Ed. Trace Elements in the Environment; 1973: 1-16.

104. Wiersma, G. B. and Davidson, C. I. Trace metals in the atmosphere of remote areas. in: Nriagu, J. O. and Davidson, C. I., Eds. Toxic Metals in the Atmosphere: Wiley; 1986: 201-266.
105. Willeke, K. and Whitby, K. T. Atmospheric aerosols: size distribution interpretation. **J. Air Pollut. Control Assoc.**; 1975; **25:** 529-534.
106. Winchester, J. W. Transport processes in air. in: Hutzinger, O., Ed. Handbook of Environmental Chemistry: Springer Verlag; 1980; **2A:** 19-30.
107. Zoller, W. H. Anthropogenic perturbation of metal fluxes into the atmosphere. In: Nriagu, J. O., Ed. Changing metal Cycles and Human Health, Dahlem Konferenzen: Springer-Verlag; 1984: 27-41.

CHAPTER 8

THE HEAVY ELEMENTS IN WATER AND SEDIMENTS

INTRODUCTION

The hydrosphere is the biggest area of the earth's surface (see Table 5.1), and is made up of a number of, and in most cases relatively homogeneous, parts (Table 8.1). Some of the significant yearly fluxes between these water systems are: 3.7×10^{19} g of river water goes into the sea, rainfall adds about 3.47×10^{20} g to the oceans and 9.9×10^{19} g to land, whereas evaporation of water is around 3.83×10^{20} g from the oceans and 6.3×10^{19} g from land areas [19]. Rivers transfer their volume of water to the oceans many times a year. The mean residence time of a molecule of water in a river is a few days, whereas in the ocean it is many hundreds of years.

The material in the hydrosphere is in three main phases, dissolved material, particulate material suspended in the water and deposited sediments. Typical concentrations of the heavy elements in the various sections of the hydrosphere have been considered in Chapter 5 (see Figures 5.4, 5.5, 5.8, 5.10 and 5.12, and Tables 5.3, 5.4 and 5.9). The product of the amount of each water pool, and the concentration of the elements in the pool, gives the amount of metal in the pool. Values for lead, cadmium and mercury are listed in Table 8.2.

In addition to the main water types (see Table 8.1) other classes of water are important, and include drinking or potable water, industrial and domestic waste water, sewage and stormwater. Some of these waters are significant sources of heavy element pollution in the hydrosphere.

We will also consider in this chapter the sediments associated with water systems. Sediments are, relative to their mass, the principal carriers of the heavy elements in the hydrosphere [179]. Sediment particles are made up from materials derived from rock, soil, biological and anthropogenic inputs. They also consist of a wide range of particle sizes, including gravels, sand, silt and clay.

TABLE 8.1 The Components of the Hydrosphere

Hydrosphere	Pool (g)	Surface Area (%)
Oceans	1.39×10^{24}	70.8
Pore water in sediments	3.20×10^{23}	
Glaciers/ice	2.92×10^{22}	10
Ground water	8.40×10^{21}	
Lakes	2.30×10^{20}	1
Atmospheric water	1.30×10^{19}	
Rivers	1.25×10^{18}	

Sources; references 19,34,154,155

TABLE 8.2 Levels of Lead, Cadmium and Mercury in the Hydrosphere

Hydrosphere	Lead (g)	Cadmium (g)	Mercury (g)
Oceans	2.7×10^{13}	8.4×10^{13}	41.5×10^{14}
Pore water in sediments	12.0×10^{15}	6.4×10^{13}	
Lakes and rivers	6.1×10^{10}	1.6×10^{9}	0.2×10^{8}
Glaciers	6.1×10^{10}	8.2×10^{10}	
Ground water	8.2×10^{10}	4.0×10^{8}	

THE PROPERTIES OF WATER

The Structure of and Bonding in Water

Molecular water contains two hydrogen atoms covalently bonded to oxygen, with the bonding electrons polarized towards the oxygen. The bent shape of water (bond angle 104.5°) is a consequence of the four valence-shell electron pairs around the oxygen atom, two bonding and two non-bonding, arranged approximately tetrahedrally (Fig 8.1a).

The hydrogen-bond Water has a wide liquid range 273-373 K (0-100 °C), which is due to a relatively strong intermolecular force between water molecules, called the hydrogen-bond. In fully developed hydrogen-bonding, as in ice each oxygen atom has four neighbouring hydrogen atoms (Fig. 8.1b). This structural feature persists, to some degree even near 100° C. The hydrogen-bonding is the reason for the elevated melting and boiling points of water (Fig 8.2) compared with the other hydrides of Group VI (Group 16).

The hydrogen-bond is an electrostatic attraction between electron lone-pairs on the oxygen atoms and the partial positive charge on the hydrogen atoms of adjacent water molecules. The hydrogen-bond energy in water is around 25 kJ mol^{-1}. Two important environmental properties of water, which are a consequence of the hydrogen-bond, are that ice floats on water between 0-4° C, and large masses of water are good heat sinks. Ice has a very open structure as shown in Fig. 8.3, with hexagonal shaped channels throughout the

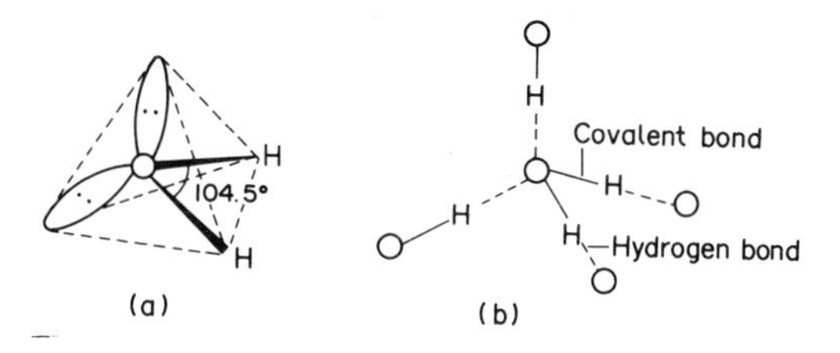

Fig. 8.1 (a) The shape of the water molecule. (b) Hydrogen bonding in water.

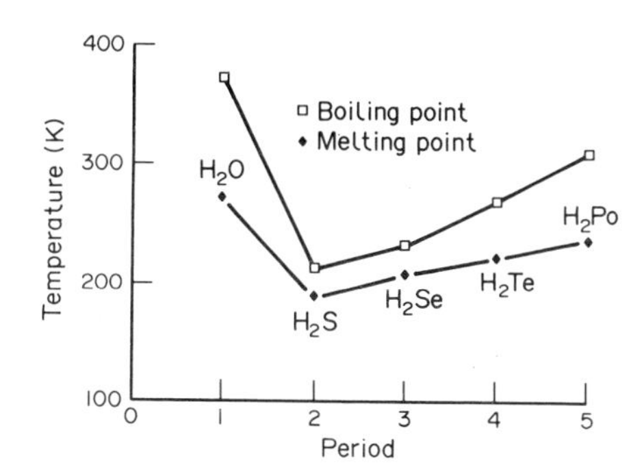

Fig. 8.2 The melting and boiling points of the hydrides of Group VI (16) elements

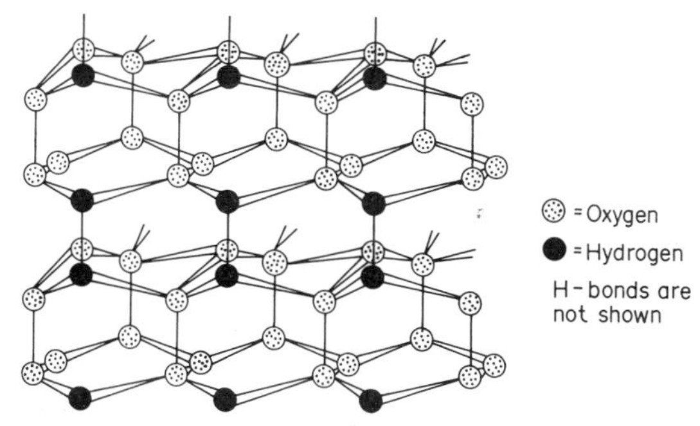

Fig. 8.3 Partial structure of ice (hydrogen atoms are not shown)

structure. When ice melts some of the water molecules become free (i.e. both hydrogen-bonds have broken) and they fall into the channels. The result is that the mass of ice remains the same, but the volume has decreased, and so the density of the water up to 4 °C is greater than the density of ice, and the ice floats. When heat energy is absorbed by large water bodies, some of the energy goes into breaking hydrogen-bonds rather than increasing the temperature of the water. Therefore water has a high heat capacity. When energy is released from water, hydrogen-bonds will reform without causing a big drop in the temperature of the water. Hence large masses of water act as heat sinks, and moderate the temperature of the earth's surface

Some Physical Properties of Water

Some of the physical properties of water, which are important in the environment, are listed in Table 8.3. Many of the properties are anomalous because of the hydrogen-bonding [106]. Water has a dipole moment of 6.1×10^{-30} Cm, and a high dielectric constant of 78.5. The polarity and high dielectric constant of water are reasons why it is a good solvent, especially towards ionic materials.

TABLE 8.3 Some Anomalous Properties of Water

Property	Comparison with other liquids	Environmental significance
State	Liquid rather than gas	Provides a media for life
Heat capacity	High	Moderates environmental temperatures, good heat sink
Latent heat of fusion	High	Moderating effect on temperatures
Latent heat of vaporization	High	Moderating effect on temperatures, affects precipitation/evaporation
Density	Maximum at 4° C	Water freezes from top, so aquatic life can survive winters
Surface tension	High	Important for surface effects, e.g. droplet formation
Dielectric constant	High	A good solvent for ionic and polar compounds
Hydration	Extensive	A good solvent, mobilization of ionic pollutants
Dissociation	Very small	Provides a near neutral medium, but a good medium for acid/base properties
Heat conduction	High	An important heat transfer material in cells

Source of data; Horne, 1978 [106].

Water as a solvent When a solid dissolves in a liquid three enthalpy changes occur: the solvent-solvent interaction is partly broken down by the solute, the lattice enthalpy of the solute is overcome and solute-solvent interactions are formed (solvation enthalpy). Because of hydrogen-bonding, water is a poor solvent for weak or non-polar solutes, but for ionic compounds, the solute-solvent interactions in water are relatively strong ion-dipole coulomb attractions, and explain why water is a good solvent in this case.

The ratio, charge/radius (z/r), is called the ionic potential, and can be used as a guide to the solubility of an ion. The greater the ratio (high charge and/or low radius) the more likely it is that a substance will be soluble. However, the lattice enthalpy will also be high, which reduces the solubility of a substance. An ionic potential diagram (i.e. plot of z against r) is given in Fig. 8.4 for some of the heavy elements. The $z/r = 0.03$ line divides the more soluble from the less soluble species.

The force of attraction (F) between oppositely charged ions (q_+ and q_-) is given by the coulomb equation:

$$F = \frac{q_+ q_-}{4\pi\varepsilon r^2}.$$

The dielectric constant (ε) of the medium between the charges, is inversely proportional to F. Since for water ε is large, the attractive forces between dissolved positive and negative ions are reduced by the water molecules interposed between them, which favours solubility.

The thermochemistry of the dissolution of a solid in water, may be represented by the following cycle (the H_2O-H_2O interaction is omitted as it is a common enthalpy term for different solutes).

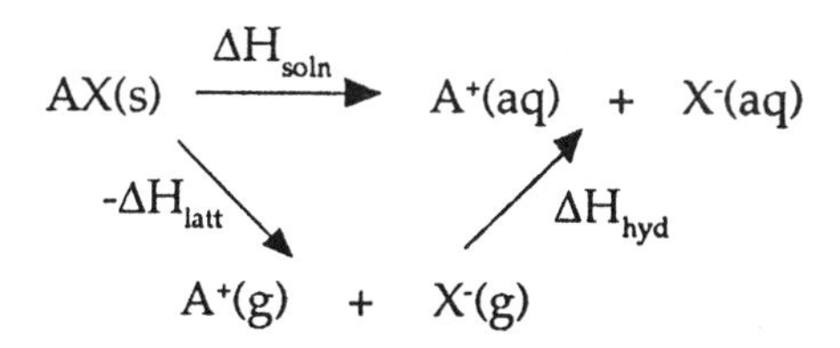

The enthalpy of solution is given by:

$$\Delta H_{soln} = -(\Delta H_{latt}) + \Delta H_{hyd}.$$

For ionic substances both these enthalpies are large, and often the difference ΔH_{soln}, is small.

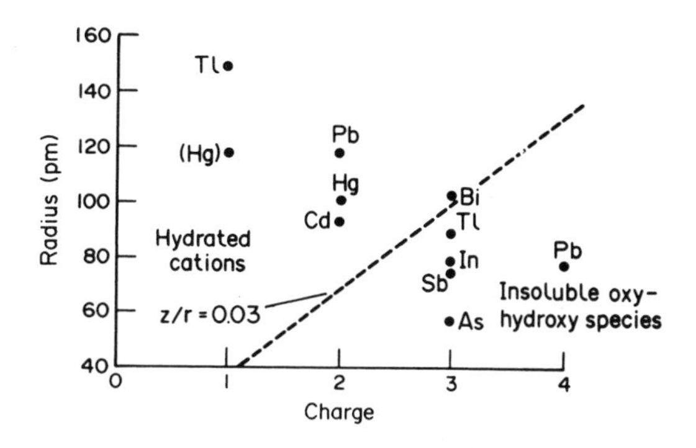

Fig. 8.4 The ionic potential for the heavy elements.

Chemical Properties of Water

Water has a standard enthalpy of formation of -286 kJ mol^{-1}, and a standard free energy of formation of -237 kJ mol^{-1}. The OH bond is strong with an energy of 464 kJ mol^{-1}. Water can be oxidized to O_2, and reduced to H_2, the process is being pH dependent (see Chapter 3). However, water has a wide stability range which is an important feature as regards the existence of species in an aqueous environment.

Acidity of water Water has a small, but important self-ionization;

$$2H_2O_{(l)} \rightleftharpoons H_3O^+_{(aq)} + OH^-_{(aq)},$$

where $K_w = [H_3O^+][OH^-]$ mol^2 l^{-2} = 10^{-14} at 298K.

By definition pure water is neutral, and $[H_3O^+] = [OH^-] = 10^{-7}$ mol l^{-1}, hence its pH = pOH = 7. Water is a good medium for acid-base reactions, strong acids are completely dissociated (0.1 mol l^{-1} HCl contains 0.1 mol l^{-1} of H_3O^+, i.e. a pH of 1), whereas weak acids undergo partial dissociation, e.g.

$$CH_3COOH + H_2O \rightleftharpoons CH_3COO^- + H_3O^+, K = 1.85 \times 10^{-5}.$$

As numerous substances are dissolved in both fresh and salt water, the pH of the water will not be 7. For example surface sea-water has a pH near 8.1 due to dissolved CO_2, existing principally as the HCO_3^- ion, as shown by the distribution diagram for CO_2 in water in Fig 8.5.

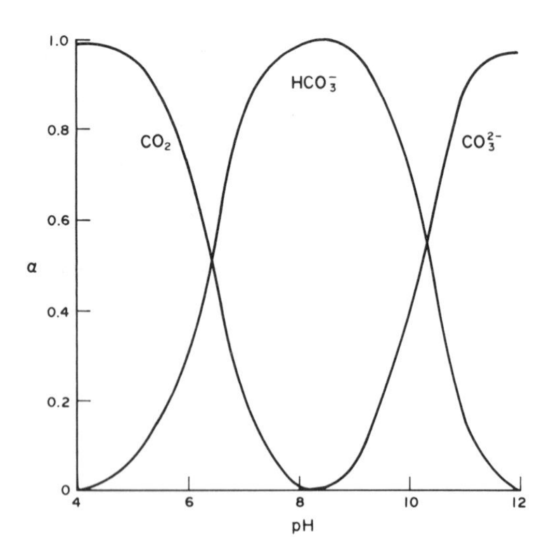

Fig. 8.5 Distribution diagram for carbon dioxide in water.

Hydrolysis Hydrolysis is an important process in the environment and is a reaction where one, or both, of the O-H bonds of water are broken. Metallic cations are generally solvated first, either through ion-dipole interaction (e.g. $[Mg(H_2O)_6]^{2+}$), or by bond formation (e.g. $[Cu(H_2O)_4]^{2+}$), followed in some cases by hydrolysis. For cations of the heavy elements if hydrolysis takes place acid solutions are produced;

$$M^{n+} + xH_2O \underset{\text{solvation}}{\rightleftharpoons} [M(H_2O)_x]^{n+} \underset{\text{hydrolysis}}{\rightleftharpoons} [M(H_2O)_{x-1}OH]^{(n-1)+} + H^+(aq)$$

$$[M(H_2O)_{x-1}OH]^{(n-1)+} \underset{\text{hydrolysis}}{\rightleftharpoons} [M(H_2O)_{x-2}(OH)_2]^{(n-2)+} + H^+(aq)$$

The process may stop at the formation of either a neutral hydroxide or a charged hydroxy or oxy-species. The extent of hydrolysis is influenced by the charge on the cation, its radius and the pH. The process may be represented as follows;

$$M^{n+} + xH_2O \xrightarrow{\text{solvation}} \underset{\text{hydrate}}{M^{n+} - - \overset{\delta-}{O}\langle^{H}_{H}} \;\; \delta+$$

$$\downarrow$$

$$[M\text{-}OH]^{(n-1)+} + H^+(aq) \xleftarrow{\text{hydrolysis}} \underset{\text{intermediate}}{M - O\langle^{H}_{H}} \;\; n+$$

The ease with which the reaction proceeds will depend on the polarizing power of the cation, i.e. on its charge density or ionic potential. Metals that exhibit variable valency hydrolyse more readily in the higher oxidation states, i.e. where the cation charge is greatest and the radius is smallest.

For a low soluble hydroxide $M(OH)_n$, the solubility product is given by:

$$K_{sp} = [M^{n+}][OH^-]^n.$$

Combing this with K_w and eliminating $[OH^-]^n$ gives;

$$K_{sp} = \frac{[M^{n+}]K_w^{\ n}}{[H_3O^+]^n},$$

which can be re-written as;

$$pH = \frac{1}{n}\log K_{sp} - \frac{1}{n}\log[M^{n+}] - \log K_w.$$

In general, the solubility of hydroxides decrease as n increases. If the concentration of M^{n+} is kept constant, the pH at which precipitation occurs will drop the higher the charge on the cation. Hence, certain cations cannot exist in natural water, at a pH around 8. This is clear from considering the distribution diagrams given in Chapter 3, and the data in Table 8.4.

Most oxides, when soluble, react with water giving either acidic or basic solutions. Basic oxides react as follows (M is a divalent metal);

$$MO + H_2O \rightarrow M(OH)_2 \rightleftharpoons M^{2+} + 2OH^-.$$

This reaction is a direct result of the instability of the oxide ion in water, the reaction being;

$$O^{2-} + H_2O \rightleftharpoons 2OH^-, K > 10^{22}.$$

Acidic oxides react to give acid solutions in water as shown for SeO_3;

$$SeO_3 + H_2O \rightarrow (HO)_2SeO_2 \rightleftharpoons HSeO_4^- + H^+ \rightleftharpoons SeO_4^{2-} + 2H^+.$$

Some oxides function as acids and bases (amphoteric) depending on the pH of the solution. A classification of the oxides of the heavy elements according to their reactions with water, or with acids or bases, is given in Table 8.5.

THE PROPERTIES OF WATER BODIES

We will consider some features of water systems that are relevant to the present study. These include the composition of natural water some physical properties, such as the temperature, and the processes of water movement.

TABLE 8.4 The pH of Precipitation of Heavy Metal Hydroxides

Hydroxide	$\log K_{sp}$	pH of precipitation*
$Cd(OH)_2$	-13.8	8.1
$Hg(OH)_2$	-25.2	2.4
$Pb(OH)_2$	-19.8	5.1
$In(OH)_3$	-33	3.7
$Tl(OH)_3$	-43.8	0.1
$Bi(OH)_3$	-30	4.7
TlOH	3.9	12

* Concentration is 0.01 mol l^{-1}.

TABLE 8.5 Acidic and Basic Properties of Heavy Element Oxides

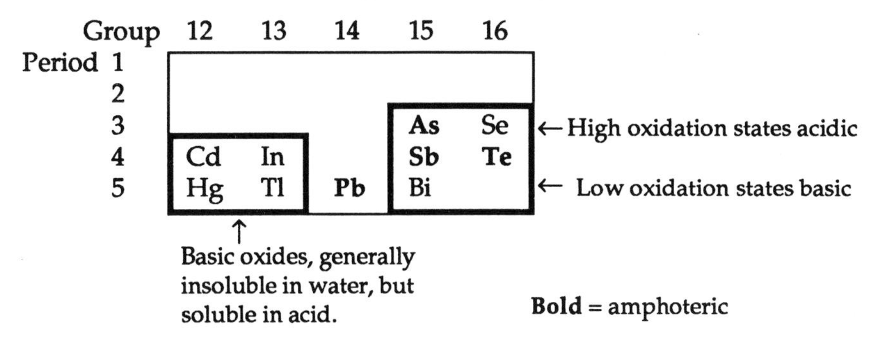

Chemical Composition

The oceans are relatively constant in their composition, as regards the major constituents. Localized differences may occur, especially near to land, due to the input from rivers and the weathering of the sea coasts. The ionic strength of the sea is approximately 0.7 M and the overall composition (i.e. total amount of dissolved material) called the salinity, is usually estimated by measuring the conductivity of the water. The salinity lies between 32–37°/$_{oo}$ with a mean of 35°/$_{oo}$ (parts per thousand). The concentrations of the major species in sea water are given in Table 8.6.

TABLE 8.6 The Composition of Seawater

Cations	Concentration g kg^{-1}	Anions	Concentration g kg^{-1}
Na^+	10.77	Cl^-	19.354
Mg^{2+}	1.29	SO_4^-	2.712
Ca^{2+}	0.412	HCO_3^-	0.142
K^+	0.399	Br^-	0.067
H_3BO_3	0.026	F^-	0.0013
Sr^{2+}	0.008		
SiO_2	0.003		

The pH of sea water is also relatively constant between 7.5-8.3, with most samples lying between 7.8-8.2. The pH is controlled by the bicarbonate ion which is the major inorganic carbon species present in the water (see Fig 8.5) [19,103,136,176,192,207,212].

Fresh water has a variable composition, as the amount of material depends on the location of the water. River water passing through limestone terrain will be high in dissolved carbonate and calcium, whereas in granite areas the levels will be much lower. The median values for the main constituents of fresh water are listed in Table 8.7, and typical concentration ranges are Na^+, 0.7-25 mg l^{-1},

TABLE 8.7 The Composition of Fresh Water

Cations	Concentration mg l^{-1}	Anions	Concentration mg l^{-1}
Ca^{2+}	15.0	HCO_3^-	58.5
SiO_2	13.1	SO_4^{2-}	11.2
Na^+	6.3	Cl^-	7.8
Mg^{2+}	4..4	F^-	0.10
K^+	2.3	Br^-	0.014
Sr^{2+}	0.07		
H_3BO_3	0.015		

Mg^{2+}, 0.4-6.0 mg l^{-1}, Cl^-, 1-35 mg l^{-1} and Br^-, 0.00005-0.055 mg l^{-1}. The pH of fresh water can also vary widely from 1.5–11, depending on the type of environment the water is in [19,103,136,176,192,207,212].

The major cations and anions in sea water are: $Na^+ >> Mg^{2+} > Ca^{2+} > K^+$ and $Cl^- >> SO_4^{2-} > HCO_3^- > Br^-$, whereas in fresh water (using the median values) they are: $Ca^{2+} >> Na^+ > Mg^{2+} > K^+$, and $HCO_3^- >> SO_4^{2-} > Cl^- > Br^-$. Some reasons for the different relative orders for sea and fresh waters are, that the majority of the earth's Na^+ and Cl^- is already in the sea, and the HCO_3^- and Ca^{2+} ions are continually removed from the sea by biological processes, such as shell and coral formation.

Water mixing Processes

Movement of water occurs in oceans, estuaries, lakes and rivers, which allows for mixing. Rivers, because of their greater movement, normally mix rapidly.

The three properties temperature, salinity and density vary with depth in the oceans. The profiles of the first two are shown in Fig 8.6. The temperature profile consists of a warm well mixed area in the top 0-200-500 m. The temperature then drops to around 4-5° C, within a zone called the thermocline, from around 500 m to 1000-1500 m. From the lower depth to the ocean floor, the water has a further slight decrease in temperature. The salinity shows a similar profile, a zone 0-100-200 m well mixed and of uniform salinity, then a decline in salinity, the halocline, occurs from 100 m down to 1000 m, followed by deep water, >1000 m, of almost uniform salinity. The density of sea water increases with an increase in salinity, decrease in temperature and an increase in pressure. The overall result is, the density tends to increase at greater depths, [17,62,176,192,207,212]. The stratification of the oceans disappears around the poles, where cold water exists throughout the profile as shown in Fig 8.7 [62].

Similar thermal and density profiles occur in lakes. Since the maximum density of water occurs at 4° C, warmer water will float on the colder water, giving a thermal stratification in the summer, as shown in Fig. 8.8a. The well mixed top layer is called the epilimnion. Below this is the thermocline, where the temperature falls about 1° C per meter to the deep cold water at a constant

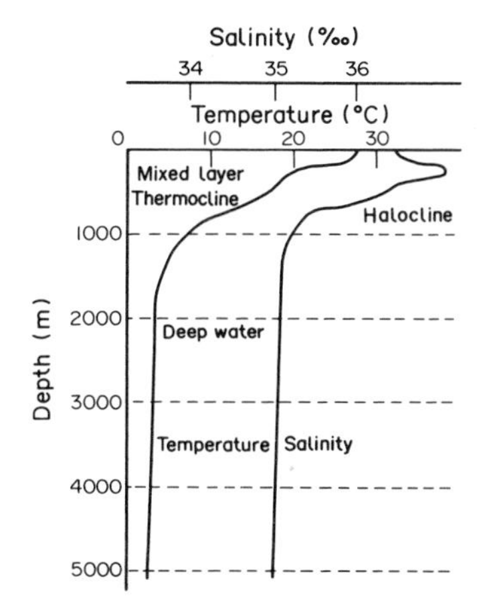

Fig. 8.6 Typical salanity and temperature profiles for the oceans.

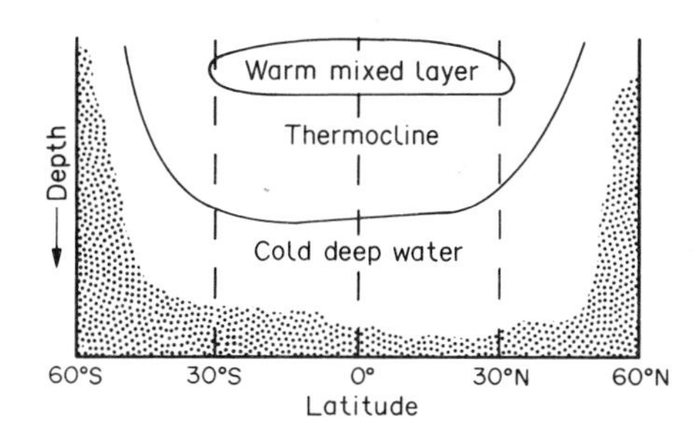

Fig. 8.7 Stratification of the oceans with respect to latitude.

temperature called the hypolimnion. The temperature profile differs in the winter (Fig. 8.8b) with the temperature on the surface falling from 0° to 4°C (if freezing occurs) down to the deeper water also at a temperature around 4° C.

The circulation of water is primarily driven by solar energy, and occurs on the surface, where mixing is activated mainly by winds, and in deep water, where circulation is affected by evaporation and chilling of the water. The top 100 m of the oceans are affected by winds and waves. The world's wind patterns produce a circular movement of currents (gyres), clockwise in the northern hemisphere, and anti-clockwise in the southern hemisphere. The horizontal circulation of main currents (e.g. Gulf Stream) move at speeds of 1 to 5 km h^{-1}. Tides and waves are influenced by the moon's gravitational pull, and wind, earthquakes, ocean land slides and volcanic activity contribute to the wave motion. Deep water circulation is controlled by the density stratification of the water and the influence of the earth's gravity. The density of a water body can alter, either as a result of dilution from a less dense source, or increase as a result of evaporation. Hence a water body will move to a level in the ocean appropriate

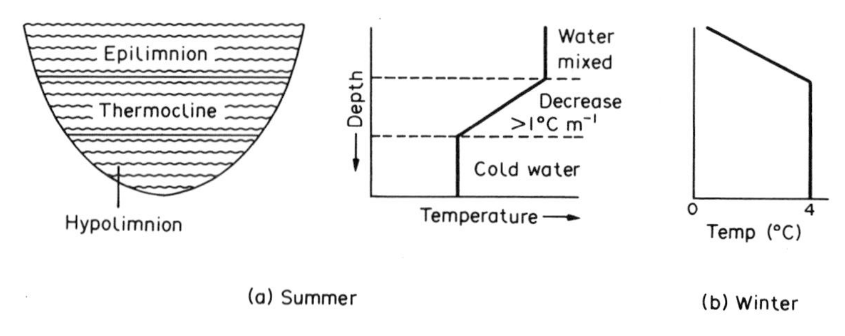

Fig. 8.8 Thermal stratification in a lake, (a) summer, (b) winter.

to its density, and in so doing displace other water. This movement produces the deep circulation, which becomes continuous as the density of water bodies change. The movement is slow and is ≤0.1 km h^{-1}, though evidence is appearing for faster deep water currents of 0.5-1 km h^{-1}. Movement also occurs by upwelling of deep water. Coastal water is blown away from the coast by land based winds and is replenished by the upwelling of cold deep water, this happens, for example, off the coast of Peru [17,62,71,176,192,207,212].

Because of different rates of movement of water bodies through the ocean, the residence time of a water molecule varies. In the mixed top layer it is around 10 years, before moving to the next layer. In the thermocline a molecule may remain for ~ 500 years before moving out of the zone, and in the deep water 2000-3000 years before moving back into the thermocline [62].

Movement of water in estuaries is influenced by tidal currents and river flow. River water will flow over the more dense saline water, whereas the tides will cause mixing to occur. There are three main categories of estuary water circulation, a salt wedge with virtually no mixing, partially mixed and well mixed. The physical dimensions of the estuary, and the earth's rotation as experienced in the Coriolis force influence the mixing. There is also a movement of the deeper saline water up to the junction of the fresh or diluted salt water in order to balance out the effective loss of salt water [17,18,179,207].

SEDIMENTS

Sediments in rivers, lakes and the oceans are important materials as they are the main location of the heavy elements in the hydrosphere. Sediments have been classified in a variety of ways according to different criteria such as their source, the particle size, and their composition. Some of the more important sources are listed in Table 8.8 [51,106,120,176,207,216].

The size of sediments range from large >256 mm, called boulders down to <2 µm (<0.002 mm) called clay. The more important size groupings are; sand 63-2000 µm, silt 2-63 µm and clay <2 µm [179,207]. The major components include clays, quartz, feldspars, various silicate minerals, gibbsite and calcium carbonate [51,120,179,216].

TABLE 8.8 Classification of Sediments According to Source

Source	Comments
Terrigenous	From weathering of terrestrial rocks, main source of inorganic sediments, ~ 206 x 10^{14} g y^{-1} of solid material
Biogenous	From living material in water systems, the sediments arise from biological skeletal material and main components are SiO_2 (~0.49 x 10^{15} g y^{-1}) and $CaCO_3$ (~1.4 x 10^{15} g y^{-1})
Organic	From the plant and animal material, i.e. soft tissue
Volcanic	From volcanism on the sea floor
Authigenic	Chemical processes and precipitation in the ocean sediments
Cosmogenic	Material from outside the earth's atmosphere

The transport of sediments can be considered under two headings; transport of the material to the oceans, estuaries and lakes, and transport of the material within the oceans, estuaries and lakes. Sediments get into the hydrosphere from the atmosphere, from rivers and streams, from glacial activity and from ground water [51,207]. Within the ocean, and to a lesser extent in lakes, the sediments move with the water, there being a critical velocity of a water current below which a particle will settle, and above which the particle will be transported [51,179]. The sediments can therefore move with the ocean's surface currents, as well as down a decline in the ocean, called a turbidity current, where a slurry of sediment is carried down a slope in the ocean. Deep ocean currents may also transport sediments [51,120,179,207]. Hence factors that influence the movement of water can also affect the movement of sediments.

The deposition of sediments occurs when the energy drops below the level necessary to sustain movement. The settling can be explained by Stokes' law described in Chapter 7 [51,168,207]. Particles of 1 μm diameter settle with a velocity of 0.00025 cm s^{-1}, and take 51 years to fall 4000 m, Whereas, 10 μm particles settle with a velocity of 0.025 cm s^{-1} and will take 185 days. However, the deposition of particles with diameters <2 μm (clays) is retarded by Brownian movement, and large particles do not follow Stokes' law well, because of the turbulence produced by the falling particle. Small clay particles may also flocculate from collisions and cohesion due to their surface charge [15,120,207]. The water content of freshly deposited sediments is high, from 40 to >80%, but water is expelled on compacting, due to the increased weight of new sediments [179].

In estuaries, the movement and deposition of sediments depends on the type of mixing process in the estuary. When a salt wedge is present, most of the solids are carried out to sea in the dilute saline layer. For a partially mixed estuary the sediments move out into the estuary and are then deposited, and for a well mixed estuary, the sediments are deposited near the shore. These different cases are depicted in Fig. 8.9 [18,120,179]. Estuaries are sinks for river

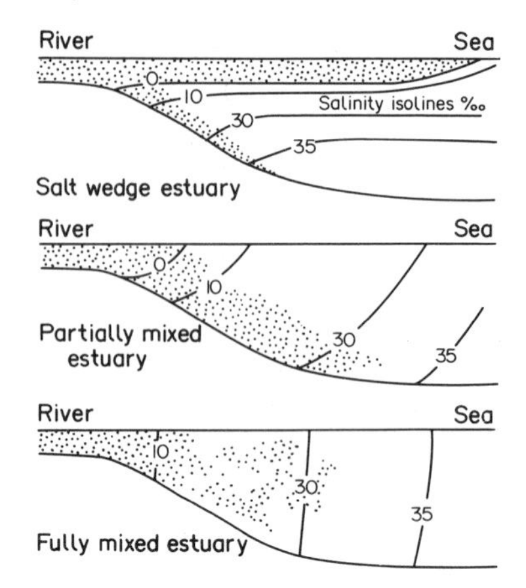

Fig. 8.9 Movement of sediments in different estuary types.

and marine sediments (that enter with the tide), and they are a sheltered environment compared with the oceans.

The transport of pollutants added to the hydrosphere follow closely the processes outlined above for the water and sediments. The biosphere component of the hydrosphere is also important in this respect. The diagram in Fig 8.10 indicates how pollutant materials move within the hydrosphere, and between the various sections [71].

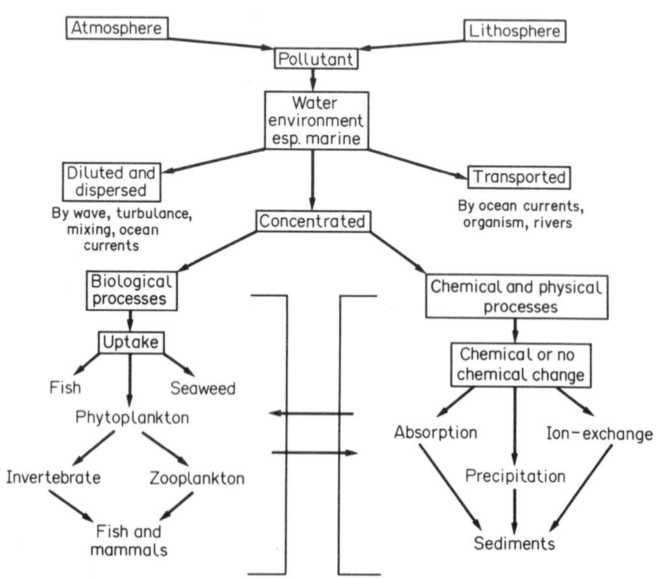

Fig. 8.10 Movement of pollutants in the hydrosphere. Source; Fergusson, 1982 [71].

SILICATES AND ALUMINOSILICATES

The basic structural material of inorganic sediments are silicates and aluminosilicates, because they are the structural materials of the earth's rocks and derived soils. Three features contribute to the wide occurrence of silicates and aluminosilicates. The first is the high abundance of silicon, aluminium and oxygen in the earth's crust (Si 27.7%, Al 8.1%, O 46.6%). The second is the strong Si-O single bond (464 kJ mol^{-1}). Lastly the absence of a Si-O π-bond, of the type found for CO_2, means that Si-O bonds are single and hence polymerisation is a result of the requirements of divalent oxygen. Some of the silicate mineral structural type are shown in Fig 8.11. The basic unit is the tetrahedral SiO_4 group and in the various silicates the oxygen atoms either bridge between the silicon atoms i.e. Si–O-Si, or carry a negative charge i.e. $Si-O^-$. The Si:O ratio indicates the degree of condensation within the Si-O system, the greater the ratio the more developed is the condensation, i.e. the degree of oxygen bridging. Complete oxygen bridging occurs in SiO_2 (silica).

The replacement of some silicon atoms with aluminium is feasible as the atoms are similar in size (metallic radii 117 and 125 pm respectively, a difference of 6.8%). The Al-O bond is also a strong interaction. Some examples of aluminosilicates are given in Fig 8.12. A consequence of replacing Si by Al is an increase in the negative charge on the anion (one per Al), because of the different oxidation states of the two metals. An important class of aluminosilicates are the feldspars which are major components of many rocks (e.g. 2/3 igneous rocks). In a feldspar, the unit $(SiO_2)_4 = Si_4O_8$ can have one or two silicon atoms replaced by aluminium atoms, to give $AlSi_3O_8^-$ and $Al_2Si_2O_8^{2-}$. As for SiO_2, the structures of feldspars are open, with cavities and channels in which the cations reside. There is some ordering of the silicon and aluminium atoms in the structures, which is probably due to the requirements arising from the difference in the Si-O and Al-O bond lengths of 160.5 and 176.0 pm respectively.

The other important aluminosilicates are the clays, in this case the silicon and aluminium atoms are in different 'sheets' of the structures, and the aluminium ions are now in their more usual octahedral coordination. Aluminium may also, however, replace some of the silicon atoms in the silicon 'sheet'. The arrangements within the silicon and aluminium sheets are shown in Fig 8.13 for a 1:1 clay kaolinite and a 2:1 clay pryophyllite. The 1:1 refers to one SiO_4 sheet and one AlO_6 sheet which gives a layer, and the 2:1 layer has two SiO_4 sheets and one AlO_6 sheet. There is also third type, a 2:2 clay. A summary is given in Table 8.9, in which is listed the species that occur between the layers.

The feature of clays which is of significance regarding the heavy elements, is their ability to absorb metals ions by their outer sheath of hydroxyl groups.

$$\text{nClay-OH} + M^{n+} \rightleftharpoons (\text{Clay-O})_n\text{-M} + nH^+.$$

The clay surface may also be negatively charged, providing absorption sites for metal ions.

Structures	Formula	Name	Example
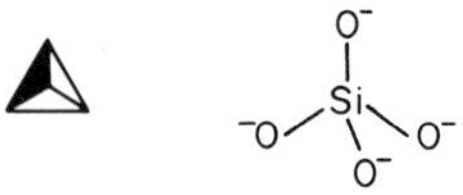	SiO_4^{4-} 1:4	Orthosilicate (neosilicate)	Olivine $(Mg, Fe)_2SiO_4$ Zircon $ZrSiO_4$
	$Si_2O_7^{6-}$ 1:3.5	Pyrosilicate (sorosilicate)	Thortvetite $Sc_2Si_2O_7$
	$Si_3O_9^{6-}$ $Si_6O_{18}^{12-}$ 1:3	Closed ring silicates 3 and 6 membered (cyclosilicate)	Benitoite $BaTiSi_3O_9$ Beryl $Be_3Al_2Si_6O_{18}$
	$(SiO_3^{2-})_n$ Chains	Pyroxenes (inosilicate)	Diopside $(Ca, Mg)SiO_3$ Enstatite $Mg_2Si_2O_6$
	$(SiO_3^{2-})_n$ Chains 1:3		Wollastonite $Ca_3Si_3O_9$
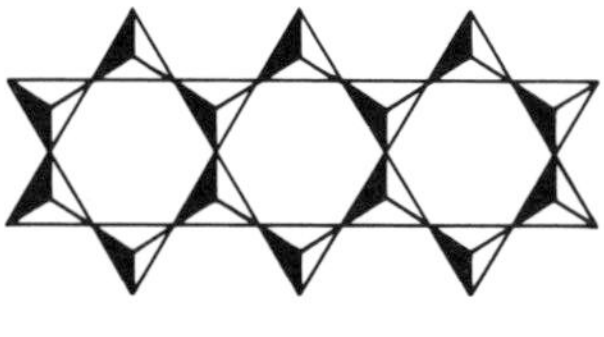	$(Si_4O_{11}^{6-})_n$ Double chains 1:2.75	Amphiboles (inosilicate)	Tremolite $Ca_2Mg_5(Si_4O_{11})(OH)_2$ Crocidolite $Na_2Fe_3^{2+}Fe_2^{3+}Si_8O_{22}(OH)_2$
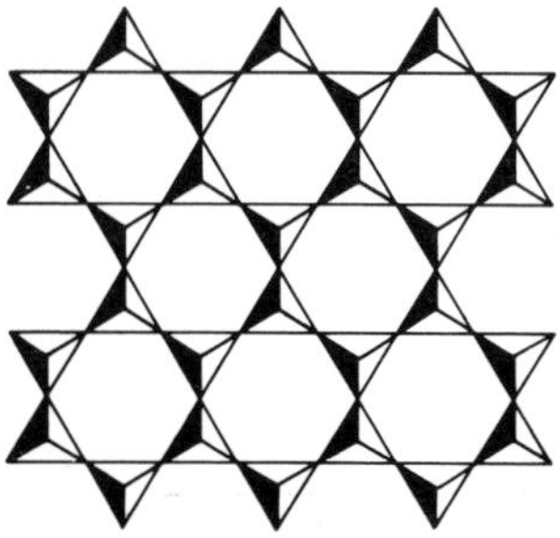	$(Si_4O_{10}^{4-})_n$ or $(Si_2O_5^{2-})_n$ Sheets 1:2.5	Infinite sheet silicate (phyllosilicate)	Talc $Mg_3(Si_4O_{10})(OH)_2$ Chrysotile $Mg_3(Si_2O_5)(OH)_4$
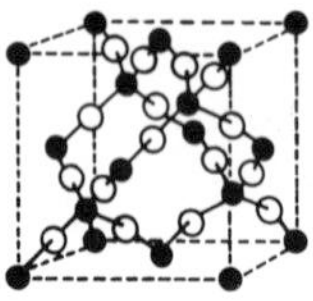 Cristobalite	$(SiO_2)_n$ 1:2	Framework silicate (tektosilicate)	Quartz Cristobalite Tridymite

Fig. 8.11 Structures of silicates.

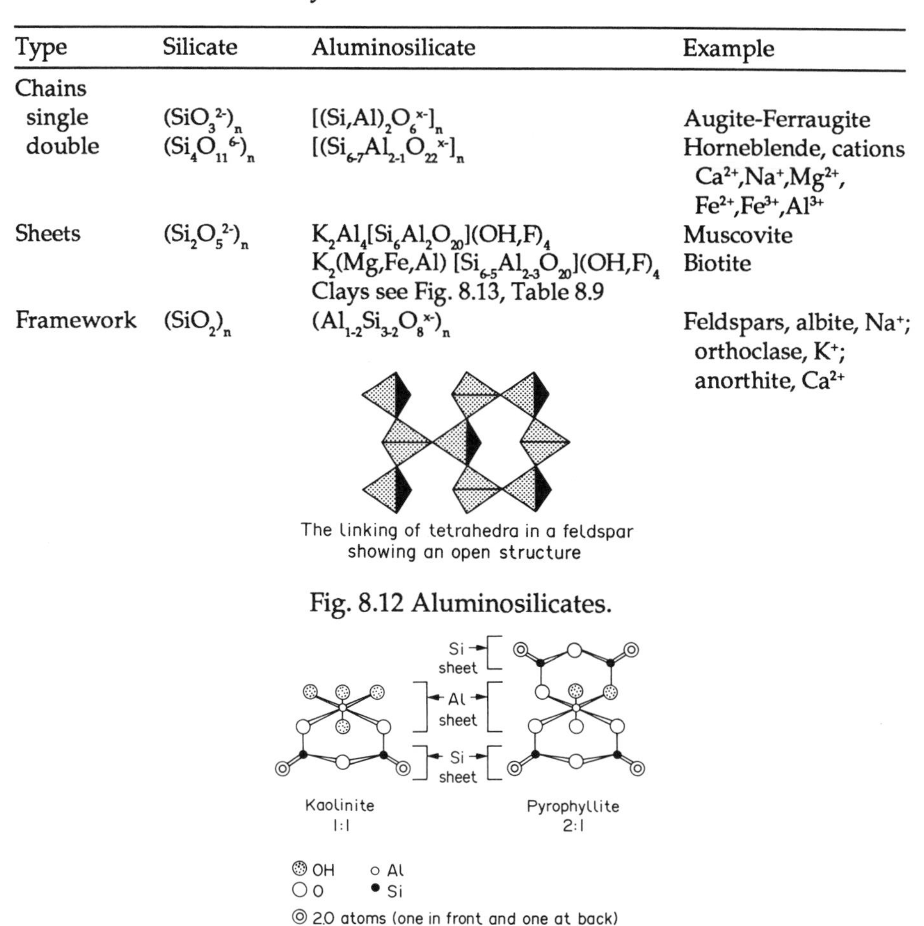

Type	Silicate	Aluminosilicate	Example
Chains			
single	$(SiO_3^{2-})_n$	$[(Si,Al)_2O_6^{x-}]_n$	Augite-Ferraugite
double	$(Si_4O_{11}^{6-})_n$	$[(Si_{6-7}Al_{2-1}O_{22}^{x-}]_n$	Horneblende, cations Ca^{2+},Na^+,Mg^{2+}, Fe^{2+},Fe^{3+},Al^{3+}
Sheets	$(Si_2O_5^{2-})_n$	$K_2Al_4[Si_6Al_2O_{20}](OH,F)_4$	Muscovite
		$K_2(Mg,Fe,Al)\ [Si_{6-5}Al_{2-3}O_{20}](OH,F)_4$	Biotite
		Clays see Fig. 8.13, Table 8.9	
Framework	$(SiO_2)_n$	$(Al_{1-2}Si_{3-2}O_8^{x-})_n$	Feldspars, albite, Na^+; orthoclase, K^+; anorthite, Ca^{2+}

Fig. 8.12 Aluminosilicates.

Fig. 8.13 A schematic represetation of 1:1 and 1:2 clays.

TABLE 8.9 The Constituents of Clay

Clay type	'SiO_4' sheet	'AlO_6' sheet	Between layers	Name
1:1	4 Si	4 Al	H-bonding	kaolinite
	4 Si	4 Al	water	halloysite
2:1	4 Si	4 Al		pyrophyllite[a]
	3 Si + Al	4 Al	K^+	illite[a]
	4 Si	5 Al + Mg'	Various cations and water	montmorillonite[b]
	3 Si + Al	6(Al,Mg,Fe)	Ca^{2+}, Mg^{2+}	vermiculite
2:2	3 Si + Al	6(Al,Mg,Fe)		chlorite

[a] non-expanding 2:1 clay, [b] expanding 2:1 clay

$$nClay^- + M^{n+} \rightleftharpoons (Clay^-)_n—M^{n+}.$$

The space between the layers in clays is also a place where heavy metal ions may reside.

SOURCES OF HEAVY ELEMENTS IN THE HYDROSPHERE

The heavy elements in the hydrosphere come from natural and anthropogenic sources. Natural materials enter the hydrosphere from the atmosphere and water run-off, and in the case of lakes and the oceans from stream and river inputs. Materials from the atmosphere include windblown silicate dust, volcanic emissions, sea spray, combustion and biological emissions [8,148,151, 220]. The material that enters by run-off comes from the weathering of geological materials, which eventually gets into rivers, lakes and oceans [81,196]

For large water systems (lakes and oceans), there has been debate over the relative strengths of atmospheric and riverine (fluvial) inputs. The atmosphere is considered to be the major source of the heavy elements to the oceans [8,33,129]. Estimates of atmospheric fluxes to the sea surface are given in Table 8.10, together with the ratio of the atmospheric flux to the dissolved riverine flux [33]. A ratio >1 suggests that the atmospheric input is the more significant, however, the particulate material in the river was not included, and if included may reduce the ratio to less than 1. In the case of lead it does appear, because of the high ratio, that the atmospheric input is the main route to the hydrosphere. A comparison of atmospheric inputs with the upwelling from the deeper ocean to the surface layer, indicates that 90% of the lead, and 50% of the cadmium comes from the atmosphere.

Atmospheric inputs to lakes are also important [99,186,188]. In a semi-arid, relatively remote lake, it appeared that river input was greater than atmospheric, on the other hand, the atmospheric input into the feeder streams from runoff from the land was also significant [13]. Atmospheric input to sediments

TABLE 8.10 Atmospheric Fluxes of Heavy Metals to the Sea Surface

Element	Sth Atlantic Bight		North Sea		W Mediteranean	
	Flux*	Ratio**	Flux	Ratio	Flux	Ratio
As	45	2.1	280	1.7	54	
Se			22	1.1	48	
Cd	9	2.7	43		13	
Sb			58		48	
Hg	24	2.2		2.1	5	0.8
Pb	660	9.5	2650	6.8	1050	6.2

* ng cm^{-2} y^{-1}, ** Ratio = (atmospheric flux/riverine flux(soluble))
Source of data; Buat-Mènard, 1984 [33].

in a subalpine pond was estimated to be 40 times the natural weathering input, and the increase had been 4-fold since 1930 [186].

Some of the anthropogenic sources of the heavy elements to the hydrosphere are: industrial, such as mining and smelting, production and use of compounds and materials containing the heavy elements, burning of fossil fuels, leaching of waste dumps, urban run-off, sewage effluent, shipping, waste dumping and agriculture run-off [8,41,79,80,81,100,101,140,144,148,151,175, 195,196].

The effects of mining, and exposing mine tailings to the atmosphere and water, has created high levels of heavy elements, such as arsenic and lead, in waterways as a consequence of subsequent weathering. The weathering of FeAsS, found in gold mine areas, has already been discussed (p148) [27,93,100, 140]. The leaching of tailings in W. Australia, elevated concentrations of lead in water close to the source to >1000 $\mu g\ l^{-1}$, and cadmium to 680 $\mu g\ l^{-1}$. The lead concentration in sediments rose to >9600 $\mu g\ g^{-1}$ [132]. The mobilization of the lead probably arises from the oxidation of the S^{2-} ion to sulphate.

$$S^{2-} + 4H_2O - 8e \rightarrow SO_4^{2-} + 8H^+,$$

$$2O_2 + 8H^+ + 8e \rightarrow 4H_2O,$$

i.e.

$$PbS + 2O_2 \rightarrow Pb^{2+} + SO_4^{2-}.$$

In restricted areas, such as bays and bights, it is possible to make reasonable estimates of the inputs to the water, as shown by the data for cadmium, mercury and lead in the New York Bight in Fig 8.14 [81,144]. The results show that barge dumps are the significant source for cadmium and lead, and municipal waster water for mercury. Urban run-off (both metered and un-metered) is also a

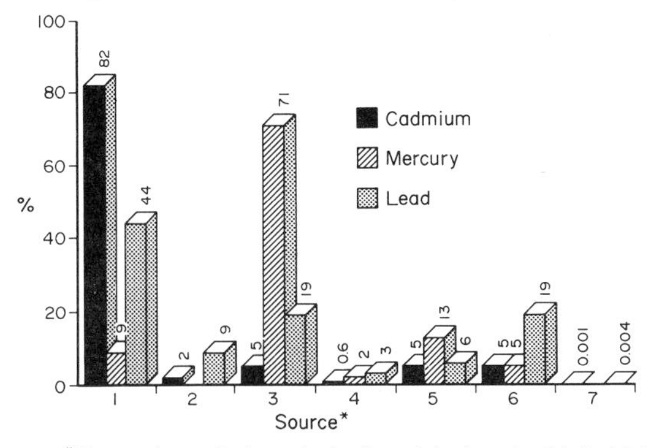

Fig. 8.14 Heavy metal sources to the New York Bight. Sources of data; Förstner, 1986 [81], Mueller et al., 1976 [144].

reasonably significant source, especially for mercury and lead. The high urban run-off for lead, 19%, is probably street dust containing deposited car lead [41,47,144]. Cadmium inputs to the south basin of Lake Michigan is made up of 7.5 tonne y^{-1} (60.9%) from the atmosphere and 3.1 tonne y^{-1} (29.3%) from run-off via tributaries. Of the total input of 10.6 tonne y^{-1} approximately 4.1 tonne y^{-1} can be accounted for in outflow and sedimentation. This means the metal is accumulating in the lake over time [145].

LEVELS OF HEAVY ELEMENTS IN THE HYDROSPHERE

We will now examine the concentrations of the heavy elements in the various sections of the hydrosphere. The concentration units used in the literature have been normalized in order for comparisons to be made. Concentrations in the water are given per litre, and 1 kg of water has been taken as 1 litre.

Oceans: Water

Lead The lead concentrations in the ocean, appear to have fallen over the years as shown in Fig 8.15 for surface lead levels. This is not, however, because of a fall in the amount of lead entering the ocean, but due to improvements in analysis, and more particularly in avoiding contamination of the samples during collection and preparation for analysis [10,35,39,181].

Surface lead levels in the open ocean are around 15 ng l^{-1} (or less) [10,34,35,39, 79,133,181,182,184,195,209]. A variation with the location of the sample has been observed, such as North Atlantic 36 ng l^{-1}, North Pacific 11-14 ng l^{-1} and South Pacific 2.5 ng l^{-1} [74,133,184]. A transect from the North to the South Pacific shows a fall in the surface lead levels from 13 ng l^{-1} at 30° N (160° W) to 8.2 ng l^{-1} at 10° N to 4.1 ng l^{-1} at 17° S [74,184]. Levels in the deep ocean fall to as low as 1 ng l^{-1} at 2000-2500 m deep [39,74,76,79181,182]. A typical profile of lead levels from the central North Pacific Ocean is given in Fig 8.16 [182].

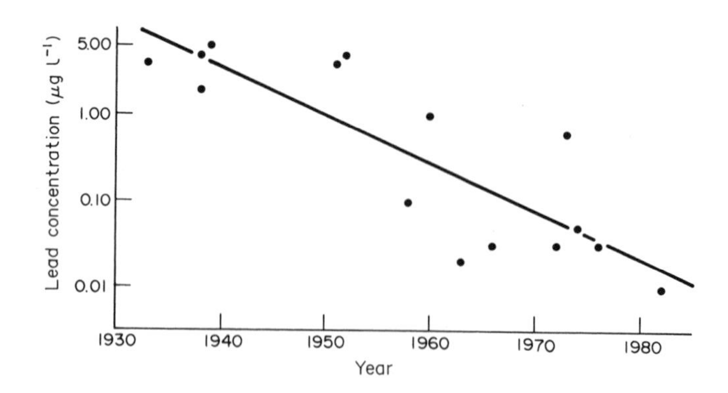

Fig. 8.15 Surface ocean lead concentrations with respect with time. Sources of data; references 10,35,39,181

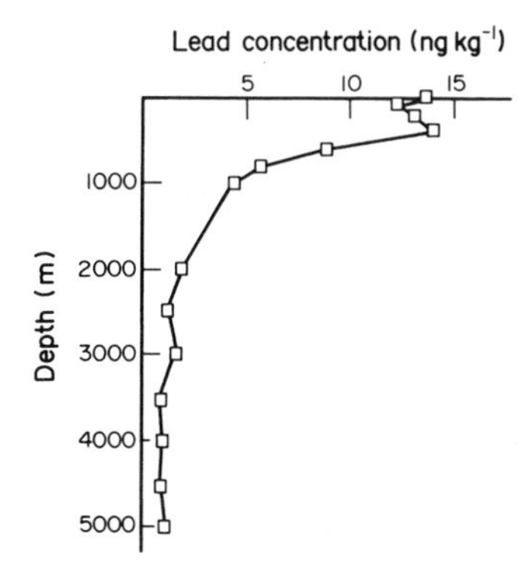

Fig. 8.16 Lead concentration profile in the central North Pacific Ocean. Source of data; Schaule and Patterson, 1981 [182].

The reason for the surface variation in lead levels and the fall in levels with depth, is because lead enters the ocean mainly from the atmosphere. A comparison of present day and prehistoric inputs of lead to the ocean given in Fig 8.17 [181], highlights the elevated present day inputs. Because of the increase in the residence time of lead with depth, the levels in the deep ocean are close to true prehistoric concentrations [181]. Estimates of the lead concentrations, and excess over prehistoric at different depths are as follows [181,182]: at the surface, 5-15 ng l^{-1} 10 fold excess; 100-900 m, 10 ng l^{-1}, 2 to 5 fold excess and 900-5000 m, 1 ng l^{-1} 2 fold excess. Estimates of present day atmospheric lead fluxes from industrial sources in different parts of the world, correlate strongly with surface lead levels as shown by the following data [74,182,184].

Location	Flux		Surface lead
	present ng cm^{-2} y^{-1}	prehistoric ng cm^{-2} y^{-1}	ng l^{-1}
Sth. Pacific westerlies	3	0.3	2.5
Nth. Pacific easterlies	6, 7	0.3	11
Nth. Pacific westerlies	50	0.3	14
Nth. Atlantic westerlies	170	3	36

Concentrations are usually higher in areas closer to the land, and in confined areas such as in harbours and channels, because of additional sources of lead such as urban run-off and sewage [10,34,39,79,81,134,140,166,175,182,195]. Concentrations of 100-200 ng l^{-1} have been found in the Ligurian and Tyrrhenian Seas, and 2000 ng l^{-1} near Genoa in shipping lanes [10,79,134]. Levels as high as 5000-10,000 ng l^{-1}, located near the Belgium and Dutch coasts, are associated with dumping grounds [10,134]. In the Red Sea brines and in hot brines, lead levels are quite elevated, 80-560 μg l^{-1} and 30,000-80,000 μg l^{-1} respectively [39].

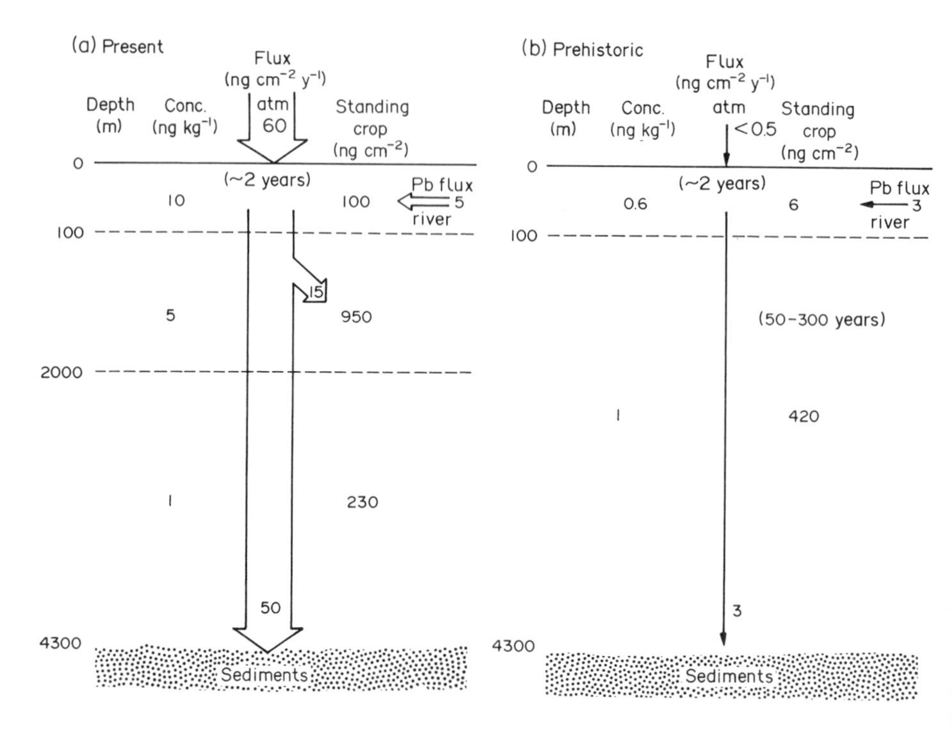

Fig. 8.17 Oceanic inputs and outputs, (a) present time, (b) estimated prehistoric. Source of data; Schaule and Patterson, 1980 [181].

Cadmium The concentrations of cadmium in the surface of the open ocean are <10 ng l^{-1} (e.g. 3.9, 6.6, 3-4 and 5-9 ng l^{-1}). The levels then increase with depth to around 500-1500 meters, to values reported as 112, 92, 110, 25 ng l^{-1} for different environments. The levels then stay relatively constant or drop slightly at greater depths [14,21,30,31,35,61,76,133,135,140,166,171,193,195,218]. A number of detailed profile studies of cadmium have been carried out confirming these changes [28,30,171], and one example is shown in Fig 8.18 [28]. Similar profiles occur for the nutrient species PO_4^{3-} and NO_3^- (Fig 8.19a, b) [28], suggesting a correlation between the metal and the nutrient species (correlation coefficients are around 0.9) [14,21,28,31,35,79,117,133,135,140,171]. The maximum levels of cadmium, phosphate and nitrate are also associated with a minimum value in the level of dioxygen dissolved in the water [135]. A weaker association exists between Cd^{2+} and silicate [171,218].

An explanation of the similar cadmium, phosphate and nitrate profiles, is that in the surface water the levels are depleted because the species are taken up by organisms and organic matter. Whether the cadmium is taken up by a growing organism, or by sorption onto the organism is not known. In deeper

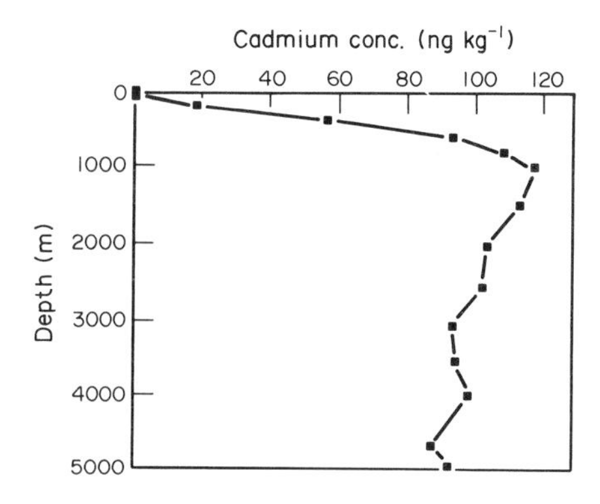

Fig. 8.18 Cadmium concentrations with depth into seawater. Source of data; Bruland, 1980 [28].

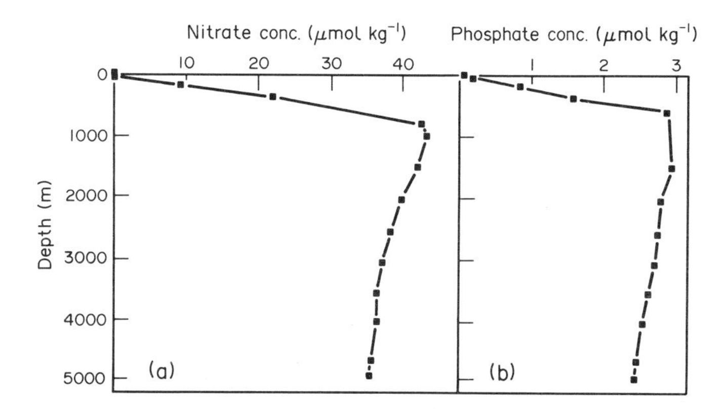

Fig. 8.19 The variation of; (a) nitrate levels with depth, and (b) phosphate levels with depth in seawater. Source of data; Bruland, 1980 [28].

water the organic detritus and debris, which is sinking, releases the cadmium and the nutrients. This corresponds to a minimum in the O_2 level, presumably because the dioxygen is being used in the decay [11,28,79,140,171]. The ratio of cadmium concentrations, deep/surface (enriched/depleted) is ~30, which is large [31]. The atomic ratios for the three elements Cd : P : N are 3.5×10^{-3} : 1 : 15.2 [218]. The higher cadmium concentration at greater depths, is evident because of its elevated levels in upwelling water, compared with non-upwelling surface water, i.e. 9 and 1 ng l^{-1} respectively [21,195].

The concentrations of cadmium tend to be higher in the Pacific than in the Atlantic, which is strange, in that the Atlantic is more associated with industrial countries. A reason is probably the lateral movement of water between the oceans, in the direction; Atlantic → Indian → Pacific for the deep ocean, and

Pacific → Indian → Atlantic in the wind mixed surface ocean [179]. Hence the more concentrated deep water moves towards the Pacific.

Levels of cadmium in the surface seas near coastlines, in shipping lanes and in confined spaces such as harbours and channels are higher. Levels reported for coastal areas are 80-230 ng l^{-1} [61], 12-410 ng l^{-1} [171] and 60-1940 ng l^{-1} [81]. In the Ligurian and Tyrrhenian seas levels are reported as: low, 5-9 ng l^{-1}; elevated, 21-50 ng l^{-1} and in shipping lanes as high 52-452 ng l^{-1} [79,81,134,171].

Mercury Levels of mercury in the open ocean lie around <0.01-0.03 μg l^{-1} (<10-30 ng l^{-1}) [26,35,73,108,140]. In polluted areas, such as in the Mediterranean Sea or New York harbour, levels can rise to >1μg l^{-1} [166,175]. In the deep ocean a wide range of values are reported i.e. 0.07-1.1 μg l^{-1} [35,73,76,166,193]. A possible reason for the high values is under-water volcanism, as mercury deposits are clustered around areas of hydrothermal activity.

Selenium The surface sea water levels of selenium are around 100 ng l^{-1} [26,35,48,76], with an increase to 200-250 ng l^{-1} at greater depths [48]. The results of a detailed study of dissolved (~80% of total) selenium levels and the speciation (organic selenium, selenite, Se(IV), and selenate, Se(VI)) are presented in Fig 8.20 [48]. It appears that organic selenium is the main species in the surface water (~63 ng l^{-1}) down to 100 m (~24 ng l^{-1}). Selenite is low in the surface water (~4 ng l^{-1}) rising to around ~40 ng l^{-1} at 600 m, and selenate is also low in surface water (~10 ng l^{-1}), but becomes the dominant form at a depth around 100 m (~79 ng l^{-1}). In the deeper ocean both selenite (~79 ng l^{-1}) and selenate (~100 ng l^{-1}) are around the same concentrations, with selenate a little higher. Particulate selenium, which decreases with depth, appears to be in the selenide (-II) form. Selenium(IV) concentrations in the Atlantic Ocean correlate more closely with silicon than phosphate, whereas the reverse is occurs for selenium(VI) [35]. Nearer to the coast the surface values are elevated, 300-600 ng l^{-1}, presumably due to pollution inputs [26].

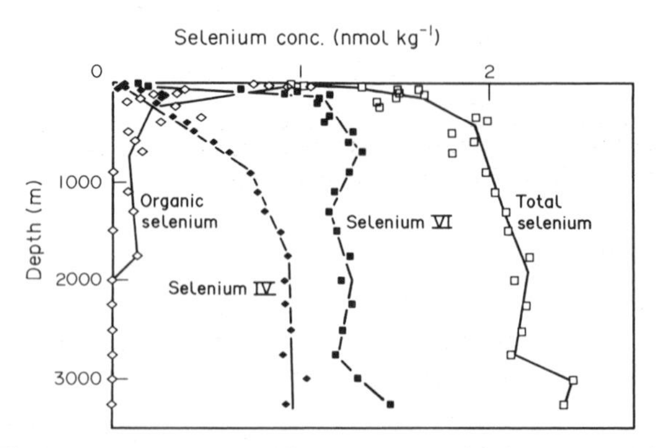

Fig. 8.20 The concentrations and speciation of selenium with depth in seawater. Source of data; Cutter and Bruland, 1984 [48].

Arsenic The concentrations of arsenic in the open ocean are higher than for the elements discussed above, and are around 2 μg l^{-1} (2000 ng l^{-1}), with a range of 0.15-6 μg l^{-1} [26,35,70,76,193,195]. The levels increase with depth, by a factor of 1.5 [26]. Arsenic is present mostly as As(V).

Remaining elements For the remaining elements, much less information is available on concentrations in the open ocean. Typical ocean levels are listed in Table 8.11, together with values for the elements discussed above. Often levels are elevated near coastlines, e.g. a concentration of 1.2 ng l^{-1} is reported for indium in the Irish Sea [26]. The concentrations of Se, Sb, Hg, and Pb in the marine atmosphere and suspended particles in the deep ocean, suggest that the elements are enriched in both materials compared with crustal abundances. The atmospheric flux may therefore control the deep water particulate enrichments of Pb, Se and Sb [8]. The concentrations of the elements in ng l^{-1} and the enrichment factors in the sea water particulate are Pb 8.6, 510; Se 0.19, 2690; Hg 0.4, 3740; and Sb 0.16, 560. For the last two elements the enrichments are said to be natural rather than anthropogenic. The four elements As, Se, Cd and Hg all show an increase in their concentrations with depth in the oceans. But in the case of lead the reverse is true, indicative of the high inputs of the metal to the surface ocean, swamping any other effects that may be occurring.

TABLE 8.11 Levels of the Heavy Elements in the Open Ocean

Element*	Surface level	Deep water level
Pb	5-15 ng l^{-1}	<1-2 ng l^{-1}
Cd	3-10 ng l^{-1}	25-112 ng l^{-1}
Hg	10-30 ng l^{-1}	70-1100 ng l^{-1}
Se	<10-100 ng l^{-1}	160-190 ng l^{-1}
As	~2000 ng l^{-1}	~3500 ng l^{-1}
Bi	15-20 ng l^{-1}	decreases
In	~0.11-0.31 ng l^{-1}	decreases
Sb	180-5600 ng l^{-1}	relatively constant
Tl	10-20 ng l^{-1} (high?)	

* Te no data? Sources of data; Brewer, 1975 [26], Burton and Stratham, 1882 [35], and in the text.

Oceans: Sediments

Concentrations of the heavy elements in ocean sediments vary considerably with geographical location, and depend on whether the sediments are deep sea or coastal. Coastal levels are often significantly elevated because of nearby land based pollution sources. For example, near an outfall drain in Los Angeles, the coastal surface sediments contain 1.1-66 μg g^{-1} Cd, 0.13-5.4 μg g^{-1} Hg and 19-576 μg g^{-1} Pb. The highest values are nearest the drain outlet, and the contamination (enrichment) factors for the three elements are 36, ~23 and 17 respectively [105]. Deep sea sediments, and sediments close to the coast, may be similar or different in composition depending on whether or not an element is deposited rapidly on entering the ocean, or is carried out into the ocean before depositing.

Lead In a similar manner to the lead profiles in the sea water, the concentrations of lead are highest near the top of coastal sediments, falling off at lower depths. A dated lead sediment profile from the Santa Monica basin in the Southern Californian bight is shown in Fig 8.21 [150]. The sediments were leached with 1M HNO_3, which was said to remove non-mineral, that is pollution lead. The concentration of the remaining lead, called natural, was around 5 to 6 $\mu g\ g^{-1}$ [2,12,34,150]. The most recent sediments (1972-76) have less lead in them then material a little lower in the profile (1962-64). The [Pb]/[Al] concentration ratio, increases towards the top of the profile. If it is assumed that the aluminium derives from the detrital minerals, and the input has remained constant over the years, it would appear that the lower levels of lead in the surface sediments are due to dilution, and not to a decrease in the lead input [12]. Approximately 16% (160 tonnes y^{-1} per 1200 km^2) of the lead input into the basin is tied up with sediments in the basin [34], and around 76% of the lead input has a pollution source [29]. Higher concentrations of lead near the coast (e.g. 20-320 $\mu g\ g^{-1}$), compared with lower values in the open ocean, suggests that lead is rapidly deposited [155,160]. An average lead level of 47 $\mu g\ g^{-1}$, reported in pelagic sediments is higher than levels in rocks and is probably due to underwater volcanic activity [157].

Cadmium Background levels of cadmium in oceanic sediments lie in the range 0.13-0.6 $\mu g\ g^{-1}$ [14,46,79,80]. Much higher levels are reported in sediments close to coasts, such as Swansea Bay, Wales, 11-25 $\mu g\ g^{-1}$, and Santa Monica Bay 1.7 $\mu g\ g^{-1}$ [29,79,80]. The high levels are due to sewage outfall, smelting/plating processes and mineralized areas.

Mercury Mercury levels in Minamata Bay sediments were highly elevated over the background, being 2010 $\mu g\ g^{-1}$ at the sewer outfall, and 130 $\mu g\ g^{-1}$ a few hundred meters away. Background levels are around 0.1-0.35 $\mu g\ g^{-1}$ [79,81,86]. High enrichments have also been reported in Santa Monica Bay, and in this case the levels fall from around 0.16 $\mu g\ g^{-1}$ at the top of the sediment profile (dated 1960-70) to 0.06 $\mu g\ g^{-1}$ (dated 1820-60) to 0.04 $\mu g\ g^{-1}$ (dated 1400-1500 BP) [2].

Other elements Non-carbonate and carbonate based sediments in the East Pacific are reported to contain 0.17 and 0.06 $\mu g\ g^{-1}$ bismuth, 4.8 and 3.0 $\mu g\ g^{-1}$ thallium and 2.6 and 1.3 $\mu g\ g^{-1}$ selenium respectively [46].

Nodules One sedimentary material which enriches trace elements is manganese/iron nodules. The nodules, which are produced by precipitation of the

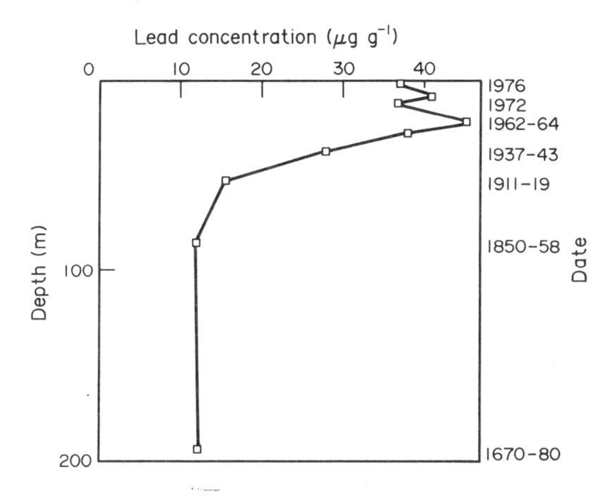

Fig. 8.21 Levels of lead down a sediment profile in the Santa Monica Basin. Source of data; Ng and Patterson, 1982 [150].

hydrous oxides of manganese and iron, are found wide spread over the shallow and deep ocean floor. The heavy metals are probably absorbed onto the oxide particles and become incorporated into the nodules. Some typical levels of the heavy elements in nodules are given in Table 8.12. It is interesting that thallium is so enriched. This maybe because it can exist in the trivalent oxidation state, and like cobalt, become incorporated into the ferric oxide lattice. Also, like cobalt, thallium is not so enriched in nodules with a low iron content [14,46,47, 52,91,157].

Bioturbation In addition to knowing the sedimentation rate in interpreting data, it is also necessary to know if the sediments have been disturbed at all, especially by living organisms - called bioturbation [121]. It is estimated that the top 60 cm of the sediments of Chesapeake Bay are homogeneous because of

TABLE 8.12 Levels of Some Heavy Metals in Manganese/Iron Marine Nodules (µg g^{-1})

Element	Pacific	Atlantic	Indian	World	Crustal abund.	Enrichment factor
Cd	7	11		8	0.11	73
Te	50					
Hg	0.82	0.16	0.15	0.5	0.05	10
Tl	170	77	100	129	0.6	215
Pb	846	127	700	867	14	62
Bi	6	5	14	8	0.05	160

Sources of data; references 14,46,47

bioturbation [12]. Sediment stratigraphy can be blurred, particle sizes altered, pore spaces and pH changed by burrowing organisms. The organism may also bio-accumulate heavy metals and remove them from the sediment profile. All these processes can alter the levels and the speciation of the heavy elements in sediments and it is necessary to be aware of this [121].

Estuary and Coastal Waters and Sediments

Estuaries tend to act as sinks for heavy elements coming from rivers and the atmosphere. Because of their sheltered environment, there is time for chemical and physical processes to occur to sediments before being swept out to sea. In the Gironde estuary around 13% of the lead input is retained, whereas in the Scheldt estuary 93% of the lead input and 90% of the cadmium input are retained [179]. The nature of the estuary can influence the level of heavy metal retention.

An important process for trace elements in estuaries is the mixing of the fluvial and marine waters and sediments. Stable particulate suspensions in fresh water become destabilized in saline water, owing to double layer compression, leading to flocculation. For this to occur the particles have to collide, which depends on the flow rate of the water, and concentration of the particulate material [57,58].

There has been controversy over what happens to the levels of trace elements in fluvial sediments and water on entering a saline environment [5,54,57,79,146,179]. It appears, that for many estuaries the levels of trace metals in river sediments decrease when mixed with marine water. For example, the fall off in concentration of As, Pb, Hg and Cd in the Rhine, with increase in salinity, is shown in Fig. 8.22 [5]. A similar situation occurs for cadmium, lead and mercury in the sediments and water that enter the Elbe estuary as shown in Fig 8.23 [146]. On the other hand mobilization of heavy elements from the sediments into solution can also occur, as suggested from the data from the Gironde estuary. The inflow of lead in tons y^{-1} is 61 and 371 for dissolved and particulate respectively whereas the outflow is 326 and 48 tons y^{-1} for dissolved and particulate respectively [179]. Such behaviour is called non-conservative and follows the negative deviation line in Fig 8.24 [179]. A number of factors operate to produce these differing results. (a) Dilution of the fluvial sediments can occur with less contaminated marine sediments. If this is all that happens the system and the heavy metals of interest are said to have conservative behaviour portrayed by the solid line in Fig 8.24 [179]. (b) Mobilization may be achieved by the mass action effect of increasing concentration of cations in the saline water, i.e.

$$M^{n+}\text{-clay} + nNa^{+}(aq) \rightleftharpoons nNa^{+}\text{-clay} + M^{n+}(aq).$$

The reaction will move to the right as the concentration of Na^+ increases [5]. (c) Decay of organic material in the estuary produces organic ligands that may extract metals from the sediments and take them into solution. It is of interest

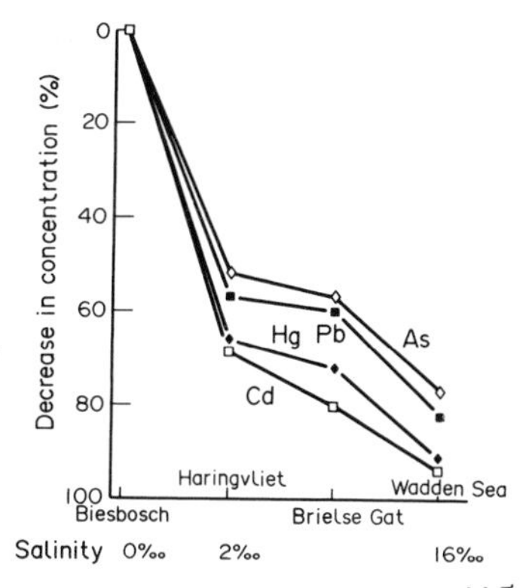

Fig. 8.22 Heavy element concentrations in sediments in relation to salinity. Source of data; Aston, 1978 [5].

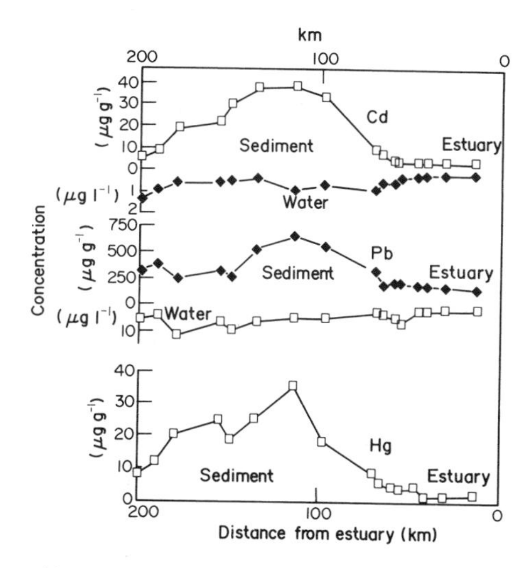

Fig. 8.23 Variation of levels of heavy metals in sediments and water, along the Elbe river, towards the estuary. Source of data; Muller and Förstner, 1975 [146].

that the Irving-Williams series of increasing stability of metal complexes, Mn < Fe < Co < Ni < Cu > Zn, is the same order as increasing mobilization [5,54,79,179]. (d) Alteration of solid species (e.g. organic and clays) on mixing, and change in grain size distribution can also influence metal mobilization. (e) A further factor that can occur is a tailing off of concentrations with distance from the source as deposition occurs. For example levels of cadmium in water in the Bristol Channel fall 5.8, 2.0, 1.0, 0.5 $\mu g\ l^{-1}$ at 4, 25, 60, and 80 km respectively, from the mouth of the Avon river [81]. The same effect is seen for the following data from New York harbour (concentrations in $\mu g\ l^{-1}$) [117].

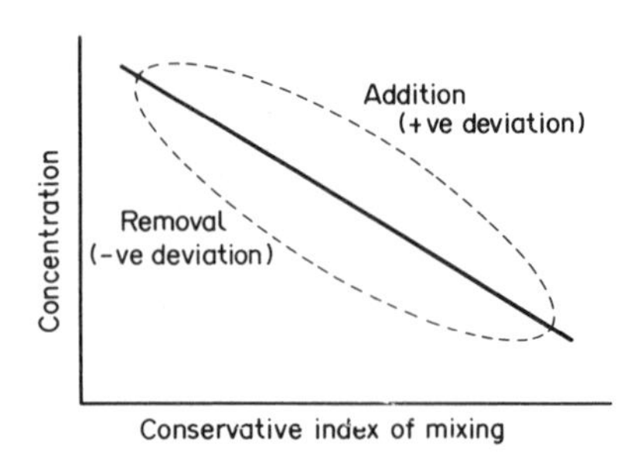

Fig. 8.24 Behaviour of heavy metals when fluvial water and sediments mix with marine water and sediments. Source of data; Salomons and Förstner, 1984 [179].

Element	Plant effluent	Harbour water	Ocean
Cd	10	2.5	0.2
Hg	6.6	4	0.03-0.05

The cadmium and lead levels fall in the sediments in Bedford harbour towards the ocean, i.e. 616, 210 and 104 $\mu g\ g^{-1}$ for lead, and 52, 6.6 and 1.4 $\mu g\ g^{-1}$ for cadmium.

It is not possible to predict or generalize as to what will happen in any individual estuary, that is, which of the above factors will dominate in a particular situation. Negative deviations from conservative behaviour for Pb, Sb and Cd frequently occurs, whereas Hg tends to display conservative behaviour and As has shown a positive deviation. In the Weser estuary the function;

$$\frac{\mu g\ g^{-1}(\text{Cd in sediment})}{\mu g\ ml^{-1}(\text{Cd in solution})}$$

first decreases with increase in salinity, indicating mobility, and then increases with increase in salinity, which may be due to either the concentration in solution decreasing due to dilution with less contaminated sea water or re-absorption on to the sediments [179].

The concentrations of the heavy elements in estuary and coastal waters and sediments vary considerably, from place to place, depending on the inputs. The data presented in Table 8.13 for water and sediments indicate the wide range of levels, especially for lead, cadmium and mercury.

The change in concentrations with time have been recorded for Foundry Cove, when during the 1951-71 period, cadmium was introduced and very high surface levels were reported, e.g. >10,000 $\mu g\ g^{-1}$ in some areas. Since 1971 the concentrations have decreased as indicated in the Table 8.14 [118]. The reason for the fall appears to be, that the contaminated sediments have been buried by

TABLE 8.13 Levels of Some Heavy Elements in Estuary and Coastal Waters and Sediments

Element	Location	Concentration		Ref.
		Water $\mu g\ l^{-1}$	Sediments $\mu g\ g^{-1}$	
Lead	Elbe Estuary	0.02-0.1		134
	Bristol Channel	0.4-5.1		39
	Firth of Forth	0.07-1.1		6
	UK Coastal waters	1.3		202
	Jervis Bay	0.4-1.9*		77
	NY Harbour	0.13-0.7	25-370	37,117
	Poole Harbour	1-5	50-190	20,205
	Conway Estuary	0.6-28	28	64,205
	Restronguet Estuary	250	684	205
	San Francisco Bay		9-174	22
	Sorfjord		720-70,000	191
	Wellington Harbour		26-6740	197
	Newport Bay		132 (max)	41
	Dewent Estuary		1000	15,79
	Rio Tinto Estuary		1600	79
Cadmium	Jervis Bay	0.19-1.1		171
	Foundry Cove	4-32	12-39,000	100,1118
	Bristol Channel	0.5-5.8		81
	NY Harbour	0.8-5.7		117
	UK Coastal waters	o.46		202
	Firth of Forth	0.03-0.25		6
	Mediterranean coast	0.005-6.9		179
	Looe Estuary	<0.1-0.54		32
	Derwent Estuary	15	up to 862	15,79
	Restronguet Estuary	13	3.4	205
	San Francisco Bay		0.5-3.3	22
	Sorfjord		16-850	191
	Bedford Harbour		0.5-130	80
	Corpus Christi		2-130	79,80
Mercury	NY Harbour	0.5-47		117
	UK Coastal waters	0.04		202
	Irish Sea	0.025-0.05		73
	Mississippi	0.08-0.37		73
	San Francisco Bay		0.04-1.05	22
	Derwent Estuary		1130	15,79
	Minamata Bay		130-2010	79,81
Arsenic	Nigeria		4.7-7.6	149
	Restronguet Estuary		900	205
	San Francisco Bay		8-12	22
Antimony	Nigeria		0.78-1.8	149
	Sorfjord		10-1080	191

Table 8.13 continued

Element	Location	Level Water μg l^{-1}	Sediments μg g^{-1}	Ref.
Bismuth	Jervis Bay	0.04-0.21*		77
	Sorfjord		BD-23	191
Indium	Jervis Bay	< 0.05*		77
	Sorfjord		BD-140	191
Thallium	Jervis Bay	< 0.2, < 0.5*		77
Selenium	San Francisco Bay		1.7-2.4	22

* Soluble fraction

TABLE 8.14 Change in Cadmium Levels with Time at Foundry Cove

Cadmium concentration	% of area 1974	1975	1983
< 1000 μg g^{-1}	67.2	67.4	92.7
1000-10,000 μg g^{-1}	25.2	26.5	6.3
> 10,000 μg g^{-1}	7.8	5.8	1.3

Source of data; Knutson et al., 1987 [118].

recent less contaminated sediments. Sediments at the 10-20 cm depth were now the most contaminated with the metal.

River and Lake Water

The levels of the heavy elements in fresh water, as for marine water, have been arbitrarily divided into two components, filterable (≤ 0.45 μm) called dissolved, and suspended matter which eventually produce sediments [112]. The transfer and fate of heavy metals in a lake system is represented in Fig. 8.25 [179,180,188]. Some factors that determine the partition of the metals between the dissolved and solid phases, are: (a) speciation, i.e. the chemical form of the heavy metal, which is influenced by pH, the redox conditions, and what other species are present, e.g. organic ligands; (b) uptake by biota by ingestion or absorption, and probable release with decay at a later stage; (c) redox conditions, for example in reducing conditions metals associated with insoluble Fe(III) and Mn(IV) oxy-hydroxy species will be released when the oxy-hydroxy species are reduced; (d) precipitation as insoluble species, particularly when concentrations of the heavy metals are high [188].

In some studies it is unclear, whether results are for dissolved, or particulate material, or both. The proportion of particulate material in fresh water tends to decrease the more polluted the water [179,182]. In the Mississippi river 89-99% of the lead and cadmium is in the particulate, whereas in polluted rivers in the

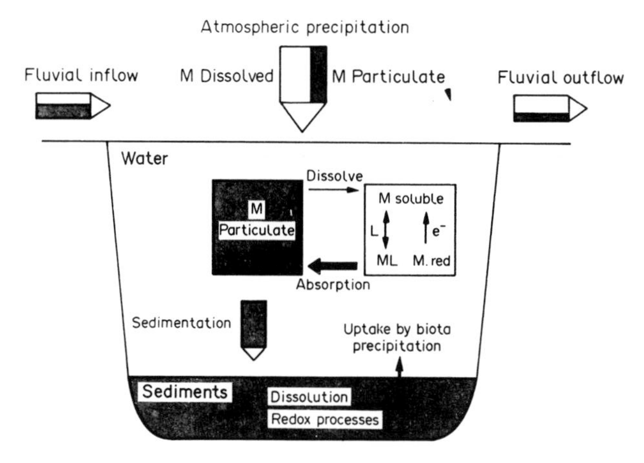

Fig. 8.25 The transfer and fate of heavy elements in a lake system. Sources of data; Salomons and Förstner, 1984 [179], Sigg, 1985 [188].

USA and FRG the percent reported is 30-84% [179]. Generally, the amount of dissolved material increases the smaller the particle size [180]. The data in Table 8.15 summarizes a lot of analytical information (especially for Cd, Hg, Pb and As), on the concentrations of the heavy elements in different fresh waters.

A major contributor to high levels of cadmium, mercury, lead and arsenic in river water is the proximity of mining and mineralized areas [27,109,132, 140,179,204]. The water is often acidic in such locations from the oxidation of sulphide from sulphide ores, which contributes to increased solubility of metal species.

TABLE 8.15 Levels of the Heavy Elements in Fresh Water ($\mu g\ l^{-1}$, ppb)

Elements	Fresh water	Rivers non-polluted	Rivers polluted	Lakes	Rivers mining areas	Geothermal water
Cd[a]	< 1	0.01-1	1- > 10	0.01-20	100-700	0.01-0.5
Hg[b]	0.02-0.1	0.0001-1	> 1	0.02		0.05-60
Pb[c]	< 10	< 1-10	20-100	0.1-30	100-1000	1-10
As[d]	< 1-5	< 1-10	10-1000	1-70	100-5000	1000-5000
Se[e]	< 0.1	0.1-0.3				
Sb[e]	< 0.3	0.3-5				
Bi[e]		0.02				
Tl[e]	0.004					

Sources of data; references a 14,16,19,79,82,102,132,135,137,145,163,166,179, 185,188,195, 196,205,210. b 19,73,79,82,108,131,140,166,179,195,196. c 16,18,39, 79,82,109,132,137, 140,155,160,163,166,179,188,195,196,205,210,213. d 19,27,38, 70,79,82,93,140,179, 195,196,205,210,214. e 19,79,114..

$$S^{2-} + 4H_2O - 8e \rightarrow SO_4^{2-} + 8H^+$$

and

$$S_2^{2-} + 8H_2O - 14e \rightarrow 2SO_4^{2-} + 16H^+$$

Because rivers transport material at a rapid rate, compared with lakes and the ocean, it is possible to identify changes in concentrations of the heavy elements with variations in population density and season. Cadmium, mercury and lead levels in the Rhine correlate well with population density [79]. The rate of flow influences the levels of trace elements, for example, cadmium concentrations in the Rhine fall with increase in the velocity of the river, presumably due to dilution [179,182]. The speed of the Rhine is lower in the summer, and cadmium levels are greater in the autumn/winter discharge than in the spring, because of the accumulation during the summer [82,179].

At Bodensee and Lake Constance, similar concentration profiles for lead and cadmium were found to those in the oceans, but no relationship exists with nutrient concentration profiles. Levels of lead are highest in the surface water, presumably due to atmospheric input, and then they fall with depth. For cadmium there is either virtually no change with depth or a slight increase [179,188].

The levels of trace elements in rivers are often higher than in the ocean, because they are often primary dumping places for waste water. As a result there is concern over the toxicity of elements to aquatic life in rivers. Suggested maximum and 24 hour mean levels which should not be exceeded are given in Table 8.16 [116,195]. Some of the data in Table 8.15 indicates that these limits have been exceeded. The situation is improving due to efforts, in some areas, to clean up materials dumped into waterways. The information plotted in Fig. 8.26 [179], shows the changes in pollution inputs to the Rhine over eight years.

Freshwater Sediments

The levels of the heavy elements in river and lake sediments vary widely with location, from what may be considered near natural concentrations, to

TABLE 8.16 Water Saftey Levels for Aquatic Life ($\mu g\ l^{-1}$)

Element	Maximum level	24 Hour mean
As	400	
Cd*	1.5	0.012
Hg	0.0017	0.00057
Pb*	74	0.75
Se	260	35

* Levels are a function of the hardness of water. Sources of data; Kirk and Lester, 1984 [116], Smith, 1986 [195].

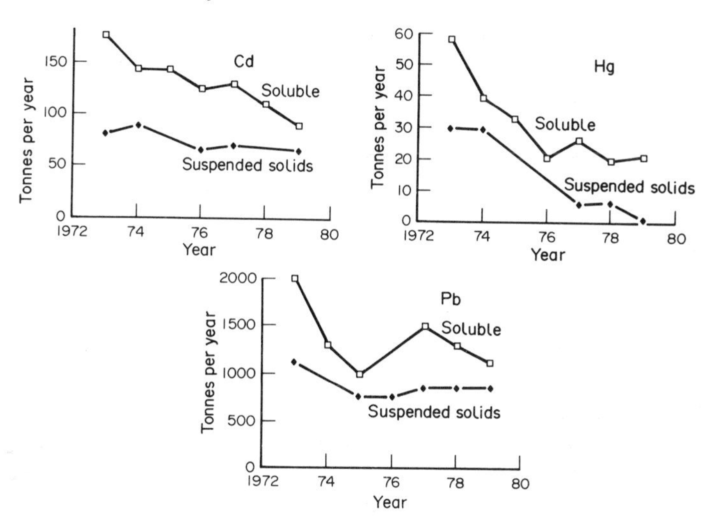

Fig. 8.26 Inputs of Cd, Hg and Pb to the Rhine, from 1972 to 1979. Source of data; Salomons and Förstner, 1984 [179].

levels many thousands times enriched close to mining and industrial sources of the metals. The wide variation is such, that tables of data are not all that informative as to typical levels. The function [79,179],

$$T = f(L,H,G,C,V,M,e)$$

summarizes the diversity of factors that can influence **T** the concentration of a trace element in a sediment. The factor **L** is the influence of the lithogenous units, **H** is the hydrologic effects, **G** is the geologic features, **C** is man made (cultural) influences, **V** is the influence of vegetation, **M** is the mineralized zone effects and **e** is the error i.e. all the factors not considered. At any one location one or more of the factors may operate, and the level of the trace metal in a sediment will also depend on the intensity of the contributing factors. The data in Table 8.17 indicate the range of values that can be expected, and a suggestion is given as to which factors dominate. Provided factor **C** (which includes mining activity) is absent, high levels of heavy elements in sediments may indicate a significant contribution from **M** . Therefore sediments can be a guide to mineralized zones of the heavy metals.

The concentrations of the heavy elements in sediments, fall with distance from the source, as material is diluted with less contaminated sediments during transport in a river [4,79,128,142,166,170,179,194]. A good example of this is shown in Fig 8.27, for the cadmium in suspended sediments from the lower Rhine River [179]. The source of the cadmium, was input from the Ruhr River, believed to come from the Duisburg copper plants. There is a rapid rise in

TABLE 8.17 Levels of Heavy Metals in Some Fresh Water Top Sediments

Location	Conc. μg g^{-1}	Comments	Ref.	Factor
Arsenic				
Canadian lakes	2.7-13.2	Near natural	140	L, G
Adirondack lakes, USA	5.3-6.5	Near natural	101	L, G
River Niger	3.9-4.3	Near natural	149	L, G
Niagara river system	2.7-14	Polluted	110,194 203	L, G, C
Lake Michigan	8-30,5	Industrial	128	L, G, C
Japanese rivers	79-760	Arsenic mining area	139	M, C
Nova Scotia streams	> 3000	Gold mining area	27	M, C
Big cedar lake, USA	150-659	$NaAsO_2$ herbicide	140	C
U.K. streams	< 50-5000	Mining areas	11	M, C
Green Bay (USA)	10-9-42.5	Pulp/paper industry	40	C
Cadmium				
Adirondack lakes, USA	1.5-2.1	Near natural	101	L, G
Thailand rivers	0.77-1.51	Near natural	138	L, G
Wellington streams NZ	< 0.3-25.8	Some pollution	197	L, G, C
Cwm river, Wales	0.2-5.5	Old mines, acid water	109	M, C
Niagara river system	0.4-5.6	Polluted	110,194 203	L, G, C
Wisconsin lakes	max 4.6	Some pollution	79	L, G, C
Illinois river	0.2-12.1	Near natural to indust.	137	L, G, C
European rivers	<0.05-95	Near natural to polluted	7,147	L, G, C
Stream, NZ	2-30	Old mining area	213	L, G, M, C
Conway river, UK	3-95	Mining area	79,80	L, G, M, C
Qishnon river, Israel	0-123	Near natural to polluted	122	L, G, C
Palestine lake, USA	3-2640	Electroplating	80,140	C
Mercury				
Canadian lakes	0.07-1.5	Near natural to polluted	79,140	L, G
Thailand rivers	0.23-1.5	Near natural to polluted	138	L, G, C
Lake Michigan	0.2	Polluted	128	L, G, C
Swedish lakes	11	Cloroalkali cell	79	C
Niagara river system	0.1-868	Polluted	110,194 142,203	L, G, C
European rivers	0.6-35	Near natural to polluted	7,147	L, G, C
Wabigoon river, Canada	8-56	Pulp and paper	4,79	C
St Clair/Erie lakes	max 88	Pulp and paper	79	C
Green Bay (USA)	0.05-0.1	Pulp/paper industry	40	C
Lead				
Arctic lakes, Canada	10-33	Near natural	140	L, G
Fresh water	20-30	Near natural	140	L, G
Illinois river, USA	3-140	Industrial, municipal	140,137	L, G, C
Thompson canyon, USA	38-42	Atmospheric pollution	186	L, G, C
Lake Michigan	88	Polluted	128	L, G, C

Table 8.17 continued

Location	Conc. $\mu g\ g^{-1}$	Comments	Ref.	Factor
European rivers	50-712	Polluted	7,147	C
Niagara river system	6-157	Polluted	194,203 142	L, G, C
Cwm Rhiedal R, Wales	100-2000	Old mining area	109	L, G, M, C
Stream NZ	500-4000	Old mining area	213	L, G, M, C
Coear d' Alene lake US	3000-6300	Mining area	140	L, G, M, C
Heathcote river,NZ	<80,000	Lead battery factory	170	C
Antimony				
Niger river	0.6-0.7	Near natural	149	L, G
Heathcote river,NZ	max 48	Lead battery factory	170	C
Selenium				
Adirondack lakes, USA	1.6-1.9	Near natural	101	L, G
Thallium				
Adirondack lakes, USA	10	Near natural	101	L, G

Other references: 43,67,131,162,166,195. Suggested background levels in $\mu g\ g^{-1}$: As 1, Cd 0.1-0.6, Hg 0.03-0.2, Pb 4-5, Sb , 0.6, Se 0.8, Tl 6-10.

cadmium concentration at the source input and then a gradual fall off over 40-50 km, but remaining high 15-20 $\mu g\ g^{-1}$ until the sea is reached. A similar situation has been observed in the sediments adjacent to a lead battery factory (Fig 8.28). In this case a slight drift of the contaminated sediments upstream was observed, probably due to tidal effects [170]. The Niagara river is a significant source of heavy metals to Lake Ontario [110,142,203,194].

Changes also occur over time, as either more or less of a material is added to a sediment. The time parameter may be studied in two ways, either surface sampling at the same spot at different times [53,179] or obtaining a sediment core and dating the sediments so the concentrations may be related to time [40,

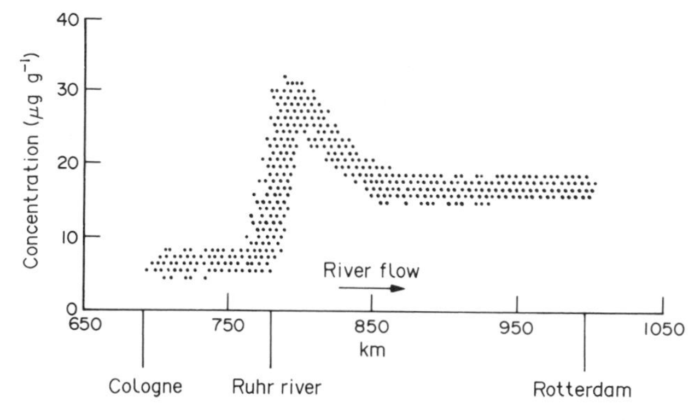

Fig. 8.27 Cadmium concentrations in suspended sediments along the lower Rhine. Source of data; Salomons and Förstner, 1984 [179].

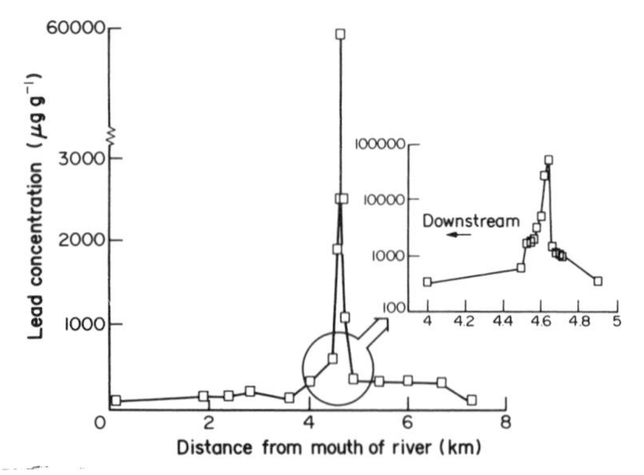

Fig. 8.28 Lead levels in river sediments in relation to a lead battery factory. Source of data; Purchase and Fergusson, 1986 [170].

44,99,101,107,156,170,179,180]. Concentrations of mercury, cadmium, arsenic and lead rose significantly in Rhine sediments over the period 1900-1958, but since then the arsenic and lead levels have fallen. Mercury continued to rise until around 1970, and cadmium to 1975 and then fell off. The changes represent a decrease in inputs, and not just improved analytical methods (Fig 8.26) [179].

Probably the most informative data comes from sediment profiles, especially if they have been dated, and if the depositional rate is known. Some results of profile studies are given in Figs 8.29, 8.30, 8.31 and 8.32. The mercury levels in Lake Windemere, Fig 8.29, show a steady increase from about 1400 AD as a result of erosion, industrial and mining activities. From around 1880 there was a more rapid increase, giving in the top sediments mercury concentrations around five time the background levels [107]. The concentration profiles of the heavy elements arsenic, cadmium, mercury, lead, selenium and thallium in Woods Lake, part of the Adirondack acid lake region (USA), are shown in Figs 8.30 and 8.31 [101]. The baseline concentrations occur at 6-9 cm depth, which is dated around 1952. Thallium shows little concentration change with depth, suggesting that it mainly arises from natural sources. The other elements all display a cultural enrichment factor, CEF;

$$\mathrm{CEF} = \frac{[\mathrm{X}]\ \text{recent sediment}}{[\mathrm{X}]\ \text{baseline level}} > 1,$$

namely, As 9, Cd 4.8, Hg 1.5, Pb 26, Se 2.8. The Woods Lake appears to have a bigger input (based on CEF's) than the nearby Sagamore Lake. This may be due to a greater sedimentation rate for the second lake (~0.4 cm y^{-1} compared with ~0.3 cm y^{-1} for Woods lake), diluting the heavy elements. A thorough study of the lead levels in a sediment profile of a subalpine pond in the Thompson Canyon (USA) was undertaken [186], and the results are depicted in Fig 8.32. The combined mean concentration of barium, strontium and calcium;

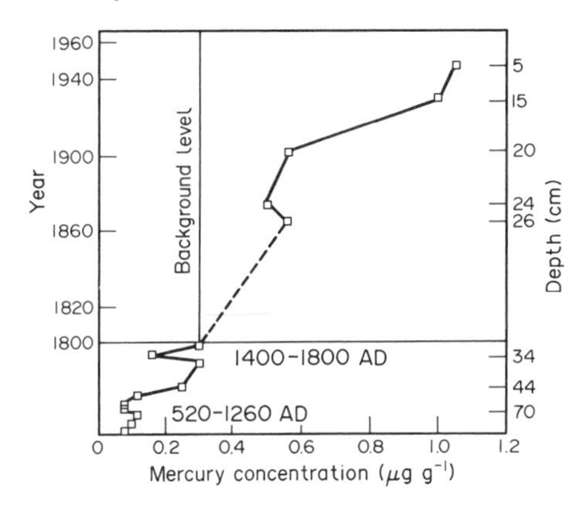

Fig. 8.29 Mercury concentrations in a sediment profile in lake Windemere, UK. Source of data; Horowitz, 1985 [107].

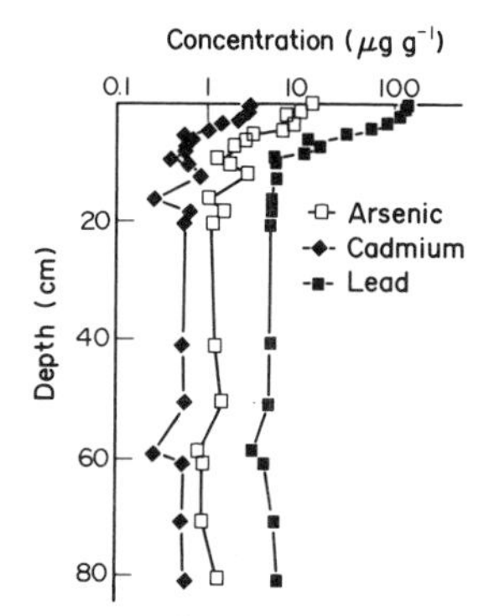

Fig.8.30 Trace element concentrations (As, Cd and Pb) In Woods Lake area, USA as a function of depth. Source of data; Heit et al., 1981 [101].

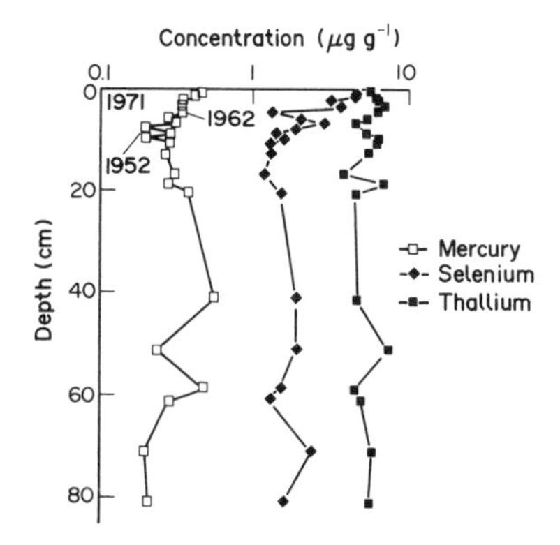

Fig. 8.31 Trace element concentrations (Hg, Se and Tl) In Woods Lake area, USA as a function of depth. Source of data; Heit et al., 1981 [101].

$$\frac{[Ba]/2+[Sr]+[Ca]/20}{3}$$

is also included, and is used as a measure of natural depositional levels at different depths. A small decrease in natural deposition had occurred from around 1900 to the present time, which would have the effect of increasing the levels from natural sources because of less dilution. The excess lead, as seen by the significantly different slopes in the two lines especially since the 1930's is due, however, to an additional input, which is from atmospheric deposition. Variations over time, which relate to depositional rates were also found in lakes near Sudbury Ontario [156]. Recent falls in lead levels in the Canadian lakes are due to reduced inputs because of higher stacks [156].

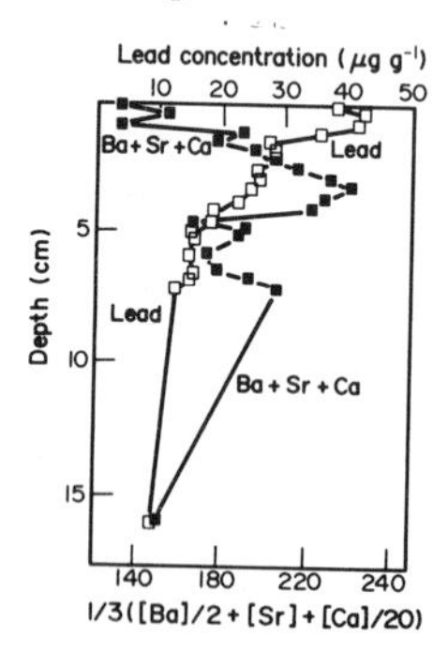

Fig. 8.32 Concentration of lead and averages for the Ba, Sr and Ca concentrations in sediments at Thompson's Canyon. Source of data; Shirahata et al., 1980 [186].

Other Waters

There are a number of other types of water, including rainwater, ice, snow, groundwater, drinking water, waste water and stormwater. In some cases the water is difficult to define, for example groundwater may be considered as soil pore water, or spring water, or well water, or the water table.

Rainwater The concentrations of the heavy elements in rain water are related to their levels in atmospheric aerosols, and their speciation. The intensity and duration of the rain also influences the levels. Generally, slow even rainfall has the best washout characteristics and produces the highest concentration of the elements in water (Chapter 7) [39,127]. Autumn and winter rainfalls usually have higher levels of the heavy metals than summer rainfall [39,127], probably because in the colder seasons winds are frequently lower and temperature inversions are more common, trapping higher levels of the heavy elements in the atmosphere.

Some typical levels of the heavy elements in rain water are listed in Table 8.18. Mercury concentrations lie in the range 0.001-4 $\mu g\, l^{-1}$ with the higher levels being associated with industrial sites. The concentration of lead in precipitate

TABLE 8.18 Levels of Some Heavy Elements in Rainwater

Element	Location	Concentration $\mu g\ l^{-1}$	Reference
Mercury		0.05-0.48	108
	Industrial sites	<0.01->1.0	140
	Remote	0.011-0.43, mean 0.08	87
	Rural	0.005-2.2, mean 0.09	87
	Coastal Pacific	0.001	73
	Urban	0.002-3.8, mean 0.75	87
Lead	Chicago	mean 18.9	39
	East Europe	1-48, mean 5.5	39
	General range	1-50	140
	Greenland	0.25	196
	High population	up to 1000	140
	Lake district, UK	39	39
	New Hampshire, USA	4-68, mean 13.4	39
	Nth. New England	0.06	73
	NZ cities	<0.8-34	196
	Remote	0.02-0.41, mean 0.09	87
	Rural	0.6-64, mean 12	87
	Southern Alps, NZ	1.6-3.1, mean 2.4	196
	UK stations	mean 27	160
	Urban	5.4-147, mean 44	87
	USA stations	mean 34	39,127,160
	USSR	mean 4.6	160
Arsenic	Greenland	0.03	196
	Near volcanoes	63-812	140
	Remote	0.019	87
	Rural	0.005-4, mean 0.29	87
	Rural Europe	0.007-0.1	140
	Seattle, USA	7.0	196
	Urban	5.8	87
Cadmium	Heathrow, UK	4-90, mean 37	153
	Lake Michigan	0.01-500	140
	London, UK	2-108, mean 30	153
	Ontario, Canada	0.01-50	153
	Remote	0.004-0.64, mean 0.008	87
	Rural	0.08-46, mean 0.5	87
	Rural, UK	<5	153
	Urban	0.48-2.3, mean 0.7	87
Antimony	Remote	0.034	87

varies from <0.1 $\mu g\ l^{-1}$ in remote areas, to >1000 $\mu g\ l^{-1}$ in populated areas, with many values around 5-30 $\mu g\ l^{-1}$. Arsenic levels vary widely, and close to volcanoes high concentrations have been observed. Cadmium levels in rain water, in some industrial areas, reach 30-500 $\mu g\ l^{-1}$, but mostly are <1 $\mu g\ l^{-1}$.

Ice and snow We have discussed in Chapters 6 and 7 the lead, content of snow and ice. The results indicate that lead levels have increased over the centuries, as demonstrated by the data presented in Table 8.19 [39,148,151]. In areas of high contamination, i.e. near industries and in cities, lead levels in snow can reach 800,000 μg l^{-1} [140]. Near a lead mine in Yugoslavia the concentrations recorded in snow 0.5 km away were 19,000 μg kg^{-1}, on the road side in Ottawa 55-410 μg kg^{-1} and in Columbus, Ohio, 50-1090 μg kg^{-1} [39]. In more isolated areas, levels of 1-100 μg l^{-1} are more common [39,140,195]. Concentrations of lead in city snow relate to traffic density [92].

Concentrations of the heavy metals (e.g. Pb and Cd), acidity and electrical conductivity, were high in the grey bands of snow profiles found in Sth. Norway. The bands correspond to times of more intense pollution (Fig 8.33) [65,97]. Mercury concentrations in ice are reported as <0.005-0.05 μg l^{-1} [108], for arsenic <10-28 pg g^{-1} and for cadmium <1-17 pg g^{-1} [49,50,97].

Groundwater Groundwater concentrations of mercury have been reported in the range 0.01-0.10 (mean 0.05) μg l^{-1} [73,108]. Levels of lead are reported to occur within the range 1-100 μg l^{-1} [39,104,155], arsenic levels in the range <1-61

TABLE 8.19 Trends in Lead Levels in Ice in Greenland and the Antartica

Greenland			Antartica		
Date	Lead levels μg kg^{-1}	Enrichment Factor	Date	Lead levels μg kg^{-1}	Enrichment Factor
Natural	0.0001	-	Natural	0.00003	-
800 BC	0.001	~10	19th C	0.002	~70
1753 AD	0.011	~100	> 1916	0.01	~330
1815	0.033	~330			
1933	0.066	~660			
1965	0.2	~2000			

Sources of data; Chow, 1978 [39], Murozumi et al., 1969 [148], Ng & Patterson 1982 [151].

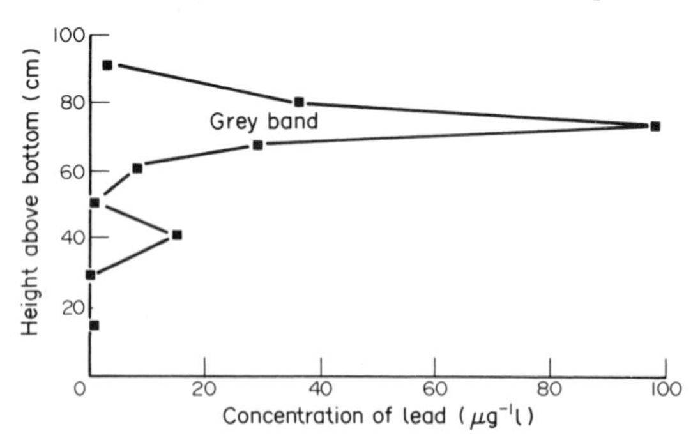

Fig. 8.33 Levels of lead in a snow core in Norway. Sources of data; Elgmark et al., 1973 [65], Hagen and Langeland, 1973 [97].

μg l^{-1} and cadmium levels in the range <0.2-17 μg l^{-1} [104]. The concentration of cadmium is highest in oxidizing conditions, whereas arsenic levels are highest in reducing conditions. Under these conditions the metals are more mobile. Lead levels do not appear to be influenced by the redox conditions, acidity is more important [104]. In a remote subalpine area [66] soil moisture lead levels were found to be 0.7 and 2.0 ng g^{-1} in meadow soils, and 8.9 ng g^{-1} in forest soil. The concentration in groundwater was found to be much lower, at 0.05 ng g^{-1}.

Drinking water Statutory bodies usually provide water of a quality which is fit for human consumption, called potable (drinkable) water. Upper allowable limits of the heavy elements vary from country to country and tend to be lowered with time. Suggested upper limits are given in Table 8.20 [39,42,55,87, 93,116,127,160,195].

The element which has received most attention is lead, because of the use of lead metal in plumbing, and in some solders. In twentyfour towns in the UK 12% of the water sampled, taken as first draw, had a lead concentration >50 μg l^{-1}, 6.9% were >50 μg l^{-1} during normal day time use, and 2.3% for flushed water [168]. Homes in Scotland were worst, with 34.4% having lead levels >50 μg l^{-1}, whereas the proportion was 7.8% in England and 8.8% in Wales [55].

TABLE 8.20 Upper Limits of Potable Water

Element	Upper limit μg l^{-1}
Antimony	10
Arsenic	50
Cadmium	5
Lead	50
Mercury	1
Selenium	10
Thallium	<10

TABLE 8.21 Effect of pH and Alkalinity of the Levels of Lead in Water

pH	Median lead level μg l^{-1}	Alkalinity mg l^{-1}	Median lead level μg l^{-1}
<6.8	170	<10	108
6.9-7.5	44	10-39	42
7.6-8.3	32	40-99	51
>8.4	21	100-199	28
		>200	28

Source of data; Pocock, 1980 [167].

It is clear from various studies on lead plumbing that acid and/or soft water have the highest levels of lead. The effect of both pH and alkalinity (ppm of calcium carbonate) may be seen from the data in the Table 8.21 [167]. Factors that increase the levels of lead in water passing through lead plumbing are: the length and diameter of the pipe, the amount of lead in the pipe, the contact time, the age of the house, the flow rate, the pH, the alkalinity, the amount of organic and colloidal material and the temperature [55,167].

Elemental lead becomes soluble in acidic conditions, due to its oxidation by dioxygen, see the E_h-pH diagram in Fig 3.17.

$$Pb - 2e \rightarrow Pb^{2+}, \; -E^{o} = 0.126 \text{ V}$$

$$O_2 + 4H^{+} + 4e \rightarrow 2H_2O, \; E^{o} = 1.229 \text{ V}$$

$$\text{i.e. } O_2 + 4H^{+} + 2Pb \rightarrow 2Pb^{2+} + 2H_2O, \; E = 1.355 \text{ V}$$

$$\text{or } Pb - 2e \rightarrow Pb^{2+}, \; -E^{o} = 0.126 \text{ V}$$

$$O_2 + 2H_2O + 4e \rightarrow 4OH^{-}, \; E^{o} = 0.401 \text{ V}$$

$$\text{i.e. } O_2 + 2H_2O + Pb \rightarrow Pb^{2+} + 4OH^{-}, \; E = 0.527 \text{ V}$$

For the first reaction acid is required, and for the second reaction acid assists in the removal of the hydroxide ions. Compounds such as $Pb_3(CO_3)_2(OH)_2$ and $PbCO_3$ that may coat the pipes, will dissolve under acid conditions [55].

Electrochemical processes may occur when lead solder has been used on copper pipes.

$$Cu^{2+} + 2e \rightarrow Cu, \; E^{o} = 0.337 \text{ V}$$

$$Pb - 2e \rightarrow Pb^{2+}, \; -E^{o} = 0.126 \text{ V}$$

$$Cu^{2+} + Pb \rightarrow Cu + Pb^{2+}, \; E^{o} = 0.463 \text{ V}$$

The influence of alkalinity on decreasing lead solubility, appears to be stabilization of the basic and normal lead carbonate coatings on the pipes, probably by a mass action effect. Lead in drinking water from lead plumbing can be reduced by making the water alkaline at around a pH of 8.0-8.5. Another method is to add a precipitating agent such as orthophosphate, to form insoluble lead compounds, such as $Pb_3(PO_4)_2$ and $Pb_5(PO_4)_3OH$. Ultimately the lead plumbing and soldering should be replaced.

Cadmium, present as an impurity in zinc in galvanized iron pipes will corrode and can be a significant source of the metal in drinking water [169]. The reactions are;

$$Cd - 2e \rightarrow Cd^{2+} \quad -E^\circ = 0.403\ V$$

$$O_2 + 4H^+ + 4e \rightarrow 2H_2O, \quad E^\circ = 1.229\ V$$

$$2Cd + O_2 + 4H^+ \rightarrow 2Cd^{2+} + 2H_2O, \quad E = 1.632\ V$$

with E = 1.63 V at pH = 0. The potential decreases as the pH increases (see the cadmium E_h-pH diagrams inChapter 3), and at a pH = 8, E = 1.16 V [169].

Waste water Wastewater, encompasses many things, such as sewage effluent, industrial waste, mining waste, agriculture land runoff and stormwater. The concentrations of some heavy elements can be quite high in such waters. In some situations, such as in sewage and industrial wastes, methods can be employed to remove the metals, but in other cases there is less control, e.g. in stormwater and agriculture runoff. It has been estimated that, in the US, when leaded petrol was widely used the stormwater runoff added lead at the rate of 8×10^9 g y^{-1} to waterways, contributing around 4 µg l^{-1} to the lead levels in rivers [39]. The few data listed in Table 8.22, indicate the heavy metal concentrations that may occur in wastewaters.

The cadmium in treated wastewater of New York, has been found to come from: electroplaters 33%, residential areas 49%, runoff 12%, and industrial 6%. The total cadmium input was around 73 kg day^{-1}. The main industrial sources were identified as fur dressing and fabric dying processes and from laundries [175].

The removal of street dust in stormwater runoff, occurs mainly during rain. The first flush, i.e. within 30 mins and often within 10 mins, carries the heaviest

TABLE 8.22 Levels of Some Heavy Elements in Waste Water

Element	Waste water	Location	Concentration µg g^{-1}
Lead	Stormwater	Durham, USA	100-12,000, mean 1500
	Stormwater	Average, USA	0-1900
	Sewage	Industrial areas	100-500
	Metal platers	Industrial areas	2000-140,000
	Mining	USSR	7000-9000
Cadmium	Industrial	NY	3-20, mean 10
	Industrial	W. Germany	220
	Mining	Sth. Africa	6-52
	Laundary effluent	NY	134
Mercury	Industrial	W. Germany	7
Arsenic	Mining	NS Canada	140

Sources of data; references 12,27,39,79,117,125,139,195,196.

load of the metals [79]. There is a corresponding fall in the levels of the metals in the remaining dust, for example the concentration of lead in street dust before a rainfall was 4820 $\mu g\ g^{-1}$, and after a heavy rain it was 2470 $\mu g\ g^{-1}$ [72].

The removal of surface contaminates with rainfall has been modelled with the equation [125];

$$L_i - L_t = L_i(1 - e^{-kR}),$$

where L_i and L_t are the surface loading initially and after time t (in mg m^{-2}), R is the cumulative rainfall from i to t (in mm) and k is the removal constant (in mm^{-1}). The constant is around 1.8 mm^{-1} for soluble species and ~0.1 mm^{-1} for insoluble materials. Soluble materials are removed in the early stages of the rainfall, and insoluble materials after a longer time. The effect of runoff on the water it enters is to increase its concentration in heavy metals. This is represented by the following equation, based on a mass balance [125].

$$C_M Q_M = C_R Q_R + C_D Q_D$$

where C = concentration, M = mixed system, R = initial concentration of receiving water and D = concentration in the drainage water, and Q is the discharge. The relative magnitudes of C_R and C_D and Q_R and Q_D will determine the effect of the storm water runoff on the final concentration, C_M, of the receiving water.

CHEMICAL SPECIATION IN WATER

The chemical speciation of an element in water is important as regards its environmental chemistry. The speciation also gives us information on the mobility and therefore availability of the metal to living things, and their potential toxicity. The elements occur in one or more of three categories as regards size. Species <0.45 µm are arbitrarily said to be soluble, and above 0.45 µm species are particulate either in suspension or as sediments. The variations in the species with size are shown in Fig 8.34 for lead and cadmium [3].

The factors that influence the speciation of the heavy elements in water are: the solubility of compounds of the elements, the oxidation state of the element, availability of complexing agents, complex formation, ion-pair formation, adsorption or desorption on to particulate material, the redox and pH conditions of the environment and biochemical processes. In order to decide the relative importance of these factors consideration needs to be given to kinetic, thermodynamic, chemical equilibria, and stability constant data [76,79,119, 126,141,198].

The experimental methods for the determination of speciation in water are principally; anodic stripping voltammetry, ion-exchange, ultrafiltration and dialysis [78,79,119,159]. Schemes may be developed around these techniques, in order to carry out progressive speciation [79]. One problem is that the

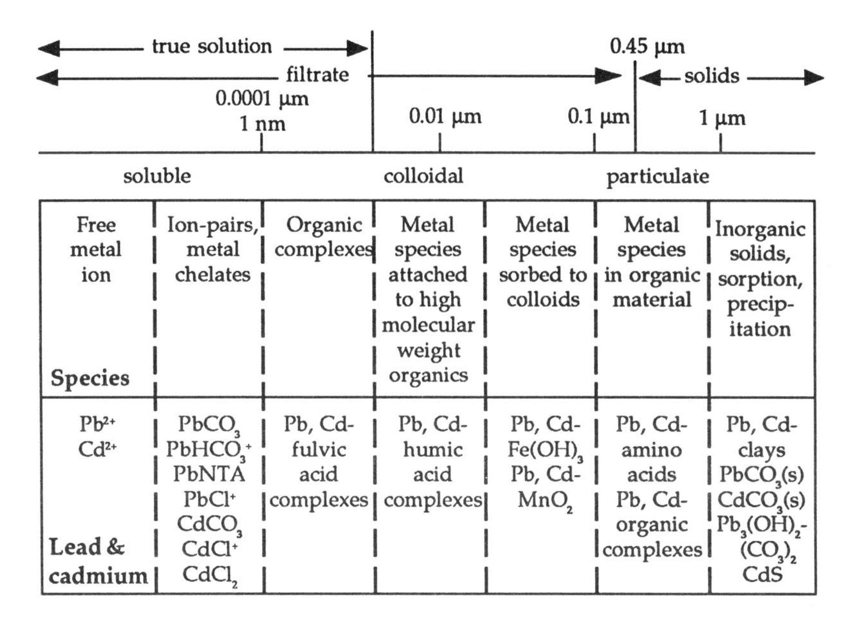

Fig. 8.34 Species in water with respect to size. Source of data; Andreae 1986 [3].

processes used, may alter the speciation and upset the equilibria in the water [76]. Also during ASV analysis there are problems, such as absorption of organic material on to the electrode, and the formation of intermetallic compounds [119].

Lead

A problem in deducing the species of an element in solution, is the availability of reliable stability constant data [221]. Estimations, in the literature, of the speciation of lead in water differ because of the different data used in the calculations. The distributions of lead with respect to pH and chloride ion concentration are given in Figs. 3.15, 3.16 and 8.35 a, b, and the E_h-pH characteristics of lead, with respect to oxy/hydroxy, carbonate and sulphur, are given in Figs. 3.17, 3.18 and 3.19. It is clear from this information, that some of the likely inorganic species of lead in freshwater, and/or seawater are, the $PbCO_3$ ion-pair, $Pb(CO_3)_2^{2-}$, $PbCl^+$, $PbCl_2$, $PbOH^+$, and $Pb(OH)_2$. Only in acid conditions, or in low chloride concentration, is the ion Pb^{2+} likely to occur in reasonable amounts. Some results of estimates of lead speciation are listed in Tables 8.23 and 8.24 for seawater and freshwater respectively. The majority of studies come up with the same overall scheme, that the dominant inorganic species in seawater are lead carbonates 40-80%, followed by chloro-species around 10-25%, and of lesser importance lead hydroxy-species and perhaps some free Pb^{2+} [3,23,36,63,71,75,76,111,158,179,198,208,215]. Less work has been carried out on

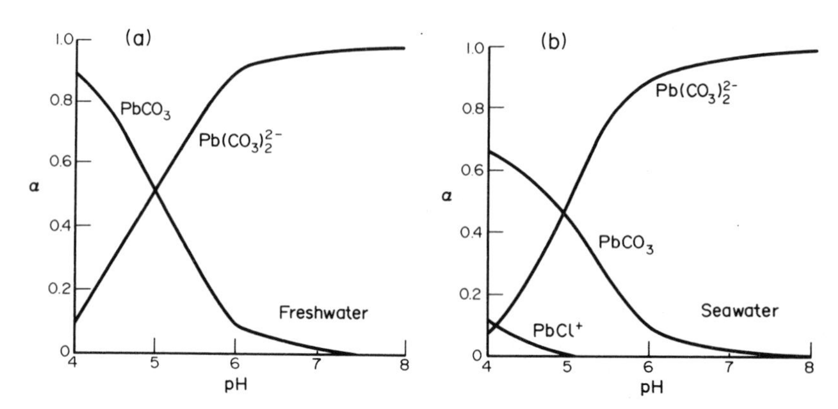

Fig. 8.35 Distribution diagrams for lead in (a) fresh water and (b) seawater.

freshwater, but it would appear that the dominant species are the lead carbonates at around 90% [23,59,63,75,76,175,187]. In thermal waters one study indicated that chloro-species dominated [39].

A change in the pH of seawater effects the relative proportions of the inorganic lead species as seen from the data in Table 8.24 [36,71,179,198,215]. From these results, it appears that as the pH increases from 6 to 8, the proportion of chloro-species falls, and at the same time the proportion of carbonate species increases. At a higher pH (i.e. > 9) the hydroxy species start to dominate. This

TABLE 8.23 Inorganic Lead Species in Seawater

Species	a	b	c	d	e	f	g	h	i	j	k	l	m	n	o
$PbCO_3$		83		80		40	68	65	55	80	40	41			43
$Pb(CO_3)_2^{2-}$															4
Carbonate species	47		83										65	77	
$PbCl^+$				11	19	8	9		7	15	15				9
$PbCl_2$		11		3	42	13	12		11	5					13
Chloro-species	21		11									47	28	21	2
Other chloro					13	3	5	29	5		9				
$PbCO_3Cl^-$									10						
$PbOH^+$					10	30					30				30
$Pb(OH)_2$											2				
Hydroxy species	7											9	6	1	
Pb^{2+}					5	4						3			

Sources; references a 158, b 76, c 75, d 215, e 215, f 215, g 215, h 215, i 215 j 71, pH = 8, k 198, pH = 8, l 179, pH = 8.2, m,n 36, o 159.

TABLE 8.24 Inorganic Lead Species in Fresh Water

Species	a	b	c	d	e	f
$PbCO_3$			main			91
Carbonate species	90				main	
$PbOH^+$		main	main			
Pb^{2+}		main		38, 71		
$PbCl^+$				60, 24		
$PbCl_2$				2, 5		

References: a 75, b 59 low alkalinity, c 59, high alkalinity, d 39, thermal waters, e 175, 187, f 76.

generalization is based on interpolation of the data given in Table 8.25, as not all studies are in agreement.

The change from freshwater to seawater, during which the chloride ion concentration increases, produces a fall in the proportion of carbonate species, with a consequential increase in the chloro-species (Fig 8.36). The point of major change occurs around the 1:1 sea and freshwater mixture. In more concentrated seawater the total amount of lead falls, but the relative proportion of each species remain unchanged [111,187]. The amount of free Pb^{2+} increases from freshwater up to the 1:1 mixture due to desorption from particulate matter (note the drop in the amount of absorbed lead as the salinity increases in the figure) owing to a decrease in sorbing surfaces, flocculation, and an increase in chloride ion concentration.

Organic ligands, either from the decay of organic materials or from added materials such as nitrilotriacetic acid $N(CH_3COOH)_3$ (NTA), can form metal

TABLE 8.25 Inorganic Species in Seawater with Respect to pH

Species	a			b			c			d		
pH	4,6	8	10	7	8	9	7.5	8	8.5	6	8.2	9
$PbCO_3$		55	15	30	80	90	51	40	24	7	41	95
$Pb(CO_3)_2^{2-}$			6									
$PbCl^+$	27	7		40	15	5	18	15	9			
$PbCl_2$	40	11		15	5		10	9	3			
Chloro-										1	47	<1
Other Cl	24	5					5	5	2			
$PbCO_3Cl^-$		10										
$PbOH^+$			13			5	12	30	51	2	5	9
$Pb(OH)_2$			29					2	10			
Pb^{2+}	6									86	3	<1
$PbCO_3(OH)^-$			25									
PbClOH			6									

References: a 215, b 71, c 198, d 179.

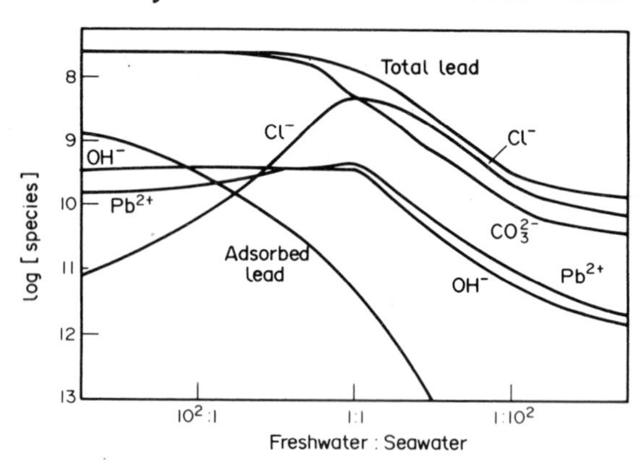

Fig. 8.36 The speciation of lead in fresh water/seawater mixtures. Sources of data; Kester, 1986 [111], Sibley and Morgan, 1975 [187].

complexes. As much as 100% of the soluble lead in surface sea water can be in organic complexes [200], but this may drop to 50-60% deep in the ocean. Lead bonds reasonably well to the donor atoms S, N and O, particularly S as it is a soft acid [140,141,198]. Ligands, such as aminoacids (e.g. cysteine), fulvic acid, humic substances, citrate, acetate, glycolate and NTA can all coordinate to lead [71,76,96,141,172,175]. Limiting factors are the amount of ligand present, and competing metals.

It is estimated that around 1×10^{-5} to 3×10^{-5} mol l^{-1} (i.e. 2-6 ppm) of NTA is required to coordinate 50% of the dissolved lead [172]. The reaction of $Pb(OH)_2$ and NTA (TH_3) is given by;

$$Pb(OH)_2(s) + HT^{2-} \rightleftharpoons PbT^- + OH^- + H_2O,$$

where

$$K = \frac{[PbT^-][OH^-]}{[HT^{2-}]} = 2.1 \times 10^{-5}$$

at a pH = 8 i.e. $[OH^-] = 1.0 \times 10^{-6}$ mol l^{-1}

$$\frac{[PbT^-]}{[HT^{2-}]} = \frac{2.1 \times 10^{-5}}{1.0 \times 10^{-6}} = 21.$$

Hence most of the lead is complexed provided sufficient NTA is present, and equilibrium has been established [71]. The distribution diagram in Fig 8.37 indicates that at pH < 7 the Pb-NTA complex can exist in preference to the lead carbonate species in freshwater [175].

Citric acid (LH_3), $HOOCCH_2$-C(OH)COOH-CH_2COOH may occur in water at concentrations 1×10^{-4} to 1×10^{-8} mol l^{-1}, and may dissolve lead carbonate at pH = 7-10. At these pH's the citrate ion L^{3-}, is present, and carbonate exists as HCO_3^-. Possible reactions are;

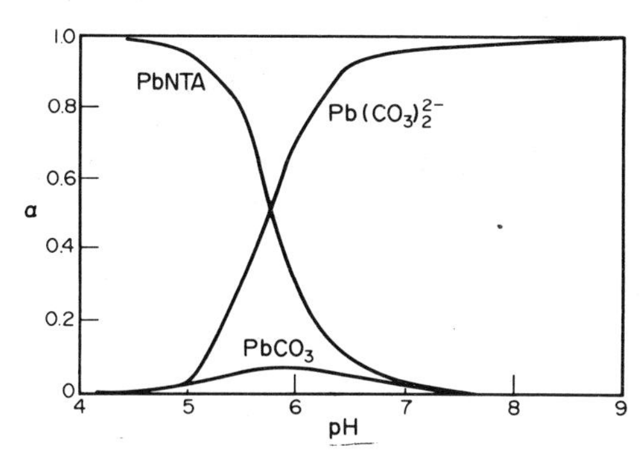

Fig. 8.37 Distribution diagram for lead with carbonate and NTA. Source of data; Rickard and Nriagu, 1978 [175].

$$PbCO_3(s) + L^{3-} + H^+ \rightleftharpoons PbL^- + HCO_3^-, K,$$

$$PbCO_3(s) \rightleftharpoons Pb^{2+} + CO_3^{2-}, K_s,$$

$$CO_3^{2-} + H^+ \rightleftharpoons HCO_3^-, K_b,$$

$$Pb^{2+} + L^{3-} \rightleftharpoons PbL^-, K_f,$$

and the equilibrium constants are;

$$K_s = [Pb^{2+}][CO_3^{2-}] = 1.5 \times 10^{-13},$$

$$K_b = \frac{[HCO_3^{2-}]}{[CO_3^{2-}][H^+]} = 2.1 \times 10^{10},$$

$$K_f = \frac{[PbL^-]}{[Pb^{2+}][L^{3-}]} = 3.2 \times 10^6,$$

$$\text{and } K = \frac{[PbL^-][HCO_3^-]}{[H^+][L^{3-}]} = K_s K_b K_f = 1.0 \times 10^4.$$

At a pH = 7 and $[HCO_3^-] = 1 \times 10^{-3}$ mol l^{-1} we have,

$$\frac{[PbL^-]}{[L^{3-}]} = \frac{1.0 \times 10^4 \times 1.0 \times 10^{-7}}{1.0 \times 10^{-3}} = 1.$$

i.e. approximately 1/2 of the lead in $PbCO_3$ could be complexed with citrate at a pH = 7 [71].

Often the characteristics of the groups bonded to the lead influence the levels and speciation of lead in water. Lead sulphide found in reducing conditions in water/sediment systems, has a low solubility, $K_s = 8.4 \times 10^{-28}$, which provides a Pb^{2+} concentration of 2.9×10^{-14} mol l^{-1}. Depending on the pH however, the sulphide ion may undergo hydrolysis (see Fig 3.14), which alters the amount of lead in solution. The reactions are:

$$PbS(s) \rightleftharpoons Pb^{2+} + S^{2-}, K_s,$$

$$S^{2-} + H_2O \rightleftharpoons HS^- + OH^-, K_h,$$

and

$$K_h = \frac{[HS^-][OH^-]}{[S^{2-}]} = 8.3 \times 10^{-2},$$

$$K_s = [Pb^{2+}][S^{2-}] = 8.4 \times 10^{-28}$$

Since $[Pb^{2+}] = [S^{2-}] + [HS^-]$ and $[HS^-] = [OH^-]$ we can re-write K_s as follows:

$$K_s = \{[S^{2-}] + [HS^-]\}[S^{2-}]$$

$$= \left(\frac{[HS^-]^2}{[S^{2-}]} + [HS^-]\right)\frac{[HS^-]^2}{K_h}$$

i.e. $K_s K_h = [HS^-]^3\left(\frac{[HS^-]^2}{K_h} + 1\right)$.

This equation solves to give $[HS^-] = 4.1 \times 10^{-10}$ mol l^{-1} and therefore $[S^{2-}] = 2.0 \times 10^{-18}$ mol l^{-1}, and $[Pb^{2+}] = 4.1 \times 10^{-10}$ mol l^{-1}. Hence there is more lead in solution than suggested by the solubility product, and an increase in acidity would increase the concentration even more [71].

Cadmium

The inorganic speciation of cadmium in water has similarities to that of lead, but is simpler in that the free cation (hydrated) exists to relatively high pH values, i.e. < 7 to 8 [42,59,63,76,88,90,102,111,112,140,171,208]. Hence, it is a major component of freshwaters, and a significant component of seawater. This can be seen from the distribution diagrams in Figs. 3.3 and 3.6. The hydroxy species only persists down to a pH of 8-9 (Fig. 3.6), below which the Cd^{2+} ion occurs [98,171]. This is also clear from the E_h-pH diagram for the Cd/H_2O system (Fig. 3.5). In seawater, with a concentration of chloride ions of 0.54 mol l^{-1} ($\log[Cl^-] = -0.27$), from the data in Fig. 3.3 it appears that cadmium chloro species are the major components, especially $CdCl_2$, $CdCl_3^-$ and $CdCl^+$, at pH's of 7-9 [59,63,75,76,98,111,112,123,171,177,208,219]. In more dilute systems however, i.e. $[Cl^-] \sim 0.01$-0.03 mol l^{-1}, the cadmium ion begins to dominate. The relative importance of different species with respect to salinity is shown in two separate diagrams Fig. 8. 38 [187] and Fig 8.39 [111,171]. There are differences

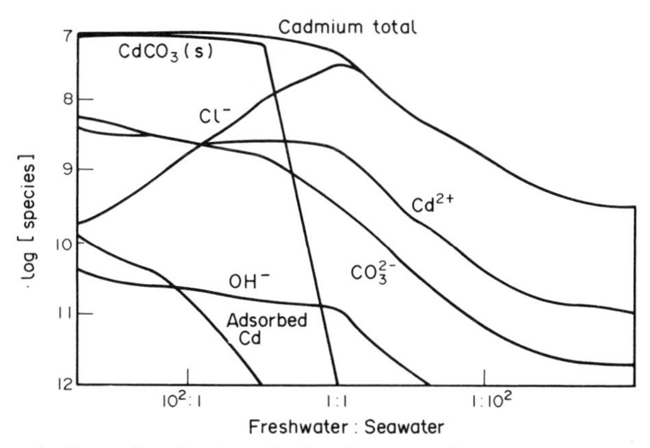

Fig. 8.38 Speciation of cadmium in fresh water/seawater mixtures. Source of data; Sibley and Morgan, 1975 [187].

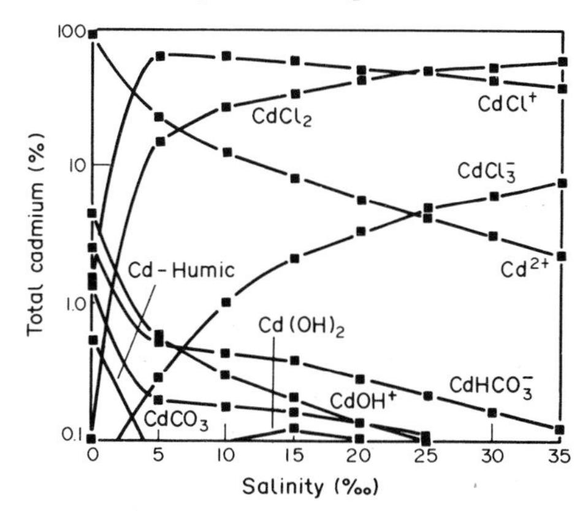

Fig. 8.39 Distribution of cadmium species as a function of salinity. Sources of data; Kester, 1986 [111], Raspor, 1980 [171].

between the diagrams, but in general they indicate the same features. The cadmium chloro-system is relatively stable, one reason for which is the strong Cd-Cl bond [140].

The $CdCO_3$ ion-pair in solution is not as significant as for the corresponding lead species. Calculated amounts, in a variety of freshwaters, range from 4 to 21% [171]. The amount of $CdCO_3$ will depend on the pH of the water [75,76,171,188] and the alkalinity, the higher the alkalinity the greater the amount of $CdCO_3$ expected [59]. The influence of pH on the carbonate is depicted in Fig 8.40, where its existence is negligible at a pH = 6, but becomes more significant at pH's > 8 [198].

Under oxidizing conditions (see Figs. 3.5 and 3.7), cadmium is mobile and present as the aquated cation [82,104], hence under normal environmental conditions the major species of cadmium in fresh to weakly saline water, is the

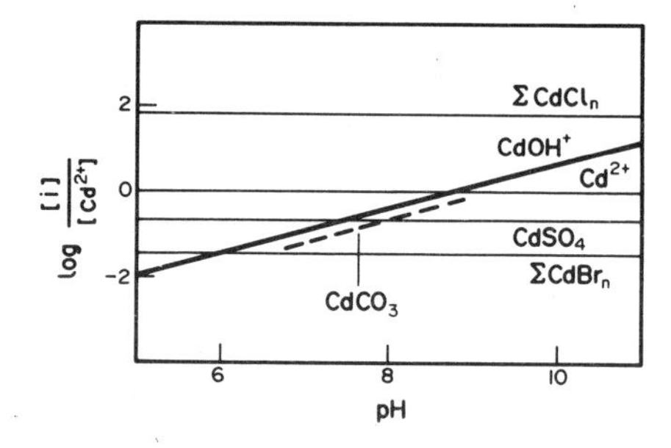

Fig. 8.40 Concentration of cadmium species in seawater. Source of data; Stumm and Brauner, 1975 [198].

hydrated Cd^{2+} ion. In reducing conditions the soluble species of cadmiu (when sulphur is present) is probably the bisulphide ion $CdHS^-$ [75,76]

In contaminated environments where other materials occur, different cadmium species may exist in solution, such as $CdSO_4$ and Cd-organic ligan complexes [90,140,185]. Cadmium, typical of a soft acid, bonds well to sulphu donors such as cysteine, but also with aminoacids, carboxylic acids, polysaccharides and organic pollutants such as NTA [140,141]. Whether or not organi complexes of cadmium occur will depend on the amount of organic materia available, and the presence of competing cations [90,198]. Estimates of th proportion of cadmium-organic complexes in fresh and seawater, range fro 10 to 80% of the total cadmium [88,90,200]. Cadmium-humic species ar reported as minor around 2.7% in fresh water, see Fig. 8.39 [42,96], on the othe hand others report more significant levels [88]. Unless the organic content of th water is high, cadmium organic complexes appear of less importance than th free ion, and inorganic species. Interestingly, in a study of the variation o cadmium levels and soluble organo-cadmium complexes with depth, it wa found that when the salinity was at a minimum so also was the proportion o cadmium complexes [200].

Mercury

The speciation of mercury is significantly different to those of cadmium and lead [98]. The dominant inorganic species in freshwater is $Hg(OH)_2$, which forms by pH ~6, a lower pH than required for the formation of the analogous cadmium and lead hydroxides. The Hg^{2+} ion hydrolyses to $HgOH^+$ and $Hg(OH)_2$ in the pH range 2-6, as shown by the distribution diagram Fig. 8.41 [9,98,140].

In seawater the dominant mercury species are chloro-complexes [3,9, 23,63,98,140,208]. Mercury chloro-complexes are quite stable and they can exist at quite low concentrations of chloride ion (see Fig. 3.8) [98]. At a concentration of chloride of 10^{-7} mol l^{-1} (0.0035 mg l^{-1}) the species $HgCl^+$ is significant, and for $[Cl^-]$ = 10^{-6} to 10^{-2} mol l^{-1} (0.035 to 350 mg l^{-1}) $HgCl_2$ is the dominant species, and

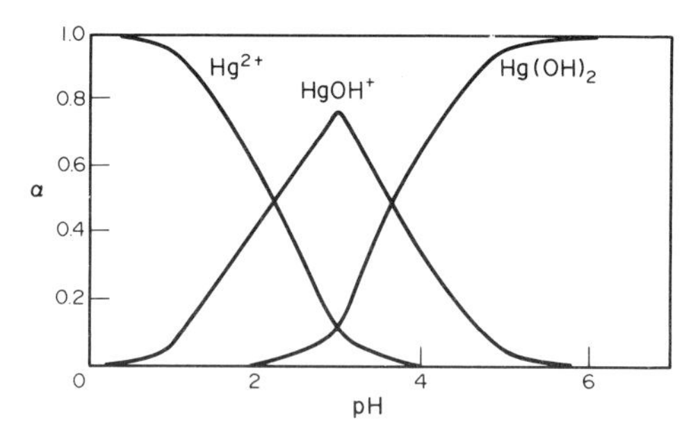

Fig. 8.41 Mercury(II)-hydroxy species distribution diagram. Sources of data; references 9,98,140.

$HgCl_3^-$ peaks at a chloride ion concentration of 10^{-1} mol l^{-1} (3550 mg l^{-1}). At the concentration of chloride ion in seawater $HgCl_4^{2-}$ is probably the main species. This information is summarized in Fig. 8.42, showing the mercury species as a function of chloride ion concentration and pH, and it is clear that in seawater $Hg(OH)_2$ has a negligible existence [9,140]. Also as shown in the Figs. 3.12 and 8.43 the chloro-species occur at low pH and hydroxy species at high pH [9,198].

Marked changes occur on mixing freshwater with seawater as in an estuary [59,76]. Because of the stability Hg/Cl complexes, a salinity of 3-5 °/$_{oo}$ is sufficient for the chloro-species to be dominant (Fig. 8.44) [9,42]. The hydroxychloride, HgOHCl, is one of the materials produced at low salinity.

As for cadmium and lead, mercury-organic ligand complexes may occur depending on the amount of organic material present, and competition with other cations [108,140]. Mercury-humic species are claimed to be the main form of mercury in freshwater [42,96], but this drops to negligible amounts in saline water as shown in Fig. 8.44.

Methylation of mercury in water is important, and can occur by either chemical or bacterial processes [3,9,23,59,76,108,217]. In freshwater the main species is likely to be CH_3HgOH, whereas in seawater it is CH_3HgCl [59]. We

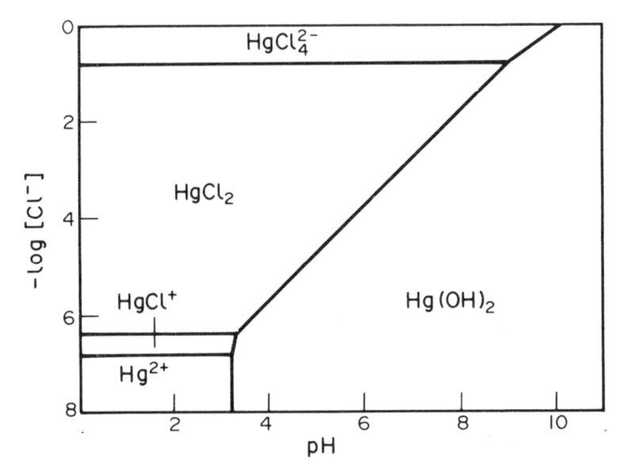

Fig. 8.42 Mercury(II) species as a function of chloride concentration and pH.

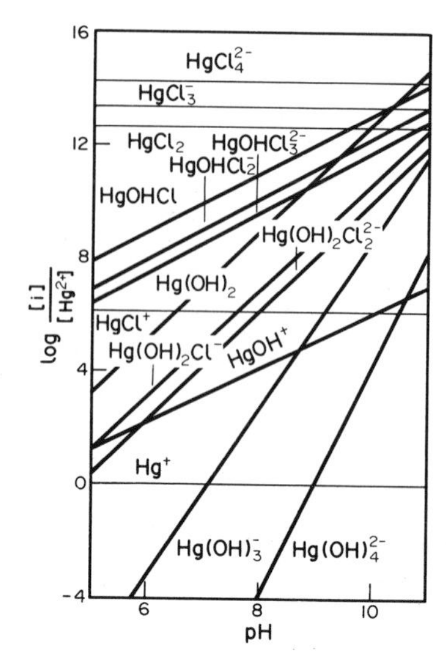

Fig. 8.43 Mercury(II) chloro- and hydroxy species as a function of pH. Sources of data; Benês and Havlìk, [9], Stumm and Brauner, 1975 [198].

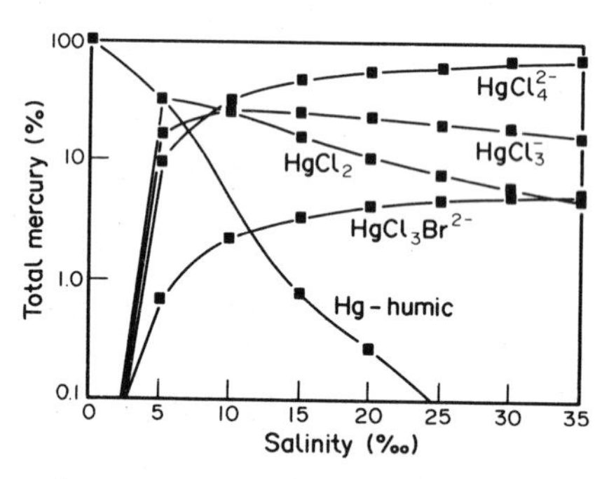

Fig. 8.44 Distribution of mercury species as a function of salanity. Sources of data; Benês and Havlìk, [9], Clarkson et al., 1984 [42].

have previously discussed the reactions leading to the production of methylmercury species in Chapter 4, and we will consider them in more detail in Chapter 12.

Arsenic

Arsenic has four oxidation states available to it +5, +3, 0 and -3. The first exists in water at high E_h values, and in oxygenated systems (see Figs. 3.28a and 8.45). The trivalent state (+3) occurs at lower E_h values, and under mild reducing conditions, whereas oxidation states 0 and -3 occur only under strongly reducing conditions, and are rare in the natural water environment [140]. The dominant species in water appears to be arsenate (arsenic(V)) as $HAsO_4^{2-}$

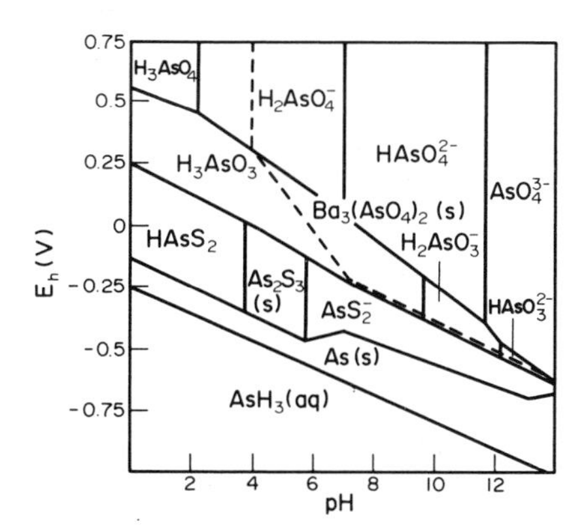

Fig. 8.45 E_h-pH diagram for arsenic, the broken line represents the boundary line for barium arsenate.

[1,3,23,59,63,70,76,177,195,197,198,208,214]. This can be removed, however, by the formation of insoluble arsenates of iron(III) and chromium(III) [1,210], but more probably as $Ba_3(AsO_4)_2$, which has a solubility product of 7.7×10^{-51} [210]. In the E_h-pH diagram given in Fig 8.45 the influence of barium arsenate is portrayed.

Arsenic mobility in water increases in the trivalent oxidation state [104] and arsenite (arsenic (III)) may be a more important state in freshwater than seawater [1,70,76,104]. Arsenic(III) is higher in the surface ocean than the deep ocean, and this may be because of biological activity in the surface ocean [3]. The speciation cycle for arsenic in a stratified lake is given in Fig. 8.46 [70]. Arsenate and arsenite are the main species in water, with arsenate more important in the oxygenated epilimnion, and arsenite in the oxygen depleted hypolimnion. Deeper into the lake further reduction, forming arsenic sulphides is possible, as well as the formation of insoluble arsenate compounds.

The production of organoarsenic species in water systems also occurs, and has been mentioned in Chapter 5, (see Fig 5.11). Their formation is biochemical, involving bacteria [123].

Remaining Elements

For the remaining elements, indium, thallium, antimony, bismuth, selenium and tellurium, much less work has been carried out on speciation. In all cases more than one oxidation state is accessible for these elements in the natural environment. The hydroxy and chloro-distribution diagrams and the E_h-pH diagrams for the elements have been given in Chapter 3.

Indium probably exists in the trivalent oxidation state, and in freshwater at a pH as low as 4 will form $InOH^{2+}$, whereas in natural waters, $In(OH)_3$ is likely to be the dominant species (Figs. 3.21 and 3.25). In seawater, the most likely

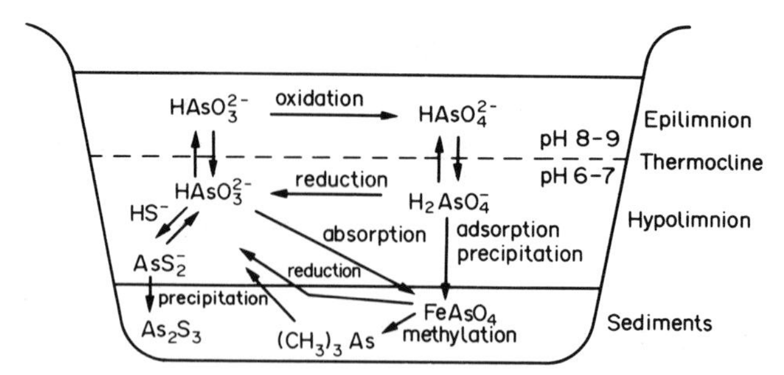

Fig 8.46 Speciation of arsenic in a stratified lake. Source of data; Ferguson and Gavis, 1972 [70].

species is $InCl_2^+$, but at lower concentrations of chloride i.e. $<10^{-2}$ mol l^{-1}, In^{3+} is the preferred species, or hydroxy species, depending on the pH (Fig. 3.22). Hydroxy species of thallium(III) form at low pH i.e. > 2 (Fig. 3.20), whereas in seawater chloro-species such as $TlCl_4^-$ may occur. Even at a chloride concentration of 10^{-7} mol l^{-1} thallium(III) chloro- and chlorohydroxy species can exist (Fig. 3.23) [59,177]. If reducing conditions prevail, thallium(I) may occur, and some TlCl may exist in seawater (Fig. 3.24) [177].

Antimony is more likely to be found as Sb(V) than Sb(III) in water, but as for arsenic, the trivalent oxidation state is more common in the surface ocean than deep water [3,23]. The element forms a wide range of chloro-species and probably the dominant one in seawater is $SbCl_4^-$ (Fig 3.29), whereas in freshwater the pentavalent oxidation state hydroxy species $Sb(OH)_6^-$ is probable (Fig 3.28b) [23,63,198]. A suggested scheme for the environmental chemistry of antimony in water, is given in Fig. 8.47, involving redox and methylation reactions [3,23]. The chloro-species of bismuth(III) start to exist at a chloride concentration of 10^{-4} mol l^{-1} and in seawater where $[Cl^-]$ = 0.54 mol l^{-1} $BiCl_4^-$ would be the dominant material (Fig. 3.30). In freshwater, hydroxy/oxy bismuth(III) compounds are probably the principal species (Fig. 3.28c), such as BiO^+, $Bi(OH)_2^+$ and $Bi_6(OH)_{12}^{6+}$ [198].

Both Se(IV) and Se(VI) exist in seawater and in freshwater (Fig. 3.32a) [3,23,63,76]. The most likely Se(IV) species is selenite, SeO_3^{2-} or $HSeO_3^-$, depending on the pH [198]. The changes in selenium speciation with depth in seawater has been studied [48] (see Fig 8.20). In the surface water organic selenium (e.g. adsorbed onto organic material or incorporated into organic systems and living organisms) dominates, but at greater depths these levels decrease, and Se(VI) is produced. At still lower depths some reduction occurs to give Se(IV), so that both Se(VI) and Se(IV) are important selenium species in the deeper ocean [48]. Tetravalent tellurium , i.e. tellurite is probably the main tellurium species in water (Fig. 3.32b), and it probably occurs as the $HTeO_3^-$ ion.

Industrial discharges can add a whole variety of species into waterways, because a great number of compounds are involved in industrial processes. The species added may change in the aqueous environment, depending on the

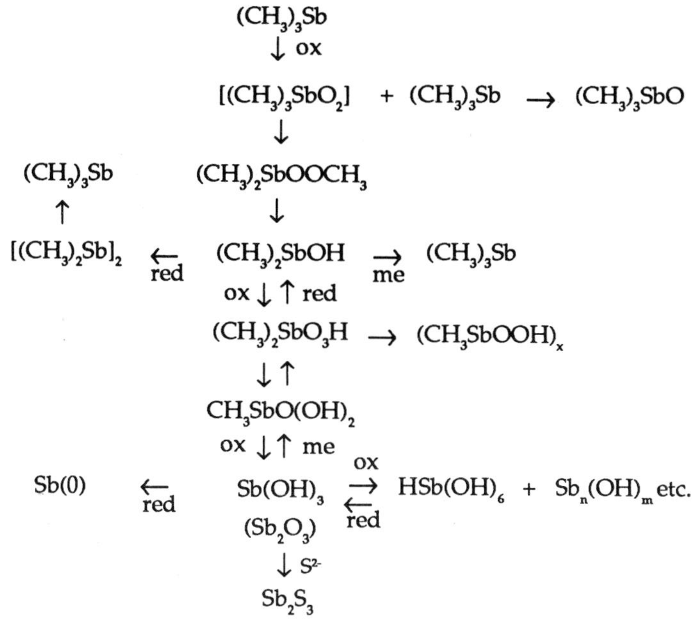

$(CH_3)_3Sb$

$\downarrow$ ox

$[(CH_3)_3SbO_2]$ + $(CH_3)_3Sb$ $\rightarrow$ $(CH_3)_3SbO$

$\downarrow$

$(CH_3)_3Sb$ $\quad$ $(CH_3)_2SbOOCH_3$

$\uparrow$ $\quad$ $\downarrow$

$[(CH_3)_2Sb]_2$ $\xleftarrow[red]{}$ $(CH_3)_2SbOH$ $\xrightarrow[me]{}$ $(CH_3)_3Sb$

ox $\downarrow\uparrow$ red

$(CH_3)_2SbO_2H$ $\rightarrow$ $(CH_3SbOOH)_x$

$\downarrow\uparrow$

$CH_3SbO(OH)_2$

ox $\downarrow\uparrow$ me

Sb(0) $\xleftarrow[red]{}$ $Sb(OH)_3$ $\underset{red}{\overset{ox}{\rightleftarrows}}$ $HSb(OH)_6$ + $Sb_n(OH)_m$ etc.

(Sb_2O_3)

$\downarrow S^{2-}$

Sb_2S_3

Fig. 8.47 Suggested environmental chemistry of antimony in water, ox, oxidation, red, reduction, me, methylation. Source of data; Bramen, 1983 [23].

conditions, or they may remain unchanged. Some examples of products getting into the water systems from industry are given in Table 8.26 [45].

CHEMICAL SPECIATION IN SEDIMENTS

The determination of the speciation of the heavy elements in sediments is more difficult than for dissolved substances. This is because the solid state speciation techniques available are not as sensitive as the methods applicable for solution. In addition, chemical methods used to determine speciation in the solid state, can alter the chemistry and speciation.

There are three main ways of investigating speciation in solids, (a) the use of direct instrumental techniques, such as XRD, electron microprobe analysis and thermal analysis, (b) fractionation of samples by physical methods such as size, density and magnetism, and (c) the use of chemical extractants.

General Features of Sediment Speciation

The heavy elements can become associated with sediments in a number of ways, and an attempt to distinguish the main components, and the ways the trace elements are associated with them is given in Table 8.27 [58,79,107,179]. In any sample a number of the species, listed in the headings of the table, exist together. The more important processes are chemisorption of heavy metals on

TABLE 8.26 Chemical Species for Lead, Cadmium, and Mercury in Industrial Discharges

Source	Species
Lead	
Scrap, ores	PbS, $PbCO_3$, $PbSO_4$, Pb
Batteries	Pb, PbO
Antiknock chemicals	$Pb(CH_3)_4$, $Pb(C_2H_5)_4$, PbX_2, PbXOH
Pigments	$PbCO_3$, $PbCrO_4$
Rust inhibitors	Pb_3O_4, Ca_2PbO_4
Glazing	$PbSi_2O_5$
PVC stabelizers	$PbSO_4$
Mercury	
Ores	HgS, Hg, HgO
Chloro-alkali cell	Hg
Electrical equipment	Hg
Paint	$C_6H_5Hg^+$
Agriculture, paper/pulp preservatives	$C_6H_5Hg^+$, RHg^+
Catalyst, e.g. for PVC acetaldehyde	$HgCl_2$, $HgSO_4$
Pharmaceutics	HgO, $HgCl_2$, Organic derivatives
Cadmium	
Ores	CdS
Electroplating	CdO/NaCN, $CdCl_2$, $CdSO_4$
Batteries	$Cd(OH)_2$, Cd, CdS, CdO
Pigments	CdS, CdSe, $Cd(NO_3)_2$
PVC stabelizers	R_2Cd, $CdCO_3$
Engraving	CdI_2, $CdBr_2$, $CdCl_2$
Semiconductors	CdS, CdSe, CdTe
TV tube phosphors	CdS

Source of data; Förstner and Wittmann, 1981 [84].

to Mn/Fe hydrous oxides, precipitation of discrete heavy metal compounds and flocculation/complexation of heavy elements associated with reactive organic materials. We will consider the major solid phases and the interaction that occurs between them and the heavy elements. Some properties of these phases are listed in Table 8.28 [79,107,179].

Incorporation into detrital minerals The incorporation of heavy metals into detrital minerals, and other minerals such as clays, may occur by replacement of the ions of one metal with another, such as Pb^{2+} replacing K^+ or Ca^{2+}. The metal may also become incorporated into the structure of minerals such as heavy metals in the shells of sea-shells.

TABLE 8.25 The Mechanism and Relative Importance of Speciation of Heavy Metals in Sediments

Mechanism	Detrital minerals (inc. clays)	Organic residues (e.g. shells)	Reactive organic material	Heavy metal precip.	Fe/Mn hydrous oxides	Carbonates phosphates
Incorporated into lattice positions	a**	a*			a*	a*
Physical sorption	b*,c	b*,c	b*		b*	
Chemical sorption or precipitation	d**,c	d**,c	d**	d*	d***	d**
Discrete insoluble compounds				e***		
Flocculation, aggregation, complexation	f**,c	f**,c	f***		f**,c	f*,c

(a) bonded in lattice positions such isomorphous replacement, (b) physical sorption such as by charge neutralization, (c) coatings on other substances, (d) involves the formation of chemical bonds between trace metal and species such as M-O-, (e) precipitation of heavy metal compound by exceeding the solubility product of the substance, (f) small particles <0.45 µm aggregating together to give insoluble larger particles. Trends in importance of the process for heavy metal speciation; *** > ** > *
Sources; references 58,79,107,179

TABLE 8.28 Some Properties of Sediment Phases that Accumulate Heavy Metals

Property	Phase		
	Mn/Fe hydrous oxides	Organic matter	Clays*
Size	Fine grained	Small to large	Fine grained
Crystallinity	Amorphous or poor crystalline	Non-crystalline	Crystalline
Surface area	Large	Large	Large
Cation exchange capacity	High	High	Moderatly high
Surface charge	High, negative	High, negative	High, negative

* Capacity of clays to concentrate heavy metals: montmorillonite (2:1 expanding clay) > vermiculite (2:1 limited expanding clay) > illite (2:1 non-expanding clay) ~ chlorite (2:2 clay) > kaolinite (1:1 clay).

Precipitation of specific compounds Provided the solubility product, K_{sp}, of a poorly soluble material is exceeded, then precipitation of specific compounds may occur.

$$MX_n(s) \rightleftharpoons M^{n+}(aq) + nX^-(aq)$$

$$K_{sp} = [M^{n+}][X^-]^n$$

Low soluble compounds of the heavy elements, associated with common environmental anions, are hydroxides, sulphides, carbonates and phosphates. Solubility product data for some of these compounds are listed in Table 8.29.

Physical sorption Sorption may occur by electrostatic attraction, which is non-specific in character. An electric double layer is set up between the charged surface (e.g. on a clay or Mn/Fe hydrous oxide), and the sorbed material, which maybe an ionic species or a polar species such as water. As a result the charged surface can alter, which can influence subsequent processes. One estimate of this is the zero point charge (ZPC) which is the pH at which the surface has zero charge. Consequentially, the surface charge alters from positive to negative at pH's below or above the pH of ZPC. The ZPC's of some common environmental materials are given in Table 8.30. Below the ZPC, the material will be an anion exchanger as the surface is positive, and above the ZPC the material will be a cation exchanger as the surface is negative. Also at pH's well away from the ZPC the charged surfaces on solids maintain dispersion of the particulate material, but as the ZPC is reached flocculation and coagulation will occur, because repulsion between the like charged surfaces is reduced.

Chemisorption and co-precipitation Sorption of heavy metals on to clays is controlled by the number of free sorption sites on the clay surface. The number are influenced by the free or broken bonding positions, and the proportion of atoms replaced with others of different valencies in the clays [164,183]. The pH, and the nature of the heavy metal species i.e. their charge and hydration, also influence the sorption. The type of clay is also significant as may be seen by the their ability to accumulate heavy metals, in the order: montmorillonite (2:1 expanding clay) > vermiculite (2:1 limited expanding clay) > illite (2:1 non-expanding clay) ~ chlorite (2:2 clay) > kaolinite (1:1 clay). The greater the ability of the clay to expand, and effectively increase its surface area, the more heavy

TABLE 8.29 Solubilities ($-\log K_{sp}$) of Some Heavy Metal Compounds

Hydroxide	pK_{sp}	Sulphide	pK_{sp}	Carbonate	pK_{sp}	Phosphate	pK_{sp}
$Cd(OH)_2$	13.8	CdS	27.2	$CdCO_3$	13.7	$Pb_3(PO_4)_2$	44
$Pb(OH)_2$	19.8	PbS	28.2	$PbCO_3$	12.8	$Hg_2(HPO_4)_2$	14
$Hg(OH)_2$	25.4	HgS	53	Hg_2CO_3	16.1		
$Tl(OH)_3$	43.8	Hg_2S	43.2	$Pb_3(OH)_2$			
$Bi(OH)_3$	30	Tl_2S	19.1	$-(CO_3)_2$	18.8		
BiOCl	34.9	Sb_2S_3	93				
		Bi_2S_3	96				

TABLE 8.30 Zero Point Charges for Some Environmental Materials

Material	pH at ZPC
α-Al_2O_3	9.1
α-$Al(OH)_3$	5.0
γ-Fe_2O_3	6.7
$Fe(OH)_3$ amorphous	8.5
MnO_2	2-4.5
SiO_2	2.0
Kaolinite	4.6
Montmorillonite	2.5
Calcite	8-9.5

Source of data; Förstner, 1980 [79].

metals it can accumulate [124,183]. Clays reduce the pH at which metal hydroxy species may precipitate [68,69].

Chemisorption, and eventual co-precipitation, of the heavy elements with the hydrous oxides of manganese and iron is a major process for incorporating heavy metals into sediments. The hydrous oxides are both amorphous (to X-rays) and crystalline. The main iron compounds are $Fe(OH)_3$, Fe_3O_4 (magnetite), amorphous FeO(OH) and α-FeO(OH) (goethite), whereas the more important manganese compounds are δ-MnO_2, $Mn_7O_{13}.5H_2O$ (birnessite), MnO(OH) (manganite) and $Na_4Mn_{14}O_{27}.9H_2O$ (mango-manganite). The two metals undergo oxidation and reduction reactions, both are mobile in the reduced state, Mn(II) and Fe(II), and form insoluble species in the oxidized states Mn(III), Mn(IV) and Fe(III). The sorption of heavy metals involves sorption to the surface of the hydrous oxides, and then exchange for protons or other metal ions.

$$FeO(OH) + M^{2+} \rightleftharpoons FeO(OH)\cdots M \rightleftharpoons (FeO(O)\cdots M)^{+} + H^{+}$$

As the particle size increases the metal ions can become incorporated into the lattice of the hydrous oxide, taking up lattice positions [52]. Sorption of the heavy metals onto the hydrous oxides and eventual sedimentation of the materials, may also be called co-precipitation.

Association with organic materials Organic material also feature prominently in the speciation of the heavy elements in sediments. Organic materials participate in the solubilization of metal species by complexing the metal ions, but also, they can take metal ions out of solution and contribute to the sediments. For example humic substances, with molecular weights ranging from 700 to > 2,000,000, contain organic groups such as polysaccharides, proteins and phenols, which readily coordinate to metal ions. Because of their size humic substances are insoluble. Fulvic acids are similar, but less condensed, and remain in solution and hence metal ions coordinated to them are solubilized. Dead organisms in sediments may carry the heavy metals with them, either taken in by the organism while alive or sorbed on to the animal before or after death. Organic compounds containing metal ions may also be sorbed onto Fe/Mn oxides.

Association with carbonates Both carbonates and phosphates are significant inorganic materials in the sediments. Metal ions may be co-precipitated with calcium carbonate, as it forms slowly in the aquatic environment. The metal ions can also alter the structural form of $CaCO_3$ from the usual calcite to the less stable aragonite structure. Sorption can be a first step to incorporation of the heavy metals in $CaCO_3$. Once the metal species is sorbed, and as the crystals grow, the metal ion will be trapped in the solid material. The heavy elements themselves form relatively insoluble carbonates (see Table 8.29), and may precipitate in their own right.

The above survey may give the impression that sediments consist of well defined separable phases. This is not the case however, as the phases are intermingled, for example one phase may coat another phase. Organic coatings occur on Fe/Mn hydrous oxides and clays and Fe/Mn hydrous oxide coatings occur on clays.

The Investigation of Sediment Speciation

Instrumental methods Instrumental methods available for solid state speciation are listed in Table 8.31 [107]. The first two methods are used on bulk materials, and give information on specific compounds which display their own characteristic X-ray diffraction pattern or differential thermal analysis curve. The other methods are principally surface analytical techniques, giving the elemental chemical composition over a small surface area, such as 1 μm^2.

Physical separation techniques Separation of solid materials based on density, particle size, surface area, and magnetism allow for some estimation of speciation, for example particles < 2-4 μm in diameter are principally clay materials [107]. Also an advantage of physical fractionation is that the heavy elements may become enriched in a particular size or density fraction. If sufficient enrichment occurs an instrumental method can be employed to elucidate the mineralogy [56,170].

Speciation through chemical extractants An attractive approach to heavy metals speciation in sediments, is to add chemical reagents that extract the elements selectively from certain phases. The known chemistry of the system permits the phase to be determined. A wide variety of methods have been

TABLE 8.31 Instrumental Techniques for the Investigation of Chemical Speciation in Sediments

Technique	Information
X-ray powder diffraction, XRD	Chemical compounds provide they occur at levels > 5%
Differential thermal analysis, DTA	DTA curves may be related to particular compounds
Electron microprobe X-ray photoelectron spectroscopy, ESCA, XPS UV photoelectron spectroscopy Auger electron spectroscopy, AES Secondary ion mass spectroscopy, SIMS Ion scattering spectroscopy, ISS	Chemical composition at a small area, (down to 1 μm^2) and therefore possible stoichiometry. Generally elemental detection limits need to be > 100 ppm.

employed and some of these are listed in Table 8.32. The chemical principal involved are described below

The addition of a high concentration of a cation such as Mg^{2+}, Ba^{2+}, NH_4^+ an Li^+ will tend to liberate ion-exchangeable cations

$$\text{Sediment-M} + Mg^{2+} \rightleftharpoons \text{Sediment-Mg} + M^{2+}$$

The pH needs to be sufficiently high so that protons do not compete, and also do not react with other phases in the sediments, but not too high so that hydroxides precipitate. A slightly more acid solution, such as provided by an acetate buffer (pH = 4.7), is used to decompose carbonates,

$$CaCO_3 + H^+ \rightleftharpoons Ca^{2+} + HCO_3^-,$$

liberating incorporated trace metals, and also dissolving any free heavy metal carbonates. Treatment with stronger acid (e.g. 0.3M HCl, pH ~ 0.5) is used to dissolve authigenic minerals (materials crystallized after sedimentation). To release heavy metals associated with iron and manganese hydrous oxides reducing agents are used, such as hydroxylamine (NH_2OH), oxalate ($C_2O_4^{2-}$) and dithionite ($S_2O_4^{2-}$), which reduce Mn(III) and Mn(IV) to Mn(II) and Fe(III) to Fe(II).

e.g. $Fe(OH)_3 + 3H^+ + e \rightarrow Fe^{2+} + 3H_2O, \quad E^\circ = 1.06\ V,$

$C_2O_4^{2-} - 2e \rightarrow 2CO_2, \quad -E^\circ = 0.49\ V,$

i.e.

$2Fe(OH)_3 + 6H^+ + C_2O_4^{2-} \rightarrow 2Fe^{2+} + 6H_2O + 2CO_2, E = 1.55V.$

TABLE 8.32 Methods for Extraction of Metals from Chemical Phases in Sediments

Classification	Reagents
Absorbates and exchangeable	Distilled, deionised water
	0.2 M $BaCl_2$-triethanolamine, pH 8.1
	1 M NH_4OAc, pH 7
	1 M $MgCl_2$, pH 7
	LiCl
	1 M NaOAc, pH 8.2
Carbonates	1M NaOAc (25% v/v HOAc)
	CO_2 treatment
	Exchange columns
Detrital, authigenic	EDTA treatment
Hydrogenous, lithogenous	0.1 or 0.3 M HCl
Reducible	0.04 or 1 M $NH_2OH.HCl$ + 25% v/v HOAc
Moderately reducible	Oxalate buffer, poorly orderd and non crystalline
	Dithionate/citrate buffer, poorly ordered and crystalline
Easily reducible	0.1 M $NH_2OH.HCl$ +0.01 M HNO_3
Organics	Sodium hypochlorite + dithionate-citrate
	30% H_2O_2 at 95°C, pH 2
	Organic solvents
	30% H_2O_2 + 0.02 M HNO_3 extracted with 1 M NH_4OAc in 6% HNO_3
	0.1 M NaOH for humic acid
	0.02 M HNO_3, + H_2O_2 pH 2 + HNO_3 at 85° C with NH_4OAc in 20% HNO_3
	30% in 0.5 M HCl, heat
Sulphides	30% H_2O_2 at 95° C extracted with 1 M NH_4OAc
	0.1 M HCl in air
Detrital silicates	$HF/HClO_4/HNO_3$
	Borate fusions/HNO_3

Sources; references 79,84,85,107,113,130,179,206

Since ferrous and manganous species are soluble, metal ions associated with the oxy-compounds are released into solution. Organic material is decomposed by oxidation using reagents, such as H_2O_2/HNO_3 and hypochlorite (OCl^-). The reactions are of the type;

$$\text{Organic (C/H/O)} + H_2O_2 \rightarrow CO_2 + H_2O,$$

$$\text{Organic (C/H/O)} + ClO^- \rightarrow Cl^- + CO_2 + H_2O,$$

the oxidizing agents undergo the half reactions such as;

$$H_2O_2 + 2H^+ + 2e \rightarrow 2H_2O, \quad E° = 1.77\ V,$$

$$ClO^- + 2H^+ + 2e \rightarrow Cl^- + H_2O, \quad E° = 1.50\ V.$$

After oxidation metal species sorbed to, or complexed by organic materials, will be released into solution. An oxidizing agent will also release heavy metals occurring as sulphides, or associated by sorption to other metal sulphides. The reaction involves the oxidation of sulphide to sulphite and/or sulphate, and freeing the metal ions.

$$S^{2-} + 3H_2O - 6e \rightarrow SO_3^- + 6H^+, \quad -E° = -0.59\ V,$$

$$SO_3^{2-} + H_2O - 2e \rightarrow SO_4^{2-} + 2H^+, \quad -E° = -0.17\ V.$$

Finally, the detrital silicates may be decomposed by treating with a reagent that destroys the silicate lattice, such as HF or borate. Two possible reactions can occur with HF;

$$\underset{\text{pyroxene}}{(SiO_3^{2-})_n} + 6nHF \rightarrow nSiF_4 + 3nH_2O + 2nF^-$$

$$\underset{\text{feldspar}}{AlSi_3O_8^-} + 18HF \rightarrow 3SiF_4 + AlF_6^{3-} + 8H_2O + 2H^+.$$

Since the silicon tetrafluoride is volatile it is lost from the solution, and metal ions are released from the lattice.

Sequential extraction schemes as described above have the appearance of obtaining good speciation data on the heavy elements in solids. Unfortunately there are serious problems with the process [24,45,83,85,86,95,107,113,115, 130,174,179,189,190,206]. The reagents are not as necessarily selective as implied by the above schemes. For example, repeated application of an extracting reagent continues to remove metal ions from the material. Once a metal ion is mobilized into solution by a reagent it may recombine, either by sorption or chemical reaction with one or more of the remaining phases in the solid, i.e. the metal can redistribute itself. Irrespective of these problems, wide use is made of selective and sequential extracting techniques, and interpretations are made of the chemical form of the metals on the basis of the results. However, some caution is necessary in interpretation because of the above problems. Because of problems resulting from the leaching of heavy metals from solid wastes in our environment, it is important to find reliable methods for the determination of solid state speciation.

Arsenic Mineral forms of arsenic, such as FeAsS occur in gold mine tailings [93] and FeAsS and As_2S_3 in sediments [1]. Arsenic levels are often elevated when iron levels are high in sediments [67], and the element is enriched in the iron phase of Fe/Mn nodules. It is likely that $FeAsO_4$ forms [177] by substitution for phosphate in the sediments. Arsenic in solution is removed from water, mainly by sorption on the hydrous iron oxides [43,140], and released again when reduction of Fe(III) to Fe(II) occurs [43]. Association with organic material appears to be a less important for arsenic, compared with other metals [140] Arsenic(V) is more strongly sorbed onto sediments [161], but interconversion between As(III) and As(V) occurs in sediments, depending on the anoxic/oxic conditions of the sediment environment [1].

Cadmium Mixed calcium and cadmium carbonates were found near to the outfall pipe of a Ni/Cd battery factory in Foundry Cove, USA, but further away the main solid species were Cd-organic materials [100,112]. Cadmium also associates with amorphous iron especially when the manganese levels were low [112,130]. Cadmium does not appear to be absorbed to colloidal material, but organic matter, such as humic substances and organic debris, appear to be the main sorption material for the metal [56,75,89,112,130,140,165]. In some localities the cadmium concentration in water is a function of the solubility of CdS [60].

Partition of sediment samples by size and density shows that cadmium levels may increase with decrease in size [56] and increase in density [107]. Separation by density allows for some mineral recognition, and therefore the phase that the metal is mainly associated with [165]. For example the most dense material (>2.95) contains the heavy minerals, whereas less dense material (2.66-2.95) contains mainly shell material, quartz is mainly associated with the fraction 2.55-2.66, and conglomerates with 2.40-2.55, and the least dense material(<2.40) contains mainly organics. Both lead and cadmium, have been found mainly in the organic debris, followed by the conglomerates [165].

The results of sequential extraction investigations of speciation for cadmium in a variety of sediments are listed in Table 8.33 [80,112,129,152]. The proportion

TABLE 8.33 Chemical Speciation for Cadmium (%) in Various Sediments

Extracted phase	German Bight Nth. Sea	Los Angeles Harbour	L Erie Ashta bula	Bridge Port Conn.	Lower Rhine	L.Moira Ontario *	San Fran -cisco Bay**
Interstitial water	-	-	0	0	-	-	0-1.6
Exchangeable	30	1	1	1	4	3-16	0-1.8
Carbonate	21	5	-	-	43	0.5-3	-
Easily reducible	-	9	5.4	4	-	0.5-3	0-3.6
Mod. reducible	11	34	0	0	6	0.5-3	0-0.7
Organic & sulphide	10	38	82.2	96	47	7-42	90-94
Residue (silicates)	28	14	12.4	1	1	49-84	1.8-8.1
Total Cd $\mu g\ g^{-1}$	1.2	2.2	1.2	15.2	28.4	1.0-8.1	0.55-1.4

* Range corresponds to different depths, ** Different locations in the bay. Sources of data; Förstner, 1987 [80], Khalid, 1980 [112], Lum and Edgar, 1983 [129].

in each fraction varies widely for the different sediments, and may reflect the different composition of the sediments. The most common feature is that organic material is a significant site for the cadmium. The sorption of cadmium to sediments, and to the clay content, increases with pH [140,164,173], and the release of cadmium from sediments is affected by acidity, redox conditions and complexing agents in the water [80]. A shift to reducing conditions, and pH 5-6.5 renders cadmium less available, in the forms of carbonate, sulphide or organic. More alkaline conditions also decrease the mobility of cadmium [113]. From reviewing a number of studies, it has been found that the affinity of materials for cadmium follows the order: Mn > $Fe_{amorph.}$ > chlorite (montmorillonite) > $Fe_{cryst.}$ = illite = humics > kaolinite > silica [124].

Lead Some of the lead compounds found in the sediment environment are $PbCO_3$, $PbSO_4$, PbS and $Pb_5(PO_4)_3Cl$ which derive from automobile aerosol lead [125] and from primary and secondary products from lead industries [170]. Lead carbonate has been found in enriched sediments, partitioned by size [197], and the compound $Pb_3(CO_3)_2(OH)_2$ is formed as a precipitate in lead water pipes [55]. The solid lead compounds in sediments alter with the redox properties of a sediment, changing from oxy-compounds to PbS as the system becomes anoxic, as shown in Fig 8.48 [170]. The less soluble $PbCO_3$ (K_{sp} = 6 x 10^{-14} compared with $PbSO_4$, K_{sp} = 1.7 x 10^8) occurs at the top of the sediment profile, even though $PbSO_4$ was one of the materials added to the river. Lower down, the reducing conditions produce the sulphide ion, and the even less soluble PbS (K_{sp} = 8.4 x 10^{-28}) starts to appear [170].

Lead species are strongly sorbed to Fe/Mn oxides, which are said to be more important than association with organic material [75]. Lead correlates best with iron when manganese levels are low, and best with manganese when iron levels

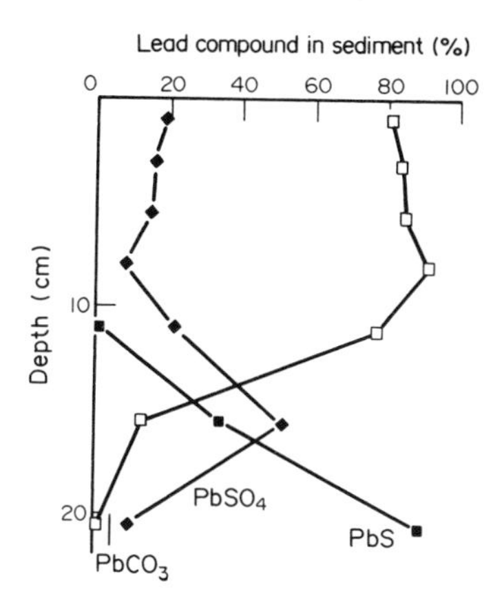

Fig. 8.48 The percentage of lead in a sediment for three lead compounds, with respect to depth. Source of data; Purchase and Fergusson, 1986 [170].

are low, suggesting competition between the two metals for the lead [130,143]. Lead is highly correlated with iron in nodules, and may occur associated with Fe/PO_4^{3-} species, as well as compounds such as $PbFe_2O_4$ ($PbO.Fe_2O_3$) [177]. The lead sorbed onto Fe/Mn oxides is unaffected by aging, but does decrease with an increase in Cl^- ion concentration, due to the formation of lead chloride complexes [201]. The process of sorption is pH dependent as indicated in the following equation, but does occur at the pH of natural water [188];

$$\text{-M-OH} + Pb^{2+} \rightleftharpoons \text{-M-O-}Pb^{+} + H^{+}$$

where M = Al, Si, Mn or Fe. In the case of the manganese and iron hydrous oxide compounds other reactions are possible [175];

$$Pb^{2+} + MnO_2 + H_2O \rightarrow \underset{\text{quenselite}}{PbMnO_2(OH)} + H^{+},$$

$$Pb^{2+} + Fe(OH)_3 \rightarrow PbFeO(OH)_2 + H^{+}.$$

If a clay is saturated with cations sorption will be less dependent on the acidity [175], e.g.

$$Pb^{2+} + \text{Clay-Ca} \rightleftharpoons Ca^{2+} + \text{Clay-Pb}.$$

Around 3-12% of lead is associated with surface active organic material in seawater [211], and organic matter and debris are reported as being the main site for lead [56,85,125,140,165]. It is claimed that mineral forms of lead are not extracted with HCl, and approx 1/3 of sediment lead in a remote subalpine pond is mineral, i.e. natural lead [186].

Separation of sediments by size and density allows for the enrichment of lead, especially in the small size and the dense fractions [56,107,165,197]. Lead levels have been found to both increase with particle size and organic content, and increase with a decrease in particle size and clay content in L. Erie [143]. Harbour and coastal sediments appear to have a significant amount of lead associated with Fe/Mn hydrous oxides [29,152,197]. However, selective extractions, do not recover all the metal added to synthetic sediment materials [115]. Montmorillonite sorbs lead over a broad pH range [183], and the sorption is dependent on amount of calcium present, owing to competition [94]. Lead sorption onto river muds increases with pH up to 6.5, with temperature and agitation, and the sorption is proportional to $(\text{time})^{1/2}$ [140,178].

Mercury The concentration of mercury in water is dependent on the solubility of HgS [60] which is increased by bicarbonate ions [108]. Mercury (4-50%) is associated with surface active organic materials in seawater [211] and mercury ions and the metal are rapidly sorbed by sediments [108]. Factors that influence the sorption of mercury, follows the order: surface area > organic > CEC > grain size, whereas in terms of the bonding of the mercury the order is: organic > grain size > CEC > surface area. Mercury is often bonded to sites containing sulphur [140].

Thallium Thallium occurs in the iron phase of iron/manganese nodules. It probably acts in a similar way to cobalt, i.e. absorbed as Tl^{+} and then oxidized to Tl^{3+} [177]. The thallic ions may then replace the ferric ions in the iron oxide lattice. It is also likely that Tl^{+} replaces K^{+} in nodules [177].

Antimony Most antimony in sediments appears to be associated with Fe/Al compounds, especially under anaerobic conditions [25]. During aerobic leaching, antimony also appears to be in a less mobile phase [25].

REFERENCES

1. Aggett, J. and O'Brien, G. A. Detailed model for the mobility of arsenic in lacustrine sediments based on measurements in Lake Ohakuri. **Environ. Sci. Techn.**; 1985; **19:** 231-238.
2. Alderton, D. H. M. Sediments. in: Historical Monitoring, MARC Report No. 31: MARC; 1985: 1-95.
3. Andreae, M. O. Chemical species in seawater and marine particulates. In: Bernhard, M., Brinckman, F. E. and Sadler, P. T., Eds. Speciation in Environmental Proceses, Dahlem Konferenzen: Springer Verlag; 1986: 301-335.
4. Armstrong, F.A.J. and Hamilton, A.L. Pathways of mercury in a polluted Northwestern Ontario Lake. in: Singer, P.C., Ed. in Trace Metals and Metal-Organic interactions in Natural Waters; 1974: 131-156.
5. Aston, S.R. Estuarine Chemistry. in: Riley, J.P. and Chester, R., Eds. **Chemical Oceanography.** 2nd Ed. ed.: Academic Press.; 1978; **7:** 362-440.
6. Balls, P. W. and Topping, G. The influence of inputs to the Firth of Forth on concentrations of trace metals in coastal waters. **Environ. Pollut.**; 1987; **45:** 159-172.
7. Banat, K., Forstner, U. and Muller, G. Schwermetalle un sedimenten von Donau, Rhein, Ems, Wezer und Elbe in Bereich der Bundesrepublick Deutschland. **Naturwiss**; 1972; **59:** 525-528.
8. Baut Mènard, P. and Chesselet, R. Variable influence of the atmospheric flux on the trace metal chemistry of oceanic suspended matter. **Earth Planet. Sci Lett.**; 1979; **42:** 399-411.
9. Benês, P. and Havlík, B. Speciation of mercury in natural waters. In: Nriagu, J. O., Ed. **The** Biogeochemistry of Mercury in the Environment: Elsevier/Nth. Holland; 1979: 175-202.
10. Bernhard, M. The relative importance of lead as a marine pollutant. in: Brancia, M. and Konrad, Z., Eds. Lead in the Marine Environment: Pergamon; 1980: 345-352.

11. Berry, P. J. An introduction to the exposure commitment concept with reference to environmental mercury. MARC Tech. Rept.: MARC; 1979; No 12.
12. Bertine, K. K. Lead in the historical sedimentary record. in: Branica, M. and Konrad, Z., Eds. Lead in the Marine Environment: Pergamon; 1980: 319-324.
13. Bertine, K. K., Walawender, S. J. and Koide, M. Chronological strategies and metal fluxes in semi-arid lake sediments. **Geochim. et Cosmochim. Acta;** 1978; **42:** 1559-1571.
14. Bewers, J. M., Barry, P. J. and MacGregor, D. J. Distribution and cycling of cadmium in the environment. **Adv. Environ. Sci.;** 1987; **19:** 1-18.
15. Bloom, H. and Ayling, G.M. Heavy metals in the Derwent Estuary. **Environ. Geol.;** 1977; **2:** 3-22.
16. Bone, K. M. and Hibbert, W. D. Solvent extraction with ammonium pyrrolidenethiocarbamate and 2,6-dimethyl-4-heptanone for the determination of trace metals in effluents and natural waters. **Anal. Chim. Acta;** 1979; **107:** 219-229.
17. Bowden, K. F. Oceanic and estuarine mixing processes. in: Riley, J. P. and Skirrow, G., Eds. Chemical Oceanography. 2nd. Ed.: Academic Press; 1975; **1:** 1-41.
18. Bowden, K. F. Physical factors: salinity, temperature, circulation and mixing processes. in: Olausson, E. and Cato, I., Eds. Chemistry and Biogeochemistry of Estuaries: Wiley; 1980: 37-70.
19. Bowen, H. J. M. Environmental Chemistry of the Elements.: Academic Press; 1979.
20. Boyden, C.R. Distribution of some trace metals in Poole Harbour, Dorset. **Marine Pollut. Bull.;** 1975; **6:** 180-187.
21. Boyle, E. A., Huested, S. S. and Jones, S.P. On the distribution of copper, nickel and cadmium in the surface waters of the North Atlantic and North Pacific oceans. **J. Geophys. Res.;** 1981; **86C:** 8048-8066.
22. Bradford, W.L. and Luoma, S.N. Some perspectives on heavy metals concentrations in shellfish and sediment in San Francisco Bay, California. In: Baker, R.A., Ed. Contaminants and Sediments: Ann Arbor Science; 1980; **2:** 501-532.
23. Bramen, R. S. Chemical speciation. In: Natusch, D. F. S. and Hopke, P. K., Eds. Analytical Aspects of Environmental Chemistry: Wiley; 1983: 1-59.
24. Brannan, J.M., Plumb, R.H. and Smith, I. Long term release of heavy metals from sediments. in: Baker, R.A., Ed. Contaminants and Sediments.: Ann Arbour Sci.; 1980; 1: 221-266.
25. Brannon, J. M. and Patrick, W. H. Fixation and mobilization of antimony in sediments. **Environ. Pollut.;** 1985; **B9:** 107-126.

26. Brewer, P. G. Minor elements in sea water. in: Riley, J. P. and Skirrow, G., Eds. **Chemical Oceanography.** 2nd Ed.: Academic Press; 1975; **1:** 415-496.
27. Brooks, R. R., Fergusson, J. E., Holzbecher, J. Ryan, D. E., Zhang, H. F., Dale, J. M. and Freedman, B. Pollution by arsenic in a gold mining district of Nova Scotia. **Environ. Pollut.;** 1982; **4B:** 109-117.
28. Bruland, K. Oceanographic distributions of cadmium, zinc, nickel and copper in the North Pacific. **Earth Planet. Sci Lett.;** 1980; **47:** 176-198.
29. Bruland, K. W., Bertine, K., Koide, M. and Goldberg, E. D. Heavy metal pollution in southern California coastal zone. **Environ. Sci. Techn.;** 1974; **8:** 425-432.
30. Bruland, K. W., Franks, R. P., Knauer, G. A. and Martin, J. H. Sampling and analytical methods for the determination of copper, cadmium, zinc and nickel at the nanogram per litre level in sea water. **Anal. Chim. Acta;** 1979; **105:** 233-245.
31. Bruland, K. W., Knauer, G. A. and Martin, J. H. Cadmium in northeast Pacific waters. **Limnol. Oceanogr.;** 1978; **23:** 618-625.
32. Bryan, G.W. and Hummerstone, L.G. Indicators of heavy-metal contamination in the Looe estuary (Cornwall) with particular regard to silver and lead. **Marine Biol. Ass. U.K.;** 1977; **57:** 75-92.
33. Buat-Mènard, P. E. Fluxes of metals through the atmosphere and oceans. in: Nriagu, J. O., Ed. Changing metal Cycles and Human Health, Dahlem Konferenzen: Springer-Verlag; 1984: 43-69.
34. Burnett, M., Ng, A., Settle, D. and Patterson, C. C. Impact of man on coastal marine ecosystems. in: Branica, M. and Konrad, Z., Eds. Lead in the Marine Environment: Pergamon; 1980: 7-13.
35. Burton, J. D., Statham, P. J. Occurrence, distribution and chemical speciation of some minor dissolved constituents in ocean waters. In: Bowen, H. J. M., Rept. Environmental Chemistry: Roy. Soc. Chem.; 1982: 234.
36. Byrne, R.H. Inorganic lead complexation in natural seawater determined by UV spectroscopy. **Nature;** 1981; **290:** 487-489.
37. Carmody, D.J., Pearce, J.B. and Yasso, W.E. Trace metals in sediments of New York Bight. **Marine Pollut. Bull.;** 1973; **4:** 132-135.
38. Cermisini, C., Dall'aligo, M. and Ghiara, E. Arsenic in Italian rivers and in some cold thermal springs. In: Management and Control of Heavy Metals in the Environment: CEP Consultants; 1979: 341-344.
39. Chow, T. J. Lead in natural waters. in: Nriagu, J. O., Ed. The Biogeochemistry of Lead in the Environment, Part A, Ecological Cycling: Elsevier/ Nth Holland; 1978: 185-218.
40. Christensen, E. R. and Chien, N. -K. Arsenic mercury and other elements in dated Green Bay sediments. In: Management and Control of Heavy Metals in the Environment: CEP Consultants; 1979: 373-376.

41. Christensen, E. R., Scherfig, J. and Koide, M. Metals fron urban runoff in dated sediments of a very shallow estuary. **Environ. Sci. Techn.**; 1978; **12:** 1168-1173.

42. Clarkson, T.W., Hamada, R. and Amin-Zaki, L. Mercury. in: Nriagu, J.O., Ed. Changing Metal Cycles and Human Health: Dahlem Konferenzen Springer Verlag; 1984: 285-309.

43. Clement, W. H. and Faust, S. D. The release of arsenic from contaminated sediments and muds. **J. Environ. Sci. Health;** 1981; **A16:** 87-122.

44. Cowgill, U.M. Mercury contamination in a 54m core from Lake Huleh. **Nature;** 1975; **256:** 476-478.

45. Craig, P. J. Chemical species in industrial discharges and effluents. In: Bernhard, M., Brinckman, F. E. and Sadler, P. T., Eds. Speciation in Environmental Proceses, Dahlem Konferenzen: Springer Verlag; 1986: 443-464.

46. Cronan, D.S. Basal metalliferous sediments from the Eastern Pacific. **Geol. Soc., Amer Bull.**; 1976; **87:** 928-934.

47. Cronon, D. S. Manganese nodules and other ferro-manganese oxide deposits. in: Riley, J. P. and Chester, R., Eds. Chemical Oceanography. 2nd. Ed.: Academic Press; 1976; 5: 217-263.

48. Cutter, G. A. and Bruland, K. W. The marine biogeochemistry of selenium: a reevaluation. **Limn. Oceanogr.**; 1984; **29:** 1179-1192.

49. Davidson, C. I., Chu, L., Grimm, T. C., Nasta, M. A. and Qamoos, M. P. Wet and dry deposition of trace elements onto the Greenland ice sheet. **Atmos. Environ.**; 1981; **15:** 1429-1437.

50. Davidson, C. I., Santhanam, S., Fortmann, R. C. and Olson, M. P. Atmospheric transport and deposition of trace elements onto the Greenland ice sheet. **Atmos. Environ.**; 1985; **19:** 2065-2081.

51. Davies, T. A. and Gorsline, D. S. Oceanic sediments and sedimentary processes. in: Riley, J. P. and Chester, R., Eds. Chemical Oceanography. 2nd. Ed.: Academic Press; 1976; 5: 1-80.

52. Dawson, B. S. W., Fergusson, J. E., Campbell, H. S. and Cutler, J. B. Distribution of elements in some Fe-Mn nodules and an iron-pan in some gley soils of New Zealand. **Geoderma.**; 1985; **35:** 127-143.

53. de Groot, A. J., Zschuppe, K. H. and Salomons, W. Standardization of methods of analysis for heavy metals in sediments. **Hydrobiologia;** 1982; **92:** 689-695.

54. de Groot, A. J. and Allersma, E. Field observations on the transport of heavy metals in sediments. In: Krenkel, P.A., Ed. in Heavy Metals in the Aquatic Environment: Pergamon; 1975: 85-94.

55. de Mora, S. J. and Harrison, R. M. Lead in tap water: contamination and chemistry. **Chem. in Brit.**; 1984; **20:** 900-906.

56. Dossis, P. and Warren, LJ. Distribution of heavy metals between the minerals and organic debris in a contaminated marine sediment. in:

Baker, R.A., Ed. Contaminants and Sediments: Ann Arbour Sci.; 1980; **1:** 119-139.

57. Duinker, J. C. Formation and transformation of element species in estuaries. In: Bernhard, M., Brinckman, F. E. and Sadler, P. T., Eds. Speciation in Environmental Proceses, Dahlem Konferenzen: Springer Verlag; 1986: 365-384.
58. Duinker, J. C. Suspended matter in estuaries: adsorption and desorption processes. in: Olausson, E. and Cato, I., Eds. Chemistry and Biogeochemistry of Estuaries: Wiley; 1980: 121-151.
59. Dyrssen, D. and Wedborg, M. Major and minor elements, chemical speciation in estuarine waters. in: Olausson, E. and Cato, I., Eds. Chemistry and Biogeochemistry of Estuaries: Wiley; 1980: 71-119.
60. Dyrssen, D., Hall, P., Haraldsson, C., Iverfeldt, A. and Westerlund, S. Trace metal concentrations in the anoxic bottom water of Framvaren. in: Kramer, C.V.M and Duinker, J.C., Eds. Complexation of Trace Metals in Natural Waters: Nijheff / Junk; 1984: 239-245.
61. Eaton, A. Marine geochemistry of cadmium. **Marine Chem.;** 1976; **4:** 141-154.
62. Ehrlich, P. R., Ehrlich, A. H. and Holdren, J. P. Ecoscience, Population, Resources, Environment: Freeman; 1977.
63. Eichenberger, B. A. and Chen, K. Y. Origin and nature of selected inorganic constituents in natural waters. In: Minear, R. A, and Keith, L. H., Eds. Water Analysis, Inorganic Species: Academic Press; 1982; 1, Part 1: 1-54.
64. Elderfield, H. and Hepworth, A. Diagenesis, metals and pollution in estuaries. **Marine Pollut. Bull.;** 1975; **6:** 85-87.
65. Elgmark, K., Hagen, A. and Langeland, A. Polluted snow in southern Norway during the winters 1968-71. **Environ. Pollut.;** 1973; **4:** 41-52.
66. Elias, R. W., Hirao, Y. and Patterson, C. C. The circumvention of the natural biopurification of calcium along nutrient pathways by atmospheric inputs of industrial lead. **Geochim. et Cosmochim. Acta;** 1982; **46:** 2561-2580.
67. Farmer, J. G. and Cross, J. D. The determination of arsenic in Loch Lomond sediment by instrumental neutron activation analysis. **Radiochem. Radioanal. Lett.;** 1979; **39:** 424-440.
68. Farrah, H. and Pickering, W.F. The effect of pH and complex formation on the uptake of heavy metal ions clays and sediments. Proc NZ Seminar Trace Elements in Health; 1979: 192-199.
69. Farrah, H. and Pickering, W.F. pH effects on the adsorption of heavy metal ions by clays. **Chem. Geol.;** 1979; **25:** 317-326.
70. Ferguson, J. F. and Gavis, J. A review of the arsenic cycle in natural waters. **Water Res.;** 1972; **6:** 1259-1274.
71. Fergusson, J. E. Inorganic Chemistry and the Earth: Pergamon Press; 1982.
72. Fergusson, J.E., Hayes, R.W., Tan Seow Yong and Sim Hang Thiew. Heavy metal pollution by traffic in Christchurch, New Zealand: Lead and

cadmium content of dust, soil and plant samples. **N. Z. J. Sci.**; 1980; **23:** 293-310.

73. Fitzgerald, W. F. Distribution of mercury in natural waters. In: Nriagu, J. O., Ed. The Biogeochemistry of Mercury in the Environment: Elsevier/ Nth. Holland; 1979: 161-173.
74. Flegal, A. R. and Patterson, C. C. Vertical concentration profiles of lead in the central Pacific at 15 °N and 20°S. **Earth Planet. Sci. Lett.**; 1983; **64:** 19-32.
75. Florence T. M. Electrochemical approach to trace element speciation in waters: a review. **Analyst**; 1986; **111.**
76. Florence, T. M. The speciation of trace elements in waters. **Talanta**; 1982; **29:** 345-364.
77. Florence, T.M. Determination of trace metals in marine samples by anodic stripping voltammetry. **Electroanal Chem.**; 1972; **35:** 237-245.
78. Florence, T.M. Development of physico-chemical speciation procedures to investigate the toxicity of copper, lead, cadmium and zinc towards aquatic biota. **Anal. Chem. Acta**; 1982; **141:** 73-94.
79. Förstner, U. Cadmium in polluted sediments. in: Nriagu, J. O., Ed. **Cadmium in the Environment, Part I,** Ecological Cycling: Wiley; 1980: 305-363.
80. Förstner, U. Changes in metal mobilities in aquatic and terrestrial cycles. In: Patterson, J. W. and Pasino, R., Eds. Metal Speciation; Separation and Recovery: Lewes Pub. Co.; 1987: 3-26.
81. Förstner, U. Chemical forms and environmental effects of critical elements in solid-waste materials: combustion residues. In: Bernhard, M., Brinckman, F. E. and Sadler, P. J., Eds. The Importance of Chemical "Speciation" in Environmental Processes, Dahlem Konferenzen: Springer Verlag; 1986: 465-491.
82. Förstner, U. Inorganic pollutants, particularly heavy metals in estuaries. in: Olausson, E. and Cato, I., Eds. Chemistry and Biogeochemistry of Estuaries: Wiley; 1980: 307-348.
83. Förstner, U. Metal pollution of terrestrial waters. in: Nriagu, J. O., Ed. Changing metal Cycles and Human Health, Dahlem Konferenzen: Springer-Verlag; 1984: 71-94.
84. Förstner, U and Wittmann, G. T. W. Metal Pollution in the Aquatic Environment. 2nd. Ed.: Springer-Verlag; 1981.
85. Förstner, U., Patchineelam, R. and Schmoll, G. Chemical forms of heavy metals in natural and polluted sediments. In: Management and Control of Heavy metals in the Environment: CEP Consultants; 1979: 316-319.
86. Fujiki, M. The pollution of Minamata Bay by mercury and Minamata disease. in: Baker, R.A., Ed. Contaminants and Sediments.: Ann Arbor; 1980; 2: 493-500.
87. Galloway, J. N., Thornton, J. D., Norton, S. A., Volchok, H. L. and McLean, R. A. N. Trace metals in atmospheric deposition: a review and assessment. **Atmos. Environ.**; 1982; **16:** 1677-1700.

88. Gardiner, J. The chemistry of cadmium in natural water I A study of cadmium complex formation using the cadmium specific-ion electrode. **Water Res.;** 1974; **8:** 23-36.
89. Gardiner, J. The chemistry of cadmium in natural water II The absorption of cadmium on river muds and naturally occurring solids. **Water Res.;** 1974; **8:** 157-164.
90. Giesy, J. P. Cadmium interactions with naturally occurring organic ligands. in: Nriagu, J. O., Ed. Cadmium in the Environment, Part I, Ecological Cycling: Wiley; 1980: 237-256.
91. Glasby, G.P., Keays, R.R. and Rankin, P.C. The distribution of rare earth, precious metals and other trace elements in recent and fossil deep sea manganese nodules. **Geochem. J.;** 1978; **12:** 229-243.
92. Grandstaff, D. E. and Meyer, G. H. Lead contamination of urban snow. **Arch. Environ. Health;** 1979; **34:** 222-223.
93. Grantham, D. A. and Jones, J. F. Arsenic contamination of water wells in Nova Scotia. **J. Amer. Water Wks. Assoc.;** 1977; **69:** 653-657.
94. Griffin, R.A. and An, A.K. Lead adsorption by montmorillonite using a competitive Langmuir equation. **Soil Sci. Soc. Am. J.;** 1977; **41:** 880-882.
95. Guy, R. D., Chakrabarti, C. L. and McBain, D. C. An evaluation of extraction techniques for the fractionation of copper and lead in model sediment systems. **Water Res.;** 1978; **12:** 21-24.
96. Haekel, W. Investigations of the complexation of heavy metals with humic substances in estuaries. in: Kramer, C.V.M and Duinker, J.C., Eds. Complexation of Trace Metals in Natural Waters: Nijheff/Junk; 1984: 229-238.
97. Hagen, A. and Langeland, A. Polluted snow in southern Norway and the effect of meltwater on fresh water and aquatic organisms. **Environ. Pollut.;** 1973; **5:** 45-57.
98. Hahne, H. C. H. and Kroontje, W. Significance of pH and chloride concentration and behaviour of heavy metal pollutants: mercury(II), cadmium(II), zinc(II) and lead(II). **J. Environ. Quality;** 1973; **2:** 444-450.
99. Hamilton-Taylor, J. Enrichment of zinc, lead and copper in recent sediments of Windermere, England. **Environ. Sci. Techn.;** 1979; **13:** 693-697.
100. Hazen, R. E. and Kneip, T. J. Biogeochemical cycling of cadmium in a marsh ecosystem. in: Nriagu, J. O., Ed. Cadmium in the Environment, Part I, Ecological Cycling: Wiley; 1980: 399-424.
101. Heit, M., Tan, Y., Klusek, C. and Burke, J. C. Anthropogenic trace elements and polycyclic aromatic hydrocarbon levels in sediment cores from two lakes in the Adirondack acid lake region. **Water, Air Soil Pollution;** 1981; **15:** 441-464.
102. Hem, J. D. Chemistry and occurrence of cadmium and zinc in surface water and ground water. **Water Resources Res.;** 1972; **8:** 661-679.
103. Henderson, P. Inorganic Geochemistry: Pergamon; 1982.
104. Hermann, R. and Neumann-Mahlkau, P. The mobility of zinc, cadmium,

copper, lead, iron and arsenic in ground water as a function of redox potential and pH. **Sci. Total Environ.**; 1985; **43:** 1-12.

105. Hershelman, G.P., Schafer, H.A., Jan T-K and Young D.R. Metals in marine sediments near a large California Municipal Outfall. **Marine Pollut. Bull.**; 1981; **12:** 131-134.
106. Horne, R. A. The Chemistry of Our Environment.: Wiley; 1978.
107. Horowitz, A. J. A primer on trace metal sediment chemistry. US Geol. Survey, Water Supply Paper, 2277; 1985.
108. Jonasson, I. R. and Boyle R. W. Geochemistry of mercury and origins of natural contamination of the environment. **Canad. Mining Met. Bull.**; 1972; **65:** 32-39.
109. Jones K. C. The distribution and partitioning of silver and other heavy metals in sediments associated with an acid mine drainage stream. **Environ. Pollut.**; 1986; **12B:** 249-263.
110. Kauss, P.B. Studies of trace contaminants, nutrients and bacteria levels in the Niagara river. **Great Lakes Res.**; 1983; **9:** 249-273.
111. Kester, D. R. Equilibrium models in seawater: applications and limitations. In: Bernhard, M., Brinckman, F. E. and Sadler, P. T., Eds. Speciation in Environmental Proceses, Dahlem Konferenzen: Springer Verlag; 1986: 337-363.
112. Khalid, R. A. Chemical mobility of cadmium in sediment-water systems. in: Nriagu, J. O., Ed. Cadmium in the Environment. Part I. Ecological Cycling: Wiley; 1980: 257-304.
113. Khalid, R. A., Gambrell, R. P. and Patrick, W. H. Chemical mobilization of cadmium in the estuarine sediment as affected by pH and redox potential. In: Management and Control of Heavy metals in the Environment: CEP Consultants; 1979: 320-324.
114. Kharhar, D. P., Turekain, K. K. and Bertine, K. K. Stream supply of dissolved silver, molybdenum, antimony, selenium, chromium, cobalt, rubidium and caesium to the oceans. **Geochim. et Cosmochim. Acta**; 1968; **32:** 285-298.
115. Kheboian, C. and Bauer, C. F. Accuracy of selective extraction procedures for metal speciation in model aquatic sediments. **Anal. Chem.**; 1987; **59:** 1417-1423.
116. Kirk, P.W. W. and Lester, J. N. Signifiance and behaviour of heavy metals in waste water treatment processes IV water quality standards and criteria. **Sci. Total Environ.**; 1984; **40:** 1-44.
117. Klein, L. A., Lang, M. Nash, N. and Kirschner, S. L. Sources of metals in New York city waste water. **J. Water Pollut. Control Assoc.**; 1974; **46:** 2653-2662.
118. Knutson, A. B., Klerks, P. L. and Levinton, J. S. The fate of metal contaminationed sediments in Foundry Cove, New York. **Environ. Pollut.**; 1987; **45:** 291-304.

119. Kramer, C. J. M., Yu Guo-hui, and Duinker, J. C. Possibilities for misinterpretation in ASV-speciation studies of natural waters. **Fresenius' Z. Anal. Chem.**; 1984; **317:** 383-384.
120. Kranck, K. Sedimentation processes in the sea. in: Hutzinger, O., Ed. Handbook of Environmental Chemistry: Springer-Verlag; 1980; **2A:** 61-75.
121. Krantzberg, G. The influence of bioturbation on physical, chemical and biological parameters in aquatic environments : A review. **Environ. Pollut.**; 1985; **39A:** 99-122.
122. Kronfeld, J. and Navrot, J. Transition metal contamination in the Qishon River system, Israel. **Environ. Pollut.**; 1974; **6:** 281-288.
123. Langston, W.J. and Bryan, G.W. The relationships between metal speciation in the environment and bioaccumulation in aquatic organisms. in: Kramer, C.V.M and Duinker, J.C., Eds. Complexation of Trace Metals in Natural Waters: Nijheff / Junk; 1984: 375-392.
124. Laxen, D. P. H. Cadmium adsorption in fresh waters - a quantitative appraisal of the literature. **The Sci. Total Environ.**; 1983; **30:** 129-146.
125. Laxen, D. P. H. and Harrison, R. M. The highway as a source of water pollution, an appraisal with the heavy metal lead. **Water Res.**; 1977; **11:** 1-11.
126. Laxen, D.P.H. The Chemistry of Metal Pollutants in water.. Harrison, R.M., Ed. in Pollution: Causes, Effects and Control: Royal Chem. Soc.; 1983: 104-123.
127. Lazrus, A. L., Lorange, E. and Lodge, J. P. Lead and other metal ions in United States precipitation. **Environ. Sci. Techn.**; 1970; **4:** 55-58.
128. Leland, H.V., Shukla, S.S. and Shimp, N.F. Factors affecting distribution of lead and other trace elements in sediments of Southern Lake Michigan. In: Singer, P.C., Ed. Trace Metals and Metal-Organic Interactions in Natural Waters; 1974: 89-129.
129. Lum, K.R. and Edgar, D.G. Determination of the chemical forms of cadmium and silver in sediments by Zeeman effect flame atomic absorption spectrometry. **Analyst.**; 1983; **108:** 918-924.
130. Luorna, S.N. and Bryan, G.W. A statistical assessment of the form of trace metals in oxidised estuarine sediments employing chemical extractants. **The Sci. Tot. Environ.**; 1981; **17:** 165-196.
131. Lutze, R. L. Implementation of a sensitive method for determining mercury in surface waters and sediments by cold vapour atomic absorption spectrophotometry. **Analyst**; 1979; **104:** 979-982.
132. Mann, A. W. and Lintern, M. Heavy metal dispersion patterns from tailings dumps Northampton district Western Australia. **Environ. Pollut.**; 1983; **B6:** 33-49.
133. Mart, L. and Nürnberg, H. W. Trace metal levels in the eastern Arctic ocean. **Sci. Total Environ.**; 1984; **39:** 1-14.

134. Mart, L., Nurnberg, H. W. and Valenta, P. Comparative base line studies on Pb-levels in European coastal waters. in: Branica, M. and Konrad, Z., Eds. Lead in the Marine Environment: Pergamon; 1980: 155-179.

135. Martin, J. H., Knauer, G. A. and Flegal, A. R. Cadmium in natural waters. in: Nriagu, J. O., Ed. **in:** Cadmium in the Environment Part 1. Ecological Cycles: Wiley; 1980: 141-145.

136. Mason, B and Moore, C. B. Principles of Geochemistry. 4th Ed. ed.: J. Wiley; 1982.

137. Mathis, B.J. and Cummings, T.F. Selected metals in sediments, water and biota in the Illinois River. **Water Poll. Control Fed.;** 1973; **45:** 1573-1583.

138. Menasveta, P. and Cheevaparanapiwat, V. Heavy metals, organochlorine pesticides and PCBs in green mussels, mullets and sediments of river mouths in Thailand. **Marine Pollut. Bull.;** 1981; **12:** 19-25.

139. Miyazaki, A., Kimura, A. and Umezaki, V. Determination of arsenic in sediments by chloride formation and DC plasma arc emission spectrometry. **Anal. Chim. Acta;** 1979; **107:** 395-398.

140. Moore, J. W. and Ramamoorthy, S. Heavy metals in Natural Waters: Applied Monitoring and Impact Assessment.: Springer-Verlag; 1984.

141. Morgan, J. J. General affinity concepts, equilibria and kinetics in aqueous metals chemistry. In: Patterson, J. W. and Pasino, R., Eds. Metal Speciation; Separation and Recovery: Lewes Pub. Co.; 1987: 27-61.

142. Mudroch, A. Distribution of major elements and metals in sediment cores from the western basin of Lake Ontario. **Great Lakes Res.;** 1983; **9:** 125-133.

143. Mudroch, A. Particle size effects on concentrations of metals in Lake Erie bottom sediments. **Water Pollut. Res. J. Canada.;** 984; **19:** 27-35.

144. Mueller, J. A., Anderson, A. R. and Jeris, J. S. Contamination in the New York Bight. **J. Water Pollut. Control Fed.;** 1976; **48:** 2309-2326.

145. Muhlbaier, J. and Tisue, G. T. Cadmium in the southern basin of Lake Michigan. **Water Soil Pollut.;** 1981; **15:** 45-59.

146. Muller, G. and Förstner, U. Heavy metals in sediments of Rhine and Elbe estuaries: mobilization or mixing effect. **Environ. Geol.;** 1975; **1:** 33-39.

147. Muller, G. and Förstner, U. Schwermetalle un den sedimenten der Elbe bei Stade : Veranderungen seit 1973. **Naturwiss.;** 1976; **63:** 242-243.

148. Murozumi, M., Chow, T. J. and Patterson, C. C. Chemical concentrations of pollutant lead aresols terrestrial dusts and sea salt in Greenland and Antarctic snow strata. **Geochim. et Cosmochim. Acta;** 1969; **33:** 1247-1294.

149. Ndiokwere, C. L. An investigation of the heavy metal content of sediments and algae from the river Niger and Nigerian Atlantic coastal waters. **Environ. Pollut.;** 1984; **7B:** 247-254.

150. Ng, A. and Patterson, C. C. Changes of lead and barium with time in California off-shore basin sediments. **Geochim. et Cosmochim. Acta;** 1982; **46:** 2307-2321.

151. Ng, A. and Patterson, C. C. Natural concentrations of lead in ancient Arctic and Antarctic ice. **Geochim. et Cosmochim. Acta;** 1981; **45:** 2109-2121.
152. Nissenbaum, A. Trace elements in Dead Sea sediments. **Israel J. Earth-Sci.;** 1974; **23:** 111-116.
153. Nriagu, J. O. Cadmium in the atmosphere and precipitation. in: Nriagu, J. O., Ed. **Cadmium in the** Environment Part I Ecological Cycling: Wiley; 1980: 71-114.
154. Nriagu, J. O. Global cadmium cycle. in: Nriagu, J. O., Ed. in: Cadmium in the Environment Part 1. Ecological Cycles: Wiley; 1980: 1-12.
155. Nriagu, J. O. Properties and the biogeochemical cycling of lead. in: Nriagu, J. O., Ed. The Biogeochemistry of Lead in the Environment, Part A, Ecological Cycling: Elsevier/Nth. Holland; 1978: 1-13.
156. Nriagu, J. O. and Rao, S. S. Response of lake sediments to changes in trace emtal emissions from smelters at Sudbury, Ontario. **Environ. Pollut.;** 1987; **44:** 211-218.
157. Nriagu, J.O. Lead in soils, sediments and major rock types . in: Nriagu, J. O., Ed. The Biogeochemistry of Lead in the Environment, Part A, Ecological Cycling: Elsevier / Nth. Holland.; 1978: 15-72.
158. Nurnberg, H. W,. The voltammetric appproach in trace metal chemistry of natural waters and atmospheric precipitation. **Anal. Chim. Acta;** 1984; **164:** 1-21.
159. Nurnberg, H.W. Potentialities of voltammetry for the study of physico-chemical aspects of heavy metal cmplexation in natural waters. in: Kramer, C.V.M. and Duinker, J.C., Eds. Complexation of Trace Metals in Natural Waters: Nijheff/Junk; 1984: 95-115.
160. O'Brien, B. J., Smith, S. and Coleman, D. O. Lead pollution of the global environment. in: Progress Reports in Environmental Monitoring and Assessment. 1 Lead. Tech. Rept.: MARC; 1980: No. 16-18.
161. Oscarson, D. W., Huang, P. M. and Liaw, W. K. The oxidation of arsenite by aquatic sediments. **J. Environ. Qual.;** 1980; **9:** 700-703.
162. Pace, C. B. and di Giulio, R. T. Lead concentration in soil, sediment and clam samples from the Pungo River peatland area of North Carolina, USA. **Environ. Pollut.;** 1987; **43:** 301-311.
163. Perhac, R. M. Distribution of Cd, Co, Cu, Fe, Mn, Ni, Pb and Zn in dissolved and particulate solids from two streams in Tennessee. **J. Hydrol.;** 1972; **15:** 177-186.
164. Pickering, W.F. Cadmium retention by clays and other soil or sediment components. in: Nriagu, J.O., Ed. in Cadmium in the Environment, Part 1 Ecological Cycling: Wiley; 1980: 365-397.
165. Pilkington, E.S. and Warren, L.J. Determination of heavy-metal distributions in marine sediments. **Environ. Sci. Techn.;** 1979; **13:** 295-299.
166. Piotrowski, J. K. and Coleman, D. O. Environmental hazards of heavy metals: summary evaluation of lead, cadmium and mercury. MARC Report No 20.; 1980.

167. Pocock, S. J. Factors influencing household water lead: a British national survey. **Arch. Environ. Health**; 1980; **35:** 45-50.
168. Pocock, S. J. Shaper, A. G., Walker, M., Wale, C. J. Clayton, B., Delves, T., Lacey, R. F., Packham, R. F. and Powell, P. Effects of tap water lead, water hardness, alcohol and cigarettes on blood lead concentrations. **J. Epidemol. Commun. Health**; 1983; **37:** 1-7.
169. Posselt, H. S. and Weber, W. J. Studies on the aqueous corrosion chemistry of cadmium. in: Aqueous Environmental Chemistry of Metals: Ann Arbor Science; 1976: 291-315.
170. Purchase, N.G. and Fergusson, J.E. The distribution and geochemistry of lead in river sediments, Christchurch, New Zealand. **Environ. Pollut.**; 1986; **B12:** 203-216.
171. Raspor, B. Distribution and speciation of cadmium in natural waters. in: Nriagu, J. O., Ed. Cadmium in the Environment Part I Ecological Cycling: Wiley; 1980: 147-236.
172. Raspor, B., Nürnberg, H. W., Valenta, P. and Brancia, M. The chelation of lead by organic ligands in sea water. in: Brancia, M. and Konrad, Z., Eds. Lead in the Marine Environment: Pergamon; 1980: 181-195.
173. Reid, J. D. and McDuffie, B. Sorption of trace cadmium on clay minerals and river sediments: efects of pH and Cd(II) concentrations in a synthetic river water river medium. **Water Air Soil Pollut.**; 1981; **15:** 375-386.
174. Rendell, P. S. Batley, G. E. and Cameron, A. J. Adsorption as a control of metal concentrations in sediment extracts. **Environ. Sci. Techn.**; 1980; **14:** 314-318.
175. Rickard, D. T. and Nriagu, J. O. Aqueous environmental chemistry of lead. in: Nriagu, J. O., Ed. in: The Biogeochemistry of Lead in the Environment, Part A, Ecological Cycling: Elsevier/Nth Holland; 1978: 219-284.
176. Ross, D. A. Introduction to Oceanography: Appleton-Century Crofts; 1970.
177. Ruppert, H. Fixation of metals on hydrous manganese and iron oxide phases in marine Mn-Fe nodules and sediments. **Chem. Erde**; 1980; **39:** 97-132.
178. Salim, R. and Cooksby, B.C. Kinetics of the adsorption of lead on river mud. **Plant and Soil**; 1980; **54:** 399-417.
179. Salomons, W. and Förstner, U. Metals in the Hydrocycle: Springer-Verlag; 1984.
180. Salomons, W. and Baccini, P. Chemical species and metal transport in lakes. In: Bernhard, M., Brinckman, F. E. and Sadler, P. T., Eds. Speciation in Environmental Proceses, Dahlem Konferenzen: Springer Verlag; 1986: 193-216.
181. Schaule, B. and Patterson, C. C. The occurrence of lead in the Northeast Pacific and the effects of anthropogenic inputs. in: Branica, M. and Konrad, Z., Eds. Lead in the Marine Environment: Pergamon; 1980: 31-43.

182. Schaule, B. K. and Patterson, C. C. Lead concentrations in the northeast Pacific: evidence for global anthropogenic pertubation. **Earth Planet. Sci. Lett.;** 1981; **54:** 97-116.
183. Scrudato, R.J. and Estes, E.L. Clay-lead sorption relations. **Environ. Geol.;** 1975; **1:** 167-170.
184. Settle, D. M. and Patterson, C. C. Magnitude and sources of precipitation and dry deposition fluxes of industrial and natural fluxes to the North Pacific at Enewetok. **J. Geophys. Res.;** 1982; **87c:** 8857-8869.
185. Shephard, B. K., McIntosh, A. W., Atchison, G. J. and Nelson, D. W. Aspects of the aquatic chemistry of cadmium and zinc in a heavy metal contaminated lake. **Water Res.;** 1980; **14:** 1061-1066.
186. Shirahata, H., Elias, R. W. and Patterson, C. C. Chronological variations in concentrations and isotopic compositions of anthropogenic atmospheric lead in sediments of a remote subalpine pond. **Geochim. et Cosmochim. Acta;** 1980; **44:** 149-162.
187. Sibley, T. H. and Morgan, J. J. Equilibrium speciation of trace metals in fresh water sea water mixtures. in: Int. Conf. on Heavy metals in the Environment, Toronto; 1975: 319-338.
188. Sigg, L. Metal transfer mechanisms in lakes. in: Stumm, W., Ed. Chemical processes in Lakes: Wiley; 1985: 283-310.
189. Sigg, L., Stumm, W. and Zinder, B. Chemical processes at the particle-water interface; implications concerning the form of occurance of solute and absorbed species. in: Kramer, C.V.M and Duinker, J.C., Eds. Complexation of Trace Metals in Natural Waters: Nijheff / Junk; 1984: 251-266.
190. Sinex, S.A., Cantillo, A.Y. and Helz, G.R. Accuracy of acid extraction methods for trace metals in sediments. **Anal. Chem.;** 1980; **52:** 2342-2346.
191. Skei, J.M., Price, N.B., Calvert, S.E. and Haltedahl, H. The distribution of heavy metals in sediments of Sorfjord, West Norway. **Water, Air and Soil Pollut.;** 1972; **1:** 452-461.
192. Skinner, B. J. and Turekian, K. K. Man and the Ocean: Prentice Hall; 1973.
193. Slavin, N. The determination of trace metals in sea water. **Atom. Spectros.;** 1980; **1:** 66-71.
194. Sly, P.G. Sedimentology and geochemistry of recent sediments off the mouth of the Niagara river; Lake Ontario. **Great Lakes Res.;** 1983; **9:** 134-159.
195. Smith, D. G. Heavy metals in the New Zealand aquatic environment: A review. N. Z. Water and Soil, Miscell. Pub. No. 100: Ministry of Works, Water Quality Centre; 1986.
196. Smith, D. G. Sources of heavy metal input to the New Zealand aquatic environment. **J. Roy. Soc. N. Z.;** 1985; **15:** 371-384.
197. Stoffers, P, Glasby, G. P., Wilson, C. J. Davis, K. R. and Walter, P. Heavy metal pollution in Wellington harbour. **N. Z. J. Marine Fresh Water Res.;** 1986; **20:** 495-512.

198. Stumm, W. and Brauner, P.A. Chemical Speciation. in: Riley, J.P. and Skirrow, G., Eds. **Chemical** Oceanography. 2nd Ed.: Academic Press; 1975; 1: 173-239.

199. Stumm, W. and Morgan, J. J. Aquatic Chemistry: Wiley-Interscience; 1970.

200. Sugimura, Y.,Suzuki, Y. and Miyake, Y. Chemical forms of minor metallic elements in the ocean. **Oceanogr. Soc. Japan;** 1978; **34:** 93-96.

201. Swallow, K.C., Hume, D.N. and More, F.M.M. Sorption of copper and lead by hydrous ferric oxide. **Environ. Sci. and Techn.;** 1980; **14:** 1326-1331.

202. Taylor, D. The distribution of heavy metals in the United Kingdom coastal waters of the North Sea. In: Management and Control of Heavy Metals in the Environment: CEP Consultants; 1979: 312-315.

203. Thomas, R.L. Lake Ontario sediments as indicators of the Niagara River as a primary source of contaminants. **Great Lakes Res.;** 983; **9:** 118-124.

204. Thorne, L. T. Thornton, I., Reynolds, B. McCalley, D. and Metcalfe, A. Geochemical investigation of the dispersal of metals in surface drainage systems derived from pollution and geological sources. In: Management and Control of Heavy Metals in the Environment: CEP Consultants; 1979: 225-228.

205. Thornton, I., Watling, H. and Darracott, A. Geochemical studies in several rivers and estuaries used for oyster rearing. **Sci. Total Environ.;** 1975; **4:** 325-345.

206. Trefry, J. H. and Metz, S. Selective leaching of trace metals from sediments as a function of pH. **Anal. Chem.;** 1984; **56:** 745-749.

207. Turekian, K. K. Oceans: Prentice Hall; 1968.

208. Turner, D.R. Speciation and cycling of arsenic, cadmium, lead and mercury in natural waters. In: Hutchinson, T.C. and Meena, K.M., Eds. Lead, Mercury, Cadmium and Arsenic in the Environment. Scope 31: Wiley; 1987: 175-186.

209. Twomey, S. Atmospheric aerosols: Elsevier; 1977.

210. Wageman, R. Some theoretical aspects of stability and solubility of inorganic arsenic in the freshwater environment. **Water Res.;** 1978; **12:** 139-145.

211. Wallace, G.T. The association of copper, mercury and lead with surface-active organic matter in coastsl seawater. **Marine Chem.;** 1982; **11:** 379-394.

212. Wangersky, P. J. Chemical oceanography. in: Hutzinger, O., Ed. Handbook of Environmental Chemistry: Springer-Verlag; 1980; **1A:** 51-67.

213. Ward, N.E., Brooks, R.R. and Reeves, R.D. Copper, cadmium, lead and zinc in soils, stream sediments, waters and natural vegetation around the Tui mine, Te Aroha, New Zealand. **N. Z. J. Sci.;** 1976; **19:** 81-89.

214. Waslenchuk, D. G. The geochemical controls on arsenic concentrations in southeastern United States rivers. **Chem. Geol.;** 1979; **24:** 315-325.

215. Whitfield, M. and Turner, D. R. The theroretical studies of the chemical speciation of lead in seawater. in: Brancia, M. and Konrad, Z., Eds. Lead in the Marine Environment: Pergamon; 1980: 109-148.
216. Windom,. Lithogenous materials in marine sediments. in: Riley, J. P. and Chester, R., Eds. **Chemical Oceanography.** 2nd. Ed.: Academic Press; 1976: 103-135.
217. Wood, J. M. Biological cycles for toxic elements in the environment. **Science;** 1974; **183:** 1049-1052.
218. Yeats, P. A. and Campbell, J. A. Nickel, copper, cadmium and zinc in the Northwest Atlantic ocean. **Marine Chem.;** 1983; **12:** 43-58.
219. Zirino, A. and Yamamoto, S. A pH dependent model for the chemical speciation of copper, zinc, cadmium and lead in seawater. **Limn. Oceanogr.;** 1972; **17:** 661-671.
220. Zoller, W. H. Anthropogenic perturbation of metal fluxes into the atmosphere. in: Nriagu, J. O., Ed. Changing metal Cycles and Human Health, Dahlem Konferenzen: Springer-Verlag; 1984: 27-41.
221. Zuehlke, R.W. and Byrne, R.H. Thermodynamic and analytical uncertainties in trace metal speciation calculations. in: Kramer, C.V.M. and Duinker, J.C., Eds. Complexation of Trace Metals in Natural Waters: Nijheff/Junk; 1984: 181-185.

CHAPTER 9

HEAVY ELEMENTS IN SOILS

The earth's crust, part of the lithosphere, is not as easily described as are the atmosphere and hydrosphere because of its heterogeneous nature. The two broad divisions of the crust are rocks and soil. Of these, soil is more directly in contact with human beings, and the part which we depend upon for most of our food.

THE NATURE OF SOIL

The basic materials of the crust are the silicates and the aluminosilicates, making up the rocks and clays. These have been described in Chapter 8 and in the diagrams in Figs. 8.11 to 8.13, and in Table 8.9. The feldspars, which are aluminosilicates, ($MAlSi_3O_8$ and $MAl_2Si_2O_8$) where M = Na^+, K^+ and Ca^{2+} are important materials in rocks.

Rocks

There are three rock types in the earth's crust, igneous, sedimentary and metamorphic. Their interrelation, in which the importance of weathering is clear, is shown in Fig. 9.1.

Igneous rocks are solidified magma and are composed of intrusive and extrusive components. Intrusive igneous rocks cooled below the earth's crust, and since the cooling was normally slow the grain or crystal size can be quite large. Extrusive igneous rocks cooled on the earth's surface and normally have a smaller crystal size because of more rapid cooling. Igneous rocks are also classified in terms of their silica (SiO_2) or silicon content. When SiO_2 > 66% the rocks are called acidic, 52-66% intermediate, 45-52% basic and < 45% ultrabasic. The most common rocks are the basalts (extrusive and basic) and granites (intrusive and acidic).

Sedimentary rocks slowly form from sediments by a process called diagenesis. The process includes compaction, loss of water under pressure, and chemi-

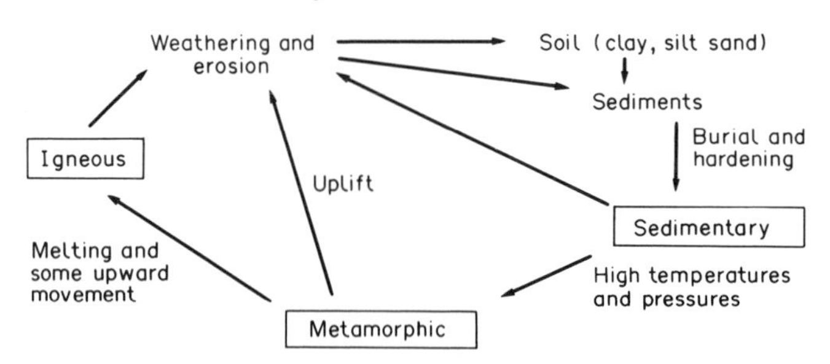

Fig. 9.1 The rock cycle.

cal breakdown of the sediments. The particles are then cemented together with materials such as gypsum ($CaSO_42H_2O$), anhydrite ($CaSO_4$), calcite ($CaCO_3$), dolomite (($Ca,Mg)CO_3$), silica, iron oxides and sulphides. Because of their origin sedimentary rocks generally contain quartz, clay, calcite and dolomite, and of lesser importance goethite (FeO(OH)), hematite (Fe_2O_3), halite and gypsum. The rocks are three main types; clastic, formed from fragments of rocks; chemical, formed from precipitates and evaporites; and organic, formed from organic material. Clastic sedimentary rocks include sandstone, mudstone and greywacke. Chemical sedimentary rocks include dolomite, rocksalt, flint, rock gypsum and organic sedimentary rocks include coal and oil shale. The most abundant sedimentary rocks are shale (> 50%), sandstone (~20%), limestone and dolomite (~15%).

Sedimentary and igneous rocks can alter under high temperatures and/or pressures, to produce metamorphic rocks. Numerous new minerals are formed under these conditions. The materials may settle into bands giving banded rocks. Typical changes are, shale to slate and limestone to marble. Metamorphic rocks are classified as slate, which is fine-grained and splits into sheets; schistose, coarser grained and has good cleavage; granular, which have no cleavage, e.g. marble and quartzite; and gneisses, which are coarse grained rocks with irregular banding. Some of the more important metamorphic minerals are: muscovite $KAl_2AlSi_3O_{10}(OH)_2$, biotite $K(Mg,Fe)_5(Al,Fe^{3+})_2$-$SiO_{10}(OH)_8$ and garnets $A_3B_2(SiO_4)_3$ where A = Ca, Mg, Fe, Mn and B = Fe^{3+}, Cr^{3+}.

Weathering

Some of the chemical materials in rocks weather more rapidly than others (see Chapter 5), and as a result the minerals in soils have either remained behind from weathering, or have been transported. Weathering is most advanced, in the humid tropics, where the basic materials (alkali and alkaline earth minerals) have weathered most rapidly, then quartz followed by iron and aluminium minerals. Some specific weathering products from primary minerals are given in Table 9.1. The extent of weathering, and the minerals present determine the soil formed and its subsequent properties. Contamination of a soil and the

TABLE 9.1 Minerals and their Weathered Products

Stage of weathering	Mineral	Released ions	Residual material	Soil type
Early	Biotite	K^+, Mg^{2+}	Clays, limonite, hematite	Dominated by these minerals and residual materials in young soils of the world, especially in arid regions.
	Calcite		Calcite	
	Gypsum	Ca^{2+}, SO_4^{2+}	Gypsum	
	Olivine	Mg^{2+}, Fe^{2+}	Clays, limonite, hematite	
	Feldspars (Na, Ca)	Na^+, Ca^{2+}	Clays	
	Pyroxene	Mg^{2+}, Ca^{2+} Fe^{2+}	Clays, limonite, hematite	
Intermediate	Clay minerals	SiO_2	Tending to baux-ite, humid tropics	Soils of temperate regions, wheat and corn belts.
	Muscovite	K^+, SiO_2	Muscovite, clays (e.g. montmorill-onite)	
Advanced	Clay minerals	SiO_2	Tending to baux-ite and iron oxide	In humid tropics, low fertility solis
	Gibbsite		Gibbsite	
	Hematite		Hematite	

Source of data; Fergusson, 1982 [39].

parent material could alter the rate of weathering, and increased acid, for example, will speed the process up. The important chemical weathering processes are: dissolution, hydration, hydrolysis, oxidation, reduction and carbonation. [39,54]

The solubilities of ionic solids are determined by their lattice energies and the hydration energies of the dissolved ions (see Chapter 8). The two energies are in opposition to each other, as the ionic radius-ionic charge diagram (Fig. 9.2) indicates. A low ionic charge and a large ionic radius, i.e. low ionic potential, is associated with more soluble species.

The solubility of many minerals is also pH dependent, for example, increase in the pH of water increases the solubility of SiO_2, by shifting the second of the reactions below, to the right.

$$SiO_2(s) + 2H_2O \rightleftharpoons H_4SiO_4(aq)$$

$$H_4SiO_4 + H_2O \rightleftharpoons H_3O^+ + H_3SiO_4^-$$

The change in solubility of three important soil species with pH is shown in Fig. 9.3. Note the amphoteric nature of Al_2O_3, and the change in solubility of SiO_2 with pH.

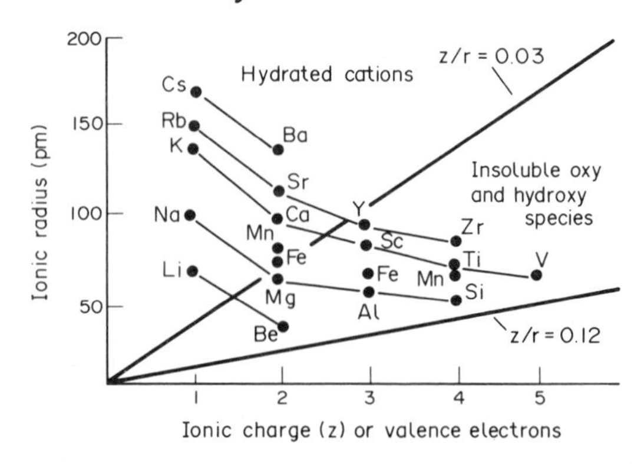

Fig. 9.2 The relationship between ionic charge, ionic radius and solubility.

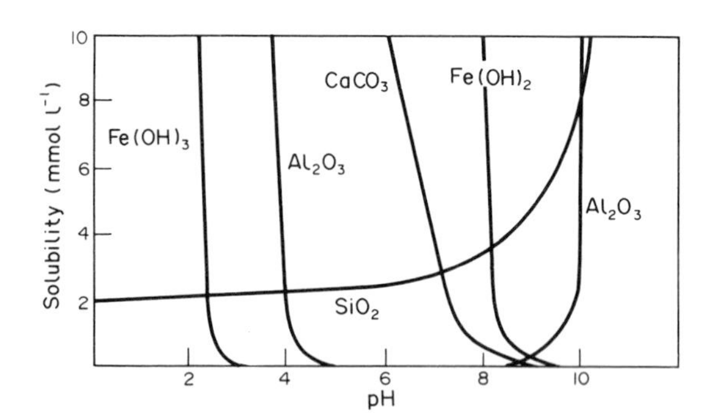

Fig 9.3 The variation in solubility of some soil constituents.

The term carbonation refers to the reaction of dissolved CO_2 with materials in the earth's crust, and in particular reaction with $CaCO_3$.

$$CaCO_3(s) + CO_2(g) + H_2O \rightleftharpoons Ca^{2+}(aq) + 2HCO_3^-(aq).$$

The formation of limestone caves is represented by the forward direction of reaction and limestone growths (stalactites and stalagmites) by the reverse reaction. Carbonation can free heavy metals from carbonates, giving concentrations in solution higher than expected, on the basis of solubility products.

For the reaction,

$$MCO_3 + CO_2 + H_2O \rightleftharpoons M^{2+} + 2HCO_3^-$$

the equilibrium constant, K, can be obtained from the following equilibria:

$$MCO_3 \rightleftharpoons M^{2+} + CO_3^{2-},\ K_{sp}$$

$$CO_2 + H_2O \rightleftharpoons H_2CO_3,\ K_p = 3.4 \times 10^{-2}$$

$$H_2CO_3 + H_2O \rightleftharpoons H_3O^+ + HCO_3^-,\ K_1 = 4.5 \times 10^{-7}$$

$$HCO_3^- + H_2O \rightleftharpoons H_3O^+ + CO_3^{2-},\ K_2 = 4.8 \times 10^{-11}$$

$$K = \frac{K_{sp}K_pK_1}{K_2}$$

$$= 318.8\ K_{sp}$$

The equilibrium constant, K, for the reaction involving lead where for $PbCO_3$ $K_{sp} = 1.6 \times 10^{-13}$ is;

$$K = \frac{[Pb^{2+}][HCO_3^-]^2}{pCO_2} = 5.1 \times 10^{-11}.$$

If we replace $[Pb^{2+}]$ by x, and since we get two moles of HCO_3^- per mole of Pb^{2+} we can write;

$$K = \frac{x.(2x)^2}{pCO_2}.$$

For $pCO_2 = 0.032$ kPa (partial pressure of CO_2 in one atmosphere at 298 K). we have;

$$5.1 \times 10^{-11} = \frac{4x^3}{0.032},$$

$$\text{i.e. } x = \sqrt[3]{\frac{5.1 \times 10^{-11} \times 0.032}{4}}$$

$$= 7.4 \times 10^{-5}\ \text{mol l}^{-1} = 15\ \text{mg l}^{-1}\ \text{(ppm)}.$$

For a partial pressure of CO_2 of 1kPa (more typical of the pressure of CO_2 in soil) the possible concentration of lead from lead carbonate would be 48 mg l^{-1}. For $CdCO_3$ which has $K_{sp} = 2.0 \times 10^{-14}$ the concentration of cadmium at the two pressures of CO_2 are 4.1 and 13 mg l^{-1} respectively.

The weathering reaction called hydrolysis is described in Chapter 8 and redox reactions are discussed in Chapter 3, in relation to E_h-pH diagrams.

Soil

Soil may be described as a mixture of inorganic and organic materials ranging from colloids to small particles, containing both dead and living materials, water and gases in variable proportions, and normally in dynamic balance. The main components are: minerals, living and dead organic material, water and air.

Sand, silt and clay The principal components of soil are the inorganic materials sand, silt and clay. The division between the three is somewhat arbitrary, and is based on particle size. Sand is mainly quartz and the particles have diameters within the range 0.063-2.0 mm. Soil particles with diameters 0.002-0.063 mm are called silt, and are also mainly quartz, with other silicate minerals of both primary and secondary origin. Finally clay particles have diameters <0.002 mm i.e. 2 μm (<0.004 mm has also been used), and consist of a wide range of secondary silicate and aluminosilicate minerals. The proportions of sand, silt and clay in soils vary widely from soil to soil. Soil texture is determined by the relative proportions of the three materials and the texture chart is portrayed in Fig. 9.4.

Organic Soil organic material may account for around 2-5% of the total soil mass, however, it has an important and active role. The materials, which are mainly in the top soil horizons, consist of dead and living organic substances such as plant litter, decaying plants and numerous organisms. The types of organisms include bacteria, fungi, algae, protozoa, worms, arthropods, mollusks and vertebrates. The amount of organic materials in a soil depends on the climatic conditions, the type of inorganic soil components and topography. An important intermediate product in the decay of organic substances in soil is humus. In the formation of humus most of the cellulose material has disappeared, lignins are modified and protein retained. Humus, containing humic acid, influences the water-holding capacity of a soil, its ion-exchange capacity and binding to metal ions.

Water and air Soil water and atmosphere fill up the interstitial spaces between soil particles. The movement of H_2O and gases is controlled by the amount of space and the size of the pores. This is in part determined by the size of the soil particles, i.e. the proportion of sand, silt and clay. The smaller the particles the greater the number that occur in a certain volume, and this generally means more pore space. The water is a solution of dissolved materials, the amount dissolved will depend, among other factors, on the pH of the soil solution.

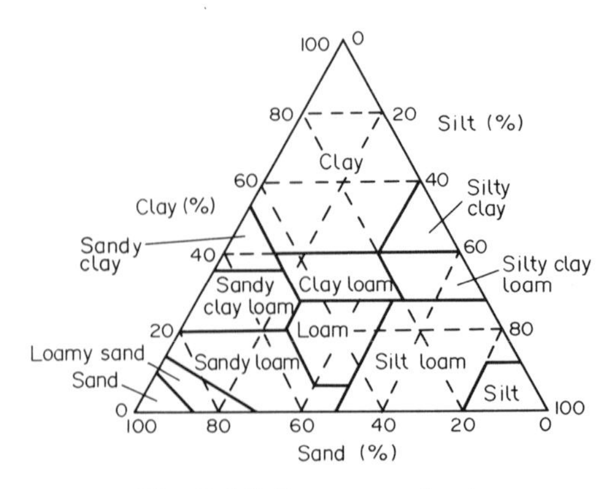

Fig. 9.4 Soil texture chart.

The soil atmosphere influences the redox processes involved in the decay of organic material and the stable oxidation state of metal ions. Absence of oxygen, as in a waterlogged soil, allows for reducing conditions to occur. The soil atmosphere and water, together fill the available space. The atmosphere is similar to, but not identical to, the normal atmosphere because of a high CO_2 concentrations from decaying materials.

Soil Minerals

Primary minerals The common primary minerals, in order of susceptibility to weathering are the felsic minerals; plagioclases (Na/Ca feldspars) > potassium feldspar > muscovite > quartz, and the mafic minerals; olivine > pyroxenes > amphiboles > biotite. The most common mineral is quartz.

Clays The clay minerals, though often a small component of soils are a significant component because of their ability to hold the heavy elements. They form from weathering of rocks, as shown in Fig. 9.5. Clays have been described briefly in Chapter 8 and in Fig. 8.13 and Table 8.9.

For the 1:1 clays, e.g. kaolinite, the repeating unit is one layer composed of one 'SiO_4 sheet' and one 'AlO_6 sheet' (Fig. 8.13). Hydrogen bonding holds the layers together and is sufficiently strong to prevent ions getting between them. The 2:1 clays, e.g. illites, have layers made up of three sheets, one 'AlO_6 sheet' sandwiched between two 'SiO_4 sheets'. In muscovite one quarter of the Si atoms are replaced by Al, and in vermiculite some of the octahedrally coordinated Al atoms are replaced by Mg and Fe. These changes alter the charge on the clays, and cations such as K^+, Ca^{2+} and Mg^{2+} fit between the layers, as well as water, all helping to hold the layers together. A 2:2 clay, e.g. chlorite, has the same layer as for a 2:1 clay but between these layers a discrete 'AlO_6 sheet' occurs, in which some of the Al atoms are replaced by Mg and Fe.

Oxides and Hydroxides Several oxides and hydroxides occur in soils, such as Al_2O_3, TiO_2, but the most important are the oxides and hydroxides of iron and

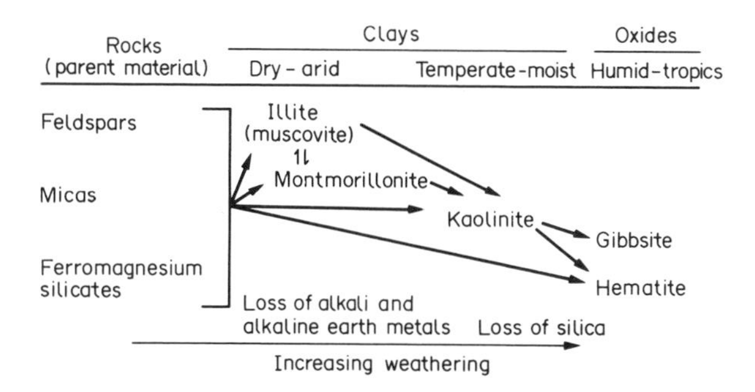

Fig.9.5 A weathering scheme for rocks and clays.

manganese. Goethite FeO(OH) is the iron species most commonly encountered, whereas for manganese the main minerals are lithiophorite $(Al_2Li(OH)_6)$-$(Mn_{0.5}Mn_{2.5}O_6)$ and birnessite $Mn_7O_{13}.5H_2O$.

Carbonates and other minerals Calcium carbonate is the dominant carbonate in soils, and normally occurs with the calcite structural form. Other minerals found in soils are phosphates of calcium, iron and aluminium, sulphides, sulphates and chlorides. A list of common primary and secondary minerals that occur in soils is given in Table 9.2 [77].

Soil forming factors

Over the centuries, various factors have determined soil type. The parent material, the rock from which the soil has developed, determines the basic mineral content and the proportions of sand, silt and clay. The typography, i.e. the slope and its exposure influences the drainage of the soil and the impact of the climate. The climate (temperature, rainfall and wind), affects the water content of, and evaporation from, the soil and leaching within the soil. Organisms influence the soils continuing fertility, as distinct from the fertility arising from the parent material. Time is important as regards continuing weathering and change. In quite recent times (relative to the time it has taken to form some soils), man has modified soils by his treatment of them and by altering the vegetation. All these factors are interrelated, and cannot be discussed in isolation. For example, a soil with a gentle undulating topography, which experiences alternating wet and dry seasons, which contains bacteria and is rich in iron and manganese, may develop iron-manganese concretions which tend to be enriched in heavy metals.

The result of soil formation can be observed from the soil profile. The profile divides into horizons which differ in the amounts of soil components. The top horizon, the O horizon, contains the majority of the organic material and humus. Next the A horizon is the main mineral zone. Downward leaching and movement of iron, aluminium and clay (eluvial zone) leaves behind the resistant minerals. The clay, iron and aluminium, accumulate in the third, the B horizon, (illuvial zone). The C horizon contains the parent material. Horizons may be further subdivided depending on the contents.

Three of the principal processes in well drained areas are, podzolization in the cool humid areas, latosolization in the humid tropics and calcification in dry areas. In podzolization the soil is leached by acid water (as low as pH 4), the acidity coming from organic materials. Soluble bases, aluminium and iron oxides are transported downwards, clays peptise (i.e. particle size reduces) and also move downwards. Hence the soils have pronounced A and B horizons. In latosolization, weaker acid (pH = 5-7) leaching occurs, or even alkaline leaching. This is probably because of the rapid mineralization of humus in the warmer climate. At higher pH SiO_2 is more soluble (Fig. 9.3) and removed from the soil, leaving behind low silicate clays (e.g. kaolinite) and iron and alumin-

TABLE 9.2 Some Common Primary and Secondary Minerals Found in Soils

Mineral group	Name	Chemical formula
a Primary minerals		
Silica minerals	Quartz, cristobalite and tridymite	SiO_2
	Opal	SiO_2nH_2O
Feldspars	Orthoclase	$KAlSi_3O_8$
	Albite	$NaAlSi_3O_8$
	Anorthite	$CaAl_2Si_2O_8$
Micas	Muscovite	$K(Si_3Al)Al_2O_{10}(OH)_2$
	Biotite	$K(Si_3Al)(Mg,Fe^{II})_3O_{10}(OH)_2$
Pyroxenes	Enstatite	$MgSiO_3$
	Diopside	$CaMgSi_2O_6$
	Augite	$Ca(Mg,Fe,Al)(Si,Al)_2O_6$
Amphiboles	Tremolite	$Ca_2Mg_5Si_8O_{22}(OH)_2$
	Actinolite	$Ca_2Fe_5Si_8O_{22}(OH)_2$
	Hornblende	$(Ca,Na)_{2(}Mg,Fe,Al)_5(Si,Al)_8O_{22}(OH)_2$
Olivines	Forsterite	Mg_2SiO_4
	Fayalite	Fe_2SiO_4
	Zircon	$ZrSiO_4$
Phosphates	Apatite	$Ca_{10}(F,OH,Cl)_2(PO_4)_6$
Oxides	Rutile & anatase	TiO_2
b Secondary minerals		
Carbonates	Calcite & aragonite	$CaCO_3$
	Dolomite	$CaMgCO_3$
	Nahcolite	$NaHCO_3$
	Trona	$Na_3CO_3HCO_32H_2O$
Sulphur containing	Gypsum	$CaSO_42H_2O$
	Pyrite	FeS_2
Halide minerals	Halite	$NaCl$
Layer silicates	Kaolinite	$Si_2Al_2O_5(OH)_4$
	Halloysite	$Si_2Al_2O_5(OH)_42H_2O$
	Chlorite	$(AlSi_3)AlMg_5O_{10}(OH)_8$
	Smectite	$M_{0.4}(Al_{0.15}Si_{3.85})(Al_{1.3}Fe^{III}{}_{0.45}Mg_{0.25})O_{10}(OH)_2nH_2O$
	Vermiculite	$M_{0.55}(Al_{1.15}Si_{2.85})(Al_{0.25}Fe^{III}{}_{0.35}Mg_{2.4})O_{10}(OH)_2nH_2O$
Oxides, hydroxides (crystalline)	Gibbsite	α-$Al(OH)_3$
	Boehmite	α-$AlOOH$
	Goethite	α-$FeOOH$
	Hematite	α-Fe_2O_3
	Lepidocrocite	γ-$FeOOH$
	Maghemite	γ-Fe_2O_3
	Pyrolusite	β-MnO_2
Oxides, hydroxides (non-crystalline)	Allophane	$Al_2O_3.SiO_2nH_2O$
	Ferrihydrate	$Fe_5HO_84H_2O$

Source of data; Paul and Huang, 1980 [77].

ium oxides which are insoluble (Fig. 9.3). In calcification, $CaCO_3$, $MgCO_3$ and $CaSO_4$ build up in top horizons because of the lack of water. A brief summary of the FAO-UNESCO classification of the world's soils is given in Table 9.3. The other well known classification is the 7th. USA Comprehensive Soil Classification System covering the main groupings; entisol, vertisol, inceptisol, aridisol, mollisol, spodosol, alfisol, ultisol, oxisol and histosol.

Soil Properties

We will consider just two important soil properties, their ion exchange capacity and soil pH, as these factors are significant as regards heavy metals in

TABLE 9.3 FAO-UNESCO Soil Map Units of the World

Soil Unit	Sign	Distri-bution %	Soil Factor **	Description	Old Names
Fluvisol	J	2.4	a	From recent alluvial deposits*	Alluvial
Gleysols	G	4.7	c	Having hydromorphic props*	Black earths
Regosols	R	10.1	a	Unconsolidated parent material*	
Arenosols	Q		a	From sand*	
Lithosols	I		a	Limited depth over rock	
Rendzinas	E	17.2	a	Shallow soil over limstone	Rendzinas
Rankers	U		a	Shallow soil over siliceous dep.*	
Andosols	T	0.8	a	From volcanic ash	Volcanic
Vertisols	V	2.4	a	Clay soils	Brown soils
Solonchaks	Z	2.0	c	With high salinity*	
Solonetz	S		c	Hydromorphic props. and salinity	
Yermosols	Y	8.9	c	Desert soil formed in arid condit.	Aridsols
Xerosols	X	6.8	c	As above more develop. horizons	Grey soils
Kastanozems	K		b	Formed under steppe vegetation	
Chernozems	C		a,b	Formed under prairie vegetation	
Phaeozems	H	3.1	b	More leached than K and C soils	
Greyzems	M		b	Under forests, cold temp. climate	
Cambisols	B	7.0	c	Highly altered, a cambic B horizon	Brown soils
Luvisols	L	7.0	c	Similar to B, more clay and leached	
Podsoluvisols	D	2.0	c	Transition between L and P soils	Grey podsols
Podzols	P	3.6	c	Highly altered due to leaching	
Planosols	W	0.9	c	Slightly leached slow permeable horizon	
Acrisols	A	8.0	c	Highly weathered & clay horizon	
Nitosols	N		c	Transition between A and F soils	Red-brown
Ferralsols	F	8.1	c	Sesquioxide clay, tropical climate	Oxisols
Histosols	O	1.8	b	Organic soils bogs, peats	

* No diagnostic horizons, ** a: parent rock, b: vegetation, c: pedogenic processes influenced by climate. Source of data; Kabata-Pendias and Kabata, 1984 [54].

soils. The properties are interrelated, and influence the availability of plant nutrients and both are susceptible to interference from pollution.

Ion exchange capacity The ion exchange capacity of a soil is its ability to hold and exchange ions. Organic matter, and clays, are effective ion-exchangers. The large surface area of clays is a reason why they have a good ion-exchanger capacity. Clays in which Si atoms are replaced by Al, are anionic, and at high pH the hydroxyl groups lose protons, or the protons are replaced by metal ions. The hydroxyl groups of the soil organic acids react in a similar way. The equations below summarize the ways clays and organic acids act as cation exchangers.

$$\text{Clay}^- + M^+ \rightleftharpoons \text{Clay-M},$$

$$\text{Clay-M}' + M^+ \rightleftharpoons \text{Clay-M} + M'^+$$

$$\text{Clay-OH} + M^+ \rightleftharpoons \text{Clay-O-M} + H^+$$

$$\text{R-C(O)-OH} + M^+ \rightleftharpoons \text{(R(O)-C-O-)M} + H^+$$

The strength of sorption of cations on to clays reflects the charge on the ions, i.e. $M^{3+} > M^{2+} > M^+$. Sometimes for cations with the same charge the order is the inverse of what is expected on the basis of ionic size. However, the cation size which is important, is not that of the free cation, but of the hydrated cation, which often is greater for the smaller cations. The 2:1 expanding clays have the best ion-exchange capacity, and the decreasing order of ion-exchange capability is given in Chapter 8, and Table 8.28 [18].

The anion exchange capacity of clays is not as significant as for cation exchange, but is important for the heavy metals that exist in anionic form in the soil, e.g. arsenate and selenate. The anionic exchange occurs according to the reaction;

$$\text{Clay-OH} + A^- \rightleftharpoons \text{Clay-A} + OH^-$$

Arsenate, like phosphate, is also strongly held by the iron and aluminium in the clay, as well as by any Mn, Fe and Al oxides present.

Soil pH Soil pH is influenced by the cations sorbed on to the clays. For example, a predominance of cations such as Ca, Mg, Na and K tends to raise the pH when they are released from a clay, as shown in the equation;

$$\text{Clay-Na} + H_2O \rightleftharpoons \text{Clay-H} + Na^+ + OH^-$$

Aluminium and iron cations, and some heavy metals, when desorbed from a clay produce acid soil solutions due to subsequent hydrolysis of the cations;

$$\text{Clay-Al} + 5H_2O \rightleftharpoons \text{Clay-H}_3 + Al(OH)_4^- + H_3O^+.$$

Acid soils can be associated with humid climates, where frequent leaching tends to produce protonated clays which lowers the pH;

$$\text{Clay-H} + H_2O \rightleftharpoons \text{Clay}^- + H_3O^+,$$

whereas arid soils where Ca, Mg, Na and K salts accumulate, have a higher pH.

SOURCE OF THE HEAVY ELEMENTS IN SOILS

There are two principal sources of the heavy elements in soils, from the weathering of the parent material and from numerous external contaminating sources. For some elements, such as lead, the levels from contamination often far exceed the levels from natural sources, whereas for the less used metals this is not the case.

Parent Material

As a consequence of weathering the heavy elements naturally occurring in the parent rocks become released into the soil. Because the particle size is now small and because chemical reactions, such as oxidation and reduction, are possible for the now accessible heavy metals, there is the chance of the metals being either retained or transported depending on their solubility. Therefore it does not necessarily follow that the relative levels of the heavy metals in a soil derived from a rock bear a relationship to the concentration of the metals in the rock. It is not possible to investigate precisely general relationships between the heavy elements in parent materials and derived soils, because of the variability in the levels that occur between different parent materials and the variability in the rock materials. Average data [14] have been used to produce Table 9.4 and Fig 9.6, which display probable background levels of the ten elements in rocks, crust and soil. There is a semblance of a relationship between the levels of the elements in the rocks (and crust in general) and the resulting levels in the soil. But this is because of the dominant influence of the high levels for Pb and As and the use of average data. If these two elements are removed from the data set there is not a statistical relationship, as may be seen from the presentation in Fig 9.6. However, in the main high levels in parent rocks tend to mean there are correspondingly high levels in the soil.

Lead Lead tends to be more concentrated in acidic magmatic rocks and argillaceous sediments and lower in ultramafic rocks and calcareous sediments. The main lead mineral PbS releases lead on weathering, as the sulphide is oxidized. Also Pb^{2+} (ionic radius 119 pm) occurs in feldspars and pegmatites by isomorphous replacement of K^+ (138 pm), Sr^{2+} (118 pm) and Ba^{2+} (135 pm), and maybe Ca^{2+} (100 pm) and Na^+ (102 pm). Like many of the heavy metals lead is more enriched in organic soils than in mineral soils, and in the latter, more

TABLE 9.4 Mean Levels of Heavy Elements in Rocks and Soil (in μg g^{-1})

Element	Basic* rocks (basalt)	Acidic* rocks (granite)	Sedimentary rocks	Crustal abundance	Range in soil	Approx. mean in soil
As	1.5	1.5	7.7	1.5	0.1-40	6
Bi	0.031	0.065	0.4	0.048	0.1-0.4	0.2
Cd	0.13	0.09	0.17	0.11	0.01-2	0.35
Hg	0.012	0.08	0.19	0.05	0.01-0.5	0.06
In	0.058	0.04	0.044	0.049	0.2-0.5	0.2
Pb	3	24	19	14	2-300	19
Sb	0.2	0.2	1.2	0.2	0.2-10	1
Se	0.05	0.05	0.42	0.05	0.01-1.2	0.4
Te			<0.1	0.005		
Tl	0.08	1.1	0.95	0.6	0.1-0.8	0.2

* Igneous rocks. Sources; references 2,6,14,32,51,54,90.

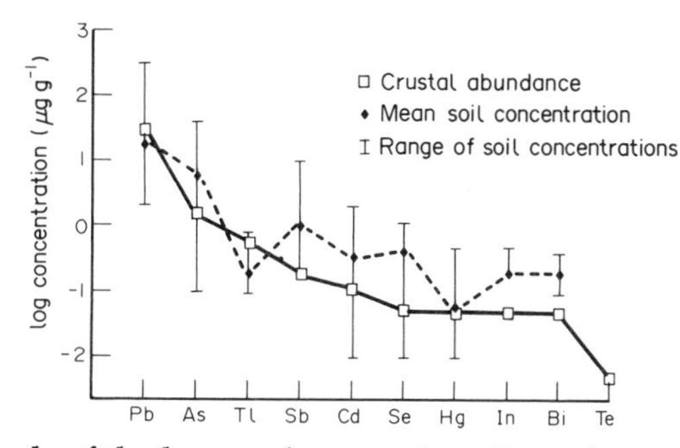

Fig. 9.6 Levels of the heavy elements in soils and average crustal concentrations.

in soils with clays such as illite and montmorillonite. Lead tends to be enriched in soils relative to the the parent material however, it is necessary to be cautious interpreting such results because of the wide spread occurrence of anthropogenic lead [2,12,32,54,55,58,73,90,94]. The drainage status of the soil is also important, and can modify the level expected from the parent material. For example, for soil derived from basic igneous parent material, the extracted lead is <0.05 μg g^{-1} in a freely drained soil, and 0.56 μg g^{-1} in a poorly drained soil. From a parent material of mica schist, the extracted lead is 0.65 μg g^{-1} in a freely drained soil, and 3.1 μg g^{-1} in a poorly drained soil [12].

Cadmium In uncontaminated soils cadmium levels tend to be determined by the parent material in the order igneous rocks (<0.1-0.3 ppm) < metamorphic rocks (0.1-1.0 ppm) < sedimentary rocks (0.3-11 ppm). The metal is more concentrated in argillaceous and shale deposits, producing high levels in the derived soils, e.g. 1.4-22 μg m^{-1} (mean 8.0 μg g^{-1}). Whereas in highly leached soils cadmium levels can be low owing to the metal ion's mobility over a wide range of pH values, it may also be enriched relative to the crustal levels. This is generally due to the soils being derived from sedimentary parent materials and from anthropogenic inputs (e.g. from fertilizers) [2,12,32,42,54,58,76,90,94]

Mercury Mercury levels in unpolluted soils probably originate from the weathering of parent rocks. Other sources may come from the degassing and thermal activity. Obviously around mineralized areas mercury levels will be high in soils [2,6,32,51,54].

Arsenic The element arsenic occurs in many minerals (>200) in the earth's crust, and around 60% of these are arsenates. The wide occurrence of arsenic means that it is present in all soils, with high levels in soils originating from sedimentary rocks, such as shales, argillaceous sediments and mica/schist. Also in some soils in Brazil higher levels occur in well weathered soils which have gibbsite as the dominant mineral [2,54,58,94]

Antimony and bismuth Since the chemistry of antimony, and to a lesser extent bismuth, is similar to that of arsenic, trends in their levels in soils tend to follow each other. This is seen for data on Scottish soils [94] presented in Fig. 9.7. The bar graphs, displayed for a range of soils from a different parent materials, show how the levels for the three elements tend to follow each other. The highest levels are in the soil derived from a quartz-mica-schist parent material. As for arsenic, both elements are enriched in soil compared with crustal abundances [2,54,58,94], even in well weathered soils [58], suggesting that they remain in residual material. Concentrations of antimony and bismuth are higher in soils from sedimentary rocks such as argillaceous sediments and shale.

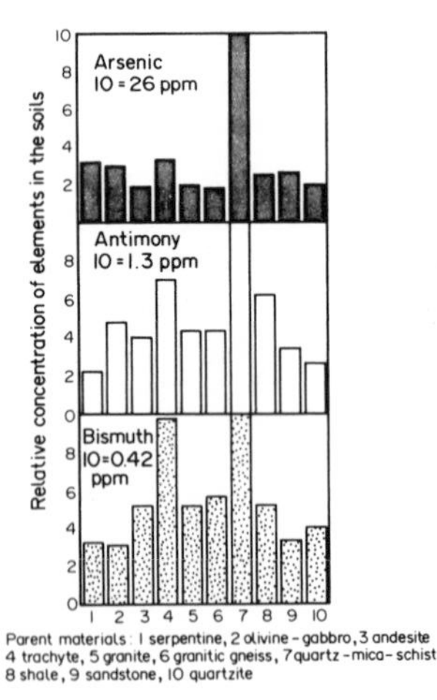

Fig. 9.7 Concentrations of As, Sb and Bi in Scottish soils in relation to parent materials. Source of data; Ure et al., 1973 [94].

Selenium and tellurium Selenium frequently follows zinc in soils as regards trends in levels, and levels are highest in Scottish soils derived from andesite parent material (for both elements). In Brazilian soils selenium is enriched in weathered soils with kaolinite as the dominant mineral. Generally the element is enriched in sedimentary rocks as a result of volcanic activity, or in black shales. Selenium distribution in Canadian soils appears to relate closely to the parent material levels, except for podzolic soils [63]. The levels in derived soils also depend on the relative solubilities of selenium compounds and redox conditions. Tellurium is poorly characterized in soils, in part due to its low abundance of ~ 0.005 ppm in the crust [2,3,54,58,94].

Indium and thallium Both indium and thallium are more concentrated in magmatic rocks, and for thallium, concentrations increase with the acidity of the rocks, and with the clay content of sedimentary rocks. Above sphalerite veins thallium levels are enriched in the derived soils [94]

Contamination of Soils

The anthropogenic sources of the heavy elements in soils are either primary sources, i.e. the heavy metals are added to the soil as an outcome of working the soil, such as fertilization or secondary sources where heavy metals are added to the soil as a consequence of a nearby activity, such as smelting or aerosol deposition. Some of the sources, and the elements that they may add to the soil are listed in Table 9.5. It appears that lead, cadmium and arsenic are the main pollutants, followed by selenium and mercury. The effects of historical mining activities remain in present day soils. In the UK, high levels of arsenic, mercury, cadmium and lead in some soils have become the heritage of earlier mining activities [91].

Some representative concentrations of the heavy elements in the primary sources, which can be added to agriculture land are given in Table 9.6 [2,20, 21,54,57,71,84,100]. Three particular sources are a cause for concern, because of the high levels of some of the trace elements in the materials. Pesticides, such as lead arsenate, calcium arsenate and arsenite, copper acetoarsenate, mercuric chloride and organomercury compounds may contribute to high levels of arsenic, lead and mercury where the pesticides have been, and are used. Fortunately the use of these materials is declining. Phosphate fertilizers are a source of arsenic, cadmium and lead (and to a lesser extent of selenium and tellurium) because of the relatively high levels of these elements in the original phosphate rock. Thirdly sewage sludge is also a significant source of cadmium and lead and to a lesser extent arsenic, bismuth, mercury, antimony and selenium. Perhaps the sludge should only be used where non-edible plants are grown, and where animals can not come into contact with it.

Since the amount of primary materials added to soils can be estimated, and their heavy element content known, it is possible to calculate the additions to a soil. The results of two such calculations are given in Table 9.7a, b [15,48]. The

TABLE 9.5 Sources of Elements that may Contaminate Soils

Source	Particular elements
Primary sources	
Fertilizers (e.g. phosphates	Cd, Pb, As
Lime	As, Pb
Pesticides	Pb, As, Hg
Sewage sludge	Cd, Pb, As
Irrigation	Cd, Pb, Se
Manure	As, Se
Secondary sources	
Automobile aerosols	Pb
Smelters	Pb, Cd, Sb, As, Se, In, Hg
Refuse incinerators	Pb, Cd
Mining areas	Pb, Cd, As, Hg
Tyre wear	Cd
Paint (weathered)	Pb, Cd
Marine	Se
Rubbish disposal	Pb, Cd, As, +
Long range aerosol	Pb, As, Cd, Se
Coal combustion	As, Se, Sb, Pb
Chloroalkali cell	Hg

Sources; references 2,13,15,19,20,21,22,25,28,34,38,42,54, 59,60,61,62,71,72,73,76,81,82,84,87,92,96,100,101.

TABLE 9.6 Levels of the Heavy Elements in Primary Sources Added to Agriculture Land (in $\mu g\ g^{-1}$)

M	Phosphate* fertilizer	Nitrogen fertilizer	Limestone	Sewage sludge	Manure	Irrigation water**	Pesti-cides¶
As	<1-1200	2-120	0.1-24	2-30	<1-25	<10	3-30
Bi				<1-100			
Cd	0.1-190	<0.1-9	<0.05-0.1	2-3000	<0.1-0.8	<0.05	
Hg	0.01-2	0.3-3		<1-56	<0.01-0.2		0.6-6
In					1.4		
Pb	4-1000	2-120	20-1250	2-7000	0.4-16	<20	11-26
Sb	<1-10			2-44	<0.1-0.5		
Se	0.5-25		≤0.1	1-17	0.2-2.4	<0.05	
Te	20-23				0.2		

*Commercial phosphates or phosphorites (phosphate rock), ** water quality criteria for short term use, ¶ % element in pesticide. Sources; references 2,17,20,21,25,42,54, 57,84,93,98,100.

TABLE 9.7 Inputs to a Soil from Primary Sources (UK)

a (in mg m^{-2} y^{-1})

Metal	Normal inputs		Abnormal inputs			
	Rain/dust	Fertilizer	Pesticides	Sewage sludge	Irrigation	Slag
As	1.6	0.3	12,000	250	0.01	15
Cd	0.5	0.6		<500	0.1	
Hg	<0.08	0.002	1.4	3-29	0.002	
Pb	37	0.8	30,000	2000	0.1	20
Se	0.4	0.03			<0.5	

Source of data; Bowen, 1975 [15].

b (tonnes y^{-1} over 5.12 x 10^6 ha)

Metal	Sewage sludge	Phosphate fertilizer*	Atmospheric deposition**	Pesticide	Total
Cd	4.9	22	15.4	0	42.3
Pb	98.3	2.4	1536	ND	1637
Hg	1.0	0.1	5.1	12	18.2
As	2.5	6.1	102.4	ND	111

* Cd 78.5, Pb 8.5 Hg 0.4, As 22.3 g t^{-1} of P_2O_5. ** Cd 3 g ha^{-1}, Pb 300 g ha^{-1}, Hg 1 g ha^{-1}, As 20 g ha^{-1}, Source of data; Hutton and Symon, 1986 [48].

three elements that tend to build up in the soil are, lead, arsenic and cadmium i.e inputs > outputs (via. cropping and drainage). The data in Table 9.7b indicates that atmospheric deposition is more important for lead, arsenic and mercury, whereas phosphate fertilizer is marginally more important for cadmium [48].

It is not as easy to quantify the contribution of the heavy elements from secondary sources as that depends on the distance and intensity of the source, and atmospheric conditions. A detailed study of the addition of secondary lead to soil in the UK has been carried out [19], where it has been estimated that since 1946 lead added from automobile generated aerosol has added ~3 $\mu g\ g^{-1}$ to top soil in rural areas and <10 $\mu g\ g^{-1}$ in urban areas. This is said to be small compared with natural levels and lead added from industrial sources over the centuries. At the time of the study automobile and industrial sources were adding lead at around the same rate in rural areas. This situation will change as less lead is used in petrol. Levels of Pb, Cd, As, Sb and Se in Norwegian surface soils show the highest levels in the south, and along the coast. The evidence suggests the elevated levels arise from long-range atmospheric transport from areas south and south-east of Norway [4].

LEVELS OF HEAVY ELEMENTS IN SOILS

The concentrations of the heavy elements in soils often cover a wide range and it is frequently difficult to determine the source of the elements. For the data presented in the Tables 9.8 to 9.18, given on pages 347 to 354, a distinction has been made between relatively uncontaminated and contaminated soils.

Levels of the Heavy Elements

Cadmium The data presented in Tables 9.8 and 9.9 for cadmium summarize some of the levels for the element in soils, which are relatively free of contamination and those which are definitely contaminated. In the former case, the mean cadmium levels lie in the range 0.1 to 1.0 $\mu g\ g^{-1}$ (Table 9.8), and for 1642 soils from over the world, a mean is 0.62 $\mu g\ g^{-1}$ has been reported [93]. The soils with high organic content e.g. histosols, tend to have the highest levels and this includes soils formed from organic containing rocks such as shales. Some soils derived from shales in California have cadmium levels of 8.0 $\mu g\ g^{-1}$. Contaminated soils can have very high levels of cadmium when close to the source, such as metal smelters (Table 9.9).

Mercury Examples of levels of mercury in relatively unpolluted and polluted soils are given in Tables 9.10 and 9.11 respectively. Mercury concentrations are usually<0.1 $\mu g\ g^{-1}$ (<100 ppb) in most soils, with levels around 0.01-0.06 $\mu g\ g^{-1}$ (10-60 ppb). The mean for 3049 soils from around the world was 0.098 $\mu g\ g^{-1}$ (98 ppb) [93]. The amount of mercury in soils derived from the parent material, may be either added to by internal degassing of the crust or thermal activity, or lost from the soil because of its volatility. Highest levels have been found in organic soils [93], some levels reaching 0.4 $\mu g\ g^{-1}$ (400 ppb). The contamination of soils by mercury, appears to be from three main sources, mining activities, the production of dichlorine and caustic soda with the mercury-chloroalkali cell, and from the use of mercury compounds in agriculture, such as fungicides [2,6,32,51,54,81,82].

Arsenic Whereas a rather wide range of levels of arsenic have been found in soils around the world, it appears that the mean level for relatively uncontaminated soils is around 5-10 $\mu g\ g^{-1}$, an overall mean of 11.3 $\mu g\ g^{-1}$ for 1193 world soils is higher than this range suggests [93] (Table 9.12). The lowest levels are normally found in sandy soils, especially those derived from granites. For example, in the USA, the mean concentrations found for sandy soils and soils over granites are 5.1 and 3.6 $\mu g\ g^{-1}$ respectively, whereas for alluvial soils and chernozems the concentrations are 8.2 and 8.8 $\mu g\ g^{-1}$ respectively. High levels of arsenic have been recorded on contaminated soils (Table 9.13), especially when associated with metal processing plants, the use of arsenical pesticides and from the mining of arsenic ores. Levels in soils reaching 600 $\mu g\ g^{-1}$ are not uncommon on soils treated with pesticides.

TABLE 9.8 Levels of Cadmium in Soils

Country	Range μg g^{-1}	Mean μg g^{-1}	Comments
Australia	0.009-0.057		Near a road
	0.013-0.56	0.28	Topsoils
Austria	0.21-0.52	0.37	Fluvisols
Brazil	<0.1-0.4		Well weathered soils
Bulgaria	0.24-0.35	0.29	Various soils
Canada		0.07-0.4	Various soils, Alberta
	0.15-0.53	0.34	Chernozem
	0.06-1.1	0.34	Luvisols
	0.19-1.22	0.57	Histosols, organic soils
	0.1-8.1	0.56	Topsoils
Denmark		0.25	Soils on glacial till
	0.8-2.2	1.05	Histosols, organic soils
England	1-17	1.6	Cities/towns/villages
England/Wales	0.08-10	1.0(med)	Topsoils
Japan	tr-0.3	0.1	
	0.03-2.53	0.44	
Poland	0.01-0.24	0.07	Podzols and sandy soils
	0.18-0.25	0.20	Loess and silty soils
	0.24-0.36	0.30	Fluvisols
	0.14-0.96	0.50	Gleysols
	0.18-0.58	0.38	Chernozem
Scotland	<0.3-1.3		Range parent materials
	0.6-1.4		Basic igneous/granite
	<0.25-2.4	0.77	Various soils
Sweden	0.03-2.3	0.22	361 samles
USA	0.005-2.4	0.27	3305 various samples, US
	<0.1-2.9	0.2	237 various samples, Ohio
		0.72	Histosols
		0.08-1.54*	Mollisols
		0.27, 1.08	Vertisols
		0.2-0.95*	Entisols
		0.27	Inceptisols
		0.21	Spodosols
		0.16-1.12*	Alfisols
		0.08	Ultisols
	1.4-22	8.0	PM** shale
	0.01-3.5	0.79	PM sandstone, basalt
	0.1-6	1.6	PM alluvial
	0.22-1.45		Near a road
	0.24-0.68	0.39	Various soils
USSR	0.01-0.07	0.06	Various soils
World	0.02-10.0	0.09-1.78*	Various soils
		0.62	Various soils

* Range of means, ** PM = parent material. Sources of data; references 4,7,11, 13,26, 27,33,54,58,60,66,70,76,80,93,94.

TABLE 9.9 Levels of Cadmium in Contaminated Soils

Country	Source of contamination	Range µg g^{-1}	Mean µg g^{-1}
Australia	Near a Pb/Zn smelter	0.03-14	
Canada	Traffic	1-14	
Eire	Industrialised	<3-3.5	
England	Old mining area	0.6-14	
	Mining area	<4-14	3
	Traffic		6
Holland	Metal processing	9-33	
	Sludged farm	15-57	
Japan	Industrial		69
	Traffic (tyres)		1.5
New Zealand	Mineralised area	15-280	
	Old mining area		12
USA	Urban soils	0.02-13.6	1.2
	One garden soil	0.48-9.53	2.45
	Near a Zn smelter (1 km)		750
	Near a smelter (300 m)		30
	Industrial		30
	Near a Pb smelter		71
	Near roads	1-10	
	Sludged	5.7-1500	185
W. Germany	Sludged farm	13-35	
Zambia	Metal processing	0.6-46	

Sources of data; references 20,25,41,54,67,69,76,81,82,84,92,95,96,97,98,101.

TABLE 9.10 Levels of Mercury in Surface Soils

Country	Range µg g^{-1}	Mean µg g^{-1}	Comments
Africa		0.023	Various soils
Austria	0.005-0.34	0.09	Various soils
Canada	0.016-0.037		
	0.020-0.20	0.05-0.07*	Various soils
	0.01-0.70	0.06	Podzols and sandy soils
	0.018-0.22	0.053	Gleysols
	0.05-1.11	0.41	Organic soils
England	0.01-0.09	0.03	Various soils
	0.008-0.19	0.04	Various soils
Holland	0.45-1.1		Loamy soils
	0.07-0.35	0.133	Topsoils
Japan	0.08-0.49	0.28	Various soils
	0.086-0.33	0.21	Rice soils
New Zealand	0.063-0.112	0.085	Surface soils
Norway	0.02-0.35	0.19	Various soils

Table 9.10 continued

Country	Range µg g⁻¹	Mean µg g⁻¹	Comments
Poland	0.02-0.16	0.06	Various soils
	0.004-0.99	0.06	Various soils
Scotland	0.09-0.2		Basic igneous/granite
	<0.01-1.71	0.13	Topsoils & profiles
Sweden	0.02-0.92	0.07	Various soils
Switzerland	0.04-0.11	0.08	Organic soils
USA	0.02-0.15	0.05	Alluvial soils
	0.01-4.60	0.28	Organic soils
		0.11	912 samples
	0.02-1.50	0.17	Various soils
W. Germany	0.025-0.35	0.09	Various soils
World		0.098	World soils

* Range of means. Sources of data; references 2,6,11,32,51,54,93.

TABLE 9.11 Levels of Mercury in Contaminated Surface Soils

Country	Source of contamination	Range µg g⁻¹	Mean µg g⁻¹
Canada	Chloroalkali cell	0.32-5.7	
	Fungicides	9.4-11.5	
England	Mining areas	0.21-3.4	
	Gardens and orchards	0.25-15	
Japan	Mercury mine		100
USA	Around a smelter		0.11
	Near chloroalkali cell		3
	Mineralized area	0.06-0.2	0.16
	Mining/mineralized area	8.2-40	
	Mining/mineralized area	0.1-2.4	
	Gardens and orchards		0.6
	Sludged land	1.93-49.0	13.6
Various	Sludged and irrigated land	0.3-10	

Sources of data; references 2,51,54,67,81,82.

TABLE 9.12 Levels of Arsenic in Surface Soils

Country	Range µg g⁻¹	Mean µg g⁻¹	Comments
Brazil	0.2-7		Well weathered soils
	0.02-2		Kaolinitic soils
Bulgaria	8.2-11.2	8.2	Chernozem
Canada	3.5-6.9		Chernozem
	0.8-6.2		Luvisol
	1.8-66.5	13.6	Histosol

Table 9.12 continued

Country	Range µg g⁻¹	Mean µg g⁻¹	Comments
Canada	1.1-16.7	6.3	Topsoils
England	20-30	25	Fluvisol
India	1.2-2.1		Various soils
Israel	0.4-2.8	1.31	Topsoils and profiles
Japan	0.4-70	11	Various soils
Korea	2.4-6.8	4.6	Podzol
Many countries		5	Various soils
	0.1-55	7.2	Various soils
Norway	0.4-7.5		Various soils
	0.7-8.8	2.5	Various soils
Scotland	4.8-8.6		Various soils
		26	Quartz/mica/schist
USA		8.7	Various soils
	<0.1-97	7.2	Various soils
	<0.1-93	3.6-8.8*	Various soils
USSR	1-10	3.6	Various soils
	5-30		Various soils
Wales	<1-46	15	Subsoils

* Range of means. Sources of data; references 2,4,33,54,58,59,72,93,94.

TABLE 9.13 Levels of Arsenic in Contaminated Surface Soils

Country	Source of contamination	Range µg g⁻¹	Mean µg g⁻¹
Canada	Metal processing	33-2000	
England	Mining area	90-900	
	Mining area	23-1080	228
Hungary	Chemical works	10-2000	
Japan	Metal processing	38-2470	
	Pesticides	38-400	
USA	Around a smelter (1.5-2 km)	89-110	80
	Pesticides	31-625	
	Pesticides	1-2550	
	Sludged land	1.13-6.21	3.08

Sources of data; references 2,54,67,82,98.

TABLE 9.14 Levels of Lead in Surface Soils

Country	Range $\mu g\ g^{-1}$	Mean $\mu g\ g^{-1}$	Comments
		44	Organic soils
		~0.2	Solonetz
		11	Various soils
Australia		12.4	Various soils
Australia	10-130		Various soils
Austria	16-22	19	Fluvisols
Brazil	5-90		Well weathered soils
	5-500		Illite/montmorillonite soil
Canada	5.2-7.3		Various soils
	12-21		Chernozems
	12-41		Luvisols
	2.3-47.5	10.4	Podzol
	1.5-50.1	16.6	Loamy and clay soils
Chad	20-50		Gleysol
China	<1-49	26	Topsoils and profiles
Denmark	43-176	50.5	Histosol
England	24-96	63	Fluvisols
	17-63	40	Gleysol
	26-142	84	Histosol
Finland	10-600	135	Topsoils
Japan	5-189	35	Various soils
Madagascar	8-175	32	Topsoils and profiles
Norway	<10-120		Various soils
		70.8	Near coast
Poland	8.5-23.5	16	Podzol
	19.5-48.5	30	Gleysol
	17-46	28.5	Rendzinas
Scotland	10-80	13-46*	Various soils
	6-20		Basic igneous or granite
	2.6-83	14	Various soils
	3.0-59	13	Mineral soils
	2-440	30	Organic soils
Sierra Leone	3-91	47	Ferralsol
Sweden	22-364	15.9	Various soils
USA		17.7	Cropland
	<10-70	17-26	Various soils
Wales	<1-356	39	Topsoils
W Germany	15-68		Various soils

* Range of means. Sources of data; references 2,4,7,11,22,27,32,33,54,55,58,62, 73,75,87, 88,93,94.

TABLE 9.15 Levels of Lead in Contaminated Soil

Country	Source of contamination	Range $\mu g\ g^{-1}$	Mean $\mu g\ g^{-1}$
	Paint	104-17,590	2010
	Paint	937-1150	
Australia	Smelter	4-2100	
Canada	Lead works	355-8750	
	Pesticides	4.4-888	123
	Traffic	400-15,000	
	Metal processing	291-12,120	
Eire	Industrial	39-540	
England	Mineralized	116-1200	
	Mineralized	1050-28,000	
	Inner city	56-1650	405
	Mining area	115-72,000	3320
	Built up areas	21-12420	671
	Traffic, inner city		1769
	Old mining area	51-21,550	
	Traffic M4	1621-6835	
	Urban gardens	270-15,240	
Holland	Sludged land	80-254	
New Zealand	Mineralized	47-12,500	
	Mineralized	20-96,000	
	Traffic	360-1210	860
	Paint		8600
Spain	Mineralized	720-30,000	
Uganda	Mineralized	500-1350	
USA	Urban gardens	1-10,900	354
	One urban garden	100-2040	740
	Smelter	560-11,450	3460
	Smelter	1700-7600	4640
	Sludged land	40-7480	1110
	Traffic	164-522	
	Paint and traffic	20-1060	
	Roadside soil	960-7000	
USSR	Metal processing		3000
Yukon	Mineralized	5-1400	
Zambia	Metal processing	92-2580	

Sources of data; references 2,20,22,25,26,29,34,38,41,54,60,61,65,67,68,69, 73,82,86, 92,96,97,98.

TABLE 9.16 Levels of Selenium in Surface Soils

Country	Range $\mu g\ g^{-1}$	Mean $\mu g\ g^{-1}$	Comments
Brazil	<0.01-0.5		Various soils
Canada	0.10-1.32	0.27	Podzols
	0.10-0.75	0.34	Histosols
	0.20-0.74	0.44	Clay and loam soils
	0.08-2.09		Various soils
Denmark	0.2-1.44	0.57	Various soils
Egypt	0.15-0.85	0.45	Fluvisols
England/Wales	0.2-1.8	0.6(med)	Various soils
England	92-230	138	Peat
Finland	0.005-1.24		Various soils
India	0.158-0.71	0.41	Various soils
Ireland	0.02-0.07	0.05	Various soils
	3-360		Peat
Japan		0.70	Various soils
	1-6		Seleniferous soils
New Zealand	0.12-2.65	0.6	Various soils
	<0.3-1.5		Various soils
Norway	0.08-1.7	0.63	Various soils
	0.07-1.35	0.42	Various soils
		2.32	Near coast
		1.05	Inland
Poland	0.06-0.38	0.14	Podzols
	0.17-0.34	0.23	Loess and silty soils
	0.24-0.64	0.44	Fluvisols
Scotland	0.02-0.36		Various soils
Sweden	0.16-0.98	0.39	Various soils
USA	<0.1-4.0	0.31	Various soils
	0.02-2.5	1.05	Lateritic soils
	<0.1-1.9	0.5	Clays and clay loam
	5.1-6.7	3.69	Near a lead smelter
	6-28	17	Mineralized area
	4.2-17.2	8.5	Sludged land
USSR	0.32-0.37	0.34	Chernozems

Sources of data; references 2,4,54,58,59,63,64,67,78,82,85,93,94,99.

TABLE 9.17 Levels of Antimony in Surface Soils

Country	Range $\mu g\ g^{-1}$	Mean $\mu g\ g^{-1}$	Comments
Brazil	0.7-20		Montmorillonite soils
	0.2-2		Kaolinite soils
	1-7		Gibbsite soils
Bulgaria		0.82	Fluvisols
		0.99	Chernozems

Table 9.17 continued

Country	Range µg g⁻¹	Mean µg g⁻¹	Comments
Bulgaria	0.82-2.32	1.32	Various soils
Canada	0.05-1.64	0.24	Various soils
	0.05-1.33	0.19	Podzols
	0.05-2.0	0.76	Loamy and clay soils
	0.08-0.61	0.28	Histosols
	1-3	1.9	Various soils
England	0.56-1.3	0.81	Various soils
Holland	0.6-2.1		Various soils
Nigeria	1-5		Various soils
Norway	0.17-2.20		Various soils
	<0.1-1.6		Various soils
		2.20	Near coast
		1.26	Inland
Scotland	0.29-1.3	0.64	Various soils
USA	<1-10	0.48(GM)	Various soils
	0.25-0.6		Various soils
	32-260	111	Near a lead smelter
	2.3-9.5		Various soils
		200	Near a copper smelter
World	0.05-2.3		Various soils
	0.05-4	0.9	Various soils

Sources of data; references 2,4,54,58,78,82,87,93,94.

TABLE 9.18 Levels of Thallium, Bismuth, Indium and Tellurium in Surface Soils

Metal	Country	Range µg g⁻¹	Mean µg g⁻¹	Comments
Tl	Canada	0.17-0.22	0.19	Garden soils
		up to 5.0		Enriched soils
		0.03-1.1		Hg mineralized area
	Scotland	0.11-0.76	0.31	Various soils
		0.1-0.8	0.2	Various soils
	USA	0.2-2.8		Various soils
			3.6	Near a cement works
Bi	Canada	1.33-1.52	1.0	Garden soils
		up to 10		Ferralitic soils
	Scotland	0.13-0.24		Various soils
		0.13-0.42	0.23	Arable land
In	Brazil	0.07-0.5	0.15	7 soils
	Canada	0.56-2.95	1.71	
Te		0.5-37		Tellurium enriched soil

Sources of data; references 2,54,93,94.

Lead The concentrations of lead in world soils vary considerably as may be seen from the data in Table 9.14, a mean of 29.2 μg g^{-1} and range of 1-888 μg g^{-1} has been reported for 4970 soils from around the world [93]. As discussed in Chapter 6 it is not possible to find a surface soil free of some lead contamination. An average figure of around 10-20 ppm is common, and levels >100 ppm indicate contamination. Organic soils tend to have higher lead levels than mineral soils, and in Scottish soils mean concentrations are 30 and 13 μg g^{-1} respectively (Table 9.14). A variety of sources including metal working, traffic, paint and smelting can produce extremely high concentrations of lead in surface soils, reaching in some cases, 1-10% lead (Table 9.15). The influence of mineralized areas on the overlying soils is shown by the data presented graphically in Fig 9.8 [73].

Selenium Some data for selenium are given in Table 9.16. An average figure for the element in soils is around 0.4 μg g^{-1}, this mean has been found for 1577 soils from around the world [50,54,63,78,93]. The element tends to follow zinc and sulphur in its variation in concentration in soils [94]. Selenium tends to be high in arid and semi-arid soils, and less high in humid areas [2]. This is probably because of the different leaching the soils have had. The element is generally enriched in soils relative to the parent material [58,63,99], and tends to build up in the organic layers [63], but in some areas it is deficient with respect to requirements for animal health (see Chapter 15). This is often the situation when the soils are derived from igneous rocks [2,99]. In a study in New Zealand low concentrations are associated with granite and rhyolitic pumic parent materials, and higher levels with andesitic ash, calcareous, argillitc and basaltic ash [99].

Antimony Antimony is generally concentrated in soils relative to parent rocks, and can be enriched as demonstrated for Brazilian soils (Table 9.17). A mean level for antimony in soils is around 0.9 μg g^{-1} or less [93]. Elevated levels have been found around smelters.

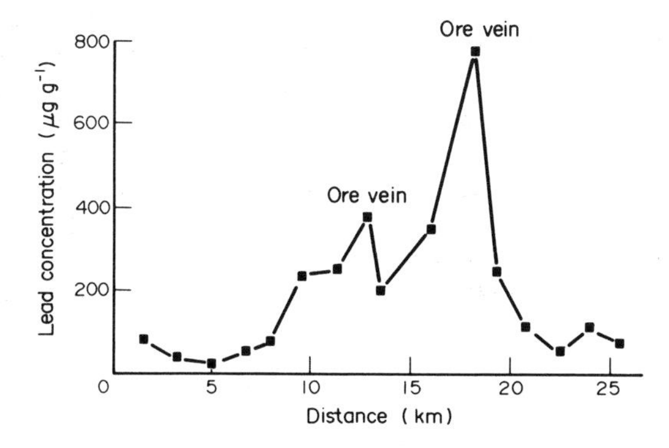

Fig. 9.8 Lead in soils in relation to an ore vein. Source of data; Nriagu, 1978 [73].

Thallium, bismuth, indium and tellurium Very much less data are available for thallium, bismuth, indium and tellurium (see Table 9.18). Mean soil concentrations of thallium are around 0.2-0.8 $\mu g\ g^{-1}$, and for bismuth around 0.2 $\mu g\ g^{-1}$, very little information is known about tellurium. For indium, levels tend to fall within the range 0.7-3.0 $\mu g\ g^{-1}$ with a mean of ~1.0 $\mu g\ g^{-1}$ [93].

Soil Profiles

In contrast to water and sediment profiles, variations in the concentrations of the heavy elements in soil profiles are not as consistent and as readily interpreted. A number of factors can influence the level of an element in a soil profile, including the nature of the parent material, pedogenic processes, the organic content of the soil, the clay content, proportion of iron containing minerals, the pH, the soil topography, the rainfall and microbial activity, and the extent of pollution. The relevance of the factors will vary from site to site.

In relatively unpolluted soils the concentrations of arsenic tend to fall with increase in soil depth, as indicated by the data plotted in Fig 9.9 [2], whereas arsenic was found to be depleted in the top horizons of Chernozem and Luvisol soils, the concentration of lead hardly changed with depth. A similar situation has been observed for cadmium in Mollisols and Alfisols. The cadmium levels fall from of 0.39 $\mu g\ g^{-1}$ on the surface to 0.23 $\mu g\ g^{-1}$ in sub-surface horizons [80]. In some cases, the cadmium concentration increases again in the horizon adjacent to the parent material [66]. It appears that the surface flux of cadmium from aerosols is greater than losses of the metal by leaching and uptake by plants in some situations and concentrations tend to be higher in the top horizon [2,33,76].

Soils derived from weathered magmatic rocks has lead concentrated in horizons rich in clay, and/or iron oxide/hydroxide compounds. The data plotted in Fig 9.10, for soils from France [73], shows how lead follows the aluminium and iron in the soil with depth. Both cadmium and lead were found

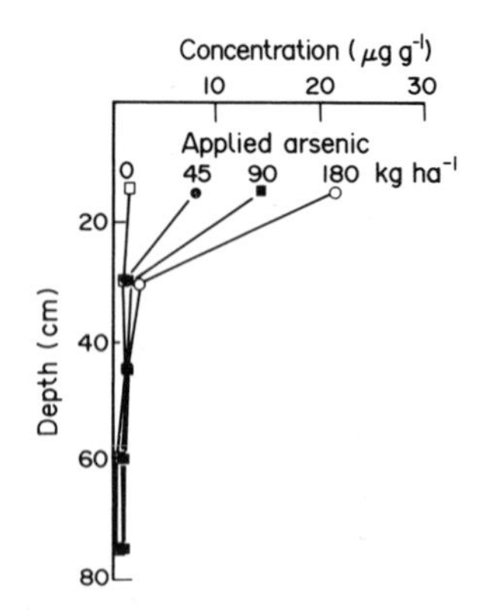

Fig. 9.9 Distribution of arsenic with soil depth. Source of data; Adriano, 1986 [2].

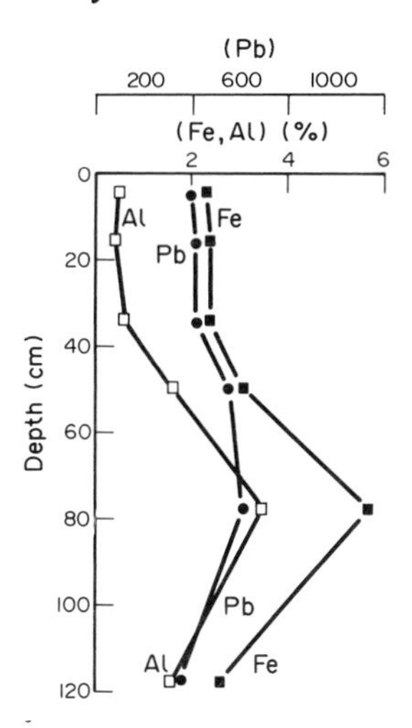

Fig. 9.10 Profiles of lead, aluminium and iron in soils formed from weathering of granites. Source of data; Nriagu, 1978 [73].

to follow the same concentration profile as the humus in four different Swedish soils, with maximum levels at the top and lower levels in deeper profiles. The results suggest an association between these two metals and the organic component of the soil [7].

The relationship between mercury levels and organic matter in a podzol (pH = 3.9-4.7) is clearly seen in the data presented in Fig 9.11 for European soils [6]. A similar situation also applies for mercury and the iron substances in a brown earth (pH = 6.2-8.2) [2,6]. The metal concentrates in the surface horizons, especially where the organic and/or clay components are high, whereas in lower horizons the metal is relatively mobile [2,93]. In mineralized zones the levels of mercury fall with soil depth [51].

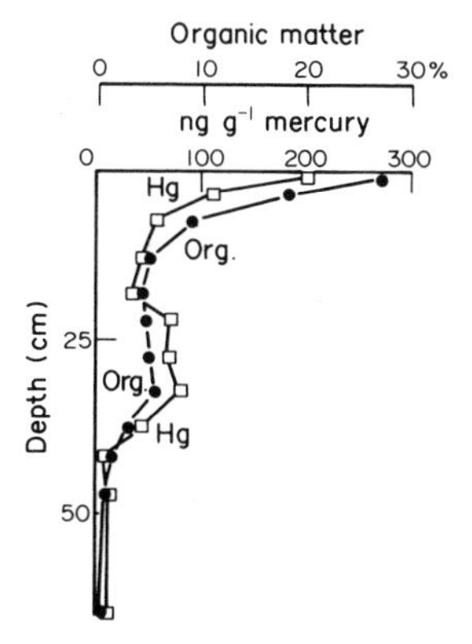

Fig. 9.11 Distribution of mercury and organic matter in a podzol. Source of data; Anderson, 1979 [6].

The levels of selenium in soil profiles, relate closely to those of organic material in the top layers and iron in the lower horizons [2]. Also the trends vary with soil type and parent material as shown for the four profiles depicted in Fig 9.12 [99]. The four parent material are schist loess over schist, greywacke, rhyolite, and andesite, and the four soils are brown grey earth, strongly weathered yellow brown earth, yellow brown pumice and brown granular clay respectively.

In the case of strongly polluted soils, from sources, such as smelters or traffic or industrial or mining areas the variation in levels of the heavy elements with depth follows a more clearly defined trend of high concentrations in the top horizons and a fall off with depth [18,27,38,40,55,60,65,82,87,95]. This is because most pollutants are added to the soil surface as shown in Fig 9.9 for arsenic [2]. Only when the levels of arsenic applied are high is there any significant transfer of the element down the profile. Similar trends apply for cadmium in contaminated soils in Japan and Canada (Fig 9.13). It appears from the data, that the cadmium has not moved very far down the soil profile [2,60]. A comparable

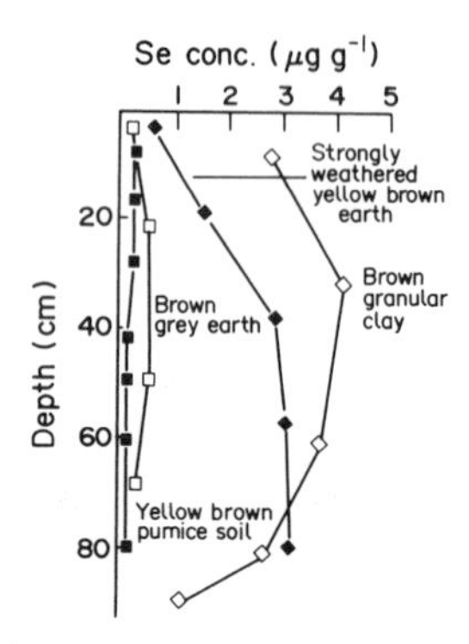

Fig. 9.12 Selenium profiles in some New Zealand soils. Source of data; Wells, 1967 [99].

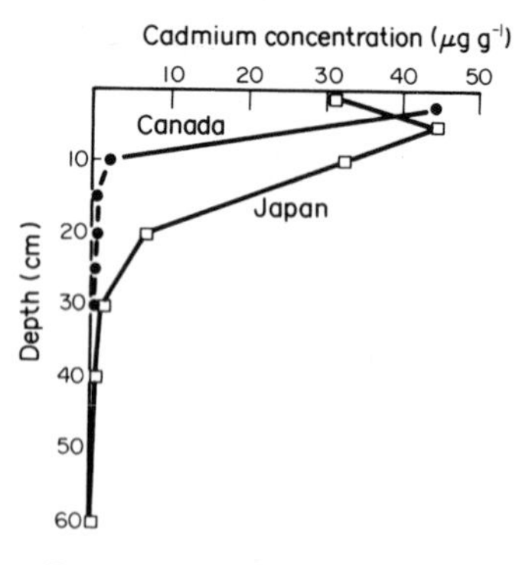

Fig. 9.13 Cadmium in polluted soil profiles. Sources of data; Adriano, 1986 [2], Lagerwerff and Specht, 1970 [60].

situation occurs for lead, this time the lead source being automobile aerosol (Fig 9.14) where a significant drop in lead down the soil profile occurs from 1130 μg g^{-1} in the top 0-2 cm down to 350 μg g^{-1} in the 8-10 cm section [38].

In soil studies the sectioning of the profile is often coarse, which masks small variations. This has been demonstrated for lead and cadmium in Israel soils, where 0.1 to 0.2 cm sections were taken at the top of the profile. In the case for lead, shown in Fig 9.15, the mean concentrations are also calculated for larger sections. It is clear that some of the finer detail is lost in making the sections too large [10].

Distance from source

One diagnostic method to identify the source of a heavy metal in a soil is to observe the change in concentration of the metal with the distance from the source. If the concentrations rise this points to the source. There is a great deal of literature available on some of the metals with respect to distance.

Cadmium from industrial sources, found in the soils on the shores of L. Michigan have been reported to be 30 μg g^{-1}, whereas 10-12 km away the levels drop to the background of 0.2 μg g^{-1} [101]. Similar fall off with distance has been

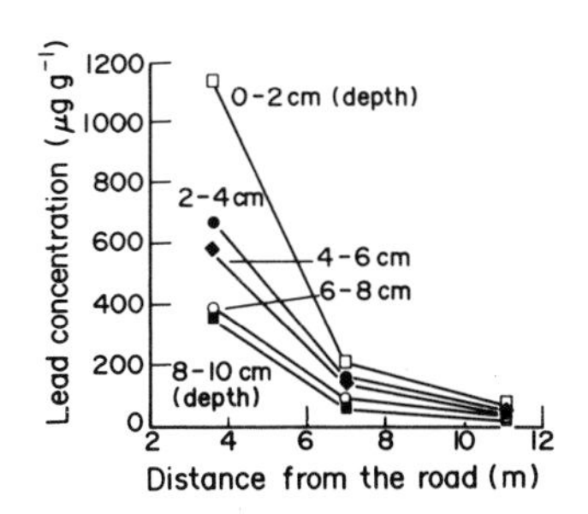

Fig. 9.14 Lead in soil close to a busy road and in relation to depth.

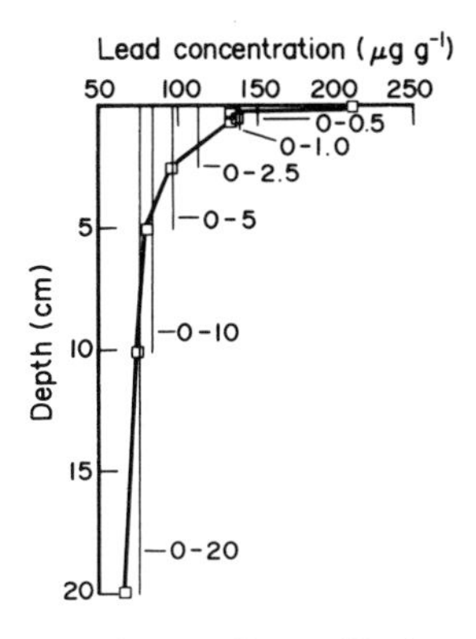

Fig. 9.15 Lead concentration in a soil profile in relation to the sampling interval. Source of data; Banin et al., 1987 [10].

reported around industries in Norway [87]. Levels of metals, such as Pb, Cd, Sb, Se and As fall off rapidly with distance from smelters, and long range sources [4,61,76,81,82], and mercury concentrations in soils around electrolytic cells drop off rapidly with distance [81]. A similar situation applies around mineralized and mining areas [73,90,95,96].

A source of selenium in Norway has been located as marine, as the levels fall with distance from the coast, on the coast the soil levels are 0.8-1.0 $\mu g\ g^{-1}$, further inland 0.4-0.6 $\mu g\ g^{-1}$ and on the Swedish border 0.3-0.4 $\mu g\ g^{-1}$. In addition, the selenium levels correlate with the iodine and bromine levels found in the same soils [59].

Numerous studies have investigated the levels of lead and cadmium in relation to traffic density, and/or distance from roads [20,22,25,27,29,31,38,40, 55,60,65,70,84,86,97]. The fall off with distance from a busy road, which is reasonably rapid, is depicted in the plotted data in Fig 9.14. Frequently there are two main sources of lead in urban areas, viz. alkyllead in petrol, and lead in paint. Hence, a typical cross section of the concentration of soil lead around painted houses along roads is high concentrations close to the road and falling off away from the road, then rising again near the house, as shown in Fig 9.16 [86]. The high levels near the building can be due to paint lead and/or lead from aerosol that has been intercepted by the wall, and then washed off with rain [86]. The influence of paint, the type and age of the house, may be seen from the data in Fig 9.17 [34,53,68,73]. Around inner city houses in Birmingham (England) the lead concentrations in soils have a significant correlation with the age of the house, suggesting a paint source. In addition a significant correlation was found with distance from roads, commercial garages, metal working places and waste land [29].

A detailed study of both lead and cadmium in soil profiles taken close to a tree trunk and extending out beyond the tree canopy indicated that closer to the tree trunk the levels tend to be higher in the top fraction and move further into

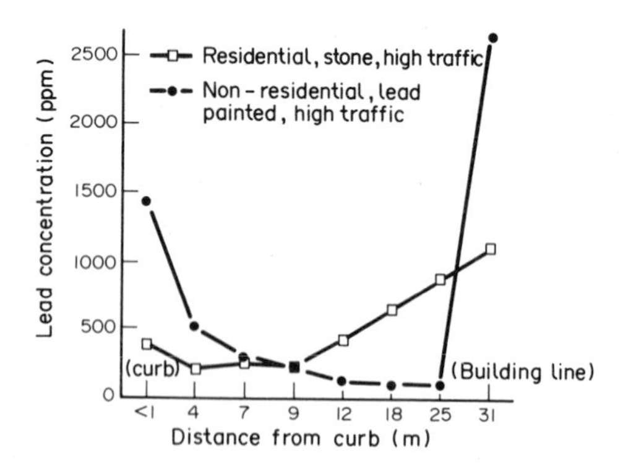

Fig. 9.16 Lead in soil in relation to distance from traffic. Source of data; Solomon and Hartford, 1976 [86].

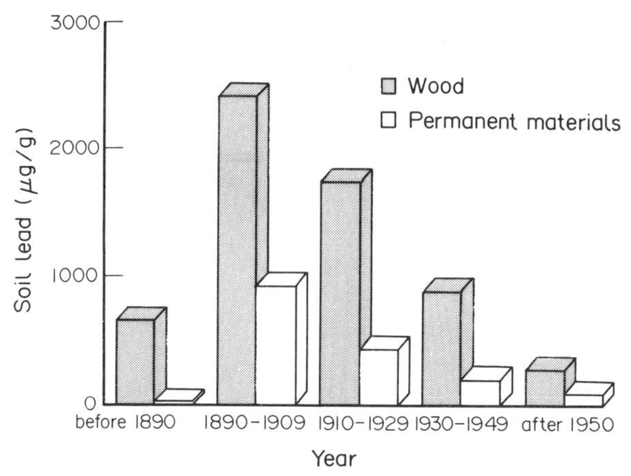

Fig. 9.17 Lead in soil in relation to house age and structural materials. Sources; references 34,53,68,73

the soil. This indicates that the tree is a collector of the metals in aerosols, which during rain are transferred to the underlying soil [10].

The concentrations of cadmium and lead in soils, in the plough layers, that have been collected since the mid 1800's to the present day and stored have been studied [17,52]. Rises of 15% for lead and 27-55% for cadmium over more than 100 years can be related to atmospheric input for both the metals. Cadmium input from phosphate fertilizer does not appear to have been retained by the soil, in the way that atmospheric cadmium has been, indicative of the different chemical forms of cadmium in the two sources.

THE CHEMISTRY OF THE HEAVY ELEMENTS IN SOILS

The chemistry of the heavy elements in soil is important area of investigation, because the chemistry will determine the availability of the elements to plants. The information that is useful to know is, the chemical forms of the elements, their solubilities, the sorption and desorption characteristics of the elements with respect to the soil components such as clay and organic matter, changes likely as a consequence of a change in soil pH and E_h and the chemical and biochemical availability of the elements. Because of the complex nature of soil, its heterogeneity and variability from place to place, it is difficult to be very precise on the soil chemistry of the heavy elements. It is quite common to come across conflicting data and interpretations. A general scheme for the reactions of heavy metals in soils is given in Fig 9.18 [18].

Arsenic

Arsenic occurs in soils, mainly as arsenate, AsO_4^{3-}, under oxic conditions. The species is strongly, and sometimes irreversibly, sorbed onto clays, iron and manganese oxides/hydroxides and organic matter. The amount of sorption relates to the arsenic concentration, to time and the iron and manganese content

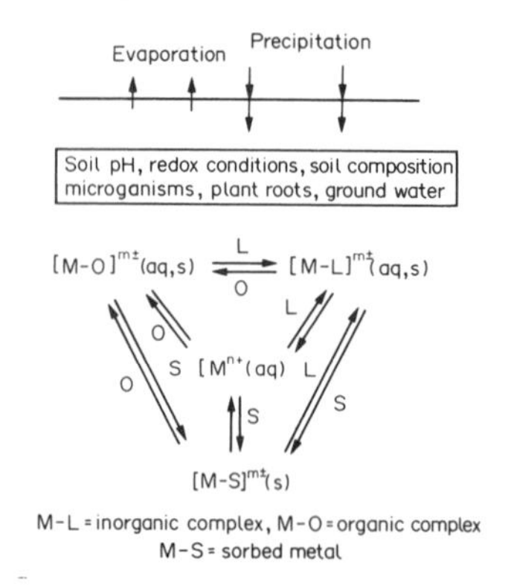

Fig. 9.18 A schematic representation of the reactions of heavy metals in soils. Source of data; Brümmer, 1986 [18].

of the soil. In acidic soils the main forms of arsenic, are aluminium and iron arsenates $AlAsO_4$ and $FeAsO_4$, whereas in alkaline and calcareous soils the main form is $Ca_3(AsO_4)_2$. The logK_{sp} of the three compounds are, -15.8, -20.2 and -18.2 respectively, but because of their stoichiometries, the aluminium and iron arsenates are less soluble than the calcium arsenate [2,33,54].

The sorption of arsenate to soil components has been studied in ways similar to that of phosphate because of related chemical properties. The results have been fitted to Langmuir and Freundlich isotherms. The removal of iron and aluminium amorphous components from soils markedly reduces the arsenate sorption.

The mobility of arsenic in soils is increased under reducing conditions, such as found in a flooded soil, because of the increase in the proportion of As(III) (arsenite). Arsenite salts are estimated to be around 5 to 10 times more soluble than the corresponding arsenates. Unfortunately, As(III) is both the more available and the more toxic form of the element. An increase in pH, e.g. by liming, also increases the mobility of the arsenic, presumably by bringing about a change from aluminium and iron arsenates to calcium arsenate. Numerous extracting agents have been used to investigate the availability of arsenic from a soil. Often the chemical significance of the results is not well understood.

Bacteria are also important in the chemistry of soil arsenic, as methylation can occur, giving As(III) methyl derivatives. The mobility of arsenic is increased by the alkylation. Bacteria also accelerate the oxidation of arsenite to arsenate.

Cadmium

The main solid cadmium species that probably occur in soils, under oxidising conditions, are CdO, $CdCO_3$ and $Cd_3(PO_4)_2$, whereas under reducing conditions ($E_h \leq -0.2$) CdS is likely to be the principal species [7,9,47,80]. The oxyanion species will exist mainly at high pH, whereas at lower pH the main mobile species of cadmium is Cd^{2+}. The pH is the single most important soil factor, as regards the speciation and mobility (availability) of cadmium in soils [2]. In acid soils, cadmium solubility and availability is controlled by the organic matter and Al, Fe, Mn hydrous oxides in the soil. In high pH soils (especially calcaerous or soils treated with lime) precipitation of cadmium compounds, e.g. $CdCO_3$ is the controlling factor as long as the solubility product of the compound is exceeded [1,9,23,49,54,74,76,90]. In soils high in chloride, cadmium chloro-species are likely to occur, which will increase the mobility of the metal [2,45].

The sorption of cadmium onto the various soil constituents has been studied by numerous people, some times with conflicting results. Sorption is said to be more important, than the solubility of cadmium compounds, in controlling the metal's mobility [54], especially at low concentrations of the metal. Sorption results usually fit either Langmuir or Freundlich isotherms [2,9]. Factors that influence the sorption are pH, ionic strength, competing cations and the constituents of the soil [2,37]. Relative sorptions, with respect to the constituents, are approximately; Al, Fe hydrous oxides, halloysite > imogalite, allophane > kaolinite, humic acid > montmorillonite > soil clay [1,5,43,79]. The relative positions of kaolinite and montmorillonite in the series is unusual, but it could be that the more acidic edge sites on kaolinite means that it absorbs Cd^{2+} better [43]. The importance of clays compared with some other inorganic species is demonstrated by the cadmium concentration ratio for clay/sand of 7 to 9 :1 [8,33]. The sorption process is probably rapid, being 95% complete within 10 minutes, and 100% within one hour [2,23]. Sorption decreases with pH [7], and is relatively reversible [24,36], and removal of organic matter from a soil reduces the sorption [35,44].

Sequential extractions on soils, to investigate the speciation of cadmium, generally indicates a high proportion of exchangeable metal, of the order of 20-40%. The other major forms are cadmium in the carbonate phase (20%), and association with the iron/manganese hydrous oxides (20%). Cadmium association with the organic matter appears to be less important, and of course depends on the amount of organic matter in the soil [2,9,46]. These results, notwithstanding the problem of the interpretation of sequential extractions [56,83,89], do suggest that compared with other heavy metals, a significant proportion of cadmium is exchangeable. In a study, with respect to distance and depth from the source of cadmium (old copper mine), the proportion of exchangeable cadmium decreased with the depth and distance whereas the metal associated with both carbonate and iron/manganese hydrous oxides increased [16]. This suggests that, either the mine source provides different cadmium speciation from other sources, or a change in speciation has occurred with distance and depth from the mine. To investigate this problem, a knowledge of the pH and E_h would be helpful.

Many studies have been undertaken to relate extractable cadmium from soil to plant availability. Reagents such as 0.5M HOAc, 1M HCl, 1M NH_4OAc, 0.05M EDTA, 1M NH_4NO_3, 0.1M HNO_3 and 0.1M $K_4P_2O_7 3H_2O$ have been used

[2,11,49,80]. The amount of cadmium extracted by 1M NH_4NO_3 appears to relate well to plant cadmium [49].

Mercury

Three aspects of mercury chemistry influences its chemistry in soil, and distinguishes it from the other heavy elements [2,6,54,93]. These are, the volatility of elemental mercury, an accessible redox chemistry whereby free mercury can be produced in soils, and the biomethylation of mercury producing very toxic, and often volatile compounds, e.g. $(CH_3)_2Hg$.

Some of the principal mercury compounds in soils of low solubility are $Hg(OH)_2$, at pH ≥ 7 $HgCO_3$, mercury phosphate and HgS in reducing conditions, such as in flooded and gleyed soils. Organomercury compounds, such as RHgOH, are likely in soils, as well as some soluble hydroxy and chloro-species depending on the pH and the chloride concentration. The strong interaction between Hg and S, means Hg-S-organic interactions will occur, for example, in mercury humic acid interactions.

As discussed in Chapter 3, metallic mercury is the oxidation state of the metal with the 'lone pair', and is accessible in soils. This can be seen from the E_h-pH diagram in Fig 3.9, and the reactions;

$$Hg(II) \rightarrow Hg(I),$$

$$2Hg(I) \rightarrow Hg(II) + Hg(0)$$

are quite possible. Microorganisms can also be involved in the redox processes. A scheme for some of the possible reactions for mercury in soils is given in Fig 9.19. The key materials are HgS, Hg/OH species and Hg-organo conpounds, and some of the more important reactions are methylation, oxidation/reduction, hydrolysis and precipitation. Other chemical cycles involving mercury have been given elsewhere in Figs. 5.8, 5.9 and 3.13.

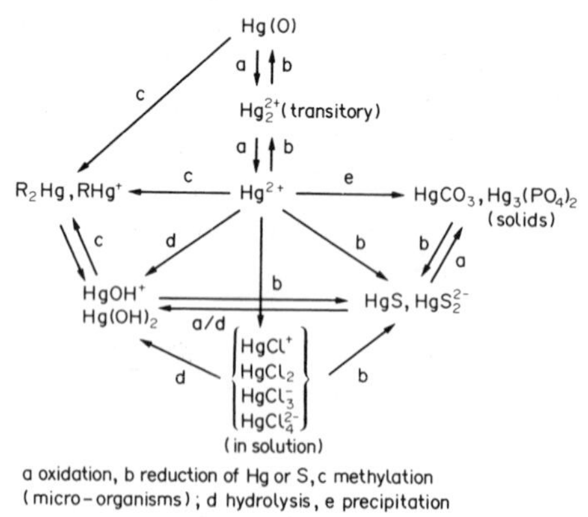

Fig. 9.19 Transformation reactions of mercury in soils.

The sorption of mercury to soil constituents depends on the chemical form of the mercury, the soil components, soil pH, other cations present and the redox potential. The main sorbing materials are clays, Fe/Mn oxides and organic matter, and the sorptions generally follow either the Langmuir or Freundlich isotherms. Maximum sorption occurs when the clay, and/or organic matter content of the soils are high. Of the clays illite sorbs mercury better than kaolinite. The availability of mercury to plants, from its various forms in soil, is said to be best measured by the amount extracted with 0.1M $CaCl_2$ solution.

Lead

Lead is probably the least mobile of the heavy elements [54], and is still found in top soils that were contaminated in the 4th century BC [28]. The half life of lead in soils lies in the range 800 to 6000 years [54]. The principal ore of lead, PbS, will produce $PbSO_4$ on weathering, as the sulphide is oxidized to SO_4^{2-}. The liberated Pb^{2+}ion is then involved in a number of chemical processes, such as sorption onto clays, organic matter and Fe/Mn oxides, precipitation of insoluble compounds and coordination to both inorganic and organic ligands [54]. Some of the insoluble lead compounds are $Pb(OH)_2$, $PbCO_3$, PbS, $PbSO_4$, PbO, $PbO.PbSO_4$, $Pb_3(PO_4)_2$, $Pb_4O(PO_4)_2$, $Pb_5(PO_4)_3(OH)$. Which species occurs, depends on the soil pH, source of the lead (e.g. natural or pollution), the anionic species present and the redox conditions [2,54]. The lead compounds in car emissions PbBrCl, PbBrOH, $(PbO)_2.PbBr_2$, are rapidly converted in soils to some of the compounds listed above. The insolubility of the compounds, their relative amounts in a soil and the pH, will regulate their soil-water concentration. This may be determined from the Pb/OH distribution diagram in Fig 3.15 and the E_h-pH diagrams in Figs 3.17 to 3.19. In saline soils lead chlorocomplexes are possible (see Fig 3.16).

Lead speciation, as determined by sequential extraction procedures, suggests that lead is tied up with Fe/Mn/Al oxides, clays and organic matter, and to a lesser extent with carbonates or as exchangeable [2,46,54]. However, the proportions vary considerably from study to study, and also vary with the distance of the soil from the source of the lead. The consistent result is that exchangeable lead is always low being $<5\%$ in most cases. The retention of lead in the surface horizons of polluted soils is said to be due to the binding of the metal to the organic matter, which is often at its highest in the top horizons. Lead moves further down a soil profile, when the organic material had been removed [38].

As for arsenic and mercury, it appears that lead in soil may also be biomethylated [2], producing volatile and toxic $(CH_3)_4Pb$, and $(CH_3)_{4-n}Pb^{n+}$. This reaction also involves the oxidation of the metal. Lead is also said to limit the enzymatic activity of microbiota, and as a result organic material, which has been incompletely decomposed, tends to accumulate in the soil [54].

The extraction of lead from soil systems has been studied, to estimate its phyto-availability. Reagents such as 0.1M HCl, 1M HNO_3, 1M NH_4OAc, 0.05M $CaCl_2$, 0.05M $BaCl_2$, organic acids and 0.02M EDTA have been used at various times. The quantity extracted is a function of a number of factors including the extractant used, soil pH and the speciation of the lead.

Lead sorbs to soil constituents, such as Fe/Mn/Al oxides, clays and organic material, and generally follows the Langmuir and Freundlich isotherms. In competition with Ca^{2+} ions, Pb^{2+} ions are generally the more strongly absorbed [2,30,38,54]. Positive correlations exist between lead and organic matter, and lead and clays. The lead ions may replace K^+ ions in both the organic and the clay materials [5.7]. Sorption follows the order; montmorillonite < humic materials, kaolinite < allophane < imogalite < halloysite, iron oxides [1,54]. The sorption tends to increase with pH up to the point where $Pb(OH)_2$ precipitates [37]. Also the species clay-Pb-OH can form [54]. Interestingly, the removal of organic matter reduces the lead sorption to soils, but the addition of organic material in the form of sewage sludge does not appear to influence the lead sorption. This may be because the sewage sludge already has a high metal content.

Selenium

Selenium is an essential element for animals, but is also toxic in excess. Because of this the soil chemistry of the element has been studied in some detail [2,3,54]. The E_h-pH diagram for the element (Fig 3.32a), shows that the accessible forms of selenium in soils are: selenium(-II), selenide HSe^- and Se^{2-}; selenium(0), Se; selenium(IV), selenite $HSeO_3^-$ and SeO_3^{2-} and selenium(VI), selenate SeO_4^{2-}. The accessibility of selenate increases the higher the pH. The interconversion of the various selenium species, and the necessary conditions are given in Fig 9.20. The divalent oxidation state, selenium(II), does not appear to occur in soils. The chemistry suggests that in gley soils, especially when

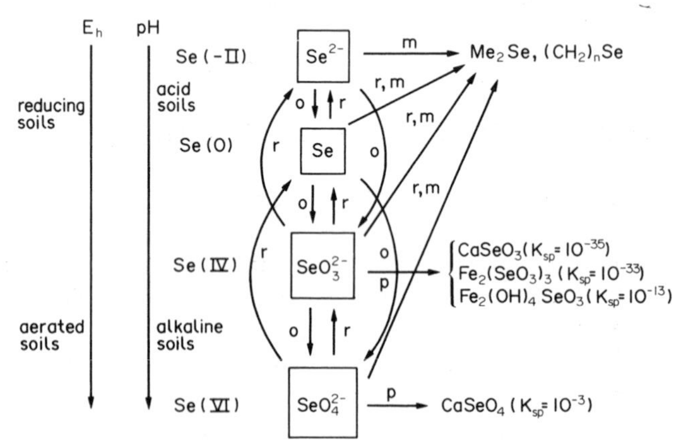

Fig. 9.20 The interconversion of selenium species in soils.

containing high organic matter, selenides and organoselenides may occur. In well drained (aerated soils), which tend towards acid, the selenite is the preferred species, and in more alkaline soils selenates occur. Selenite forms very insoluble compounds with iron such as $Fe_2(SeO_3)_3$ and $Fe_2(OH)_4SeO_3$, whereas selenate in alkaline soils does not form such insoluble compounds. Hence in acid soils with high selenium, the element is relatively immobile and unavailable to plants, on the other hand in alkaline high selenium soils, it is more mobile and available to plants. In the acid soils of Hawaii, where the selenium content can be high at 20 $\mu g\ g^{-1}$, the element is not very available to plants. In alkaline soils, even with a low selenium content, plants, especially accumulator plants (e.g. *Astragalus sp*.) can attain high levels of the element because the more soluble SeO_4^{2-} form is present.

The selenite species appears to be more firmly sorbed to iron/manganese oxides, maybe because of the formation of insoluble iron selenites on the surface of the oxide. The selenate ion is also strongly sorbed to the iron/manganese oxides. The addition of fulvic acid to soils increased the water solubility of the selenium, but not its availability to plants [64], presumably because of the selenium-fulvic acid interaction.

Selenium is often found associated with sulphur around volcanic areas and in rocks and minerals. Selenium species are less readily oxidized and are less mobile than sulphur, so it is not as readily removed from the environment by weathering. The differences in the redox chemistry of sulphur and selenium helps to explain their distinct chemistry in soils.

$$H_2S \rightarrow S + 2H^+ + 2e, \ -E^\circ = -0.14V,$$

$$H_2Se \rightarrow Se + 2H^+ + 2e, \ -E^\circ = -0.40V,$$

$$S + 3H_2O \rightarrow H_2SO_3 + 4H^+ + 4e, \ -E^\circ = -0.45V,$$

$$Se + 3H_2O \rightarrow H_2SeO_3 + 4H^+ + 4e, \ -E^\circ = -0.74V,$$

$$H_2SO_3 + H_2O \rightarrow SO_4^{2-} + 4H^+ + 2e, \ -E^\circ = -0.17V,$$

$$H_2SeO_3 + H_2O \rightarrow SeO_4^{2-} + 4H^+ + 2e, \ -E^\circ = -1.15V.$$

Relative to sulphur, elemental selenium is the more redox stable element, and for the oxyanions, selenite is the favoured form for selenium whereas sulphate is the favoured form for sulphur.

The biomethylation of selenium occurs in soils. The methylation is achieved by microorganisms, in plants, fungi and bacteria. The selenium accumulator species *Astragalus* produces $(CH_3)_2Se$ when exposed to selenite. The biomethylation is enhanced by the addition of organic matter to the soil.

Other Elements

Very little data are available on the soil chemistry of the remaining elements indium, thallium, antimony, bismuth and tellurium. This is probably because of their lower abundance, and that they are used less and are not so polluting.

Indium and thallium Indium(III) is the stable form in soils (see the E_h-pH diagram Fig 3.25) and appears to follow Fe^{3+}, Mn^{4+} and Al^{3+} in soils. Therefore it will occur associated with the oxides of these metals. At higher pH's $In(OH)_3$ may well precipitate, and other species such as $In(OH)_4^-$, $InOH^{2+}$, and $InCl^{2+}$ may occur depending on the pH and soil conditions. Thallium as Tl^+ is the stable form in soils (see E_h-pH diagram Fig 3.26) and is highly associated with K^+ (and Rb^+) and where it occurs, may replace K^+ in soil, such as in clays and where K^+ is sorbed [2,54].

Antimony bismuth and tellurium Whereas little data are available on these elements, it is possible to predict the most likely forms in soils from the appropriate E_h-pH diagrams. For antimony, the stable form would probably be Sb(III) (Fig 3.28b), whereas for bismuth the elemental form Bi(0) is also likely to have some existence (Fig 3.28c). The most likely oxidation state for tellurium is Te(IV) ((Fig 3.32b), but Te(VI), Te(0) and Te(-II) are also possible as may be seen by comparing the E_h-pH diagrams of selenium and tellurium (Figs 3.32a, b).

Mobility of the Heavy Elements

The mobility of elements in soils is one of the key properties that determines their concentrations and their position in a soil profile. A summary of the relative mobilities of the heavy elements is given in Table 9.19 [54]. The attempt to generalize in this way means there will be exceptions. It would appear, however, that the more mobile of the ten elements in soils are Se and Hg followed by Cd and As and Tl(I) (depending on the conditions). The less mobile elements are Sb, Tl, Bi, Te, In and Pb.

TABLE 9.19 Relative Mobilities of the Heavy Elements in Soil

Relative	Conditions			
Mobility	Oxidising	Acid	Neutral to alkaline	Reducing
Very high			Se	
High	Se	Se, Hg		
Medium	Hg, As, Cd	As, Cd	As, Cd	Tl?
Low	Pb, As, Sb, Tl	Pb, Bi, Sb, Tl	Pb, Bi, Sb, Tl, In	
Very low to Imobile	Te	Te	Te, Hg	Te, Se, Hg, As Cd, Pb, Bi, Tl

Source of data; Kabata-Pendias and Kabata, 1984 [54].

REFERENCES

1. Abd-Elfattah, and Wada, K. Adsorption of lead, copper, zinc, cobalt and cadmium by soils that differ in cation exchange materials. **J. Soil Sci.;** 1981; **32:** 271-283.
2. Adriano, D C. Trace Elements in the Terrestial Environment: Springer Verlag; 1986.
3. Allaway, W H. Control of the environmental levels of selenium. **Trace Subst. in Environ. Health;** 1968; **2:** 181-206.
4. Allen, R. O. and Steinnes, E. Contribution from long range atmospheric transport to the heavy metal pollution of surface soil. In: Management and Control of Heavy metals in the Environment: CEP Consultants; 1979: 271-274.
5. Anderson, A. The distribution of heavy metals in soils and soil material as influenced by the ionic radius. **Swedish J. Agric. Res.;** 1977; **7:** 79-83.
6. Anderson, A. Mercury in soils. In: Nriagu, J. O., Ed. The Biogeochemistry of Mercury In the Environment: Elsevier/North Holland; 1979: 79-112.
7. Andersson, A. Heavy metals on Swedish soils; on their retention distribution and amounts. **Swed. J. Agric. Res.;** 1877; **7:** 7-20.
8. Andersson, A. On the distribution of heavy metals as compared to some other elements between grain size fractions in soils. **Swed. J. Agric. Res.;** 1979; **9:** 7-13.
9. Asami, T. Pollution of soils by cadmium. In: Nriagu, J. O., Ed. in: Changing Metal Cycles and Human Health, Dahlem Konferenzen; 1984: 95-111.
10. Banin, A., Navrot, J. and Perl, A. Thin horizon sampling reveals highly localised concentrations of atmophile heavy metals in a forest soil. **The Sci. Total Environ.;** 1987; **61:** 145-152.
11. Berrow, M. L. and Mitchell, R. L. Location of trace elements in soil profiles total and extractable contents of individual horizons. **Trans. Roy. Soc. Edinburgh, Earth Sci.;** 1980; **71:** 103-121.
12. Berrow, M. L. and Burridge, J. C. Sources and distribution of trace elements in soils and related crops. In: Management and Control of Heavy Elementsin the Environment, Int. Conf.: CEP Consultants; 1979: 304-311.
13. Bewers, J. M., Barry, P. J. and MacGregor, D. J. Distribution and cycling of cadmium in the environment. **Adv. Environ. Sci.;** 1987; **19:** 1-18.
14. Bowen, H. J. M. Environmental Chemistry of the Elements: Academic Press; 1979.
15. Bowen, H. J. M. Soil pollution. **Educ. Chem.;** 1975; **12:** 72-74.
16. Bradley, S. B. and Cox, J. J. Heavy metals in the Hamps and Manifold valleys, North Staffordshire, UK: I Partitioning of metals in flood plain soil. **The Sci. Total Environ.;** 1987; **67:** 75-89.

17. Bradley, S. B. and Cox, J. J. Heavy metals in the Hamps and Manifold valleys, North Staffordshire, UK: II Cadmium. **The Sci. Total Environ.**; 1987; **67**: 75-89.
18. Brümmer, G. W. Heavy metal species, mobility and availability in soils. In: Bernhard, M., Brinckman, F. E. and Sadler, P. J., Eds. The Importance of Chemical Speciation in Environmental Processes, Dahlem Konferenzen: Springer Verlag; 1986: 169-192.
19. Chamberlain, A.C. Fallout of lead and uptake by crops. **Atmos. Environ.**; 1983; **17**: 693-706.
20. Chaney, R. L., Sterrett, S. B. and Miekle, H. W. The potential for heavy metal exposure from urban gadens and soils. In: Preer, J. R., Ed. Proc. Symp. Heavy Metals in Urban Soils: Univ. District Columbia, Washington D.C.; 1984: 37-84.
21. Chaney, R. L. Potential effects of sludge borne heavy metals and toxic organics in soils, plants and animals and related regulatory guidelines. In: Report Workshop Int. Transportation, Utilization and Disposal of Sewage Sludge Including Recommendations, PNSP/85-01: Pan American Health Organisation, Washington, D.C.; 1985: 1-56.
22. Chaney, R. L. and Miekle, H. W. Standards for soil lead limitations in the United States. **Trace Subst. Environ. Health**; 1986; **20 (in press).**
23. Christensen, T. H. Cadmium soil sorptions at low concentrations: 1 Effect of time, cadmium load, pH and calcium. **Water, Air Soil Pollut.**; 1984; **21**: 105-114.
24. Christensen, T. H. Cadmium soil sorptions at low concentrations: II Reversibility, effect of changes in solute composition and effect of soil aging. **Water, Air Soil Pollut.**; 1984; **21**: 115-125.
25. Cool, M., Marcoux, F. Paulin, A. and Mehra, M. C. Metallic contaminants in street soils of Moncton, New Brunswick, Canada. **Bull. Environ. Contam. Toxicol.**; 1980; **25**: 409-415.
26. Culbard, E., Thornton, I., Watt, J. Moorcroft, S., Brooks, K. and Thompson, M. A nation wide reconnaissance survey of the United Kingdom to determine metal concentrations in urban dusts and soils. **Trace Subst. Environ. Health**; 1983; **XVI.**
27. David, D. J. and Williams, C. H. heavy metal contents of soils and plants adjacent to the Hume highway near Marulan, New South Wales. **Aust. J. Expt. Agric. Animal Husb.**; 1975; **15**: 414-418.
28. Davies, B. E. Trace metals in soils. **Chem. Brit.**; 1988; **24**: 149-154.
29. Davies, D. J. A., Watt, J. M. and Thornton, I. Lead levels in Birmingham dusts and soils. **The Sci. Total Environ.**; 1987; **67**: 177-185.
30. Dawson, B.S.W., Fergusson, J.E., Campbell, H.S. and Cutler, J.B. Distribution of elements in some Fe-Mn nodules and an iron-pan in some gley soils of New Zealand. **Geoderma**; 1985; **35**: 127-143.

31. Deroanne-Bauvin, J., Delcarte, E. and Impens, R. Monitoring of lead deposition near highways in a ten year study. **The Sci. Total Environ.**; 1987; **59:** 257-266.
32. Dudas, M. J. and Powluk, S. heavy metals in cultivated soils and in cereal crops in Alberta. **Can. J. Soil Sci.**; 1977; **57:** 329-339.
33. Dudas, M. J. and Pawluk, S. Natural abundances and mineralogical partitioning of trace elements in selected Alberta soils. **Can. J. Soil Sci.**; 1980; **60:** 763-771.
34. Duggan, M. J. Childhood exposure to lead in the surface dust and soil: a community health problem. **Pub. Health Rev.**; 1985; **13:** 1-54.
35. Elloitt, M. R. Liberati, M. R. and Huang, C. P. Competitive adsorption of heavy metals by soils. **J. Environ. Qual.**; 1986; **15:** 214-219.
36. Farrah, H. and Pickering, W. F. Extraction of heavy metal ions sorbed on clays. **Water, Air Soil Pollut.**; 1978; **9:** 491-498.
37. Farrah, H. and Pickering, W. F. Influence of clay-solute interactions on aqueous heavy metal ion levels. **Water, Air Soil Pollut.**; 1977; **8:** 189-197.
38. Fergusson, J.E., Hayes, R.W., Tan Seow Yong and Sim Hang Thiew. Heavy metal pollution by traffic in Christchurch, New Zealand: Lead and cadmium content of dust, soil and plant samples. **N. Z. J. Sci.**; 1980; **23:** 293-310.
39. Fergusson, J. E. Inorganic Chemistry and the Earth: Pergamon Press; 1982.
40. Fergusson, J. E. Lead: petrol lead in the environment and its contributions to human blood lead levels. **The Sci. Total Environ.**; 1986; **50:** 1-54.
41. Fleming, G. A., and Parle, P. J. Heavy metals in soils, herbage and vegetables from an industrialised area of West Dublin city. **Irish J. Agric. Res.**; 1977; **16:** 35-48.
42. Förstner, U. Cadmium in polluted sediments. In: Nriagu, J. O., Ed. Cadmium in the Environment, Part I, Ecological Cycling: Wiley; 1980: 305-363.
43. Garcia-Miragaya, J. and Cardenas, R. Surface loading effect on Cd and Zn sorption by kaolinite and montmorillonite from low concentration solutions. **Water, Air Soil Pollut.**; 1986; **27:** 181-190.
44. Haas, C. N. and Horowitz, N. D. Adsorption of cadmium to kaolinite in the presence of organic material. **Water, Air Soil Pollut.**; 1986; **27:** 131-140.
45. Hahne, H. C. H. and Kroontje, W. Significance of pH and chloride concentration and behaviour of heavy metal pollutants: mercury(II), cadmium(II), zinc(II) and lead(II). **J. Environ. Quality**; 1973; **2:** 444-450.
46. Harrison, R. M., Laxen, D. P. H. and Wilson, S. J. Chemical associations of lead, cadmium, copper and zinc in street dusts and roadside soils. **Environ. Sci. Techn.**; 1981; **15:** 1378-1383.
47. Hermann, R. and Neumann-Mahlkau, P. The mobility of zinc, cadmium, copper, lead, iron and arsenic in ground water as a function of redox potential and pH. **Sci. Total Environ.**; 1985; **43:** 1-12.

48. Hutton, M. and Symon, C. The quantities of cadmium, lead, mercury and arsenic entering the UK environment from human activities. **The Sci. Total Environ.;** 1986; **57:** 129-150.

49. Jastow, J. D. and Koeppe, D. E. Uptake and effects of cadmium in higher plants. In: Nriagu, J. O., Ed. Cadmium in the Environment, Part 1. Ecological Cycling: Wiley; 1980: 607-638.

50. Johnson, C. M. Selenium in soils and plants; contrasts in conditions providing safe but adequate amounts of selenium in the field chain. In: Nicholas, D. J. D. and Egan, A. R., Eds. Trace Elements in Soil-Plant-Animal Systems: Academic Press; 1975: 165-180.

51. Jonasson, I. R. and Boyle R. W. Geochemistry of mercury and origins of natural contamination of the environment. **Canad. Mining Met. Bull.;** 1972; **65:** 32-39.

52. Jones, K. C., Symon, C. J. and Johnston, A. E. Retrospective analysis of an archived soil collection I Metals. **The Sci. Total Environ.;** 1987; **61:** 131-144.

53. Jordan, L. D. and Hogan, D. T. Survey of lead in Christchurch soils. **N. Z. J. Sci.;** 1975; **18:** 253-260.

54. Kabata-Pendias, A. and Pendias, H. Trace elements in soils and plants: CRC Press; 1984.

55. Khan, D. A. Lead in the Soil Environment. MARC Tech. Rept.: MARC; 1980.

56. Kheboian, C. and Bauer, C. F. Accuracy of selective extraction procedures for metal speciation in model aquatic sediments. **Anal. Chem.;** 1987; **59:** 1417-1423.

57. Kloke, A., Sauerbeck, D. R. and Vetter, H. The contamination of plants and soils with heavy metals and the transport of metas in terrestrial food chains. In: Nriagu, J. O., Ed. Changing Metal Cycles and Human Health, Dahlem Konferenzen.; 1984: 113-141.

58. Kronberg, B. I., Fyfe, W. S. Leonardos, O. H. and Santos, A. M. The chemistry of some Brazilian soils: element mobility during intense weathering. **Chem. Geol.;** 1979; **24:** 211-229.

59. Lag, J. and Steinnes, E. Regional distribution of selenium and arsenic in humus layers of Norwegian forest soils. **Geoderma;** 1978; **20:** 3-14.

60. Lagerwerff, J. V. and Specht, A. W. Contamination of roadside soil and vegetation with cadmium, nickel, lead and zinc. **Environ. Sci. Techn.;** 1970; **4:** 583-586.

61. Landrigan, P. J., Stephen, H. Gehlbach, M. D., Rosenblum, B. F., Shoults, J. M., Candelaria, R. M., Barthel, W. F., Liddle, J. A., Smrek, A. L., Staehling, N. W. and Sanders, J. F. Epidemic lead absorption near an ore smelter: the role of particulate lead. **New Engl J. Med.;** 1975; **292:** 123-129.

62. Le Riche, H. H. The distribution of minor elements among the components of a soil developed in loess. **Geoderma;** 1973; **9:** 43-57.

63. Lévesque, M. Selenium distribution in Canadian soil profiles. **Canad. J. Soil Sci.**; 1974; **54:** 63-68.
64. Lévesque, M. Some aspects of selenium relationships in Eastern Canadian soils and plants. **Canad. J. Soil Sci.**; 1974; **54:** 205-214.
65. Little, P. and Wiffen, R. D. Emissions and deposition of petrol engine exhaust lead II Airborne concentration, particle size and deposition of lead near motorways. **Atmos. Environ.**; 1978; **12:** 1331-1341.
66. Lund, L. J., Betty, E. E., Page, A. L. and Elliott, R. A. Occurrence of naturally high cadmium levles in solis and its accumulation in vegetation. **J. Environ. Qual.**; 1981; **10:** 551-556.
67. Mahler, R. J., Ryan, J. A. and Reed, T. Cadmium sulphate application to sludge amended soils I Efect on yield and cadmium availability to plants. **The Sci. Total Environ.**; 1987; **67:** 117-131.
68. Markunas, L. D. Barry, E. F. Guiffre, G. P. and Litman, R. An improved procedure for the determination of lead in environmental samples by atomic absorption spectroscopy. **J. Environ. Sci. Health**; 1979; **A14:** 501-506.
69. Mielke, H. W., Anderson, J. C., Berry, K. J. Mielke, P. W. Chaney, R. L. and Leech, M. Lead concentrations in inner city soils as a factor in the child lead problem. **Amer. J. Pub. Health**; 1983; **73:** 1366-1369.
70. Minami, K. and Araki, K. Distribution of trace elements in arable soil affected by automobile exhausts. **Soil Sci. Plant Nutr.**; 1875; **21:** 185-188.
71. Mortvedt, J. J. and Osborn, G. Studies on the chemical form of cadmium contaminants in phosphate fertilizers. **Soil Sci.**; 1982; **134:** 185-192.
72. Murti, C.R.K. The cycling of arsenic, cadmium, lead and mercury in India. In: Hutchinson, T.C. and Meena, K.M., Eds. Lead, Mercury, Cadmium and Arsenic in the Environment. Scope 31: Wiley; 1987: 315-333.
73. Nriagu, J.O. Lead in soils, sediments and major rock types. In: Nriagu, J. O., Ed. The Biogeochemistry of Lead in the Environment, Part A, Ecological Cycling: Elsevier / Nth. Holland.; 1978: 15-72.
74. O'Connor, G. A., O'Connor, C. and Cline, G. R. Sorption of cadmium by calcareous soils: influence of solution composition. **Soil Sci. Soc. Amer. J.**; 1984; **48:** 1244-1247.
75. Pace, C. B. and di Giulio, R. T. Lead concentration in soil, sediment and clam samples from the Pungo River peatland area of North Carolina, USA. **Environ. Pollut.**; 1987; **43:** 301-311.
76. Page, A. L., Chang, A. C. and El-Amamy, M. Cadmium levels in soils and crops in the United States. In: Hutchinson, T.C. and Meena, K.M., Eds. Lead, Mercury, Cadmium and Arsenic in the Environment. Scope 31: Wiley; 1987: 119-148.
77. Paul, E. A. and Huang, P. M. Chemical aspects of soil. In: Hutzinger, O., Ed. The Handbook of Environmental Chemistry. Vol 1. Part A, The Natural and Biogeochemical Cycles: Springer Verlag; 1980; 1: 69-86.

78. Peterson, P. J. Benson, L. M. and Zieve, R. Metalloids. In: Lepp, N. W., Ed. Effect of Heavy Metal Pollution on Plants. Vol 1, Effects of Trace Metals on Plant Function: Applied Science Publishers; 1981: 279-342.
79. Pickering, W.F. Cadmium retention by clays and other soil or sediment components. in: Nriagu, J.O., Ed. Cadmium in the Environment, Part 1 Ecological Cycling: Wiley; 1980: 365-397.
80. Pierce, F. J., Dowdy, R. H. and Grigal, D. F. Concentrations of six trace metals in some major Minnesota soil series. **J. Environ. Qual.**; 1982; **11:** 416-422.
81. Piotrowski, J. K. and Coleman, D. O. Environmental hazards of heavy metals: summary evaluation of lead, cadmium and mercury. MARC Report No 20.; 1980.
82. Ragaini, R. C., Ralston, H. R. and Roberts, N. Environmental trace metal contamination in Kellog, Idaho, near a lead smelting complex. **Environ. Sci. Tech.**; 1977; **11:** 773-781.
83. Rendell, P. S. Batley, G. E. and Cameron, A. J. Adsorption as a control of metal concentrations in sediment extracts. **Environ. Sci. Techn.**; 1980; **14:** 314-318.
84. Sharma, R. P. Soil-plant-animal distribution of cadmium in the environment. In: Nriagu, J. O., Ed. Cadmium in the Environment, Part 1. Ecological Cycling: Wiley; 1980: 587-605.
85. Soil Bureau N. Z. Single factor map. Soil Bureau, DSIR, N. Z.: 89, 90; 1967.
86. Solomon, R. L. and Hartford, J. W. Lead and cadmium in dust and soils in a small urban community. **Environ. Sci. Techn.**; 1976; **10:** 773-777.
87. Steinnes, E. Impact of long range atmospheric transport of heavy metals to the terrestrial environment in Norway. in: Hutchinson, T.C. and Meena, K.M., Eds. Lead, Mercury, Cadmium and Arsenic in the Environment. Scope 31: Wiley; 1987: 107-117.
88. Taylor, R. M. and McKenzie, R. M. The association of trace elements with manganese minerals in Australian soils. **Aust. J. Soil Res.**; 1966; **4:** 29-39.
89. Tessier, A., Campbell, P. G. C. and Bisson, M. Sequential extraction procedure for the speciation of particulate trace metals. **Anal. Chem.**; 1979; **51:** 844-851.
90. Thornton. I. Geochemical aspects of the distribution and forms of heavy metals in soils. In: Lepp, N. W., Ed. Effect of Heavy Metal Pollution on Plants. Vol 2, Metals in the Environment: Applied Science Publishers; 1981: 1-33.
91. Thornton, I. and Abrahams, P. Historical record of metal pollution in the environment. In: Nriagu, J. O., Ed. Changing Metal Cycles and Human Health, Dahlem Konferenzen: Springer Verlag; 1984: 7-25.
92. Tiller, K. G. and de Vries, M. P. C. Contamination of soils and vegetables near the lead-zinc smelter, Port Pirie, by cadmium, lead and zinc. **Search**; 1977; **8:** 78-79.

93. Ure, A. M. and Berrow, M. L. The elemental constituents of soils. In: Bowen, H. J. M., Reporter. Environmental Chemistry, Vol. 1, Roy. Soc Special Period. Repts.: Royal Society; 1982: 142-204.
94. Ure, A. M., Bacon, J. R., Berrow, M. L. and Watt, J. J. The total trace element content of some Scottish soils by spark source mass spectrometry. **Geoderma**; 1979; **22:** 1-23.
95. Ward, N. I., Brooks, R. R. and Robets, E. Contamination of a pasture by a New Zealand base metal mine. **N. Z. J. Sci.**; 1977; **20:** 413-418.
96. Ward, N.E., Brooks, R.R. and Reeves, R.D. Copper, cadmium, lead and zinc in soils, stream sediments, waters and natural vegetation around the Tui mine, Te Aroha, New Zealand. **N. Z. J. Sci.**; 1976; **19:** 81-89.
97. Warren, R. S. and Birch, P. Heavy metal levels in atmospheric particulates, roadside dust and soil along a major urban highway. **The Sci. Total Environ.**; 1987; **59:** 253-256.
98. Webber, J. Trace metals in agriculture. In: Lepp, N. W., Ed. Effect of Heavy Metal Pollution on Plants. Vol 2, Metals in the Environment: Applied Science Publishers; 1981: 159-184.
99. Wells, N. Selenium in horizons of soil profiles. **N. Z. J. Sci.**; 1967; **10:** 142-179.
100. Woodis, T. C., Hunter, G. B. and Johnson, F. J. Statistical studies of matrix effects on the determination of cadmium and lead in fertilizer materials and plant tissue by flameless atomic absorption spectrometry. **Anal. Chim. Acta**; 1977; **90:** 127-136.
101. Yost, K. J. Some aspects of the environmental flow of cadmium in the United States. In: Mennear, J. H., Ed. Cadmium Toxicity: Dekker; 1979: 181-206.

CHAPTER 10

HEAVY ELEMENTS IN PLANTS

When considering the heavy metals in plants there is an extra dimension compared with the heavy elements in air, water and soil, which is the wide range of species. Therefore it is not as straight forward to generalize over concentrations and effects. For this reason only selected and limited data will be described, and information on many species will be omitted. The situation is more complex for animals, because they have greater access to the heavy elements because of their mobility. We will not deal particularly with animals in this book.

PLANTS

A very brief description will be given of the structure of plants, and the mechanism by which minerals are transported into and within plants. The nature and the extent of world vegetation will also be outlined briefly.

Plant Structure

A basic function of plants is the conversion of carbon dioxide and water into carbohydrates through the use of solar energy [42]. The process called photosynthesis is catalysed by chlorophyll, the green colouring material in leaves.

$$CO_2 + H_2O \xrightarrow[\text{chlorophyll}]{h\upsilon} \underset{\text{carbohydrate}}{\{CH_2O\}} + O_2$$

For most species in the plant world, the carbon dioxide is the sole source of plant carbon, but some species such as fungi also need organic compounds for their carbon. Another activity of plants is transpiration, the transport of water from the soil via the roots and stem of the plant to the leaves, where the photosynthesis occurs, and then emission into the atmosphere.

Plants consist of two main structural components, the root system, which anchors the plant in the soil, and is the site for absorption of materials from the soil, and the shoot system consisting of the stem and leaves. The stem supports the plant and is the place where the minerals and water are conveyed to the stem and leaves, and where the carbohydrates produced in the leaves are transported to the roots.

The carbohydrate produced by the plant, is used for the manufacture of the plant cells and their repair, particularly in the root system. As a consequence respiration occurs, which is the reverse of photosynthesis, and the energy evolved is utilized in the manufacture and repair processes.

$$\{CH_2O\} + O_2 \rightarrow CO_2 + H_2O + \text{energy}$$

The system that extends from roots to stem and leaf veins, and which provides for transport of material, is called the vascular system. The system contains the xylem in which water and minerals are transported, and the phloem, in which the carbohydrates are transported. In a herbaceous stem there are bundles of phloem and xylem in a soft tissue called the cortex and pith. In a woody stem the central pith is surrounded by the woody xylem, and then by the bark which consists of the phloem and cork.

World Vegetation

Because of agricultural practices, it is not always obvious what are the basic vegetation types of the different areas of the world. A list of the some of the main vegetations and some of the areas where they may be found is given in Table 10.1 [11,43]. Agriculture and forestry have produced significant changes to this pattern, and in particular the growing of food. The production and consumption of food, is one principal means by which people come into contact with the heavy elements.

PLANT SOIL INTERACTIONS

The incorporation of the heavy elements into plants is mainly achieved by uptake from the soil through the roots. Uptake may also occur from deposits of the heavy elements on the leaves from soil or aerosol [3,4,24,42,47]. Many factors influence the uptake and include the growing environment (such as temperature, soil pH, soil aeration, E_h conditions and fertilization), competition between the plant species, the type of plant, its size, the root system, the availability of the elements in the soil or foliar deposits, the type of leaves, soil moisture and the plant energy supply to the roots and leaves.

The uptake can be estimated by the biological absorption coefficient (BAC),

$$BAC = \frac{[M_P]}{[M_S]},$$

where $[M_p]$ is the concentration of the element in the plant and $[M_s]$ is the concentration in the soil. Alternatively the relative absorption coefficient (RAC),

$$RAC = \frac{[M_p]}{[M_{rp}]},$$

may be used, $[M_{rp}]$ is the concentration of the element in a reference plant [4].

Root Absorption

An element in a soil may either diffuse to a root surface through the soil solution, provided there is a concentration gradient, or be involved in ion-exchange between a clay, for example, and a root in contact with the clay. Movement of the element into the root occurs either by passive diffusion through the cell membrane, or by the more common process of active transfer against a concentration gradient and an electrochemical potential (as calculated from the Nernst equation). In some respects the root acts like a selective ion electrode where, in most cases, the concentration of the ion is greater inside the root than outside. Hence energy is required to facilitate the element's movement across the cell membrane.

The transfer of an element (e.g. sorbed onto a clay) from the soil to the root surface, can be by either contact exchange or proton exchange as demonstrated

TABLE 10.1 World Vegetation

Vegetation type	Some locations
Deciduous woodlands	N. Europe, E. USA
Coniferous forests	Canada, USSR
Tropical rain forests	N. South America, Equatorial Africa, S.E. Asia
Temperate rain forests	E. China, E. Australia, North Island N.Z.
Savannas	Mid South Africa, S.E. India, Mid E. South America
Grasslands	Mid USA & Canada, South Africa, S.W. USSR
Sclerophyllous vegetation (e.g. eucalyptus)	Mediterrean area, S.W. Australia
Thorn scrub	Southern Africa, W. & N. Australia, Mid South America
Cactus scrub	S.W. USA, Mexico, South America
Semi-desert vegetation	Central Asia
Desert	N. Africa, Central Australia, W. USA
Tundra	N. USSR, N. Canada, Alaska
Mountain vegetation	Himalays, Andes

Sources of data; Eyre, 1968 [11], Riley and Young, 1968 [43].

in Figs 10.1a, b. The energy for the active transfer comes from the metabolic process of respiration (see Fig 10.1b), and is called a pump. The carbon dioxide formed, with water, produces the necessary protons from the dissociation of carbonic acid for the exchange process. Another possibility is that the metal species moves across the membrane into the root system assisted by a carrier. The carrier could be a complexing agent, such as an organic acid or protein, that bonds to the metal species, transports it across the membrane and then dissociates freeing the species to move into the cell vacuoles. Roots may exude compounds, which help in making metal species available to the plant. Materials produced are organic acids and ionophores, which are chelating agents that solubilize metal species. Symbiotic fungi, called mycorrhizae, also are involved in this process. Other uptake mechanisms have also been proposed, such as pinocytosis, where the species binds to the membrane surface, becomes incorporated into the membrane (invagonation) and released on the other side.

Foliar uptake of elements occurs either through the stomata or leaf cuticle or both, the principal route is probably through the leaf cuticle. The entry may be by either non-metabolic or metabolic processes. Entry of elements to plants via the leaves is more significant for pollution elements because of aerosol deposits.

Once inside a plant a heavy element species may be transported through the xylem, but this varies from element to element. For example cadmium appears to be quite mobile, whereas lead tends to remain fixed in the roots. The elements may exist as free ions, or as complexes, such as citrate and oxalate, which influence the elements mobility. The metal species may also occur in the phloem (used to distribute the carbohydrate through the plant) by lateral transfer from the xylem, and by remobilization of species from the leaves into the phloem at times of exfoliation. The movement in the xylem and phloem are partly dependent on the transpiration intensity, i.e. the movement of water through a plant.

Plant varieties differ in their ability to absorb and accumulate heavy metals in their tissues. The plot in Fig 10.2, for the average BAC for green plants and fungi, shows that elements such as cadmium and mercury are more strongly

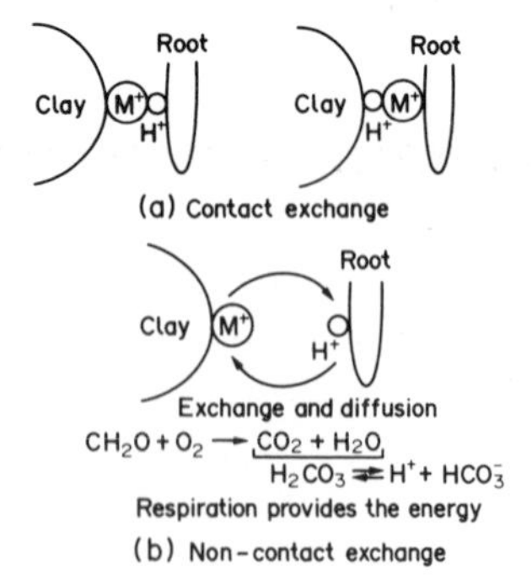

Fig. 10.1 The transfer of metal species to a root, (a) contact exchange, (b) non-contact exchange

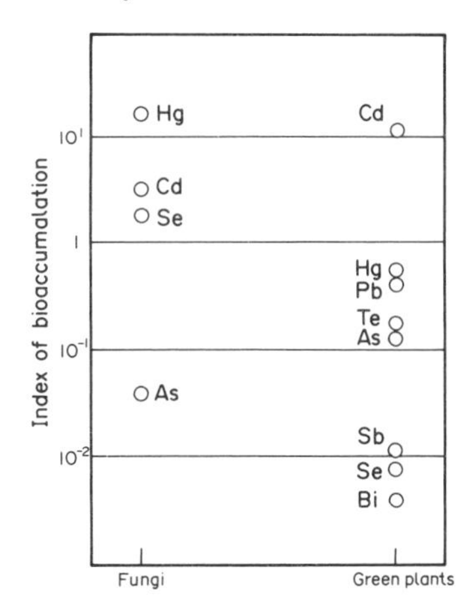

Fig. 10.2 The bioaccumulation of heavy metals by plants and fungi from soil. Source of data; Kabata-Pendias and Kabata, 1984 [24].

bioaccumulated than elements such as bismuth and antimony [24]. An exclusion mechanism appears to function for plants to avoid excess uptake of toxic metals. Concentrations in a plant may remain relatively constant, while the levels in the soil change. This means that for the equation;

$$[M_p] = [M_s](BAC),$$

the factor BAC varies as $[M_s]$ varies, so that $[M_p]$ remains constant. At very high levels of an element in the soil the exclusion mechanism can be overcome, and concentrations may increase in a plant producing toxicity. This point has been considered earlier (Chapter 6), where comparison is made between a pollution and a non-pollution element, with the biopurification factor (BPF). In these investigations it is necessary to distinguish between natural and pollution contributions of a metal taken in from the soil, as well as atmospheric additions to the foliage.

LEVELS OF THE HEAVY ELEMENTS IN PLANTS

A number of factors influence the actual level of an element found in a plant, these include; the type of plant, the particular plant tissue, level of the element in the soil, the availability of the element in the soil, the distance a plant is away from the source of the element, the season, the climatic conditions and the foliar uptake from settled aerosol. Because of these factors, the concentrations of the heavy elements vary widely, and it is not possible to give an adequate mean value. However, some attempt to do this is given in Table 10.2, whereas particular data, mainly for food plants, are listed in Tables 10.3 to 10.9 on pages 382 to 387. For plants, not growing on polluted soils or mineralized areas, the

TABLE 10.2 Average Levels of the Heavy Elements in Plants

Element	Land plants µg g^{-1*}	Edible plants µg g^{-1*}
As	0.02-7	0.01-1.5
Bi	< 0.02	~0.06
Cd	0.1-2.4	0.05-2.0
Hg	0.005-0.02	0.013-0.17
In	0.03-0.7(FW)**	0.001
Pb	1-13	0.2-20
Sb	0.002-0.06	0.0001-0.2
Se	0.01-1.0	0.001-0.5
Te	0.7-6	0.013-0.35(FW)
Tl	0.008-1.0	0.03-0.3

* Concentrations on a dry weight basis unless stated otherwise, ** FW = fresh weight. Sources; references 3,24

levels of the heavy elements in plants tend to be low and generally <1 µg g^{-1}, however, there are wide variations in the data listed in the Tables. The method of expression of the amount of a heavy element in a plant can be made in terms of the plant's wet, dry or ash weight.

Estimates of a transfer coefficient from soil to plant for some heavy elements, are given in Table 10.10. The figures have an order of magnitude accuracy, rather than exact values [26], and they depend on the plant species and soil properties. It does appear that the more available elements, i.e. the ones more readily taken up by plants, are cadmium and thallium, and to a lesser extent selenium. We will consider some of the factors listed above, that influence the levels of the heavy metals in plants.

TABLE 10.3 Levels of Cadmium in Plant Tissue

Plant type	Plant tissue	Concentration µg g^{-1} (ppm) DW	Location/ comments
Grasses and pasture			
Grasses		0.05-1.26*, 0.27**	E. Germany
		0.05-0.20, 0.08	Poland
		0.03-0.3, 0.16	USA
		<0.5	
Clover		0.28	Canada
		0.07-0.3, 0.10	Poland
		0.02-0.35, 0.16	E. Germany
Hay		0.14-0.33, 0.18	
Native vegetation		0.11-7.6	USA

Table 10.3 continued

Plant type	Plant tissue	Concentration µg g⁻¹ (ppm) DW	Location/ comments
Cereals			
Barley	grain	0.006-0.044, 0.022	Norway
Oats	grain	0.21	Canada
Rice	grain (unpol.)	0.05-0.11, 0.08	Japan
Various	grain	0.01-0.75, 0.22	W. Germany
	grain	0.1-0.5	USA
	grain	0.014-0.21, 0.047	USA
	grain	0.39-0.12	
	grain	0.015-0.08, 0.38	India
Wheat	grain	0.012-0.036, 0.022	Australia
Vegetables			
Bean	pods	0.29	
Cabbage	leaves	0.05	
Carrot	roots	0.07, 0.24, 0.2	
General		0.017-0.98, 0.044	India
Lettuce		0.5-3.3	Ireland
	leaves	0.12, 0.4, 0.66, 0.062	
Potato	tuber	0.03, 0.23	
Radish	tuber	0.1-14	USA
	tops	0.2-58	USA
Various		0.2-3.3	Ireland
	leaves	0.93-0.88, 0.056	USA
Trees			
Spruce	various parts	<0.1-5.4	Sweden
Mixed deciduous	various parts	0.2-0.4	USA
Various	leaves(unwash)	30-55	

* range, ** mean. Sources of data; references 1,10,14,24,31,32,34,36,37,44,48.

TABLE 10.4 Levels of Mercury in Plant Tissue

Plant type	Plant tissue	Concentration ng g⁻¹ (ppb) DW	Location/ comments
Cereals			
Barley	grain	5-17*, 12**	Canada
General	grain	5.3-12	Canada
	grain	30	
	grain	0.75-31.1, 7.0	India
Oat	grain	4-19, 9	Canada
	grain	<4-45, 14	Sweden
Rye	grain	3-18, 9	Poland
Wheat	grain	<10	W. Germany
	grain	0.2-2.7, 0.9	Norway
	grain	10-16, 14	USA

Table 10.4 continued

Plant type	Plant tissue	Concentration µg g^{-1} (ppm) DW	Location/ comments
Wheat	grain	7-12, 10	USSR
Vegetables			
Cabbage	leaves	6.5	
Carrot	roots	5.7, 86	
General		0.75-36.3, 6.5	India
Lettuce	leaves	8.3	
Potato	tuber	<10, 47, 20	
Sweet corn	grain	4.6	
Tomato	fruit	31, 34	

*range, ** mean. Sources of data; references10,24,26,34.

TABLE 10.5 Levels of Lead in Plant Tissue

Plant type	Plant tissue	Concentration µg g^{-1} (ppm) DW	Location/ comments
Grasses and pasture			
Clover		1-3*, 1.3**	UK
		3.3-4.7, 4.2	W. Germany
		1.6-4.0, 2.7	Poland
Grasses		<1.2-3.6, 1.8	Canada
		1-9, 2.1	UK
		5-6	Sweden
		<0.8-5.6, 1.6	USA
Cereals			
Barley	grain	<1.25-1.5	UK
General	grain	0.1-0.2	Canada
Oat	grain	2.28	Canada
Rye	grain	0.64	Austria
Wheat	grain	0.13-0.28	Finland
	grain	0.2-0.8	Poland
	grain	0.42-1.0	USA
	grain	0.4-0.6	USSR
Vegetables			
Cabbage	leaves	1.7, 2.3	
		1.4-26	Ireland
Carrot	roots	3, 1.5, 0.5	
Leeks		8.3	Ireland
Lettuce	leaves	0.7, 2, 3.3, 3.6	
		2.1-67	Ireland
Potato	tuber	0.5, 3	
Red beet	roots	2, 0.7	
Swede		12.4	Ireland
Sweet corn	grain	<0.3, 3, 0.88	

Table 10.5 continued

Plant type	Plant tissue	Concentration µg g^{-1} (ppm) DW	Location/ comments
Tomato	fruit	1, 1.2, 3	
Various		0.01-3.85, 0.05	
Trees			
Various	leaves	3.7-13.8	USA
	twigs	12.2-68.8	USA
	wood	0.35-1.1	USA
	roots	10.2-32.6	USA

* range, ** mean. Sources of data; references 1,10,14,24,25,28.

TABLE 10.6 Levels of Selenium in Plant Tissue

Plant type	Plant tissue	Concentration ng g^{-1} (ppb) DW	Location/ comments
Grasses and pasture			
Clover/alfaalfa		18-40*	Sweden
Grass		5-23, 13**	Canada
		19-134, 47	France
		10-40, 32	USA
Cereals			
Barley	grain	27-42, 33	France
	grain	2-110, 18	Denmark
Oat	grain	4-43, 28	Canada
	grain	150-1000, 480	USA
	grain	3-54, 16	Denmark
Various	grain	2-29,	Norway
	grain	4-46	Sweden
Wheat	grain	1-117, 23	Australia
	plant	420-4000, 640	on aeolian sand
	plant	880-11200, 2180	on glacial till
Vegetables			
Cabbage	leaves	150	USA
Carrot	roots	64	USA
Kale		130	
Lettuce	leaves	57	USA
Onion	bulbs	42	USA
Sweet corn	grain	11	USA
Tomato	fruit	36	USA
Various		1-500	

* range, ** mean. Sources of data; references 1,3,24,39.

TABLE 10.7 Levels of Arsenic in Plant Tissue

Plant type	Plant tissue	Concentration ng g^{-1} (ppb) DW
Grasses and pasture		
Clover	tops	280-330
Grass	tops	20-160
Cereals		
Barley	grain	3-18
General		1-88*, 25**
Oat	grain	10
Rice (brown)	grain	110-200
Wheat	grain	70, 50, 10
Vegetables		
Cabbage	leaves	20-50
Carrot	roots	40-80
General		1-8, 3.1
		260
Lettuce	leaves	20-250
Potato	tuber	30-200

* range, ** mean. Sources of data; references 24,34,39.

TABLE 10.8 Levels of Thallium in Plant Tissue

Plant type	Plant tissue	Concentration μg g^{-1} (ppm) DW
Grasses and pasture		
Clover		0.008-0.01*
	whole	0.4-0.5
Fescue		0.2-0.6 (ash weight)
herbage		0.02-1.0
Meadow hay		0.02-0.025
Various		0.01
Cereal		
Rye		1.0
Vegetables		
Beet	leaves	0.025-0.030
Cabbage	leaves	0.04, 0.125
Carrot	leaves	0.30
	roots	0.10
General		0.02-0.125
Lettuce	leaves	0.021
Potato	tops	0.025-0.03
Spinach		1.0
Turnip	leaves	5.9
	roots	0.040

Table 10.8 continued

Plant type	Plant tissue	Concentration µg g^{-1} (ppm) DW
Trees		
Birch	leaves	0.5
Larch		0.5-5.0 (ash weight)
Pine		2-100 (ash weight)

* range. Sources of data; references 1,3,24,45.

TABLE 10.9 Levels of Antimony, Bismuth, Indium and Tellurium in Plant Tissue

Plant type	Plant tissue	Concentration
Antimony		**ng g^{-1}**
Barley	roots	122
	leaves	10
Corn	grain	<2 (DW)
Edible plants		0.02-4.3* (FW)
Forage crops		100 (DW)
Grass		29
Potato	tubers	<2 (DW)
Rice		960
Bismuth		µg g^{-1}
Trees		1-15 (AW)
Vegetables		0.06 (DW)
Indium		**ng g^{-1}**
Beets		80-300 (DW)
Friut trees	leaves	0.64-1.8 (DW)
Tomato	leaves	0.64-1.8 (DW)
Vegetables		30-710 (FW), 210
Tellurium		µg g^{-1}
General		0.7-6 (DW)
Vegetables		<0.013-0.35(DW)

* range, ** mean, AW = ash weight, DW = dry weight, FW = fresh weight. Sources of data; references 24,39.

Plant Species

Provided other factors are constant, it is possible to compare the uptake of the heavy elements by different species. With respect to any metal, plants may be classified as either accumulators or indicators or excluders of the element [4,26]. Accumulator plants have the ability to take up large concentrations of certain metals without toxic effects to the plant. Indicator plants are ones whose metal uptake is in response to the amount of metal in the soil environment. They can therefore be used as indicators of the source and the intensity. There are also

TABLE 10.10 Transfer Coefficients

Metal	Transfer coefficient
Cd	1-10
Hg	0.01-0.1
Pb	0.01-0.1
Tl	1-10
As	0.01-0.1
Se	0.1-10

Source of data; Kloke et al., 1984 [26].

plants that discriminate against metal ions [4], and can in fact biopurify with respect to the element.

Most work has been carried out on arsenic, lead and cadmium for plants used in human consumption. Because of the difficulty in obtaining comparable data from different studies, it is best to give qualitative estimates for the uptake of the heavy elements by plants. The information in Table 10.11 is a summary, from a number of sources, where uptake is considered as high, medium or low. It is necessary to ensure the same sections of plants are compared. This is clear

TABLE 10.11 Relative Uptake of some Heavy Elements by Selected Plants

Plants	As	Cd	Pb	Hg	Se
Vegetables					
Beans	low	low-med.			
Cabbage	high	med.-low	low		
Lettuce		high	high	high-med.	
Peas	low	med.-low	low		
Radish	med.	low-med.	high		
Spinach	low	high			
Tomato	high	med.			
Turnip	med.	high	high		
Cereals					
Clover	low	low	med.-low		
Corn	med.	med.	low		
Oats	high	low	low		
Rice	low	low			
Wheat	high	low	low		med.
Fruit					
Apples	high				
Cherries	med.				
Peach	low				
Pears	high				

Sources of data; references 1,14,22,24,26,36,37.

from the data given in Table 10.12 for the uptake of cadmium. Genetic factors are quite relevant to metal uptake, as shown by the data and the root/shoot concentration ratios in Table 10.12a. The figures in parenthesis in Table 10.12b are the normalized data, where the cadmium concentration in the plant, at no extra application of the metal, is taken as 1. The figures highlight the differences that occur between plants, and the effect of increased cadmium. Doubling the cadmium added generally doubles the the cadmium in the plant roots of some plants but not in others. The genetic effect of plants are compared with respect to the uptake of cadmium and lead in Table 10.13. From the study of different plants from the same site, of a pine and birch forest ecosystem in Poland, it was found that lead accumulated in the mosses and lichens, whereas cadmium accumulated in the fungi [24].

Plant Tissue

It is necessary to take note of the plant tissue when comparing the uptake of the heavy elements by plants. Since the metals mostly enter the plant through

TABLE 10.12 The Uptake of Cadmium in Different Parts of a Plant

(a) From a flowing solution*

Plant	Shoots μg g^{-1} (DW)	Roots (fibrous) μg g^{-1} (DW)	Ratio roots/shoots
Maize	11.5	57.5	5
Lucerne	3.1	33.4	10.8
Fodder beet	21.1	151.1	7.2
Radish	1.8	33.6	18.7
Lettuce	15.0	48.3	3.2
Tomato	5.6	121.6	21.

* Cadmium concentration 0.01 μg ml^{-1}

(b) Cadmium added to soil

Plant	Added cadmium to soil μg/g		
	0	2.5	5.0
Oats grain	0.21 (1)**	1.50 (7.1)**	2.07 (9.9)**
straw	0.29 (1)	2.30 (7.9)	3.70 (12.6)
Soybean roots	0.99 (1)	6.09 (6.2)	11.77 (11.9)
Corn roots	0.73 (1)	10.47 (14.3)	17.02 (23.3)
Potato tubers	0.18 (1)	0.89 (4.9)	1.09 (6.1)
Lettuce tops	0.66 (1)	7.72 (11.7)	10.36 (15.7)
roots	0.40 (1)	2.96 (7.4)	5.60 (14.0)

** Values normalised with respect to levels when no cadmium was added. Source of data; Jastow and Koeppe, 1980 [22].

TABLE 10.13 Levels of Cadmium and Lead in Plants in the Same Forest Ecosystem

Plant	Cadmium	Lead
	$\mu g\ g^{-1}$ (DW)	
Grass	0.6	1.2
Clover	0.7	2.8
Plantain	1.9	2.4
Mosses	0.8, 0.7	22.4, 13.0
Lichens	0.4, 0.5	17.0, 28.0
Edible fungi	1.0, 2.7	1.2, <0.1
Inedible fungi	1.6, 1.0	0.4, 1.0

Source of data; Kabata-Pendias and Kabata, 1984 [24].

the root system it is not surprizing that the heavy elements concentrations are frequently highest in the root. This is clear from the data for cadmium in Table 10.12a. For twentythree different vegetable plants it was found that $[Cd]_{roots} > [Cd]_{shoots}$ [22,36,37]. In general levels decrease in the order roots > stems > leaves > fruit > seeds, when the source of the heavy metals is from the soil. Foliar deposits and intakes may change this order, especially for lead.

There are divergent views concerning the transport of cadmium in plants [1,10,24]. It is said that 99% of cadmium in rice roots gets through to the shoots, on the other hand cadmium appears to be held on cell walls, probably through Cd-S interactions. The element is mainly present as the Cd^{2+} ion, and no deposits of cadmium compounds have been found in plants. For lead, movement beyond the roots appears to be very limited, and there is evidence for deposits of lead compounds (e.g. lead pyrophosphate and lead orthophosphate) in cell walls [24,27]. For a larch tree the ratios of lead levels in various sections roots: bark: needles: twigs: wood are 6: 5: 2: 1: 1 respectively [24]. Data on the lead concentrations in different sections of lettuce, oats and carrots are given in Table 10.14. As the amount of lead in the soil increases, the relative order of lead concentrations in the parts of the plant (cf. oats) changes. The results suggest that for the controls another source of lead (perhaps aerosol) was contributing to higher levels in the husk and leaves. The poor translocation of lead from the roots is obvious, particularly to the oat grains. Mercury appears to be relatively readily translocated in some plants, and less so in others. Mercury movement appears greater when the metal has entered the plant by the stems or leaves. As for cadmium, the formation of a Hg-S bond, can restrict the element's movement in plants. Thallium, as Tl^{+}, is readily translocated in plants, and in pine trees more thallium is found in the needles than the stems. The movement of arsenic is more restricted, and levels are higher in the roots of plants than in other tissues. The order for arsenic levels is usually roots > vegetative growth > seeds, fruit [39].

TABLE 10.14 Lead in Various Tissues of Plants (ppm)

Plant	Plant tissue	Control	200 ppm Pb in soil	1000 ppm Pb in soil
Lettuce	leaves	2.5	3.0	54.2
	roots	5.8	84.5	867.7
Oats	grain	3.2	4.4	4.9
	husk	11.1	11.8	16.4
	leaves	6.0	6.8	20.1
	stalk	1.6	2.5	9.2
	roots	4.5	82.0	396.6
Carrot	tops	2.3	8.0	17.6
	tuber	1.9	5.3	41.0
	roots	8.9	241.7	561.4

Source of data; Kabata-Pendias ans Kabata, 1984 [24].

If selenium in soils is soluble, the element is readily taken up by plants. Selenium can, in part, replace sulphur in various compounds and up to eight organoselenium compounds have been isolated or identified in plants (Table 10.15) [39]. The element tends to be concentrated in the plant roots, growing points and seeds [39]. Tellurium, is less easily absorbed by plants.

TABLE 10.15 Selenium Compounds Found or Identified in Plants

Selenium compound
Selenocystathionine[a]
Se-methylselenocysteine[a]
Selenohomocysteine[b]
Selenomethionine[b] and oxide
Se-methylselenomethionine[b]
Selenocystine[b]
Selenocysteic acid[b]
Se-propenylselenocysteine[b] and oxide

[a] isolated and characterised, [b] identified (chromatography and electrophoresis)
Source of data; Peterson et al., 1981 [39].

Levels of Heavy Elements in Soils

An obvious influence on the concentrations of the heavy elements in plants is the concentration of the element in the soil. For all of the heavy metals, it is usual to see an increase in their concentration in plants as the level increases in the soil (Table 10.14). This relationship also applies for extractable metal from soil with reagents such as EDTA [1,7,13,22,24,32,37]. Examples of the influence of cadmium levels in soil on the levels in swiss chard and radish are shown in Fig 10.3. The correlations between plant and soil cadmium are significant at the 1% level of probability. A similar situation occurs for arsenic in lettuce, pasture and barley grains (Fig 10.4). The results also highlight the genetic difference between plants. Levels of mercury in rice roots correlate highly (r = 0.934, n = 20) with the square root of the soil mercury concentration.

In some situations there is a decrease in plant levels for some heavy elements as the soil levels increase. Both a limit on how much the plant may take in, and the toxicity of the metals, which may cause a plant to deteriorate, are the reasons. As shown in Fig 10.5 at the highest soil concentration, the levels of thallium in spinach and chives begins to decrease, whereas the concentration in rape continues to increase [1].

Availability of the Heavy Elements in Soil

More critical than the total amount of an element in a soil, is its availability as determined by the soil properties. Since the speciation of the heavy metals in soils is largely controlled by the soil pH and E_h, these two factors also influence the availability of the elements to plants. For arsenic, cadmium, selenium and lead the effect of these parameters are summarized in Table 10.16 [1,2,10,22,24, 27,30,33,36,37,39] The generalizations inherent in deriving such a table means that there are exceptions.

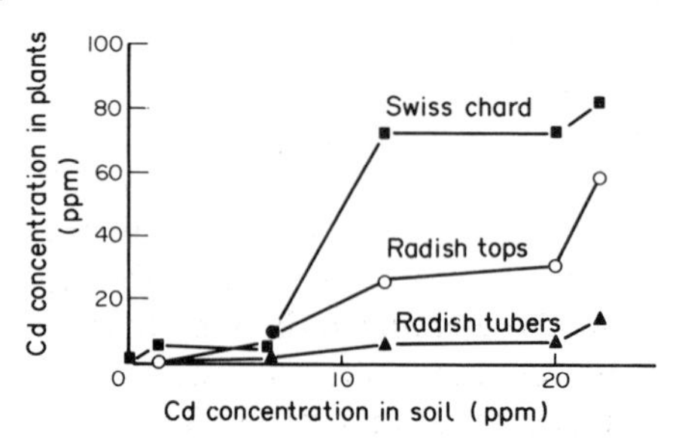

Fig. 10.3 Cadmium in soil and plants. Source of data; Adriano, 1986 [1].

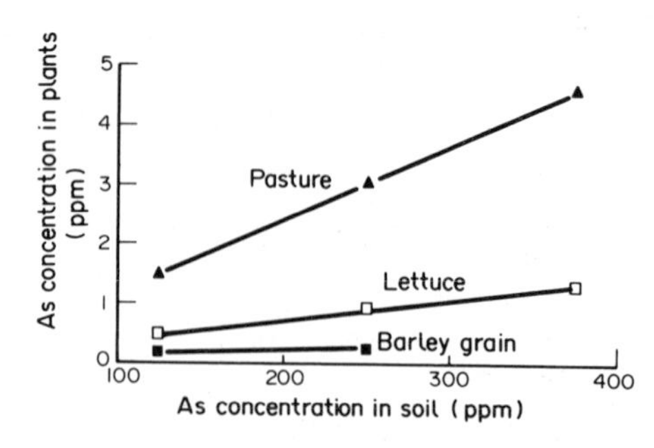

Fig. 10.4 Arsenic in soil and plants. Source of data; Adriano, 1986 [1].

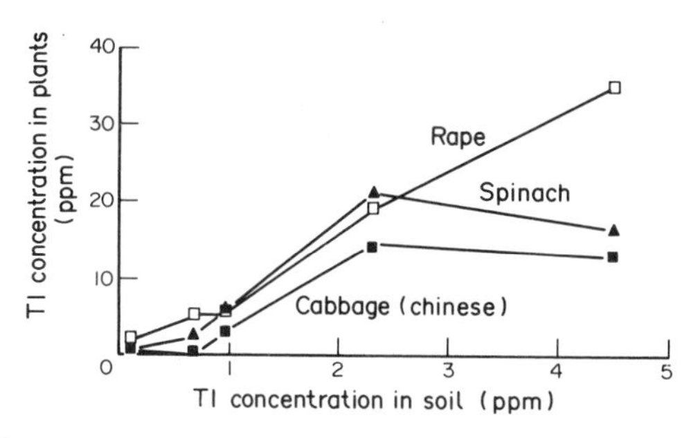

Fig. 10.5 Thallium in soil and plants. Source of data; Adriano, 1986 [1].

The uptake of selenium with change in pH and E_h has been discussed previously in terms of the soil chemistry of selenite and selenate (see Chapter 9). The data plotted in Fig 10.6, demonstrates the effect of pH on the selenium content of rye grass [24], and a good example of the influence of pH is seen from the data for selenium in herbage in relation to the yearly rainfall in Fig 10.7 [16]. The results indicate that the lower the rainfall, (and therefore probably more alkaline soil), the greater the selenium uptake, whereas in the high rainfall areas (where the soil may be more acidic and reducing), the selenium uptake is less. Except for lead, increasing the oxidizing characteristics of the environment, increases the element uptake. For lead the reverse occurs, and is presumably related to the insolubility of Pb^{2+}salts, such as $PbSO_4$.

The conditions used for growing rice makes it possible to investigate the effect of E_h because in flooded conditions the environment will be reducing. The data presented in Fig 10.8 shows clearly the effect of the redox conditions on the uptake of cadmium [1]. In the wet and reducing conditions the intake of the metal is quite significantly less than in the dry and oxidizing conditions. The reverse occurs for lead, and direct comparison of lead and cadmium concentrations with respect to pH and E_h are shown in Figs 10.9a, b. The difference

TABLE 10.16 Influence of pH and E_h on the Availability of Arsenic, Cadmium Selenium and Lead to Plants

Factor	As	Cd	Se	Pb
pH increase (more alkaline)	dec.	dec.	inc.	dec.
E_h increase (more oxidising)	inc.	inc.	inc.	dec.

dec = decrease in availability, inc = increase in availability.

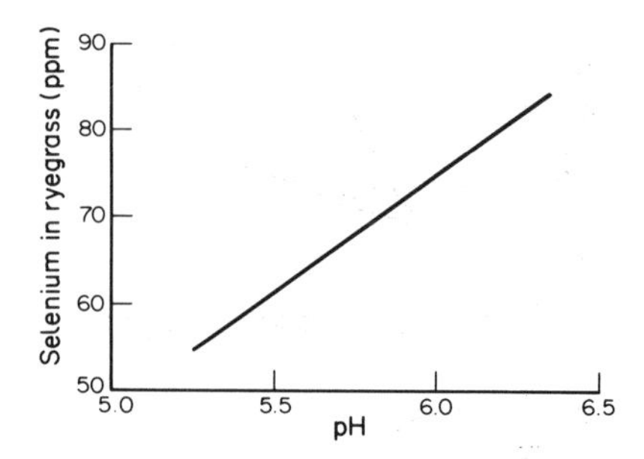

Fig. 10.6 The selenium content of rye grass as a function of pH. Source of data; Kabata-Pendias and Kabata, 1984 [24].

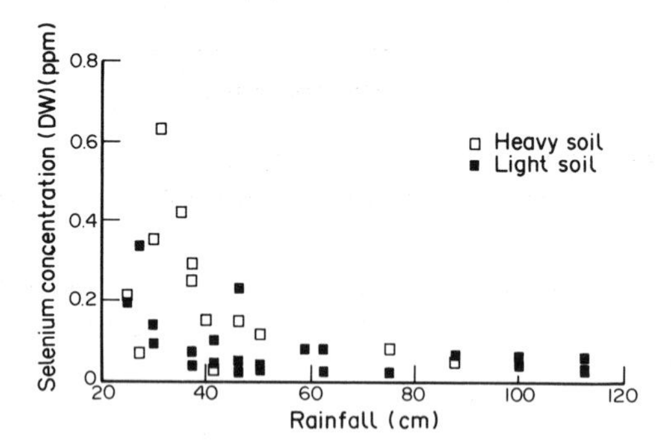

Fig. 10.7 Relationship between selenium in herbage and rainfall. Source of data; Gardiner and Gorman, 1963 [16].

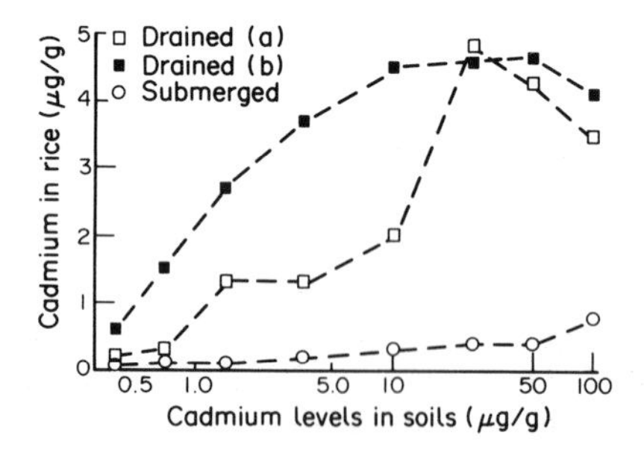

Fig. 10.8 Concentration of cadmium in rice in relation to the water status of the soil. Source of data; Asami, 1984 [2].

probably arises because of the lower solubility of lead salts, compared with cadmium salts.

The increase in arsenic levels in tree rings were related to emissions from a sulphur recovery plant. The increase in acidity in the soil resulting from the acid emissions, improved the availability of the soil arsenic [30].

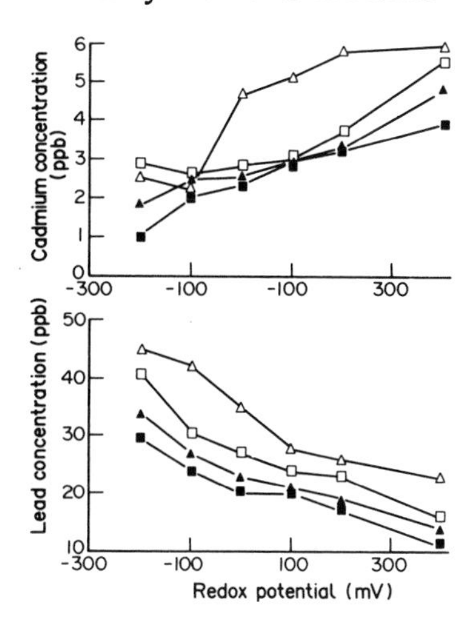

Fig. 10.9 Influence of pH and redox potential on the uptake of (a) lead and (b) cadmium by plants. Source of data; Adriano, 1986 [1].

Other soil factors that influence the availability of heavy elements to plants are listed in Table 10.17. For the first four factors an increase in the number of sites that can bind the heavy metals reduces the metal's availability for plant uptake. An increase in the concentration of other metals, because they compete for binding sites, may release the heavy metals making them more available. The extent of this, however, does depend on their relative binding energies, and the influence this has on the chemical equilibrium of the reaction;

$$(S\text{-}M)_s + Z^{n+}_{aq} \rightleftharpoons (S\text{-}Z)_s + M^{n+}_{aq}$$

where the equilibrium constant, K is;

$$K = \frac{[M^{n+}_{aq}]}{[Z^{n+}_{aq}]},$$

and where S is the binding sites in the soil material, M is the heavy metal and Z is the competing metal. An increase in $[Z^{n+}_{aq}]$ produces an increase in $[M^{n+}_{aq}]$ in order for the equilibrium to be maintained.

Distance from the Source

A number of studies have been carried out where the levels of heavy metals in plants are related to their distance from the source of the metals [8,9,16,20,24,25,41,44,46,49,50,52]. The data in Table 10.18 indicates the influence of heavy metals from a lead smelter on the plant levels, and how the levels change with distance [41]. For lead and cadmium, a fall off in concentrations with a fall in traffic density also occurs [9,13,29,44,50]. This is demonstrated in Fig 10.10, for lead levels in unwashed vegetation [50].

TABLE 10.17 Soil Factors that Influence the Availability of Heavy Metals to Plants

Soil factor	Change in factor	Effect on element's avaiability
Clay content	increase	decrease, due to more binding sites in soil
Organic matter	increase	decrease, due to more binding sites in soil
Fe/Mn oxides	increase	decrease, due to more binding sites in soil
CEC*	increase	decrease, due to more binding sites in soil
Temp	increase	increase, due to improved kinetics
Water stress	increase	decrease, due to reduced uptake of water
Other elements	increase	decrease, due to competition
Added phosphate	increase	decrease, due to increased vegatative growth

* CEC = cation exchange capacity is related to the three factors above it
Sources of data; references 1,22,24,27,36,37,39.

TABLE 10.18 Levels of Some Heavy Metals in Plants Growing Near a Lead Smelter ($\mu g\ g^{-1}$)

Metal	Distance from the smelter		
	0.27-2 miles	0.27 miles	3.3 miles
Cd	200-9480	9500	17
Sb	6.2-111	110	4.5
Pb	540-9990	10,000	320
As	4.9-59	59	3
Se	0.13-3.2		
In	0.078-2.1		
Hg	(6.0-77) x 10^{-3}		

Source of data; Ragaini et al., 1977 [41].

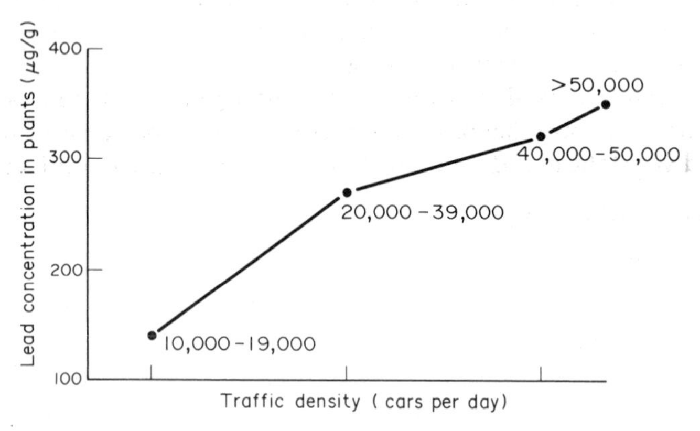

Fig. 10.10 Relationship btween traffic density and lead in plants. Source of data; Ward et al., 1977 [50].

The use of lead arsenate as an insecticide in agriculture, means that both lead and arsenic can get into crops and plants. Increases in the lead levels by approximately three times, and arsenic by about fourteen times, compared with untreated areas have been reported [8]. The addition of sewage sludge increases cadmium levels in plants. For example for oats and rye, after the application of sludge at the rate of 502 MT per hectare, the cadmium concentration increased from 0.4 to 4.8 $\mu g\ g^{-1}$ (dry weight) [15].

The distribution of lead, cadmium, arsenic and antimony in ambrotrophic bogs and in twigs in Norway indicate a decrease with depth for the bogs and between north and south Norway for the twigs. Both results point to a pollutant source [46].

Seasonal Effects

The concentration of the heavy elements in plants sometimes show a seasonal effect, for example, lead levels in plants tend to increase in the autumn and winter. There are a number of reasons for this, such as a loss of vegetative material during the winter, older leaves taking in aerosol lead more readily than new leaves, reduced wind speed giving more opportunity for aerosol deposition to occur, and the amount and intensity of rainfall varies with the season [7,13,35]. Since plant growth and decay affect the concentration of the heavy metals, it is necessary to be careful when comparing different studies and interpreting results.

Foliar Uptake of Heavy Metals

As discussed earlier, heavy metals can also enter plants through the foliage, and also foliage will have external deposits. Therefore, whether or not to wash plants before analysis will depend on the information required [12,13,18,31]. The amount of particulate material removed, will depend on the washing agent and the surface texture of the leaves. Hence there is always some doubt as to interpretation of the results of plant analysis because the variable amount of deposit and the effectiveness of washing.

Foliar uptake of lead by plants, may be more important than root uptake [6]. From data from a number of studies correlation has been observed between air lead and vegetation lead. The slope of the line in Fig. 10.11 [6], given by;

$$\frac{[Pb]_{foliage}(DW)\ (\mu g\ g^{-1})}{[Pb]_{air}\ (\mu g\ m^{-3})},$$

is 20 $m^3\ g^{-1}$, and the correlation between the two parameters is 0.805.

The lead particles on leaves have been found to be associated with sulphur and phosphorus, and the particles may be $PbSO_4$ and $Pb_3(PO_4)_2$ [28]. The material probably came from aerosol, but there is also a soil contribution in the leaf deposit. There is disagreement over whether or not lead is taken in through the leaves, but is does appear that the lead particles may clog the stomata.

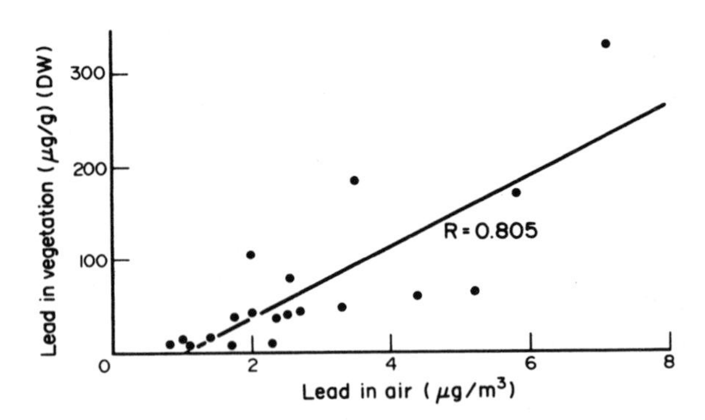

Fig. 10.11 The relationship between lead in air and plants. Source of data; Chamberlain, 1983 [6].

Heavy Metal Accumulators

Some plants are able to accumulate some heavy metals to high concentrations. A particularly good example is the accumulation of selenium, up to 1% in some cases, by *astragalus machairanthera, haplopappus* and *tanleya* species. The unpleasant odour from such plants arises from the formation of volatile compounds such as $(CH_3)_2Se_2$ [4,24,39]. The plants *andropogan scaporius* and *agrostis* are both found to accumulate arsenic when growing in arsenic rich soil. The *agrostis* species are found to be tolerant to arsenate, but not arsenite [38].

Mosses and lichens are particularly good in accumulating lead (see the data in Table 10.13) [24] and other heavy metals. The species *alyssum wulfenianum* and *thlaspi rotundifolium* [4] have been found to accumulate 1000 $\mu g\ g^{-1}$ and 7000 $\mu g\ g^{-1}$ of lead respectively. Since some plants are good accumulators of metals use may be made of this in mineral prospecting. The plant is used as an indicator of metal concentrations in the underlying rocks and soils.

TOXICITY AND BIOCHEMISTRY OF THE HEAVY ELEMENTS IN PLANTS

Estimation of the toxicity of the heavy elements on plants is not straight forward. The essential and toxic effects of elements on plants can be described by response versus concentration curves, of the type shown in Fig. 15.1. The diagrams when used for plants indicate that there is an optimum level for a nutrient element, below which the element is deficient, and above which the element is toxic. Selenium is considered essential for some bacteria, such as for the activity of formate dehydrogenase in *E. coli* and *closteridium thermoactium* and glycine reductase in *C. stricklandii* [3,23,24,39]. Deficient levels of selenium in a nutrient solution are <0.02 mg l^{-1}, optimum levels are <1 mg l^{-1} and toxic levels are 1-2 mg l^{-1}. For a non-essential element there is a concentration below which the element does not appear to have any significant effect on the plant growth, and a level above which the element becomes toxic (see Fig 15.1b). It has been reported that arsenic may slightly stimulate growth at low concentrations [3]. The data in Table 10.19 is a collection of some available information on levels of toxicity of the heavy metals for plants. The first three columns refer to concentrations in the nutrient solution and the last two refer to the concentrations in the plant leaf [3,24,26]. It may be judged, from the data, that in general the toxicity of the heavy metals to plants lie in the order As(III) ~Hg > Cd > Tl > Se(IV) > Te(IV) > Pb > Bi ~ Sb. The order is very dependent however, on soil properties and the type of plant.

An example of the influence of the toxicity of cadmium on rice, is seen from the data presented in Fig 10.12 [2]. The results demonstrate that the yield in the rice begins to fall at a cadmium concentration of around 0.1-0.2 $\mu g\ g^{-1}$ in the nutrient solution, but the concentration of the metal in the various sections of the plant continues to increase.

The effect of toxicity may be seen by physical signs on the plants, and some of these are listed in Table 10.20 [1,2,22,24,36]. These are relatively general effects and could well occur for other reasons, such as the deficiency of an essential element. On the whole, lower plants, such mosses tend to be more tolerant to higher levels of the heavy elements, and can be used as pollution monitors [19].

The biochemical processes that the heavy elements may be involved in within the plants are quite numerous, and not necessarily specific for the metals. Some which have been observed for the heavy elements, are listed in Table 10.21. Reactions can occur, for the heavy elements, which involve the element bonding to a reactive site and/or replacement of an essential element. The replacement occurs because of similar chemistry, such as selenium and tellurium for sulphur, thallium(I) for the potassium ion and arsenate for phosphate.

TABLE 10.19 Toxic Levels for Some Heavy Elements Towards Plants

Element	Nutrient solution mg l^{-1}			Concentration in leaves $\mu g\ g^{-1}$	
	Deficient	Normal	Toxic	No effect	Toxic
As(III)		<0.02	0.02-7.5	1-1.7	5-20
Bi			27		
Cd		<0.05	0.2-9	0.05-0.2	5-30
Hg					1-3
Pb		<3	3-20	5-10	30-300
Se(IV)	<0.02	<1	1-2	0.01-2	5-30
Sb				7-50	150
Te(IV)		<6	>6		
Tl		<1	1		20-30

Sources of data; Bowen, 1979 [3], Kabata-Pendias and Kabata, 1984 [24], Kloke et al., 1984 [26].

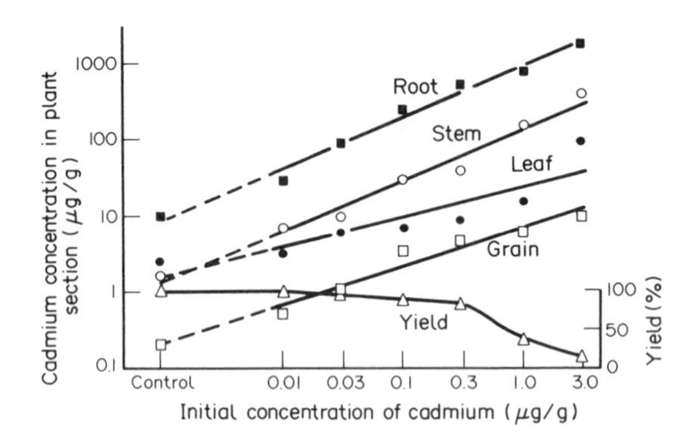

Fig. 10.12 Uptake of cadmium by rice plants and yield of unpolished rice. Source of data; Asami, 1984 [2].

TABLE 10.20 Toxic Effects of Some Heavy Elements on Higher Plants

Element	Visible signs of toxicity
Arsenic	Red brown necrotic spots on old leaves, yellow browning of roots, growth reduction.
Cadmium	Brown margin to leaves, chlorosis, curled leaves. brown stunted roots, redish veins and petioles, reduction in growth.
Mercury	Severe stunting of seedlings and roots, chlorosis, browning of leaf points, reduction in growth.
Lead	Dark green leaves, stunted foliage, increased amounts of shoots
Selenium	Interveined chlorosis, black spots at high selenium, bleaching and yellowing of young leaves, pink spots on roots.
Thallium	Reduction in growth

The element that has been studied in most detail is selenium and possible (but not always verified) reactions that occur with selenium in plants are listed out in Fig 10.13 [1,39]. It is likely that other elements, such as arsenic and mercury, also have an organic chemistry in plants.

Competition between metals is quite likely to occur and the competition may be either antagonistic, that is the combined effect of the two elements is less than the sum of their separate effects, or synergistic where the combined effect is greater than the sum of the two element's individual effects. The more common interaction is antagonistic. A list of some of the interactions for the heavy elements is give in Table 10.22. Some elements can be both antagonistic and synergistic to an element, presumably in different biochemical processes. When a heavy toxic element is antagonistic to an essential element, the toxic effect could be due to a deficiency of the essential element. For example

TABLE 10.21 Some Biochemical Processes Affected by the Heavy Elements

Process	Elements
Change in the permeability of the cell membrane	Cd, Hg, Pb
Inhibits protein synthesis	Hg
Bonding to thiol and SH groups	Hg, Pb, Cd, Tl, As(III)
Competition for sites with essential metabolites	As, Sb, Se, Te
Affinity for phosphate groups	most heavy metals
Replacement of essential atoms	Se, Tl
Occupation of sites for essential groups, such as PO_4^{3-} and NO_3^-	arsenate, selenate tellurate
Inhibition of certain enzymes	Tl. Pb, Cd
Affects respiration	Cd, Pb
photosynthesis	Cd, Pb, Hg, Tl
stomatal opening	Cd, Tl, Pb
transpiration	Cd, Pb, Hg, Tl, As

Sources of data; references 1,22,24,27,39,40.

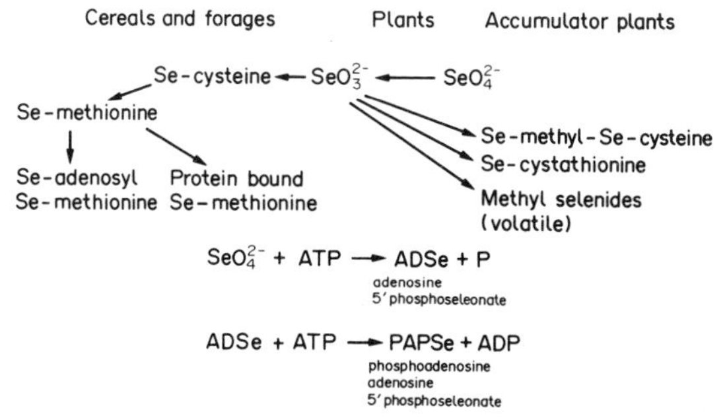

Fig. 10.13 Selenium chemistry in plants. Sources of data; Adriano, 1986 [1], Peterson et al., 1981 [39].

TABLE 10.22 The Interaction of Elements with Each Other in Plants

Element	Antagonistic to	Synergistic to
Cadmium	Ca, P, K, Zn, Al, Se, Mn	Pb, Mn, Fe, Ni, Cu*, Zn* Fe, Ni, Mg, Cu*
Lead	Ca, P, S, Zn	Cd
Arsenic	P, Mn, Zn, Mn*	
Mercury	P, K	
Selenium	P, S, Mn, Cu, Zn, Cd, W*	V
Thallium	K	

*Outside plant and close to the roots. Sources of data; references 1,24,39,40.

chlorosis of the plant leaves is due to a deficiency of iron, and maybe produced by some heavy metals being taken up by the plant in preference to iron, or the heavy metal being involved in the plant biochemistry rather than iron.

Plants which are tolerant to heavy metals have been used for the reclamation of waste land, such as mine tailings [17,21,51]. For example the two grasses *festuca rubra* (in neutral to calcareous Pb/Zn wastes) and*agrostis tenuis, arctagrostis latifolia* and *deschampsia cespitosa* (in acidic lead wastes) have been used as vegetation on certain soil types with a high heavy metal content.

REFERENCES

1. Adriano, D C. Trace Elements in the Terrestial Environment: Springer Verlag; 1986.
2. Asami, T. Pollution of soils by cadmium. In: Nriagu, J. O., Ed. in: Changing Metal Cycles and Human Health, Dahlem Konferenzen; 1984: 95-111.
3. Bowen, H. J. M. Environmental Chemistry of the Elements: Academic Press; 1979.
4. Brooks, R. R. Biological Methods of Prospecting for Minerals: Wiley; 1983.
5. Bryce-Smith, D. Heavy metals as contaminats of the human environment: Chem. Soc. Chem. Cassetts; 1975.
6. Chamberlain, A.C. Fallout of lead and uptake by crops. **Atmos. Environ.;** 1983; **17:** 693-706.
7. Chaney, R. L., Sterrett, S. B. and Miekle, H. W. The potential for heavy metal exposure from urban gadens and soils. In: Preer, J. R., Ed. Proc. Symp. Heavy Metals in Urban Soils: Univ. District Columbia, Washington D.C.; 1984: 37-84.
8. Chisholm, D. Lead, arsenic and copper content of crops grown on lead arsenate treated and untreated soils. **Canad. J. Plant Sci.;** 1972; **52:** 583-588.
9. David, D. J. and Williams, C. H. heavy metal contents of soils and plants adjacent to the Hume highway near Marulan, New South Wales. **Aust. J. Expt. Agric. Animal Husb.;** 1975; **15:** 414-418.
10. Dudas, M. J. and Powluk, S. Heavy metals in cultivated soils and in cereal crops in Alberta. **Can. J. Soil Sci.;** 1977; **57:** 329-339.
11. Eyre, S. R. Vegetation and Soils- a World Picture. 2nd Ed.: Arnold; 1968.
12. Fergusson, J.E., Hayes, R.W., Tan Seow Yong and Sim Hang Thiew. Heavy metal pollution by traffic in Christchurch, New Zealand: Lead and cadmium content of dust, soil and plant samples. **N. Z. J. Sci.;** 1980; **23:** 293-310.
13. Fergusson, J. E. Lead: petrol lead in the environment and its contributions to human blood lead levels. **The Sci. Total Environ.;** 1986; **50:** 1-54.
14. Fleming, G. A., and Parle, P. J. Heavy metals in soils, herbage and vegetables from an industrialised area of West Dublin city. **Irish J. Agric. Res.;** 1977; **16:** 35-48.

15. Freedman, B. and Hutchinson, T. C. Sources of metal and elemental contamination of terrestrial environments. In: Lepp, N. W., Ed. Effect of Heavy Metal Pollution on Plants. Vol 2, Metals in the Environment: Applied Science Publishers; 1981: 35-94.
16. Gardiner, M. R. and Gorman, R. C. Further observations on plant selenium in Western Australia. **Aust J. Exptl. Agric. Animal Husb.**; 1963; **3:** 284-289
17. Gemmell, R. P. Colonization of Industrial Wasteland. Studies in Biology, No. 80: E. Arnold; 1977.
18. Hassett, J. M., Jennett, J. C. and Smith, J. E. Heavy metal accumulation by algae. In: Baker, R. A., Ed. Contaminants and Sediments (Vol 2): Ann Arbor Science; 1980: 409-424.
19. Hawksworth, D. L. and Rose, F. Lichens as Pollution Monitors. Studies in Biology, No. 66: E. Arnold; 1976.
20. Hemphill. D. D. and Clevenger, T. C. Dispersion of heavy metals from primary smelters and consequent plant uptake. In: Management and Control of Heavy metals in the Environment: CEP Consultants; 1979: 267-270.
21. Hutchinson, T. C. and Kuja, A. Selection and use of multiple-metal tolerant native grasses for re-vegetation of mine tailings. In: Management and Control of Heavy metals in the Environment: CEP Consultants; 1979: 191-197.
22. Jastow, J. D. and Koeppe, D. E. Uptake and effects of cadmium in higher plants. In: Nriagu, J. O., Ed. Cadmium in the Environment, Part 1. Ecological Cycling: Wiley; 1980: 607-638.
23. Johnson, C. M. Selenium in soils and plants; contrasts in conditions providing safe but adequate amounts of selenium in the field chain. In: Nicholas, D. J. D. and Egan, A. R., Eds. Trace Elements in Soil-Plant-Animal Systems: Academic Press; 1975: 165-180.
24. Kabata-Pendias, A. and Pendias, H. Trace elements in soils and plants: CRC Press; 1984.
25. Khan, D. A. Lead in the Soil Environment. MARC Tech. Rept.: MARC; 1980.
26. Kloke, A., Sauerbeck, D. R. and Vetter, H. The contamination of plants and soils with heavy metals and the transport of metas in terrestrial food chains. In: Nriagu, J. O., Ed. Changing Metal Cycles and Human Health, Dahlem Konferenzen.; 1984: 113-141.
27. Koeppe, D. E. Lead: understanding the minimal toxicity of lead in plants. In: Lepp, N. W., Ed. Effect of Heavy Metal Pollution on Plants. Vol 1, Effects of Trace Metals on Plant Function: Applied Science Publishers; 1981: 55-76.
28. Koslow, E. E., Smith, W. H. and Staskawicz, B. J. Lead containing particles on urban leaf surfaces. **Environ. Sci. Techn.**; 1977; **11:** 1019-1021.

29. Lagerwerff, J. V. and Specht, A. W. Contamination of roadside soil and vegetation with cadmium, nickel, lead and zinc. **Environ. Sci. Techn.;** 1970; **4:** 583-586.
30. Legge, A. H., Kaufmann, H. C. and Winchester, J. W. Tree-ring analysis by PIXE for a historical record of soil chemistry response to acidic air pollution. **Nucl. Instr. Methods, Phys. Res.;** 1984.
31. Little, P. A study of heavy metal contamination on leaf surfaces. **Environ. Pollut.;** 1973; **5:** 159-172.
32. Lund, L. J., Betty, E. E., Page, A. L. and Elliott, R. A. Occurrence of naturally high cadmium levels in soils and its accumulation in vegetation. **J. Environ. Qual.;** 1981; **10:** 551-556.
33. Mortvedt, J. J. and Osborn, G. Studies on the chemical form of cadmium contaminants in phosphate fertilizers. **Soil Sci.;** 1982; **134:** 185-192.
34. Murti, C.R.K. The cycling of arsenic, cadmium, lead and mercury in India. In: Hutchinson, T.C. and Meena, K.M., Eds. Lead, Mercury, Cadmium and Arsenic in the Environment. Scope 31: Wiley; 1987: 315-333.
35. O'Toole, J. J., Wessels, T. E. and Malaby, K. L. Trace element levels in native vegetation as a measure of aerosol deposition. In: Management and Control of Heavy metals in the Environment: CEP Consultants; 1979: 222-225.
36. Page, A. L., Bingham, F. T. and Chang, A. C. Cadmium. In: Lepp, N. W., Ed. Effect of Heavy Metal Pollution on Plants. Vol 1, Effects of Trace Metals on Plant Function: Applied Science Publishers; 1981: 77-109.
37. Page, A. L., Chang, A. C. and El-Amamy, M. Cadmium levels in soils and crops in the United States. In: Hutchinson, T.C. and Meena, K.M., Eds. Lead, Mercury, Cadmium and Arsenic in the Environment. Scope 31: Wiley; 1987: 119-148.
38. Peterson, P. J.. Benson, L. M. and Porter, E. K. Biogeochemistry of arsenic in polluted sites in S. W. England. In: Management and Control of Heavy metals in the Environment: CEP Consultants; 1979: 198201.
39. Peterson, P. J. Benson, L. M. and Zieve, R. Metalloids. In: Lepp, N. W., Ed. Effect of Heavy Metal Pollution on Plants. Vol 1, Effects of Trace Metals on Plant Function: Applied Science Publishers; 1981: 279-342.
40. Peterson, P. J. and Girling, C. A. Other trace metals. In: Lepp, N. W., Ed. Effect of Heavy Metal Pollution on Plants. Vol 1, Effects of Trace Metals on Plant Function: Applied Science Publishers; 1981: 213-278.
41. Ragaini, R. C., Ralston, H. R. and Roberts, N. Environmental trace metal contamination in Kellog, Idaho, near a lead smelting complex. **Environ. Sci. Tech.;** 1977; **11:** 773-781.
42. Ray, P. M. The Living Plant. 2nd Ed.: Holt Rinehart and Winston; 1972.
43. Riley, P. M. and Young, A. World Vegetation: Cambridge Univ. Press; 1968.
44. Sharma, R. P. Soil-plant-animal distribution of cadmium in the environment. In: Nriagu, J. O., Ed. Cadmium in the Environment, Part 1. Ecological Cycling: Wiley; 1980: 587-605.

45. Smith, I. C. and Carson, B. L. Trace Metals in the Environment Vol. I Thallium: Ann Arbor Science; 1977.
46. Steinnes, E. Impact of long range atmospheric transport of heavy metals to the terrestrial environment in Norway. in: Hutchinson, T.C. and Meena, K.M., Eds. Lead, Mercury, Cadmium and Arsenic in the Environment. Scope 31: Wiley; 1987: 107-117.
47. Sutcliffe, J. F. and Baker, D. A. Plants ans Mineral Salts. Studies in Biology No. 48: Arnold; 1974.
48. Thomas, B., Roughhan, J. A. and Watters, E. D. Lead and cadmium content of some vegetable foodstuffs. **J. Sci. Food Agric.**; 1972; **23:** 1493-1498.
49. Ward, N. I., Brooks, R. R. and Robets, E. Contamination of a pasture by a New Zealand base metal mine. **N. Z. J. Sci.**; 1977; **20:** 413-418.
50. Ward, N. I., Brooks, R. R., Roberts, E. and Boswell, C. R. Heavy metal pollution from automotive emissions and its effect on roadside soils and pasture species in New Zealand. **Environ. Sci. Techn.**; 1977; **11:** 917-920.
51. Williamson, A. and Johnson, M. S. Reclamation of metalliferous mine wastes. In: Lepp, N. W., Ed. Effect of Heavy Metal Pollution on Plants. Vol 2, Metals in the Environment: Applied Science Publishers; 1981: 185-212.
52. Windom, H. L. and Kendall, D. R. Accumulation and biotransformation of mercury in coastal and marine biota. In: Nriagu, J. O., Ed. The Biogeochemistry of Mercury in the Environment: Elsevier/Nth. Holland; 1979: 303-323.

CHAPTER 11

ADVENTITIOUS SOURCES OF THE HEAVY ELEMENTS

Air, water, soil, plants and animals are the clear routes by which human beings come into contact with the heavy elements. There are, however, less obvious sources, called adventitious sources, which include surface dust, paint and consumer items such as heavy metals in glazed pottery and cadmium, and sometimes selenium, in red coloured plastic food containers.

The most studied adventitious material is dust, particularly urban dust in streets and houses, and the majority of the work has been carried out on lead and cadmium. Some of the sources that contribute to dust and the heavy elements in them are given in Table 11.1. In this chapter we will concentrate on urban street and house dust.

LEAD IN DUST

The two significant urban sources of lead in dust are paint and automobile emissions. Since children come into contact with dust, most information available is on the concentrations, and chemistry of the metal in street dust, and to a lesser extent house dust.

Concentrations of Lead in Dust

The best estimate of the concentrations of the heavy elements in natural dust is the dust arising from the deposition of natural atmospheric emissions (see Chapter 7). The concentrations are similar to the natural concentrations that occur in soil for, as we will see, soil is one of the main components of dust. The data in Table 11.2 [32], is for representative ranges of lead levels in contaminated dust and paints. The difference in the concentrations for rural, suburban and inner city areas relates mainly to closeness to roads, automobile lead emissions, but also to lead painted buildings, there being more in the city.

TABLE 11.1 Sources of Dust (Principally Urban)

Sources	Probable heavy element content
Soil	All heavy metals
Automobile aerosol	Lead
Industrial aerosol	Many of the elements
Organic litter	Most of the elements at low concentration
Weathered buildings	Many of the elements
Transported industrial dust (e.g. in clothes)	Many of the elements
Weathered paint	Lead, cadmium, selenium
Corroded metals:	
car parts	Cadmium, lead, antimony
galvanized iron	Cadmium
iron structures	
Rubber and plastic	Lead, cadmium

TABLE 11.2 Typical Concentrations of Lead in Dust ($\mu g\ g^{-1}$)

Sample	Rural	Suburban	Inner city	Around smelters	Mineralized areas
House dust (no paint)	20-100	50-500	200-6000	up to 3-5%	up t0 1-2%
Street dust	30-500	250-2000	250-10,000		
House dust + sanded paint	1 to > 10%				
Lead paint flakes	0.25 to many % or <1 to 5 mg cm^{-1}				
Soil + lead paint flakes	0.1 to > 5%				

Source; reference 32

Some specific data are given in Tables 11.3 and 11.4. One feature that emerges from the data, is the relatively consistent levels found in different parts of the world. This is irrespective of different conditions, including weather, geography, lead levels in sources and time of study. This suggests that a steady state exists between lead input to dust and lead output through washout and cleaning processes. The average lead in urban house dust is around 400-800 $\mu g\ g^{-1}$, but in urban street dust is around 1000-10,000 $\mu g\ g^{-1}$. There are situations, however, where very high lead levels occur [3,5,12,18,47,55,67] due to the proximity of smelters, lead based industries, and buildings with lead based paints, particularly if it has been removed by dry sanding.

House dust has been sampled by some workers by wiping a surface or collecting dust fall in certain areas. The results are expressed variously as μg/

TABLE 11.3 Lead Concentrations in Street Dust

Location	Mean µg g-1	Range µg g-1	Comments
Belgium	3541		1 km from smelter
	5466		1 km from smelter
	397		2.5 km from smelter
	152		Rural
Birmingham (UK)	2350	160-10,000	A-class roads
	932	200-5800	Residential pavements
	506	62-5100	Pavements
	805	80-2100	Roadway
Christchurch (NZ)	3020	200-7050	City intersections
	2400	200-7840	Between intersections
	1220	175-2790	City wide
		10,700, 1290	Central city
	6340	4310-9630*	Main road
		2270-5700	Side roads
	3100		Heavy traffic
	830		Light traffic
Derbyshire (UK)	4394	1190-13,352	Playgrounds
Detroit		324-1213	City streets
England	2291	1643-3268	Near major roads
	2100		Motorway gutters
Glasgow	960	150-2300	City area
	960	150-8900	National survey
Halifax (Canada)		1919, 674	Central city
Hobart	4148	120-30,100	Industrial areas
Holland	859	77-2667	Urban
	1144	457-8097	Urban
Hong Kong	7427	1570-19,073	City streets
	1102	271-1987	City streets
Kingston (Jamaica)		909, 817	Central city
Lancaster (UK)	2250		City centre
	305		Rural
	46,300	39,700-51,900	Car parks
	2130	840-4530	Town centre
	570	410-870	Rural
London (UK)	1840		Central city
	920		Outer urban
		170-3700*	Playground dust
		2008, 4053	Central city
	2937	1041-6508	Near secondary smelter
		3010-4420*	By busy roads
Manchester (UK)	1001		Moderate/heavy traffic
	993		Light traffic
New York		2213, 2952	Central city

Table 11.3 continued

Location	Mean µg g⁻¹	Range µg g⁻¹	Comments
Rotterdam	532	168-2308	Inner city
	318	113-1155	Suburbs
Swansea (Wales)	634	164-3984	
United Kingdom	1188	140-7720	Nationwide
	486	93-1828	Playgrounds
	35		Rural
Wales	173	22-1341	Urban
	269	83-870	Cul-de-sac

* Range of means. Sources of data; references 3,4,5,7,8,9,12-15,17,18,20,22,24, 26,27,29-32,36,37,39,40,52,55,57,65,67,69.

TABLE 11.4 Lead Concentrations in House Dust

Location	Mean µg g⁻¹	Range µg g⁻¹	Comments
Baltimore (USA)		100-1600 µg*	
Birmingham (UK)	2780	100-280,000	Internal surfaces
	615	79-15,000	Doormat
	430	111-3110	
Christchurch (NZ)	460		Post 1950 houses
	830		Pre 1950 houses
	727		Wood houses
	457		Brick houses
	734	287-1408	City survey
Connecticut	11,000		
Derbyshire	1870	606-7019	Playgrounds
		531-2582**	
El Paso (USA)	22,191	2800-103,756	<1.6 km from a smelter
	973	200-22,700	>6.4 km from a smelter
Glasgow	716	510-970	National survey
Holland	1144	457-8097	Urban
	1054	463-4741	Urban (fine dust)
Illinois (USA)	600	170-1440	Urban residential
		410-3380	Urban, non-residential
Lancaster (UK)	716	510-970	
London (UK)	3366	276-8619	Near secondary smelter
	1540	368-3057	Controls
New York	3310		Lead workers homes
	1240		Controls
New York State		7300, 6268	Sills in houses
Omaha (USA)		76-5571	Different sites
		76-860	
		18-845	

Table 11.4 continued

Location	Mean $\mu g\ g^{-1}$	Range $\mu g\ g^{-1}$	Comments
Rochester (USA)		33-486 $\mu g\ ft^{-2}$	Old housing
		2-24 $\mu g\ ft^{-2}$	New housing
		0-60 $\mu g\ ft^{-2}$	Suburban
		149-205 μg*	High blood lead group
		55-123 μg*	Low blood lead group
Rotterdam	81	5-740 $\mu g\ m^{-2}$	Inner city
	30	1-410 $\mu g\ m^{-2}$	Suburban
Shipham (UK)	1014	132-26,760	National survey
United Kingdom	1263	39-30,060	National survey
Wales	209	44-998	Urban
	259	62-1075	Near motorway
	177	36-851	Village
	330	28-3790	Mining village

* Amount of lead per sample or site, ** Range of means. Sources of data; references 1,2,3,8,10-13,17,18,19,25,27,28,39,47,50,55,56,58,60,62,66,67,68.

towel or μg/wipe or μg/area [8,10,11,18,58,68]. Such data are not easily related to concentrations in $\mu g\ g^{-1}$. However, lead levels as μg/wipe, can be related to levels on the hands of children, expressed as μg/hand or μg/child (two hands). Data on hand wipes are given in Table 11.5, and the relationship found between lead on children's hands and lead in playground dust is shown in Fig 11.1 [22]. There is an increase in both hand lead and dust lead, though it is not clear if the relationship is linear ($R = 0.717$) or curved. A curved relationship suggests other factors may also influence the amount of lead children get on their hands. One obvious factor, is the size of the particles. Up to 90% of the particles in dust on hands have been shown to be <10 μm, with a geometric mean of 4.5 μm [22]. It is difficult to remove all the dust, (and hence lead) from the hand of a child by one wipe, and it has been estimated that around 60% of the lead is removed [22].

TABLE 11.5 The Amounts of Lead on Children's Hands

Location	Mean/median	Range	Comments
Belgium		20-436 μg/hand	Near a smelter
Connecticut	2400 $\mu g\ g^{-1}$	650-4100 $\mu g\ g^{-1}$	Used sticky tape
London (UK)		12-43 μg/child	City schools
Rochester (USA)	5 μg/towel	<5-20 μg/towel	Suburbs
	20 μg/towel	<5-.160 μg/towel	City
	49 μg/sample		High blood lead group
	21 μg/sample		Low blood lead group
Various (USA, Belgium)		5-142 μg/hand	

Sources of data; references 8,11,18,22,50,58,59.

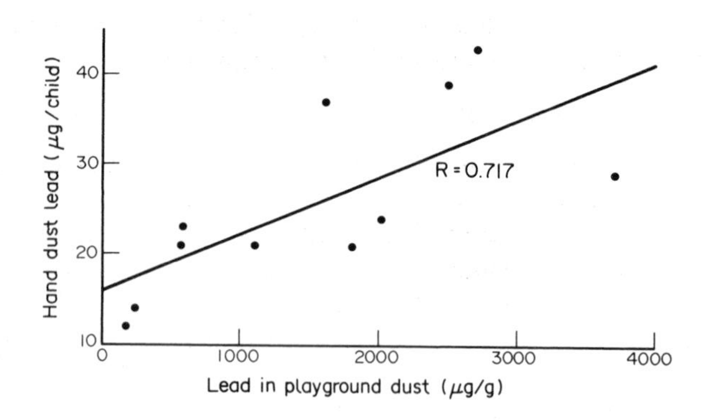

Fig. 11.1 Relationship between lead in dust and lead on children's hands. Source of data; Duggan et al., 1985 [22].

Sampling of Dust

Dust sampling methods can influence the final analytical result. For street dust, a brush and pan is frequently used, whereas for house dust a variety of methods have been employed [28,33,62], including brush and pan [2,61,62], vacuum cleaner [2], suction on to filter paper [28,62], sticky tape [50], wiping [10,11,56,58,59,60,68], and dust fall on to oiled plates [7,17]. The concentration of lead is also influenced by the location of the dust sample, such as from gutters, pavements and ledges [3,20,24,65]. Dust that has collected on high ledges and in sheltered places, may have a lower soil component, and therefore be higher in lead. Therefore a bias can occur in the sample [3], because of some fractionation of the dust, according to particle size and density. Fractionation may also occur during sampling, often the smaller particles, where the lead concentrations are highest, are left behind. Re-sampling the same spot on a carpet has demonstrated that whereas the amount of dust collected decreased the concentration of lead increased. Presumably the smaller particles lie deeper in the carpet pile [28]. House dust lead levels also depend on where the sample comes from, such as carpeted or non-carpeted areas. It has been found, that on a mass basis, the concentrations are higher from the non-carpeted areas, but, on an area basis they are lower [62].

The mean concentration of lead, in street dust samples taken from the same area (over 1 m^2) at the same time, often have reasonably low coefficients of variation, (CV),

$$CV\% = \left(\frac{\text{mean}}{\text{sd}}\right) \times 100,$$

of around 10-15% representing the analytical and sample variation [15,30,33]. The CV for repeated analysis on one sample (analytical variation) is <10%, and frequently <5% [33]. When the CV >20% this probably represents a real difference (e.g. spatial or temporal) between samples [23,33]. Because of a lot of factors it is not surprising that a wide range of lead levels occur in dust. There is a need to have standardized sampling techniques [9].

Lead in Dust in Relation to Distance

The concentration of lead in both street and house dusts is a function of the distance of the sample from the source of the lead. It has been observed frequently that the street dust lead levels fall off with distance from busy roads, with traffic density and distance from the central city area [13,15,20,30,31, 36,39,62]. For example, in London (UK) a lead concentration of 1840 $\mu g\ g^{-1}$ was recorded for the central city, 920 $\mu g\ g^{-1}$ in mid-urban areas and 35 $\mu g\ g^{-1}$ in rural areas [20]. Areas including car parks and garages, can have very high lead concentrations in the surface dust [30,39]. Similar trends occur for house and street dusts with distance from smelters, as shown in Fig 11.2 [37,47].

A study of the lead levels in dust in vacant homes demonstrated that lead had entered the house through gaps in the window systems. Levels fell off with distance from the windows, on the sill a level of 1144 µg/towel was found, 1189 µg/towel on the adjacent floor and 214 and 208 µg/towel at 2 and 3 m into the room respectively [60]. On the other hand for house dust and street dust, some investigators have found only small effect with location and source of the dust [14,25,37].

The washoff of lead aerosol, attached to the sides of buildings contributes to the dust lead near buildings [52,62,65]. A typical profile of lead concentration between roads and houses is shown in Fig 11.3, The lead levels close to the road are high from automobile emissions, then fall off away from the road, but increase closer to the house especially if lead paint has been used.

Influence of Particle Size and Density on Lead Levels

Many workers have found that, for both street and house dust, the smaller particles have higher concentrations of lead, and other heavy metals [4,7,17,

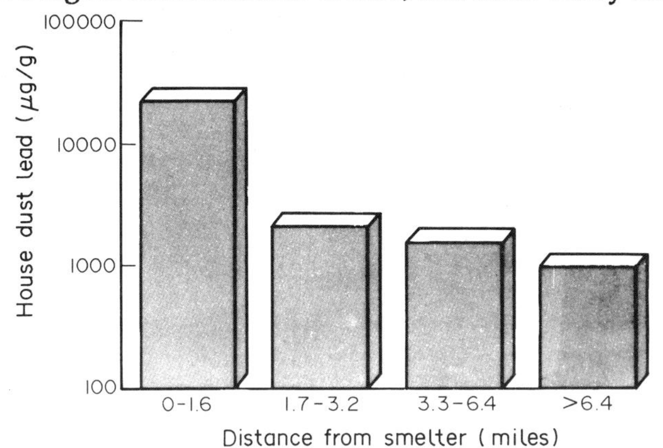

Fig. 11.2 House dust lead in relation to distance from a smelter. Sources of data; Harper et al., 1987 [37], Landrigan et al., 1975 [47].

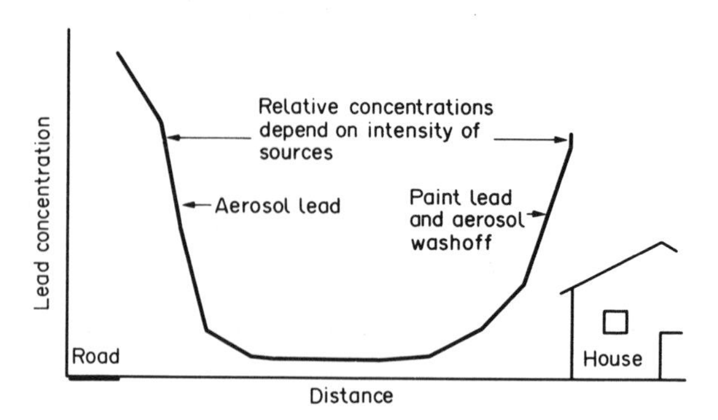

Fig. 11.3 Lead in dust with respect to distance from roads and houses. Source of data; Solomon and Hartford, 1976 [62].

18,22,24,29,30,41,44,52]. In many cases there is a fall in lead concentration with particle size, from 1000 µm (1 mm) down to 100-200 µm, and then an increase in concentration as the size decreases further (Fig. 11.4) [29]. The smaller size particles, have less silica and feldspar in them, and therefore a higher proportion of material containing lead, raising the concentration of the metal. Also in street dusts the particles emitted with car emissions are very small, most <1 µm in size. Most of the lead, however, occurs in the larger dust particles because of the greater quantity of that size dust [4]. The more dense fraction and the non-magnetic fraction of dust usually have the higher lead concentrations. Both these features are demonstrated by the data in Table 11.6 [52]. However, the location of the dust is also important, as lead occurring in ferrous materials may occur with higher concentrations in the magnetic fraction [18,29,30,41,52]. For house dust it has also been observed, that the finer particles contain the higher lead concentrations [7,17,44]. The smaller dust particles are also the ones that become attached to the hands of children.

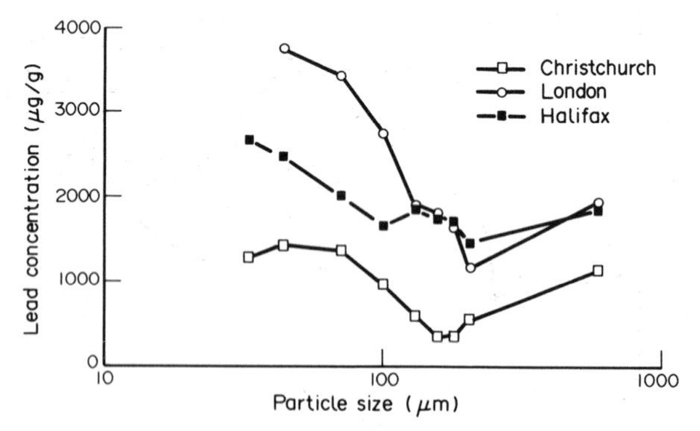

Fig. 11.4 Relation between particle size and lead concentration in dust. Source of data; Fergusson and Ryan, 1984 [29].

TABLE 11.6 The Relationship Between Lead Concentration* in Dust and Particle Size, Density and Magnetism

Size	Density (g ml^{-1})					
μm	≤ 1.5		1.5-3.3		> 3.3	
	non-mag	mag	non-mag	mag	non-mag	mag
250-600	1400*	6400	540	2700	780,000	41,000
45-250	2700	6400	1100	5800	270,000	70,000
< 45	6800	-	3500	20,000	550,000	120,000

*Concentrations in μg g^{-1}. Source of data; Linton et al., 1980 [52].

Other Factors

The weather, and mode of operation of the car also influence the levels of lead in street and house dusts [30,31,32,61]. This is clearly seen in the data plotted in Fig 11.5, for the concentration of lead at 14 sites, at an intersection on a busy urban road, studied over 13 years [31]. The effect of stopping the traffic on the main road (by installing traffic lights) was to increase the dust lead levels. The influence of the weather, particularly when there has been a lot of rain or long dry periods, is evident from the figure. Reduction of the lead content of petrol also shows up in a drop in the lead concentration in the dust. The same trends in lead concentrations, but less well developed, and at lower levels occur in the dust along the side streets at the intersection.

Lead in Dust in Relation to Sources

The sources of street dust have been investigated by statistical analysis of analytical data. Methods include factor and principal component analysis, chemical element balances, estimation of enrichment factors, metal concentra-

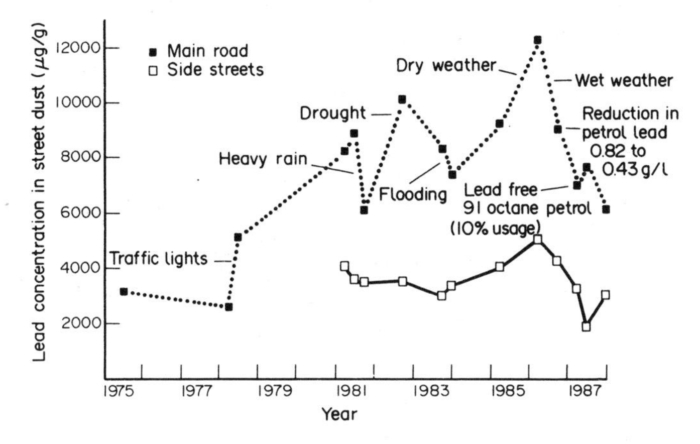

Fig. 11.5 Changes in lead concentrations in street dust with time. Source of data; Fergusson and Simmonds, 1983 [31].

tion ratios and correlation analysis from which dendograms may be produced [17,27,29,41]. The results suggest that the important components of street dust are soil (~75%) followed by iron materials (~8%), automobile tyre wear (~7%), cement (weathered concrete building materials) (~5%), automobile exhaust particles (~1.5%) and sea salt spray (~0.3%) [41]. The proportion of soil varies with the location. In central London and New York soil contributes ~57-60% to the dust, whereas in smaller cities such as Christchurch (New Zealand), Kingston (Jamaica) and Halifax (Canada) the contribution is ~75-90% [29].

The main sources of the lead in the street and playground dust, are car emissions, [3,15,22,39,41,52,62 65], paint, especially near old painted buildings [52,65], industry [3], smelters [57] and coal combustion [15]. The sources of house dust are not so easily determined, as a considerable proportion is internally produced. Estimates have pointed to soil being a major component, and may account for 30-40% of the inorganic component of the dust. It is estimated that around 60% of the total dust is inorganic, the rest being organic material [28,39,44]. Other sources of the house dust are street dust (15-30% of the inorganic material) and outside aerosol (1-2% of the inorganic material) [17,27,28,39,44]. One study found house dust was high in silica, confirming soil as a source, and a high calcium content, which appears to have been internally produced [44].

The high concentrations of lead in house dust suggest that the lead comes from a source additional to soil [12,13,27,39]. The two main sources of the lead in urban house dust have been identified as lead paint, and aerosol from car emissions [12,21,28,37,60,62]. Houses close to smelters, will receive a lot of their dust lead from that source [47,55]. Debate exists over the relative importance of paint lead and car generated aerosol lead, and the situation differs with location and circumstances. Trends in house dust lead with the age of the houses, or type of building material and with distance from roads classified according to their car density can be used to help identify the sources. Such trends are shown in Figs 11.6 and 11.7 respectively. It appears that for these houses, in Christchurch

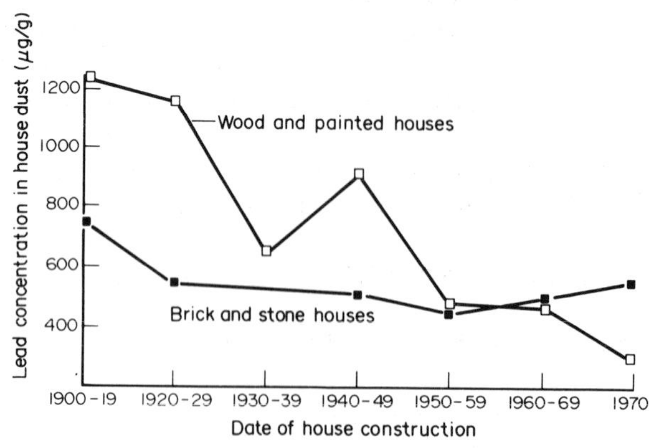

Fig. 11.6 Lead in house dust and the age of the houses.

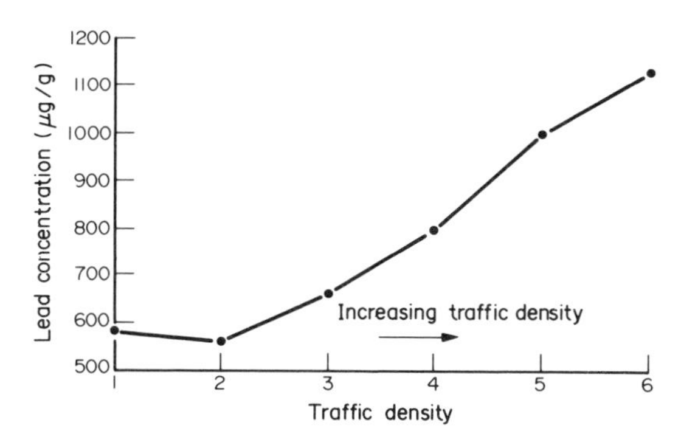

Fig. 11.7 Lead in house dust with respect to nearby traffic density.

(New Zealand), that around 400-500 μg g^{-1} of the lead comes from car aerosol, and for levels higher than that the extra comes from paint lead.

Another source of lead in house dust, but limited to certain homes, is the lead carried home in the clothing of workers in lead based industries [3,9,56]. A study of lead carried home in the socks of workers (Table 11.7) clearly demonstrates this method of adding lead to house dust [3].

Speciation of Lead in Dust

The lead species in dust are either in crystalline or amorphous forms [4,44,64], and are present as low soluble lead compounds or sorbed or bonded to organic materials, iron/manganese oxides and clays.

Some of the individual lead compounds identified in street dust using XRD are listed in Table 11.8 [4,31,63]. Because of the poor sensitivity of XRD (~5%) a compound needs to be present in large amounts, or pre-concentrated according to size, density and magnetic properties. Most of the compounds are weathered materials, but some, such as $PbSO_4.(NH_4)_2SO_4$, $2PbCO_3.Pb(OH)_2$,

TABLE 11.7 Lead in the Socks of Workers in a Lead Based Factory

Employment	Lead in socks* mg Pb/sock
Administration staff	0.36-26.7
Supervisory staff	4.2-28.5
Manual inside working staff	4.3-102.0
Manual outside working staff	5.4-73.5
Control group	0.09-0.36
Blank	0.03-0.06

* Worn for two days. Source; Archer & Barratt, 1976 [3].

TABLE 11.8 Lead Compounds in Dust

Compounds	Proportion Present*
Pb	m-t
$PbSO_4$	m
$PbSO_4{\cdot}(NH_4)_2SO_4$	m-t
Pb_3O_4	m
$PbO{\cdot}PbSO_4$	m
$2PbO{\cdot}PbSO_4$	
$2PbCO_3{\cdot}Pb(OH)_2$	t
$PbCO_3$	t
PbO	
PbO_2	
PbS	

*m = moderate, t = trace. Sources; references 4,31,63

PbO, $2PbSO_4$.PbO and $PbCrO_4$ derive directly from their sources, i.e. car emissions and paints [63]. Lead deposits on leaves appear associated with sulphate and phosphate, suggesting the species are $PbSO_4$ and $Pb_3(PO_4)_2$ [45].

The extraction of lead from dust has also been used to investigate speciation. Up to 65-88% of lead in dust is soluble at pH 2 and 30-50% at pH 3 [15,16]. Sequential extractions suggest that the main lead species are in the carbonate fraction (30-40%) and in the iron/manganese oxide fraction (25-40%) [29,36,38]. Lesser amounts occur as exchangeable lead or associated with organic material or residual silicate material [26,29,38]. The NBS urban particulate matter, which is aerosol in character, had most of the lead in the exchangeable form (46%) [53], suggesting that significant changes occur to the aerosol by weathering when it becomes part of the dust.

CADMIUM

Levels of Cadmium in Dust

Some measured levels of cadmium in street dust and house dust are given in Tables 11.9 and 11.10 respectively. Many values lie in the range <1-2 $\mu g\ g^{-1}$, and are elevated over soil levels. In some cases, very high concentrations have been reported, especially when the dust is located close to cadmium sources, such as mining areas, smelters, industry and busy roads. Though the data is limited, it appears that levels of cadmium in house dust tend to be higher than in the street dust, suggesting an internal source.

Cadmium in Dust in Relation to Distance from the Source and Particle Size

As for lead, cadmium concentrations in dust decline with the distance from sources such as smelters, electro-platers and rubber tyre industries [30,36,37, 47,55]. Cadmium occurs in rubber because of the mineralogical association of

TABLE 11.9 Cadmium Concentrations in Street Dust

Location	Mean $\mu g\ g^{-1}$	Range $\mu g\ g^{-1}$	Comments
Christchurch (NZ)	2.0	0.3-11.7	Road intersections
	2.0	0.6-5.7	Between intersections
	6.8	1.1-15.9	Near rubber factories
	1.5	0.5-4.3	
England	3.1	1.5-4.3	Near busy roads
	2.6	1.0-4.2	Near busy roads
	2.2	1.8-2.4	Near busy roads
Halifax (Canada)		0.6, 1.4	Urban
Hobart	180	6-1450	Industrial dustfall
Kingston (Jamaica)		0.8, 0.8	Urban
London (UK)	7.7 (GM)	1-336	Urban
		5.2, 7.9	Urban
	4.0	<1-280	Urban
		3.3-4.3	Busy road
New York		4.6, 11.1	Urban
Swansea (Wales)	8.3 (Med)	2.0-32	Urban
UK		5.3-6.8	Urban (<250 µm)
		1.7-4.2	Urban (>250 µm)
	1.3		Country lane
	1.7, 5.1	<1-28, 1-210	Streets, national survey
	1.8, 2.2	<1-68, 1-8	Playgrounds, nat. survey
USA	1.6		

Sources of data; references 5,12,24,29,30,31,36,37,41,67,69.

TABLE 11.10 Cadmium Concentrations in House Dust

Location	Mean $\mu g\ g^{-1}$	Range $\mu g\ g^{-1}$	Comments
Christchurch (NZ)	5.2	0.6-21.0	Urban, residential
Illinois	18	7-48	Residential
		1-1060	Non-residential
	70		Rugs
	19		Non-rugs
London	7.7	<1-336	Urban
	193	12-387	Near a smelter
	15.7	9-26	Away from a smelter
Nth. Petherton(UK)	11	0.8-150	Old mining area
Shipham (UK)	27	1-273	Old mining area
UK	6.8	1-804	National survey
	10.5	1-450	National survey

Sources of data; references 12,37,55,62,66.

cadmium with zinc, and that zinc compounds are used in vulcanizing. Since lead comes from car emissions, and cadmium from the wearing of car tyres, it is not unexpected that lead and cadmium are positively correlated in street dust.

Cadmium concentrations are also higher in the smaller particles [24,29,36]. The plot of cadmium levels in dust, fractionated into different sizes, is shown in Fig 11.8, which has similar features, to what has been observed for lead (Fig 11.4) [29].

Sources of Cadmium in Dust

Sources of the street and house dusts have been considered above. The cadmium contribution to street dust appears, from the limited studies, to arise from the wear of rubber car tyres [24,29,30,37]. Other more localized sources are smelters, mining areas, metal platers, and waste tips [37,60]. The mean levels in New York and London of 7.3 μg g^{-1} are significantly higher than the mean concentration of 0.95 μg g^{-1} for smaller cities (Christchurch, Halifax and Kingston) and may be due to higher traffic density in the larger cities [29].

Whereas the higher concentrations of lead in internal dust came from non-rug areas compared with rug areas (1500 compared with 860 μg g^{-1}), the reverse was true for cadmium (19 compared with 70 μg g^{-1}) [62]. This suggests a source of cadmium in rugs (carpets), which could be the rubber underlays used in the carpeted areas. Cadmium in house dust has also been linked with material with a silica matrix, suggesting soil and/or street dust are also sources [44]. The high cadmium concentrations in the dust from houses in Shipham (England) [66] were significantly but weakly related to zinc in the dust and with cadmium in the garden soil. The heavily contaminated soil was clearly contributing to the house dust cadmium.

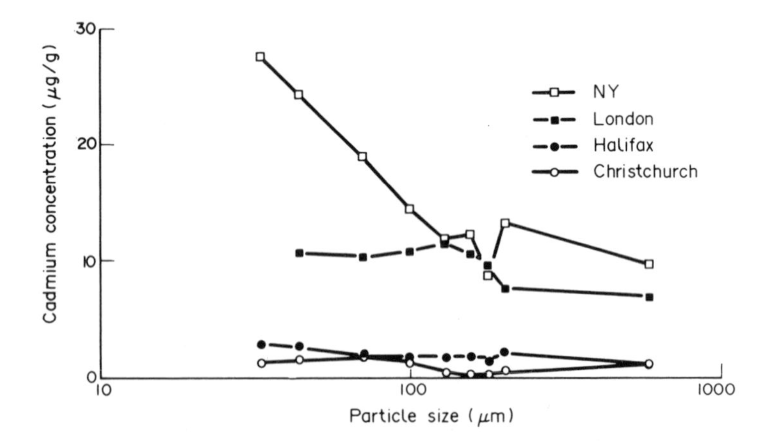

Fig. 11.8 The relationship between cadmium concentrations in dust and particle size. Source of data; Fergusson and Ryan, 1984 [29].

Cadmium Speciation

Sequential extractions have demonstrated that around 20% of cadmium is exchangeable, and the other two main species are cadmium associated with carbonate (30-40%) and iron/manganese oxides (30-40%) [29,38]. Cadmium in urban aerosol was found to be almost completely exchangeable [53]. Hence the cadmium has become less available, or diluted with the other forms, in dust. The greater availability of cadmium is similar to its greater mobility in water, sediments and soils.

OTHER ELEMENTS

Reported levels of arsenic, selenium, antimony and mercury in street dust and house dust are given in Table 11.11. A striking feature is the high levels of antimony in the London dust, but since only two samples were analysed nothing can be inferred from this [29]. All the elements are enriched in the dust compared with soil levels, indicating other sources for the elements. As for lead and cadmium, the levels of arsenic and antimony increase as the particle size of the dust decreases, but not necessarily in a regular manner. Some typical plots are given in Fig 11.9.

TABLE 11.11 Arsenic, Antimony, Selenium and Mercury Concentrations in Street and House Dusts

Element	Location	Mean $\mu g\ g^{-1}$	Range $\mu g\ g^{-1}$
Street dust			
Arsenic	Christchurch	14.5	1.24-29.5
	Halifax		5.9, 8.9
	Kingston		3.5, 6.7
	London		27, 12.9
	New York		4.0, 5.4
	USA	1.6	
Antimony	Christchurch	4.69	2.61-6.78
	Halifax		2.1, 1.9
	Kingston		1.7, 1.6
	London		15.8, 14.6
	New York		3.6, 8.4
	USA	2.2	
Selenium	USA	1.0	
Mercury	USA	0.09	
House dust			
Arsenic	Christchurch	15.8	7.95-23.4
Antimony	Christchurch	10.0	1.83-30.6

Sources of data; references 27,29,41

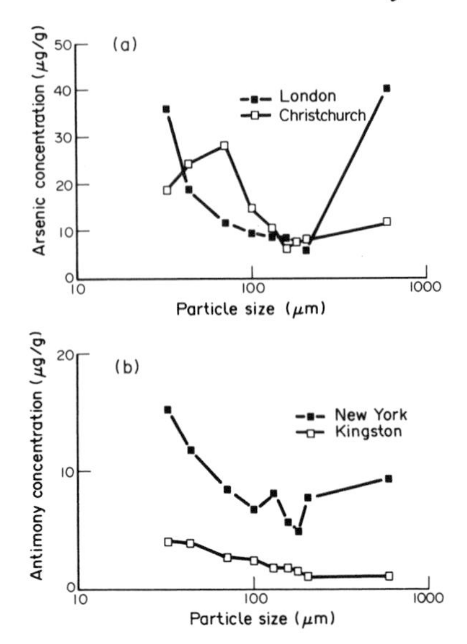

Fig. 11.9 The variation of concentration of (a) arsenic and (b) antimony in street dust with particle size. Source of data; Fergusson and Ryan, 1984 [29].

OTHER ADVENTITIOUS SOURCES OF HEAVY METALS

Whereas dust is widespread and is an important source of heavy elements, in particular lead and cadmium, other adventitious sources are more localized and not always apparent.

Some of the common adventitious sources of the heavy metals are given in Table 11.12. Printing inks and coloured sections of paper, can be a high source of lead if the paper is consumed. Levels as high as 3600 $\mu g\, g^{-1}$ have been reported for coloured pages [6,49]. For different coloured pieces extracted at a pH of 1 the levels of lead reported are blue 521 $\mu g\, g^{-1}$, yellow 1442 $\mu g\, g^{-1}$, dark green 992 $\mu g\, g^{-1}$ and red 1057 $\mu g\, g^{-1}$ [6].

Certain cosmetics may contain lead compounds, for example lead sulphide is used to darken the area of skin around the eyes and suma, used in India, can contain more than 50% PbS. Lead acetate added to hair, darkens it by forming Pb-S bonds with the sulphur in the hair [32,49].

Lead glazes are less used today, however, lead glazed vessels still exist, and acid solutions such as fruit juices may leach the lead out of the glaze. This is more likely if the glaze formula was made up incorrectly and too much PbO used, or the firing temperature was too low, i.e. less than 1200° C [49].

The removal of lead paint from buildings is a problem which can add large quantities of lead to the surrounding environment [43,48]. The results of a study, given in Table 11.13, indicates that the preferred way to remove paint is by a hot air gun.

TABLE 11.12 Adventitious Sources of Heavy Metals Other than Dust

Source	Potential source for
Sanded paint	Pb, Cd
Peeling paint	Pb, Cd
Plaster	Pb
Glazed pottery	Pb
Eye cosmetics	Pb
Medicinal compounds	Pb, As, Bi, Hg, Se
Metals	Pb, Cd, Hg, Sb
Printing inks	Pb
Alcoholic drinks	Pb
Battery casings	Pb, Sb
Coloured paper	Pb, Cd
Plastic consumer items	Pb, Cd, Se
Cigarette smoke	Pb, Cd, As, Hg, Se, Sb
Incinerated ash	Most metals

Sources; references 1,6,32,34,42,43,46,48,49,51,54,64.

TABLE 11.13 Methods for the Removal of Lead Paint and Resulting Levels

Method of removal	Air lead* $\mu g\ m^{-3}$	
	Air	Mask filter
Propane gas burner	280	460
Hot air gun	97	270
Dry sanding	10,100	4730

* In an enclosed area

Situation	Indoor dust lead $\mu g\ g^{-1}$	
	Mean	Range
Immediately after paint removal	-	5000-100,000
Recent restorative work	3350	1650-5000
No recent restorative work	515	50-2270

Sources of data; Inskip, 1984 [43], Landrigan et al., 1982 [48].

The use of CdS and CdSe pigments for the yellow, orange and red colours in various plastic consumer items exposes people to a potential hazard. This could be dangerous when the cadmium and selenium pigments are used in containers for storing food [35]. Brightly coloured plastics are also given to children to play with. The pigments are used in car paints, and in artists paints.

REFERENCES

1. Ahmad, S, Chaudhry, M. S. and Qureshi, I. H. Determination of toxic elements in tobacco products by instrumental neutron activation analysis. **J. Radioanal. Chem.**; 1979; **54:** 331-3412.
2. Angle, C. R., Marcus, A., Cheng, I-H. and McIntire, M. S. Omaha childhood blood lead and environmental lead: a linear total exposure model. **Environ. Res.**; 1984; **35:** 160-170.
3. Archer, A. and Barratt, R. S. Lead levels in Birmingham dust. **The Sci. Total Environ.**; 1976; **6:** 275-286.
4. Biggins, P. D. E. and Harrison, R. M. Chemical speciation of lead compounds in street dust. **Environ. Sci. Techn.**; 1980; **14:** 336-339.
5. Bloom, H and Ayling, G.M. Heavy metals in the Derwent Estuary. **Environ. Geol.**; 1977; **2:** 3-22.
6. Bogden, J. D., Joselow, M. M. and Singh, N. P. Extraction of lead from printed matter at physiological values of pH. **Arch. Environ. Health**; 1975; **30:** 442-444.
7. Brunekreef, B., Veenstra, S. J., Biersteker, K. and Boleij, J. J. M. The Arnhem lead study I lead uptake by 1 to 3 year old children living in the vicinity of a secondary lead smelter in Arnhem, The Netherlands. **Environ. Res.**; 1981; **25:** 441-448.
8. Brunekreef, B., Noy, D., Biersteker, K. and Boleij, J. Blood lead levels of Dutch city children and their relationship to lead in the environment. **J. Air Pollut. Control Ass.**; 1983; **33:** 872-876.
9. Brunekreef, B. Childhood Exposure to Environmental Lead. Technical Report: MARC; 1986; No 34.
10. Charney, E., Kessler, B., Farfel, M. and Jackson, D. Child lead poisoning: a controlled trial of the effect of dust control measures on blood lead levels. **New Eng. J. Med.**; 1983; **309:** 1089-1093.
11. Charney, E., Sayre, J. and Coulter, M. Increased lead absorption in inner city children: where does the lead come from? **Pediatrics**; 1980; **65:** 226-231.
12. Culbard, E., Thornton, I., Watt, J. Moorcroft, S., Brooks, K. and Thompson, M. A nation wide reconnaissance survey of the United Kingdom to determine metal concentrations in urban dusts and soils. **Trace Subst. Environ. Health**; 1983; **XVI.**
13. Davies, D. J. A., Watt, J. M. and Thornton, I. Lead levels in Birmingham dusts and soils. **The Sci. Total Environ.**; 1987; **67:** 177-185.
14. Day, J. P., Hart, M. and Robinson, M. S. Lead in urban street dust. **Nature**; 1975; **253:** 343-345.
15. Day, J. P. Lead pollution in Christchurch. **N. Z. J. Sci.**; 1977; **20:** 395-406.
16. Day, J. P., Fergusson, J. E. and Tay Ming Chee. Solubility and potential toxicity of lead in urban dust. **Bull. Environ. Contam. Toxicol.**; 1979; **23:** 497-502.

17. Diemal, J. A. L. Brunekreef, B., Boleij, J. S. M., Biersteker, K. and Veinstra, S. J. The Arnhem lead study II indoor pollution and indoor/outdoor relationships. **Environ. Res.;** 1981; **25:** 449-456.
18. Duggan, M. J. Childhood exposure to lead in the surface dust and soil: a community health problem. **Pub. Health Rev.;** 1985; **13:** 1-54.
19. Duggan, M. J. Contribution of lead in dust to children's blood lead. **Environ. Health Perspec.;** 1983; **50:** 371-381.
20. Duggan, M. J. and Williams, S. Lead in dust in city streets. **The Sci. Total Environ.;** 1977; **7:** 91-97.
21. Duggan, M. J. Lead in urban dust: an assessment. **Water, Air Soil Pollut.;** 1980; **14:** 309-321.
22. Duggan, M. J., Inskip, M. J., Rundle, S. A. and Moorcroft, J. S. Lead in playground dust and on the hands of school children. **The Sci. Total Environ.;** 1985; **44:** 65-79.
23. Duggan, M. Temporal and spatial variations of lead in air and in surface dust - implications for monitoring. **The Sci. Total Environ.;** 1984; **33:** 37-48.
24. Ellis, J. B. and Revitt, D. M. Incidence of heavy metals in street surface sediments: solubility and grain size. **Water, Air Soil Pollut.;** 1982; **17:** 87-100.
25. Elwood, P. C., Gallacher, J. E. J., Phillips, K. M., Davies, B. E. and Toothill, C. greater contribution to blood lead from water than air. **Nature;** 1984; **310:** 138-140.
26. Farmer, J. G. and Lyon, T. D. B. Lead in Glasgow street dust and soil. **The Sci. Total Environ.;** 1977; **8:** 89-93.
27. Fergusson, J. E. Forbes, E. A. and Schroeder, R. J. The elemental composition and sources of house dust and street dust. **The Sci. Total Environ.;** 1986; **50:** 217-222.
28. Fergusson, J. E. and Schroeder, R. J. Lead in house dust of Christchurch, New Zealand: sampling, levels and sources. **The Sci. Total Environ.;** 1985; **46:** 61-72.
29. Fergusson, J. E. and Ryan, D. E. The elemental composition of street dust from large and small urban areas related to city type, source and particle size. **The Sci. Total Environ.;** 1984; **34:** 101-116.
30. Fergusson, J.E., Hayes, R.W., Tan Seow Yong and Sim Hang Thiew. Heavy metal pollution by traffic in Christchurch, New Zealand: Lead and cadmium content of dust, soil and plant samples. **N. Z. J. Sci.;** 1980; **23:** 293-310.
31. Fergusson, J. E. and Simmonds, P. R. Heavy metal pollution at an intersection involving a busy urban road in Christchurch, New Zealand: I Levels of Cr, Mn, Fe, Ni, Cu, Zn, Cd and Pb in street dust. **N. Z. J. Sci.;** 1983; **26:** 219-228.
32. Fergusson, J. E. Lead: petrol lead in the environment and its contributions to human blood lead levels. **The Sci. Total Environ.;** 1986; **50:** 1-54.

33. Fergusson, J. E. The significance of the variability in analytical results for lead, copper, nickel and zinc in street dust. **Canad. J. Chem.;** 1987; **65:** 1002-1006.

34. Fraser, J. L. and Lum, K. R. Availability of elements of environmental importance in incinerated sludge ash. **Environ. Sci. Techn.;** 1983; **17:** 52.

35. Funk, W. Inorganic pigments. In: Hutzinger, O., Ed. Handbook of Environmental Chemistry: Springer Verlag; 1980; **3A:** 217-229.

36. Hamilton, R. S., Revitt, D. M. and Warren, R. S. Levels and physico-chemical associations of Cd, Cu, Pb and Zn in road sediments. **The Sci. Total Environ.;** 1984; **33:** 59-74.

37. Harper, M., Sullivan, K. R. and Quinn, M. J. Wind dispersal of metals from smelter waste tips and their contribution to environmental contamination. **Environ. Sci. Techn.;** 1987; **21:** 481-484.

38. Harrison, R. M., Laxen, D. P. H. and Wilson, S. J. Chemical associations of lead, cadmium, copper and zinc in street dusts and roadside soils. **Environ. Sci. Techn.;** 1981; **15:** 1378-1383.

39. Harrison, R. M. Toxic metals in street and household dusts. **The Sci. Total Environ.;** 1979; **11:** 89-97.

40. Ho, Y. B. Lead contamination in street dust in Hong Kong. **Bull. Environ. Contam. Toxicol.;** 1979; **21:** 639-647.

41. Hopke, P. K., Lamb, R. E. and Natusch, D. F. S. Multielement characterization of urban roadway dust. **Environ. Sci. Techn.;** 1980; **14:** 164-172.

42. Hutton, M. and Symon, C. The quantities of cadmium, lead, mercury and arsenic entering the UK environment from human activities. **The Sci. Total Environ.;** 1986; **57:** 129-150.

43. Inskip, M. J. Lead based paints: potential hazards associated with their presence and removal. London Environ. Suppl.: The Greater London Council; 1984; **6.**

44. Johnson, D. L., Fortmann, R. and Thornton, I. Individual particle characterization of heavy metal rich household dusts. **Trace Subst. Environ Health;** 1982; **16:** 116-123.

45. Koslow, E. E., Smith, W. H. and Staskawicz, B. J. Lead containing particles on urban leaf surfaces. **Environ. Sci. Techn.;** 1977; **11:** 1019-1021.

46. Landrigan, P. J. and Bridbord, K. Additional exposure routes. In: Nriagu, J. O., Ed. **Changing Metal Cycles and Human Health:** Springer Verlag; 1984: 169-186.

47. Landrigan, P. J., Stephen, H. Gehlbach, M. D., Rosenblum, B. F., Shoults, J. M., Candelaria, R. M., Barthel, W. F., Liddle, J. A., Smrek, A. L., Staehling, N. W. and Sanders, J. F. Epidemic lead absorption near an ore smelter: the role of particulate lead. **New Engl J. Med.;** 1975; **292:** 123-129.

48. Landrigan, P. J., Baker, E. L. Himmelstein, J. S. Stein, G. F., Wedding, J. P. and Straub, W. E. Exposure to lead from the Mystic river bridge: the dilemma of deleading. **New Eng. J. Med.;** 1982; **306:** 673-676.
49. Landsdown, R. and Yule, W. The Lead Debate: The Environment, Toxicology and Child Health: Croom Helm; 1986.
50. Lepow, M. L., Bruckman, L., Gillette, M., Markowitz, S., Robino, R. and Kapish, J. Investigations into sources of lead in the environment of urban children. **Environ. Res.;** 1975; **10:** 415-426.
51. Lin-Fu, J. S. Children and lead; new findings and concerns. **New Eng. J. Med.;** 1982; **307:** 615-617.
52. Linton, R. W., Natusch, D. F. S., Solomon, R. C. and Evans C. A. Physicochemical characterization of lead in urban dusts: a micoanalytical approach to lead tracing. **Environ. Sci. Techn.;** 1980; **14:** 159-164.
53. Lum, K. R., Betteridge, J. S. and Mcdonald, R. R. The potential availability of P, Al, Cd, Co, Cr, Cu, Fe, Mn, Ni, Pb and Zn in urban particulate matter. **Environ. Techn. Lett.;** 1982; **3:** 57-62.
54. Markunas, L. D. Barry, E. F. Guiffre, G. P. and Litman, R. An improved procedure for the determination of lead in environmental samples by atomic absorption spectroscopy. **J. Environ. Sci. Health;** 1979; **A14:** 501-506.
55. Muskett, C. J., Roberts, L. H. and Page, B. J. Cadmium and lead pollution from secondary metal refining operations. **The Sci. Total Environ.;** 1979; **11:** 73-87.
56. Rice, C. M., Lilis, R., Fischbein, A. and Selikoff, I. J. Unsuspected sources of lead poisoning. **New Eng. J. Med.;** 1977; **296:** 1416.
57. Roels, H. A., Buchet, J-P., Lauwerys, R., Braux, P., Claeys-Thoreau, F., Lafontaine, A., van Overschelde, J. and Verduyn, G. Lead and cadmium among children near a non-ferrous metal plant. **Environ. Res.;** 1978; **15:** 290-308.
58. Sayre, J. Dust lead contribution to lead in children. In: Lynam, D. R., Piananida, L. G. and Cole, J. F., Eds. Environmental Lead: Academic press; 1981: 23-40.
59. Sayre, J. W., Charney, E., Vostal, J. and Pless, J. B. House and hand dust as a potential source of childhood lead exposure. **Amer. J. Dis. Child.;** 1974; **127:** 167-170.
60. Sayre, J. W. and Katzel, M. D. Household surface dust; its accumulation in vacant homes. **Environ. Health Perspec.;** 1979; **29:** 179-182.
61. Schwar, M. J. R. Sampling and measurement of environmental lead present in air, surface dust and paint. **London Environ. Supple. No. 1;** 1983.
62. Solomon, R. L. and Hartford, J. W. Lead and cadmium in dust and soils in a small urban community. **Environ. Sci. Techn.;** 1976; **10:** 773-777.
63. Sturges, W. T. and Harrison, R. M. An assessment of the contribution from paint flakes to the lead content of some street and household dusts. **The Sci. Total Environ.;** 1985; **44:** 225-234.

64. Swaine, D. J. The fate of trace elements in coal after combustion. Int. Clean Air Conf. Brisbane: Ann Arbor; 1978.
65. Ter Haar, G. and Aronow, R. New information on lead in dirt and dust as related to the childhood lead problem. **Environ. Health Perspec.**; 1974; **7**: 83-89.
66. Thornton, I. Metal content of soils and dust. **The Sci. Total Environ.**; 1988; **75**: 21-39.
67. Thornton, I., Culbard, E., Moorcroft, S., Watt, J., Wheatley, M., Thompson, M. and Thomas, J. F. A. Metals in urban dusts and soils. **Environ. Techn. Lett.**; 1985; **6**: 137-144.
68. Vostal, J. J., Taves, E., Sayre, J. W. and Charney, E. Lead analysis of house dust; a method of detection of another source of lead exposure in inner city children. **Environ. Health Perspec.**; 1974; **7**: 91-97.
69. Warren, R. S. and Birch, P. Heavy metal levels in atmospheric particulates, roadside dust and soil along a major urban highway. **The Sci. Total Environ.**; 1987; **59**: 253-256.

CHAPTER 12

METHYL COMPOUNDS OF THE HEAVY ELEMENTS

Some of the alkyl derivatives of the heavy metals relevant to environmental chemistry are listed in Tables 12.1 and 12.2 [30]. In this chapter we will consider the methyl derivatives, $(CH_3)_nM$, and their salts, $(CH_3)_{n-z}MX_z$. The environmental interest in these compounds, is in their toxicity, and that methylation occurs for many of the heavy elements in the environment. The word 'methylation' refers to the transfer of a methyl group from one compound to another. The process may occur biologically (called biomethylation) or abiotically which does not directly involve a biological system.

MICROORGANISMS

The two main groups of microorganisms that are linked with biomethylation of the heavy metals are bacteria and fungi (moulds). Some brief descriptive details on these two groups are listed in Table 12.3 [2,36,39,48,60,69,70].

TABLE 12.1 Some Organometallic Compounds of the Heavy Elements

Periodic Group	IIb 12	IIIb 13	IVb 14	Vb 15	VIb 16
	Cd Hg	In Tl	Pb	As Sb Bi	Se Te
Alkyl Organo-Metallic Compounds	R_2M RMX	R_3M R_2MX RMX_2	R_4M R_3MX R_2MX_2	R_3M R_2MX RMX_2	R_2M R_2MX_2
Oxidation	2	3	4	5	2, 4

TABLE 12.2 Some Properties of Organometallic Compounds of the Heavy Elements

Metal	Compound	Melting Point °C	Boiling Point °C	Mean bond energy kJ mol^{-1}	ΔH_f kJ mol^{-1}	Stability ¶	
In	$(CH_3)_3In$	88.4	135.8	163, 197*	172	u	h
Tl	$(CH_3)_3Tl$	38.5	147 (est)	152*		u	
	$(CH_3)_2TlCl$	>280				s	s
Cd	$(CH_3)_2Cd$	-4.5	105.6	141, 189*, 92**	110,74	u	h
Hg	$(CH_3)_2Hg$		92.5	60, 121, 123 218*, 29**	93	s	s
	CH_3HgCl	170		269		s	s
Pb	$(CH_3)_4Pb$	-30.2	110, 9/13 mm	160.5, 155 138†	137	s	s
	$(C_2H_5)_4Pb$	-130.5	200, 80/13 mm			s	s
	$(CH_3)_3PbCl$	190 dec				s	h
	$(C_2H_5)_3PbCl$	155				s	h
As	$(CH_3)_3As$	-87	50	229, 238	15	u	h
	$(CH_3)_2AsCl$		107				h
	CH_3AsCl_2		131			u	h
Sb	$(CH_3)_3Sb$	-62	80	215, 224	31	u	h
	$(CH_3)_2SbCl$		157/>50 mm				h
	CH_3SbCl_2		115/60 mm			u	
Bi	$(CH_3)_3Bi$	-86	109	143, 140	192	u	
	$(CH_3)_2BiCl$	116					
Se	$(CH_3)_2Se$			247	-57	s	

* first M-C bond dissociation energy. ** second M-C bond dissociation energy, † thermochemical bond energy. ¶ Stability: first column towards oxygen, second column towards water, s = stable, u = unstable, h = hydrolyses.

TABLE 12.3 Some Details on Bacteria and Fungi

Micro-organism	Classification	Diameter	Shapes	Chlorophyll	Nutrients
Bacteria	Prokaryote Unicellular not bound by a membrane	Most around 0.2-1.5 μm	Spheres straight rods curved rods	Absent or different to plant chlorophyll	Heterotrophic some autotrophic
Fungi (mould)	Eukaryota one or more discrete cells, bounded by membranes	Large, μm to cm/meters	Variable	Absent	Heterotrophic

Bacteria are very small, spherical, linear or curved in shape and are mostly heterotrophic (dependent on organic nutrient). They contain both DNA and RNA, and do not have their nucleus surrounded by a membrane. Fungi are frequently much bigger in size, with diameters from a few μm to cm and meters. They are heterotrophic, and have one or more nuclei surrounded by membranes [2,36,39,48,60,69,70].

Examples of bacteria and fungi reported to methylate the heavy metals Hg, As, Se, Te, Pb, Cd, Tl and In are tabulated in Table 12.4. The majority of the bacteria in the list are unicellular, rod-like in shape, and heterotrophic. They are usually aerobic except *clostridium sp* and *methanobacterium* which are anaerobic [2,36,39,48,60,69,70]. It appears from the list that mercury, arsenic and selenium (in decreasing order) are the elements most frequently methylated, followed by tellurium, lead and cadmium. There is good evidence for the biomethylation of mercury, arsenic, selenium and tellurium, but some doubts exist over the biomethylation of the other elements. Methylation can also be selective, e.g. *A. versicolor* and *A. glaucus* methylate arsenic but not selenium and tellurium, whereas *P. chrysogenium*, *P. notatum* and *A. niger* methylate all three elements.

HISTORY OF METHYLATION OF THE HEAVY ELEMENTS

The two discoveries of environmental methylation of the heavy elements, arose from evidence of poisoning of human beings. These were arsenic poisoning, in the rooms of European houses in the 19th century, and mercury poisoning of people in Minamata and Niigata in Japan in the 1950's [73].

As early as 1815 arsenic poisoning was encountered in German houses, and in 1839 Gmelin commented on the emanations coming from moulds on wall paper as having a garlic odour. Gasio [40] demonstrated in 1901 that the bacteria *Penicillium brevicaulis* (now called *Scopulariopsis brevicaulis*) together with As_2O_3 gave the same gas which was identified as containing arsenic. The source of the arsenic in the wall papers was the green copper arsenite pigments Scheele's green (approximately $CuHAsO_3$) and Paris green, $Cu_2(CH_3COO)AsO_3$ [15,29,73]. In 1932 Challenger and his colleagues identified the gas as trimethylarsine [16,17,19], the first evidence for biomethylation of a heavy element. Soon after Challenger and his coworkers demonstrated the biomethylation of selenium and tellurium oxy-compounds to $(CH_3)_2Se$ and $(CH_3)_2Te$ [8,18].

People in Minamata and Niigata in Japan, whose staple diet was fish, were found to be suffering from mercury poisoning in the 1950's [73]. It was revealed in 1966 that the main form of mercury in the fish was CH_3Hg^+ [77]. Three years later it was discovered that the mercury species was produced from inorganic mercury when added to aquarium sediments [29,54]. The cobalt complex ion, $[CH_3Co(CN)_5]^{3-}$, which has been used as a model for Vitamin B_{12}, was shown in 1964 to be able to methylate mercury [42]. The first suggestion that the methylating agent, associated with methane producing bacteria, was methylcobalamin (Fig 12.1) (the methyl derivative of vitamin B_{12} where the CN^- group is replaced by CH_3^-), was made by Wood and coworkers in 1968 [89]. Thus began

TABLE 12.4 Bacteria and Fungi Involved in the Methylation of the Heavy Elements

Microorganism	Metals methylated
Bacteria	
Acinetobacter	Pb
Aeromonassp	Se, As, Pb
Aeromonashydrophila	Hg
Alcaligenessp	Pb, As
Bifidobacterium	Hg
Chromobacterium	Hg
Clostridium cochlearium	Hg
Clostridium thermoaceticium	Hg
Clostridium sticklandii	Hg
Enterobacter aerogenes (methanobacter)	Hg
Escherichia coli	Hg, As, Pb
Flavobacterium sp	Se, As, Pb
Lactobacilli	Hg
Methanobacterium sp	Hg, As
Methanobacterium MoH	Hg, As
Methanobacterium smeliansky	Hg
Pseudomonas sp	Hg, Se, Pb, Cd
Pseudomonas aeruginosa	Te
Pseudomonas fluorescenis	Hg
Staphyloccus aureus	As, Cd
Stibiobacter senarmontii	Sb
Fungi (moulds)	
Aspergillus niger	Se, Te, As, Hg
Aspergillus glaucus	As
Aspergillus versicolor	As
Candida humicola	As, Se, Te
Gliocladium roseum	As(organo)
Neurospora crassa	Hg
Penicillium chrysogenium	Se, Te, As
Penicillium notatum	Se, Te, As
Scopulariopsis brevicaulis	As, Se, Te, Hg

Sources of data; references 4,6,14,15,20,23,28,29,53,55, 59,62,63,66,71,73,79,80,82,88,90.

an extensive study of the methylation of mercury, and the interpretation of a number of poisonings by mercury around the world [35].

The first suggestion of methylation of inorganic lead was reported in 1975 [81]. The work has been challenged, reinterpreted and also confirmed, leaving some doubt as to whether biomethylation occurs. The weight of evidence however, tends to favour the process occurring.

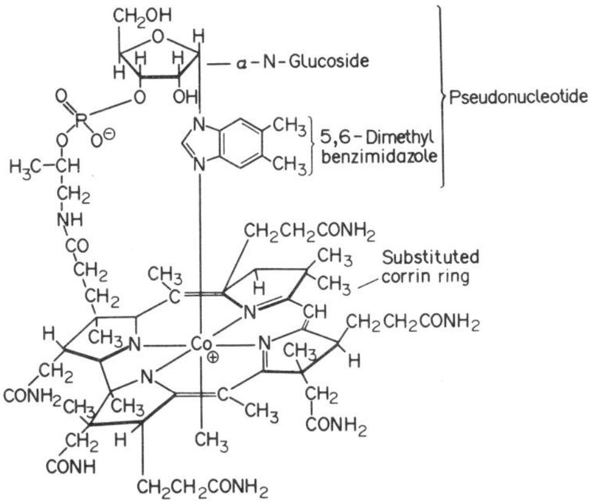

Fig. 12.1 The structure of methylcobalamin

BIOLOGICAL METHYLATION

The three biological methylating agents associated with the heavy elements are methylcobalamin (Fig 12.1), S-adenosylmethionine (Fig 12.2) and N^5-methyltetrahydrofolate (Fig 12.3). The methylcobalamin produces CH_3^- and can methylate cationic species, and is the reagent considered to be involved in the methylation of mercury. The other two reagents produce CH_3^+ and methylate anionic species, such as arsenite AsO_3^{3-}. The S-adenosylmethionine is the reagent implicated in the production of $(CH_3)_3As$ from moulds and arsenic containing pigments [6,9,10,28,29,66,73,78,82].

Methylcobalamin

Vitamin B_{12} is a cyanocobalamin [26], whose structure is similar to that given in Fig 12.1, with the exception that the methyl group is replaced by cyanide. Other species may also replace the CN^-, and other bases can replace the 5,6-dimethylbenzimidazole. The vitamin occurs in animals and microorganisms as the coenzyme (Fig 12.4), which is unstable with respect to reagents such as cyanide. The methyl group in methylcobalamin can come from a methyl donor, such as the N^5-methyltetrahydrofolate (Fig 12.3), i.e.

$$\text{cobalamin coenzyme} + N^5\text{-MeTHF} \rightarrow \text{methylcobalamin.}$$

The methylcobalamin is involved in the following biochemical reactions;

$$\text{methylcobalamin} + \text{homocysteine} \rightarrow \underset{\text{methionine}}{CH_3\text{-S-}(CH_2)_2\text{-CH}(NH_2)\text{-COOH}},$$

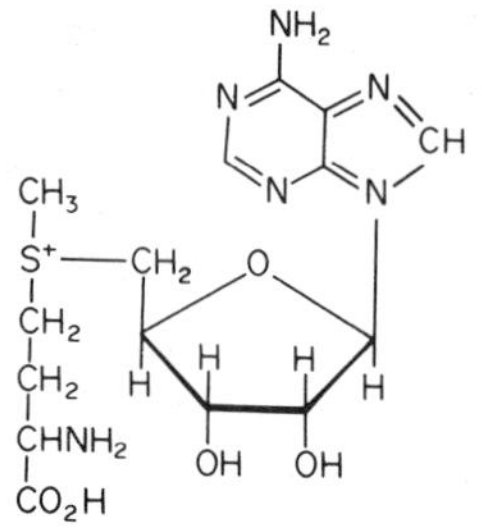

Fig. 12.2 The structure of S-adenosyl methioine.

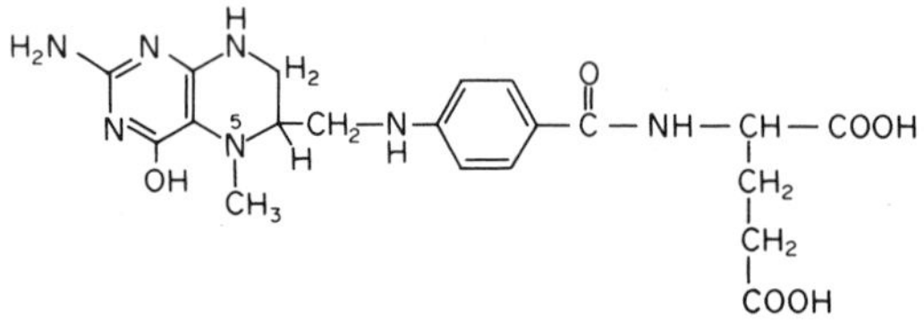

Fig. 12.3 The structure of N^5-methyltetrahydrofolate.

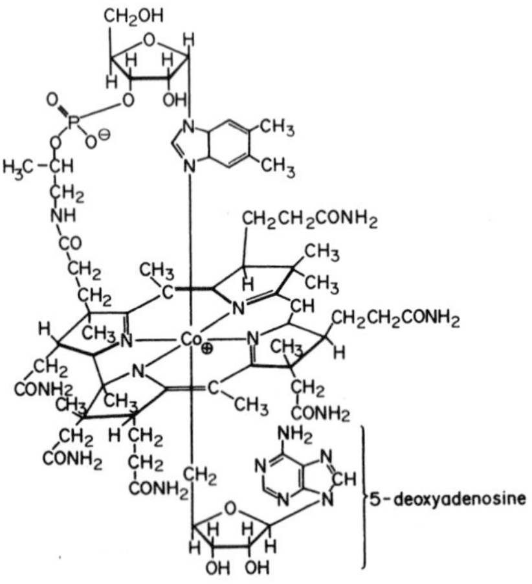

Fig. 12.4 The structure of the coenzyme-B_{12}.

methylcobalamin + reducing agent → CH_4,

methylcobalamin + reducing agent + CO_2 → CH_3COOH.

The cobalt in vitamin B_{12} is in the trivalent oxidation state, but it may be reduced to Co(II) (B_{12r}) and Co(I) (B_{12s}), using various reducing agents, such as

NADH [52,57,84]. The cobalt(I) complex is a strong nucleophile and is probably five coordinate and therefore reactive. It will readily accept a methyl group with oxidation to Co(III). For convenience we will write the reagent as $MeCoB_{12}$. The material is excreted by bacteria living in sediments and sludges [71].

The $MeCoB_{12}$ is a source of the CH_3 group, as CH_3^+ or CH_3^- or $\cdot CH_3$, as portrayed in Fig 12.5. Route 2 appears the most likely for the methylation of mercury, and route 4 for tin [28,29,30,37,65]. The route depends on the nature of the metal species, e.g. oxyanions of arsenic may well react by route 6, where the carbocation CH_3^+ is produced.

Methylcobalamin methylates either enzymatically or non-enzymatically [9,28,30,55,79,82]. In the first process, which is less common of the two, the $MeCoB_{12}$ is associated with enzymes, such as methionine synthetase, acetate synthetase and methane synthetase, and it is likely that the metal intervenes in the cell's metabolic pathway [30]. A scheme for methylation in the methylcobalamin-acetate synthetase system, under anaerobic conditions, is shown in Fig 12.6. For non-enzymatic methylation the $MeCoB_{12}$ is free from other materials, and acts as a chemical methyl transfer agent. The process can occur by two routes described as 'base on' and 'base off' [6,30,37,55,66,73,82,83,86,87,90]. The

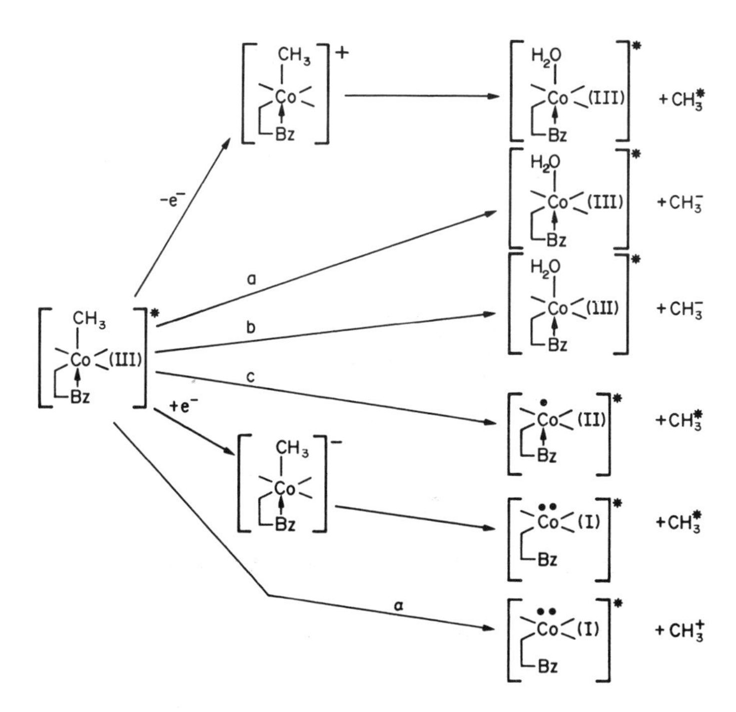

Fig. 12.5 Reaction mechanisms for $MeCoB_{12}$ producing CH_3 species. a, electrophilic attack; b, redox switch for metal being methylated; c, free radical attack; d, nucleophilic attack. Sources; references 28,29,30,37,65.

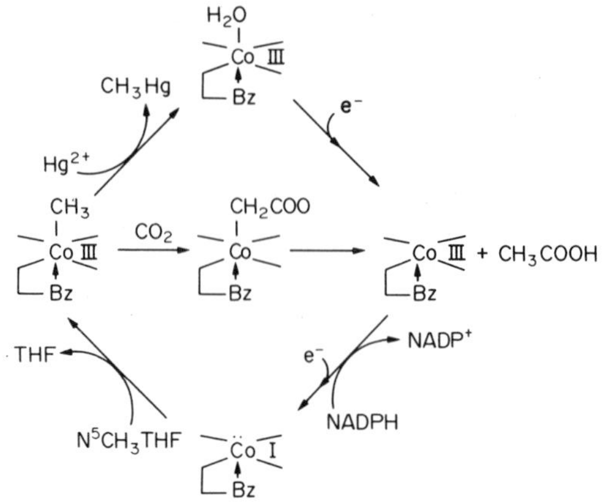

Fig. 12.6 Mercury methylation in the $MeCoB_{12}$ acetate synthetase system in anaerobic conditions. Sources; references 9,28,30,55,79,82.

two routes are given in Fig 12.7, the methyl derivatives may then be regenerated enzymatically [59]. In the 'base on' process the benzimidazole remains attached to the cobalt, whereas in the 'base off' process it becomes detached and is mercurated, and then after methylation becomes attached again. The 'base on' reaction is around 1000 times faster than the 'base off' process.

Methylation may occur through a bridged intermediate, because the high reactivity of the $\cdot CH_3$, CH_3^+ and CH_3^- species means that they are unlikely to exist by themselves in reactive solvents [73]. The most likely process would be the formation of a bimolecular transition state, as follows;

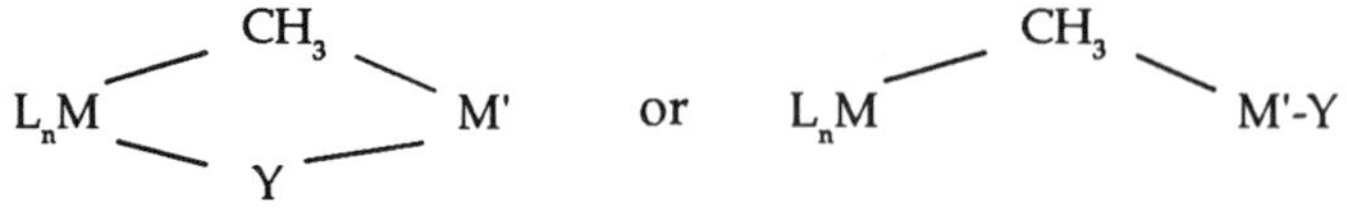

L_nM — CH_3 — M', L_nM — Y — M' or L_nM — CH_3 — M'-Y

S_E2 (closed) stereochemical retention of configuration

S_E2 (open) inversion

The reduction potential of the metal being methylated, may determine in the type of the methyl transfer. It appears that if the reduction potential E° >+0.8V then CH_3^- transfer occurs but if E° <+0.6V the transfer is oxidative addition of a free radical with homolytic cleavage of the Co-C bond [5,29,30,37,73]. Relevant reduction potentials are listed in Table 12.5, together with a suggested

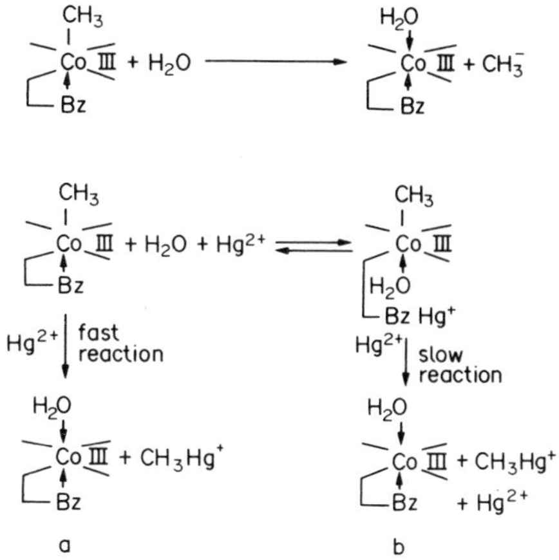

Fig. 12 7 Methylation of mercury (a) base on, (b) base off. Sources of data; references 6,30,37,55,59,66,73,82,83,86,87,90

methyl transfer process. The elements considered to be methylated by $MeCoB_{12}$ are Hg, Pb, Tl, As, Se, and Te [29,87,90].

Abiotic Methylation

Methylation by non-enzymatic $MeCoB_{12}$ may be treated as abiotic, except that the reagent is produced biotically and maybe re-methylated biotically. The two main abiotic methylation processes are transmethylation, and to a lesser extent photochemical.

TABLE 12.5 Reduction Potentials and Possible Methyl Transfer Processes for the Heavy Elements

Couple	Reduction Potential (V)	Possible process
PbO_2/Pb^{2+}	+1.46	Transfer of CH_3^-
Tl^{3+}/Tl^+	+1.26	Transfer of CH_3^-
Se(VI)/Se(IV) (in acid)	+1.15	Transfer of CH_3^-
Hg^{2+}/Hg	+0.85	Transfer of CH_3^-
$Sb_2O_5/SbO^+Sb(V)/Sb(III)$	+0.68	Oxdn. metal or CH_3 transfer
As(V)/As(III) (in acid)	+0.66	Oxdn. metal or CH_3 transfer
Se(VI)/Se(IV) (in base)	+0.05	Oxdn. metal or CH_3 transfer
Pb(II)/Pb(0)	-0.13	Oxdn. metal or CH_3 transfer
As(V)/As(III)	-0.67	Oxdn. metal or CH_3 transfer

Source of data; Craig and Brinckman, 1986 [30].

In transmethylation a methyl group is transferred from one metal to another, for example;

$$(CH_3)_nSn^{(4-n)+} + Hg^{2+} \rightarrow (CH_3)_{n-1}Sn^{(5-n)+} + CH_3Hg^+.$$

A list of some of the methylating agents and their substrates are given in Table 12.6 [1,6,9,10,29,28,56,59,63,64,73,85,87]. The reagent CH_3I, which is a good methylating agent, occurs naturally in the environment. It may form from methylation of diiodine (I_2) by methylcobalamin [31,65,87] according to the reaction;

$$MeCoB_{12} + I_2 \rightarrow ICoB_{12} + CH_3I.$$

In photochemical methylation the CH_3 group is produced photochemically, when for example acetate, methanol, ethanol and aliphatic α-amino acids are exposed to intense UV radiation [29,63,71].

METHYLATION OF THE HEAVY ELEMENTS

Mercury

Much of the discussion above relating to $MeCoB_{12}$ covers aspects of the methylation of mercury, see Figs. 12.6 and 12.7. A number of features of the chemistry of mercury facilitates its existence in organo-species. Both Hg^{2+} and CH_3Hg^+ are soft acids and bond well to soft bases such as S^{2-} and SH^-. The cation is large and polarizable, and because of the dipositive charge, it is itself a good polarizing cation and tends to form covalent bonds. The Hg-C bond (ca. 60-120 kJ mole^{-1}), though not thermodynamically strong, is stronger than the Hg-O bond, and therefore persists in the environment. The bond is also non-polar.

TABLE 12.6 Methylating Agents in Trans-methylation Reactions

Methylating agent	Substance methylated	Product
$(CH_3)_nSn^{4-n}$	Hg^{2+}	CH_3Hg^+
$(CH_3)_3Si$	Hg^{2+}	CH_3Hg+
CH_3Hg^+	Se(II)	CH_3Se^+
CH_3I	Pb(II), CdS	$(CH_3)_4Pb$, ?
$(CH_3)_3Pb^+$	$HgCl_2$	CH_3HgCl
CH_3CoB_{12}	Hg^{2+}	CH_3Hg^+
	Tl(III)	$(CH_3)_2Tl^+$
$(CH_3)_3S^+PF_6^-$	As(III)	CH_3/As species
CH_3COO^-	Hg^{2+}	CH_3Hg^+, $(CH_3)_2Hg$

Sources; references 1,6,9,10,28,29,56,59,63,64,73,85,87.

The bacteria associated with the methylation of mercury, are located in the bottom sediments of rivers, estuaries and the oceans, in intestines and faeces, in soils and yeast [3,28,29,53,59,68,71]. The factors that influence the formation of CH_3Hg^+ are many and include the temperature, mercury and bacterial concentrations, pH and types of soil or sediment, the sulphide concentration and the redox conditions [6,9,28,59,73,82]. Seasonal variations in methylation, in esturaine sediments, relate to the bacterial activity. This may also apply to organic sediments, compared with sandy sediments [73].

Methylation of mercury occurs in both aerobic and anaerobic conditions, but the latter are best, with maximum methylation occurring in the E_h range +0.1 to -0.2 V [5,28,29]. The pH of the system also has a significant effect, as CH_3Hg^+ is more stable in neutral to acid conditions as shown in the distribution diagram in Fig. 12.8 [75], and $(CH_3)_2Hg$ is more stable in basic conditions. The binuclear species $(CH_3Hg)_2OH$ appears to be of minor importance at low concentrations of organomercury species. The influence of the sulphide ion depends on the redox conditions. If the conditions are anaerobic, the low soluble HgS will remove much of the mercury from being methylated [5,6,28,55]. In aerobic conditions, however, the S^{2-} ion can be oxidized to SO_4^{2-}, freeing the mercury ion for methylation. In addition, the following reaction is possible, especially at S^{2-} ion concentrations >160 μg g^{-1} [5,28].

$$2CH_3Hg^+ + S^{2-} \rightarrow (CH_3Hg)_2S \xrightarrow{h\nu} (CH_3)_2Hg + HgS.$$

At a constant mercury concentration the production of methylmercury is reported to increase as the sulphide concentration increases, and then falls at higher sulphide concentrations [31].

The rate of production of CH_3Hg^+ depends on the above conditions and the reaction matrix. Values in the range 15-140 ng g^{-1} d^{-1} and 17-700 ng m^{-2} d^{-1} have

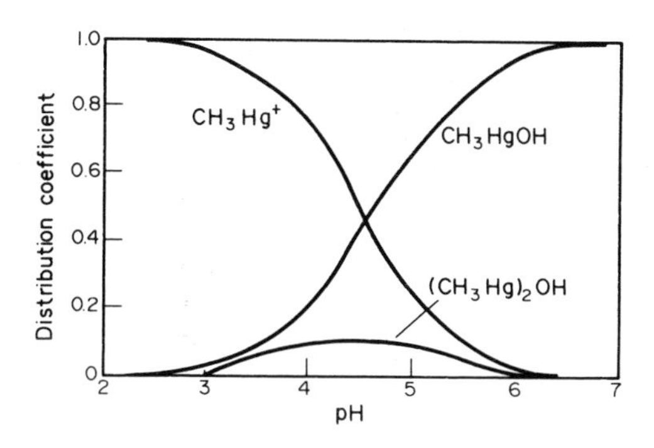

Fig. 12.8 Distribution diagram for the CH_3/Hg/OH system. Source of data; Tobias, 1978 [75].

been quoted [28,79]. The rate is generally greater in saline water than fresh [28]. The rate is affected by reverse reactions i.e. degradation or demethylation. In fresh sediments the addition of Hg^{2+} is associated with an increase in CH_3Hg^+, but then it falls due to the degradation reaction, and finally a steady state is produced [9,28,29,79]. Addition of the second methyl group to give $(CH_3)_2Hg$ is around 6000 times slower than the the production of CH_3Hg^+ [31,87].

Methylmercury accounts for approximately 0.1-1.5% of the total mercury in sediments, and around 2% of the total mercury in seawater, but in fish it is >80% of the total mercury. It is not clear if the CH_3Hg^+ is taken in by the fish from the seawater or formed within the fish. The evidence tends to suggest the former, though in rotting fish $(CH_3)_2Hg$ is formed [7,28,38,71]. In shellfish the methylmercury species found is CH_3HgSCH_3. A possible mechanism for the production of this compound is given in Fig 12.9 [6,13,28,73,87].

A number of organomercury/inorganic mercury cycles within the sediment/water/air system have been proposed. One that includes much of the known data is given in Fig 12.10 [6,12,28,31,58,71,83,84,87,88,90].

Arsenic

Bacteria associated with the methylation of arsenic are found in the soil, moulds, fungi, sewage fungi, sediments and also in animals and human beings [3,23,28,29,63,76]. Organoarsenic compounds are found in water, fish, seashells, mammal tissues and marine plants [7,29,76].

The principal methylating agent appears to be S-adenosylmethionine (see Fig 12.2), but methylcobalamin with the cofactor 2,2' dithiodiethanesulphonic acid may also methylate, as it can be methylated [29,88].

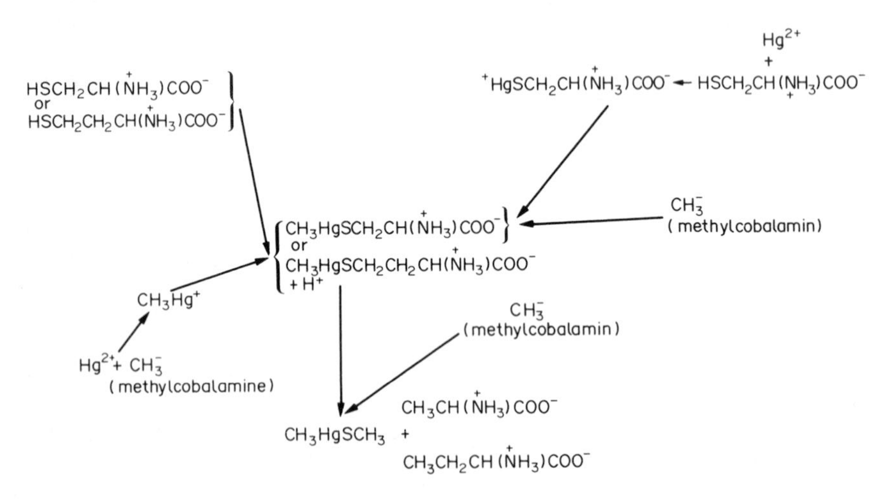

Fig. 12.9 Possible synthetic routes for CH_3HgSCH_3 in the environment. Sources of data; references 6,13,28,73,87.

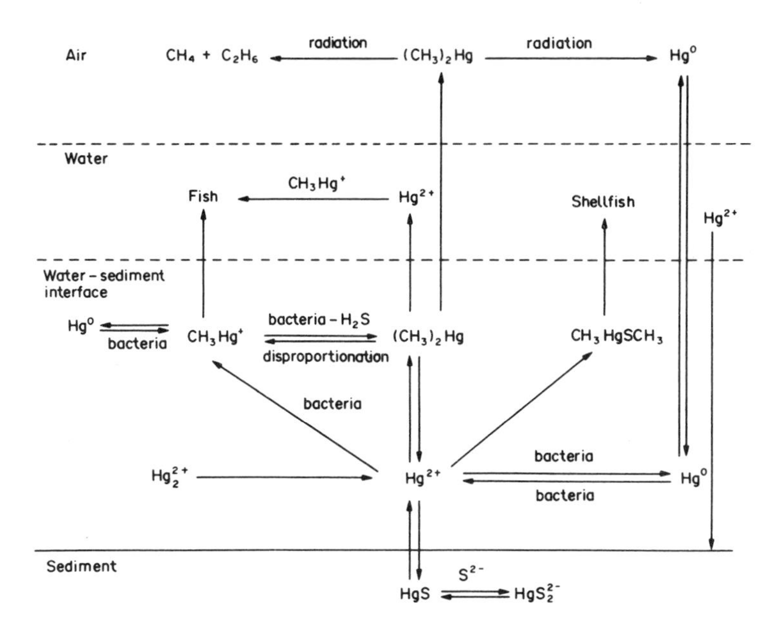

Fig. 12.10 An environmental mercury cycle. Sources of data; references 6,12,28,31,58,71,83,84,87,88,90

$$[SCH_2CH_2SO_3^-]_2 \xrightarrow[CH_3]{} CH_3\text{-}SCH_2CH_2SO_3^-$$

The methyl group transferred is probably the carbocation CH_3^+, and it has been shown, using ^{14}C labeling, that the CH_3 group bonded to sulphur in methionine bonds to arsenic. The reaction occurs with trivalent arsenic, which has a lone pair of electrons, so the transfer reaction probably proceeds according to the following reaction [3,15,28,29,62].

$$CH_3 + :As\lessdot \longrightarrow H_3C \leftarrow :As^+\lessdot$$

↑
lone pair

In the reaction the arsenic(III) is oxidized to arsenic(V), and the CH_3^+ is effectively reduced to a bonded CH_3^- group. Stable arsenic oxy-species are also formed in the reaction [3,15,55,62,73,76].

$$As(OH)_3 \xrightarrow{a} H^+ + AsO(OH)_2^- \xrightarrow{CH_3^+} CH_3\text{-}AsO(OH)_2 \xrightarrow{a}$$

As(III) As(III) As(V) methylarsonic acid

$$CH_3AsO_2(OH)^- \xrightarrow{b} CH_3AsO(OH)^- \xrightarrow{CH_3^+} (CH_3)_2AsO(OH)^- \xrightarrow{a\ \&\ b}$$

As(V) As(III) As(V) cacodylic acid

$$(CH_3)_2AsO^- \xrightarrow{CH_3^+} (CH_3)_3AsO \xrightarrow{b} (CH_3)_3As$$

As(III) As(V) trimethylarsine oxide As(III) trimethylarsine

a = ionization
b = reduction

The S-adenosylmethionine, CH_3S^+RR', being the source of CH_3^+ ends up as S-adenosylhomocysteine i.e. RSR′.

An organoarsenic cycle in sediments, water and air in which the above reactions occur is given in Fig 12.11 [3,28,29,55,62,63,83] The points at which bacteria are involved in the cycle are noted. The speciation of arsenic in systems such as water, phytoplankton or seaweeds is the result of the interplay of methylation and redox processes [3], as shown in the Table 12.7.

Lead

The weight of evidence suggests that both inorganic and organic lead are methylated in the environment. But some lingering doubts remain [27,28,29,33, 49,72,73]. The possible routes for the methylation of lead are biomethylation, chemical disproportionation and with abiotic reagents [20,23,29,33,49,71, 73,80]. For example in the reaction,

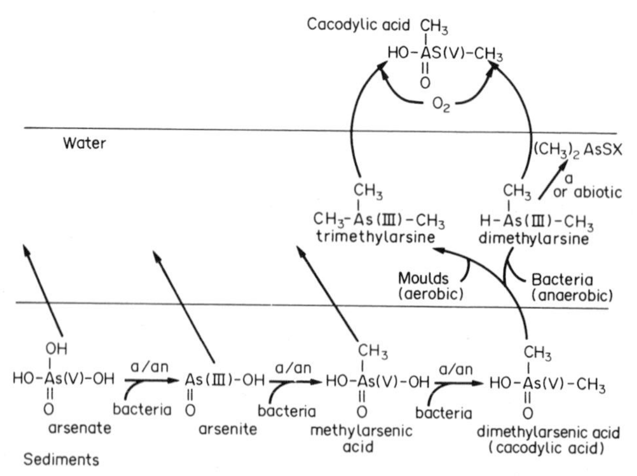

Fig. 12.11 The arsenic cycle involving methyl arsenic compounds, a = aerobic, an = anaerobic. Sources of data; references 3,28,29,55,62,63,83.

TABLE 12.7 Arsenic Speciation in Various Environmental Samples

Sample	As(III)	As(V)	CH_3/As*	$(CH_3)_2/As$*	As (total)
Seaweed (Argentina)					
Macrocysticpyrifera	0.07**	0.42	0.01	2.0	33.3
Phytoplankton	0.26	0.094	0.020	1.29	3.52
Seawater					
McMurdo Sound 25 m	0.15¶	18.3	0.13	0.11	18.7
75 m	0.08	18.3	0.11	0.03	18.5
Sth. Calif. Blight 1 m	-	18.4	0.24	1.64	20.3
67 m	-	18.3	<0.07	0.15	18.4

* Organoarsenic compounds containing one or two CH_3 groups, ** $\mu g\ g^{-1}$ dry weight, ¶ nmol dm^{-3}. Source of data; Andreae, 1986 [3].

$$(CH_3)_3PbOAc + \text{lake sediment} \rightarrow (CH_3)_4Pb,$$

the extra methyl group could arise from any one of the three processes. Estimations of the proportion of $(CH_3)_4Pb$ from each route vary, 15-19% from biomethylation [28,63] to 15-20% from chemical methylation (such as disproportionation) [20,23]. Relatively high levels of methyl lead have been found in fish, i.e. 10-40% [27,41].

It is likely that $MeCoB_{12}$ will methylate lead, according to the reaction [85],

$$(CH_3)_2PbX_2 + MeCoB_{12} \rightarrow (CH_3)_3PbX + H_2OCoB_{12}^+ + X^-$$

A problem in proposing the methylation of inorganic lead, however, is that the monomethyllead product, CH_3Pb^{3+}, is very unstable. Unless it can undergo further methylation before decomposition, it is doubtful if methylation would proceed any further [1,13,23,27]. Oxidative addition is the possible mechanism for the reaction [1,13],

$$Pb(II)X_3^- + CH_3^+ \rightarrow CH_3Pb(IV)X_2^+,$$

$$[Pb(II)L_n]^{(2-n)+} + CH_3^+ \rightarrow [CH_3Pb(IV)L_n]^{(3-n)+},$$

with the CH_3^+, or CH_3^-, coming from $MeCoB_{12}$.

A biological cycle for both methylation and demethylation of lead in the tetravalent state is given in Fig 12.12. The process which includes the methylation of mercury by the methyllead [31,85] avoids forming the very unstable CH_3Pb^{3+}. Methylation also occurs for $(C_2H_5)_nPb^{(4-n)+}$ species.

Disproportionation reactions produce $(CH_3)_4Pb$, as the series of (unbalanced) reactions in Fig 12.13 indicate [37,41,51,63,65]. The overall reactions are;

$$3(CH_3)_3PbX \rightarrow 2(CH_3)_4Pb + CH_3X + PbX_2,$$

and

$$2(CH_3)_2PbX_2 \rightarrow (CH_3)_3PbX + CH_3X + PbX_2$$

from combining equations 1, 2 and 3 in Fig. 12.13. The second reaction is the fastest, and therefore $(CH_3)_3PbX$ is the major product. Disproportionation to give $(CH_3)_4Pb$, may also involve a sulphide intermediate [1,28,29,49,65,73];

$$2(CH_3)_3Pb^+ + S^{2-} \rightarrow ((CH_3)_3Pb)_2S$$

$$3((CH_3)_3Pb)_2S \xrightarrow[h\upsilon]{} 3(CH_3)_4Pb + ((CH_3)_2PbS)_3$$

Some $(CH_3)_2S$ is also formed and this has been identified. Chemical methylation can also occur, for example the Pb^{2+} ion is methylated to $(CH_3)_4Pb$ with CH_3I [73].

The effect of pH on the methyllead species is portrayed in the distribution diagram given in Fig 12.14. The $(CH_3)_3Pb^+$ ion can exist up to a pH around 9.

It seems very likely that methyllead species are produced in the environment from one of the three routes or a combination of them. It is, however, necessary to verify unequivocally, that laboratory reactions do occur in the environment.

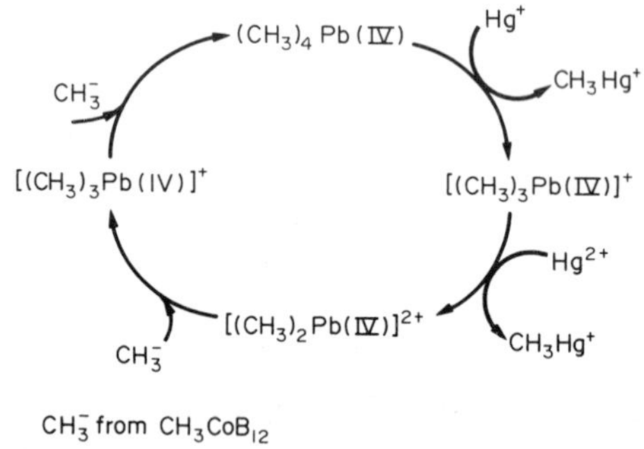

$(CH_3)_4Pb(IV)$
Hg^+
CH_3^-
CH_3Hg^+
$[(CH_3)_3Pb(IV)]^+$
$[(CH_3)_3Pb(IV)]^+$
Hg^{2+}
$[(CH_3)_2Pb(IV)]^{2+}$
CH_3^-
CH_3Hg^+
CH_3^- from CH_3CoB_{12}

Fig. 12.12 Methylation and demethylation cycle for lead in the environment. Sources of data; Craig, 1986 [31], Wood, 1980 [85].

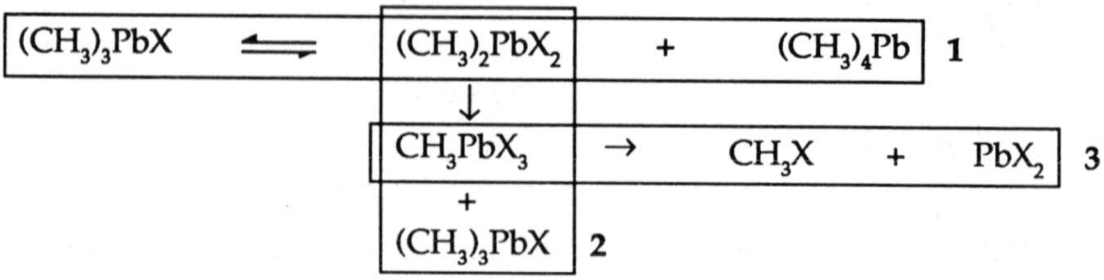

$(CH_3)_3PbX \rightleftharpoons (CH_3)_2PbX_2 + (CH_3)_4Pb$ **1**

$(CH_3)_2PbX_2 \downarrow CH_3PbX_3 + (CH_3)_3PbX$ **2**

$CH_3PbX_3 \rightarrow CH_3X + PbX_2$ **3**

Fig. 12.13 Disproportionation reactions for methyllead compounds (unbalanced).

Distribution coefficient
1.0
0.8
0.6
0.4
0.2
0
7
9
11
pH
(CH3)3Pb+
(CH3)3 PbOH
[(CH3)3 Pb]2OH+

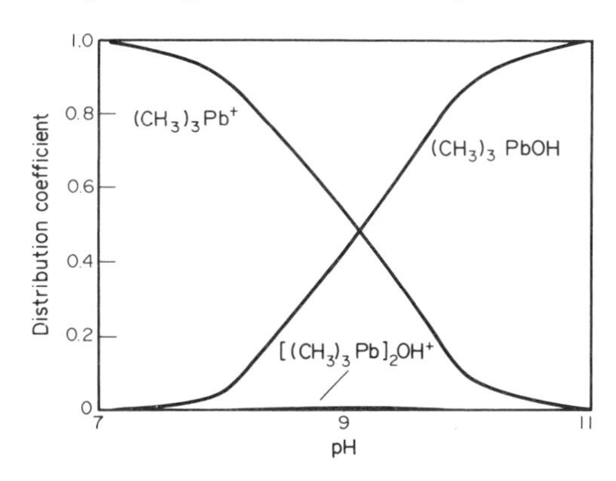

Fig. 12.14 Distribution diagram for methyllead compounds.

Selenium and Tellurium

The two Group 16 (VIb) elements, selenium and tellurium, are biomethylated by a number of bacteria (see Table 12.4). The methylating agent is S-adenosylmethionine [15,24,28,29,50,55,71,88]. The products $(CH_3)_2M$ (M = Se and Te) and $(CH_3)_2Se_2$ are produced in sediments and soils [23,28] and in animals and human beings [29]. A mechanism for the methylation similar to that given for arsenic, is shown in Fig 12.15 [24,55,73]. A similar process is suggested for tellurium, though it is possible that selenium is also required through transmethylation from selenium to tellurium [24].

Selenite (Se(IV)), selenate (Se(VI)) and tellurite (Te(IV)) interfere in the formation of arsenic methyl derivatives, presumably because the Se or Te are preferentially methylated [23,73]. Also, selenium reduces the toxicity of

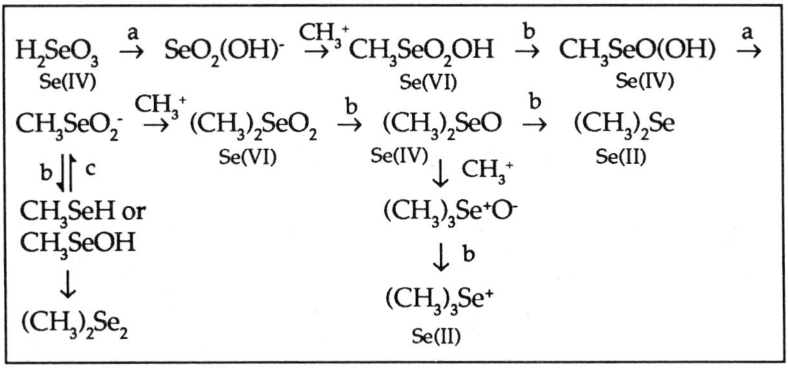

H2SeO3 →a SeO2(OH)- →CH3+ CH3SeO2OH →b CH3SeO(OH) →a
Se(IV) Se(VI) Se(IV)
CH3SeO2- →CH3+ (CH3)2SeO2 →b (CH3)2SeO →b (CH3)2Se
Se(VI) Se(IV) Se(II)
b ↓↑ c
CH3SeH or
CH3SeOH
↓
(CH3)2Se2
↓ CH3+
(CH3)3Se+O-
↓ b
(CH3)3Se+
Se(II)

Fig. 12.15 Suggested mechanism for the methylation of selenium. a = ionization, b = reduction, c = oxidation, CH_3^+ comes from S-adenosylmethionine. Sources of data; references 24,55,73.

CH_3Hg^+ as the methyl group is transferred from mercury to selenium, and the $(CH_3)_2Se$ is lost from the body [29,31] This role of selenium may arise from the exothermic heat of formation of $(CH_3)_2Se$ compared with the endothermic values for the arsenic and mercury compounds (see Table 12.2).

Thallium

There is good laboratory evidence for the biomethylation, by CH_3CoB_{12}, of both Tl(I) and Tl(III) to give species such as $(CH_3)_2Tl^+$ and CH_3Tl^{2+} from Tl(III). A number of chemical reactions are then possible to produce CH_3TlX^+, CH_3TlX_2 and $(CH_3)_2TlX$ [28,29,32,51,65,73,88]. The trimethylthalium(III) compound is unstable in water. It is not possible to extrapolate the laboratory results to the environmental situation, because as yet there is no evidence for biomethylation of thallium in the environment.

Cadmium, Indium, Antimony and Bismuth

There is evidence that a volatile cadmium compound is produced from the interaction of cadmium salts and bacteria associated with methylcobalamin [28,29,32,71,73]. More work is necessary to identify the product. The instability of $(CH_3)_2Cd$ in water, rules out its detection or existence in the environment.

Attempts to biomethylate antimony have been unsuccessful, though some early work did suggest methylation takes place [15,32]. Methylantimony(V) species such as $(CH_3)_2SbO(OH)$ and $CH_3SbO(OH)_2$, have been identified in water [32]. There is some evidence that Bi(III) and $InCl_3$ react with methylcobalamin, but there is little known about the natural or laboratory methylation of the four metals.

A summary of the bio- and chemical methylation of the ten heavy elements is given in Table 12.8.

TABLE 12.8 Summary of the Methylation of the Heavy Elements

Element	Biomethylation		Biomethyl-ating agent	Chemical methylation
	Laboratory	Environ.		
Hg	yes	yes	$MeCoB_{12}$	yes
As	yes	yes	SAM, $MeCoB_{12}$	yes
Se	yes	yes	SAM, $MeCoB_{12}$	yes
Te	yes	yes	SAM, $MeCoB_{12}$	
Pb	yes	probably	$MeCoB_{12}$	yes
Tl	yes	-	$MeCoB_{12}$	yes
Cd	maybe	-	$MeCoB_{12}$	yes
In	maybe	-	$MeCoB_{12}$	
Sb	maybe	probably		
Bi	maybe	-	$MeCoB_{12}$	

SAM = S-adenosylmethionine, $MeCoB_{12}$ = methylcobalamin

Other Methylating Reactions

Two other processes involving bacteria also occur for the heavy elements, viz. oxidation/reduction and demethylation. Bacterial reductions occurs for As(V), Hg(I), Hg(II), Se(IV) and Te(IV) [64,71] and bacterial oxidations occur for As(III), Sb(III) [71]. During soil degassing, *pseudomanas* can convert Hg(II) to Hg(0) which is then lost to the soil [29,87].

The bacterial demethylation of CH_3Hg^+ is probably faster than the methylation process [4,6,9,14,28,59,71,79,87],

$$CH_3Hg^+ \xrightarrow[\text{bacteria}]{} CH_4 + Hg(0)$$

This process also detoxifies a system from CH_3Hg^+. A non-bacterial demethylation takes place in the atmosphere [9,59,87].

$$(CH_3)_2Hg \xrightarrow{h\upsilon} CH_4 + C_2H_6 + Hg$$

Some of these process have the effect of detoxifying systems, either by increasing the volatility (Hg(II) → Hg(0)) [14,68], or producing a less toxic material (arsenite → arsenate) [14,68]. Also methylating agents such as $MeCoB_{12}$ may detoxify bacteria [79,90].

Methylation can increase the solubility of poorly soluble metal compounds, such as FeTe, FeSe, PbSe, PbTe, PbS and Sb_2S_3. Methylation of a metal atom from the surface of a crystal by CH_3I and methylcobalamin weakens its bond in the crystal lattice, and it is eventually taken into solution [74].

LEVELS OF METHYL DERIVATIVES OF THE HEAVY METALS IN THE ENVIRONMENT

The concentrations of the methyl derivatives of the heavy metals in the environment have been frequently measured and many results are available. Except for most of the lead, and in earlier times mercury, the methyl compounds arise from the methylation processes described above. In the case of lead the major environmental sources are $(CH_3)_4Pb$ and $(C_2H_5)_4Pb$ used in petrol. Methylmercury has been used as a seed protective coating from fungicides, but this use is now discontinued.

Mercury

The important feature of CH_3Hg^+, is its concentration in fish tissue, especially muscle. This is a significant source of CH_3Hg^+ for human beings that consume a lot of fish. The total concentration of mercury in the environmental materials and the percent that is methylmercury are listed in Table 12.9 [28,79]. The few results presented, indicate that in fish muscle the proportion of CH_3Hg^+ is quite high, reaching over 90%. Methylmercury bioaccumulates in the food

TABLE 12 .9 Levels of Methylmercury in the Environment

Material	Total mercury μg g^{-1}	% Methylmercury
Esturary invertebrates		
poluted	0.13-4.5	4-57
controlls	0.05-0.61	trace-21
Sea trout, muscle	2.3	99
liver	3.2	92
Black tip shark, muscle	0.75	69
liver	1.3	4.6
Gray pilot whale		
muscle	2.42	95
liver	9.56	33
Fish	0.16	85
Seed-eating birds		
liver		50-100
Duck (NW Ontario)		
liver/breast	1-10	10-40
Shellfish 0.14-0.75	40-90	
polluted	11-40	40-90
Urban air	0.5-50 ng m^{-3}	0-5
Sediments	0-100	up to 1.5
Bed sediments	0.081	3.8

Sources of data; references 28,31,79.

chain, and for example the relative levels found in lake water, macrophytes, algae and fish, are 1 : 1000 : 2800 : 23200 respectively [28].

Arsenic

Similar high concentrations of organoarsenic compounds occur in marine animals. Compared with a seawater concentration of arsenic of 2 ng g^{-1} the bioaccumulation factor for marine plant life is up to 71,000 (inorganic arsenic); algae 4-17; snail 1-23 and crayfish 1-16 [28]. The use of organic arsenicals as pesticides, in the past, means high levels of the element occurs in some soils. The concentrations of methylarsenic compounds in some environmental materials are listed in Table 12.10.

Lead

Probably most organolead compounds in the environment, come from the use of the methyl and ethyl compounds in petrol. The compounds get into the environment by volatilization from the petrol or emission from car exhausts, especially under certain driving conditions. This is the major source in the urban areas, but in rural areas it is possible that environmental methylation is

TABLE 12.10 Levels of Methylarsenic Compounds in the Environment

Substance	CH_3/As* ng g^{-1}	$(CH_3)_2/As$* ng g^{-1}	$(CH_3)_n/As$* %	Inorganic As ng g^{-1}
Lake water	0.2	0.25	18	2.0
Surface seawater	0.004	0.02	10-90	1.02
Ocean (Pacific)	0.1-0.2	1		
Diatoms	0-0.03	0.13-0.24	10-90	0.02-0.28
Kelp			4	
Macroscopic algae	0.4-5.3	1.1-10.6	0-80	2-28.4 x 10^3
Shark muscle			80	37.4 x 10^3
Flyash slurry	24	109		20

* Corresponding to substances such as $CH_3AsO(OH)_2$, $(CH_3)_2AsO(OH)$.
Source of data; Craig, 1982 [28].

more significant [45,47], as shown by the data in Table 12.11. The air sample data are dated, as the levels of alkyllead have changed with the lowering of the amount of alkyllead in petrol. The proportion of organolead in urban aerosol lead samples lies in the range 1-10%, and a background level of aerosol lead may contain 1-6% alkyllead [47]. The amount of alkyllead emitted from vehicle exhausts decreases as the car engine warms up (see Table 12.11). For example for 97 octane petrol, 0-1, 1-2, 2-3, and 3-4 minutes after starting, the amount of alkyllead falls from 1565, 905, 640 to 525 μg m^{-3} [41].

Organolead compounds are also bioaccumulated in fish relative to the levels in sediments and water. The organolead compounds that have been found in fish and in sediments are; $(CH_3)_4Pb$, $(C_2H_5)_4Pb$, $(CH_3)_3C_2H_5Pb$, $(CH_3)_2(C_2H_5)_2Pb$, $CH_3(C_2H_5)_3Pb$, $(CH_3)_3Pb^+$, $(CH_3)_2Pb^{2+}$, $(C_2H_5)_3Pb^+$ and $(C_2H_5)_2Pb^{2+}$. The most common species are; $(C_2H_5)_4Pb$, $(C_2H_5)_3Pb^+$ and $CH_3(C_2H_5)_3Pb$ presumably because more $(C_2H_5)_4Pb$ is used in petrol compared with $(CH_3)_4Pb$.

In addition to the disproportionation reactions for methyllead compounds discussed above [1,33,41,51,63], other reactions are possible. These include, attack by hydroxyl radicals and photolysis, and to a lesser extent attack by ozone and the ^3P oxygen atom [33,47]. The decomposition rate estimated for the photochemical route is 8% hr^{-1} for $(CH_3)_4Pb$ and 31% hr^{-1} for $(C_2H_5)_4Pb$ [41]. Other rates for breakdown in the atmosphere are as 21% hr^{-1} and 88% hr^{-1} respectively [28]. The stability of the alkyllead compounds to radiation decreases, with the number of alkyl groups attached to the lead and the increased size of the alkyl groups [51]. In seawater dealkylation occurs readily to produce the stable trialkyllead cations, $(CH_3)_3Pb^+$ and $(C_2H_5)_3Pb^+$ [85].

Selenium and Antimony

Organoselenium compounds have been identified both in the atmosphere and in sediments. Some data are given in Table 12.12 [24], some recent results for methylantimony levels in seawater are listed in Table 12.13 [32,65].

TABLE 12.11 Levels of Alkyllead in the Environment

Location/sample	Organic lead		Mean	Comments
	Mean	Range	%*	
Air (ng m^{-3})				
Antwerp 1981	239	210-260	-	petrol station
	24	14-44	1.7	highway
	39	12-162	0.5	tunnel
Frankfurt 1980	24	1-90	7.8	residential
	3	1-7	3.5	rural
Glasgow 1981	16	1.5-54	3.0	residential
	3.9	1.6-6.5	14	rural
Lancaster 1978		<6-206	0.3-4	urban
		210-590	4-10	petrolstation
		0.5-230	1.5-33	rural**
London 1974	70		4.0	Heathrow
London 1980	94	24-190	6.6	at 5 m height
	65	16-130	7.2	at 14 m height
Urban air		50-1240	0.7-13	
Vehicle exhaust				
cold choked	5×10^6		10-30	
warm idling	$0.5\text{-}1 \times 10^6$		1-5	
warm steady speed	$0.05\text{-}0.1 \times 10^6$		0.1-1	
Marine samples (μg g^{-1})				
Flounder meal	4800		90	
Lobster	162		81	
Mediterranean sea mussel	22	9.7-48	13	
Ontario Lakes fish	0.022	0.021-0.023	23	
Rainbow trout				
lipid	416			
gills	55			
kidney	9			
St. Lawrence River				
carp	5.6			
pike	1.5			
sediment	1.5			

* Mean % organic lead/particulate lead, ** High due to environmental methylation.
Sources; references 22,28,33,34,41,43-47,61,80.

ANALYSIS AND IDENTIFICATION OF ORGANOMETALLIC SPECIES

The detection and estimation of the organo-derivatives of the heavy elements is a two stage process, firstly, separation from the inorganic species of the metal, and secondly, analysis either for the total organic component or the individual species (see Chapter 4).

TABLE 12.12 Levels of Methylselenium in the Environment

Location/sample	$(CH_3)_2Se$	$(CH_3)_2Se_2$	$(CH_3)_2SeO_2$
Atmosphere (ng m^{-3})			
Lake Wilrij k (Belgium)	0.47-0.84	0.35-0.62	<0.10-0.43
Sediment (ng Se from 50g of sediment)*			
Kelly Lake (Canada)	2.7	3.3	2.3

* Sediment incubated for 7 days. Source of data; Chau and Wong, 1986 [24].

TABLE 12.13 Levels of Antimony in the Environment (ng l^{-1})

Location	Sb(III)	Sb(V)	MSA*	DMSA*
Mississippi River	<0.3	148.0	2.3	ND**
Main River (FDR)	0.3	311	1.8	ND
Gulf of Mexico	4.4	149	5.3	3.2
Baltic Sea***	0.001-0.1	0.01-0.7	0-0.05	<0.1

* MSA = monomethylantimony species, DMSA = dimethyl-antimony species, ** not detected, *** concentration nmol dm^{-3}.
Source of data; Craig, 1986 [32].

Methods used to separate the organic from inorganic compounds are chelation, solvent extraction, gas liquid chromatography, or specific chemical reactions. Estimation of the amounts recovered are achieved with flame or flameless AAS. The method is frequently associated with gas chromatography or gas liquid chromatography for separation of the various organometallic compounds prior to entering the AAS [3,11,21,22,25,28,31,34,41,43,44,49,67].

REFERENCES

1. Ahmad, I., Chau, Y. K. Wong, P. T. S. Carty, A. J. and Taylor, L. Chemical alkylation of lead(II) salts to tetraalkyllead(IV) in aqueous solution. **Nature**; 1980; **287**: 716-717.
2. Anderson, D. A. and Sobieski, R. J. Introduction to Microbiology. 2nd ed.: C. V. Mosby Co.; 1986.
3. Andreae, M. O. Organoarsenic compounds in the environment. In: Craig, P. J., Ed. Organometallic Compounds in the Environment: Longmans; 1986: 198-228.
4. Barkay, T., Olson, B. H. and Colwell, R. R. Heavy metal transformations mediated by estuarine bacteria. In: Management and Control of Heavy Metals in the Environment, International Conference, London: CEP Consultants; 1979: 356-363.
5. Bartlett, P. D. and Craig, P. J. Methylation processes for mercury in esturar-ine sediments. In: Management and Control of Heavy Metals in the Environment, International Conference, London: CEP Consultants; 1979: 354-355.

6. Beijer, K. and Jernelöv, A. Methylation of mercury in aquatic environments. In: Nriagu, J. O., Ed. The Biogeochemistry of Mercury in the Environment: Elsevier/North Holland; 1979: 203-210.

7. Bernhard, M. and George, S. G. Importance of chemical species in uptake, loss and toxicity of elements for marine organisms. In: Bernhard, M., Brinckman, F. E. and Sadler, P. T., Eds. The Importance of Chemical Speciation in Environmental Processes, Dahlem Konferenzen: Springer Verlag; 1986: 385-422.

8. Bird, M. L. and Challenger, F. The formation of organo-metalloidal and similar compounds by micro-organisms. Part VIII Dimethyltelluride. **J. Chem. Soc.;** 1939: 163-168.

9. Bisogni, J. J. Kinetics of methylmercury formation and decomposition in aquatic environments. In: Nriagu, J. O., Ed. The Biogeochemistry of Mercury in the Environment: Elsevier/North Holland; 1979: 211-230.

10. Brinckman, F. E. and Olson, G. J. Global biomethylation of the elements: its role in the biosphere translated to new organometallic chemistry and biotechnology. In: Craig, P. J. and Glockling, F., Eds. The Biological Alkylation of Heavy Metals: Roy. Soc. Chem.; 1988: 168-196.

11. Bye, R. and Paus, P. E. Determination of alkylmercury compounds in fish tissue with an atomic absorption spectrometer used as a specific gas chromatographic detector. **Anal. Chim. Acta;** 1979; **107:** 169-175.

12. Carty, A. J. and Malone, S. F. The chemistry of mercury in biological systems. In: Nriagu, J. O., Ed. The Biogeochemistry of Mercury in the Environment: Elsevier/North Holland; 1979: 433-479.

13. Carty, A. J. Mercury, lead and cadmium complexation by sulfhydryl containing amino acids: implications for heavy metal synthesis transport and toxicology. In: Brinckman, F. E. and Bellama, J. M, Eds. Organometals and Organometalloids: Occurrence and Fate in the Environment, ACS Symposium: Amer. Chem. Soc.; 1978: 339-358.

14. Chakrabarty, A. M. Microbial interactions with toxic meats in the environment. In: Bernhard, M. Brinckman, F. E. and Sadler, P. J., Eds. The Importance of Chemical Speciation in Environmental Processes, Dahlem Konferezen: Springer Verlag; 1986: 513-531.

15. Challenger, F. Biosynthesis of organometallic and organometalloid compounds. In: Brinckman, F. E. and Bellama, J. M, Eds. Organometals and Organometalloids: Occurrence and Fate in the Environment, ACS Symposium: Amer. Chem. Soc.; 1978; No. 82: 1-22.

16. Challenger, F. Biological methylation. **Chem. Revs.;** 1945; **36:** 315-361.

17. Challenger, F., Higinbotom, C. and Ellis, L. The formation of organo-metalloidal and similar compounds by micro-organisms. Part I Trimethylarsine and dimethylarsine. **J. Chem. Soc.;** 1933: 95-101.

18. Challenger, F. and North, H. E. The formation of organo-metalloidal and similar compounds by micro-organisms. Part II Dimethylselenide. **J. Chem. Soc.;** 1934: 68-71.

19. Challenger, F. Biological methylation. **Quart. Revs. Chem. Soc.**; 1955; **9:** 255-286.
20. Chau, Y. K. and Wong, P. T. S. Biotransformation and toxicity of lead in the aquatic environment. In: Branica, M. and Konrad, Z., Eds. Lead in the Marine Environment; 1980: 225-231.
21. Chau, Y. K., Wong, P.T.S. and Kramar, O. The determination of dialkyllead, trialkyllead, tetraalkyllead and lead(II) ions in water by chelation/ extraction and gas chromatography/atomic absorption. **Anal Chim. Acta**; 1983; **146:** 211-217.
22. Chau, Y. K. Wong, P. T. S., Bengert, G. A. and Dunn, J. L. Determination of dialkylead, trialkyllead tetraalkyllead and lead(II) compopunds in sediments and biological samples. **Anal. Chem.**; 1984; **56:** 271-274.
23. Chau, Y. K. and Wong, P. T. S. Occurrence of biological methylation of elements in the environment. In: Brinckman, F. E. and Bellama, J. M, Eds. Organometals and Organometalloids: Occurrence and Fate in the Environment, ACS Symposium: Amer. Chem. Soc.; 1978: 39-53.
24. Chau, Y. K. and Wong, P. T. S. Organic Group VI elements in the environment. In: Craig, P. J., Ed. Organometallic Compounds in the Environment: Longmans; 1986: 254-278.
25. Coker, D. T. A simple sensitive technique for personal and environmental sampling and analysis of lead alkyl vapours in air. **Annals Occup. Hyg.**; 1978; **21:** 33-38.
26. Conn, E. E. and Stumpf, P. K. Outlines of Biochemistry. 4th ed.: Wiley; 1976.
27. Craig, P. J. and Wood, J. M. The biological methylation of lead: an assessment of the present position. In: Lynam, D. R., Piantanida, L. G. and Cole J. F., Eds. **Environmental Lead**; 1981: 333-349.
28. Craig, P. J. Environmental aspects of organometallic chemistry. In: Wilkinson, G., Stone, F. G. A. and Abel, E. W., Eds. Comprehensive Organometallic Chemistry; 1982; 2.
29. Craig, P. J. Metal cycles and biological methylation. In: Huntzinger, O., Ed. Handbook of Environmental Chemistry: Springer Verlag; 1980; **1A:** 169-227.
30. Craig, P. J. and Brinckman, F. E. Occurrence and pathways of organometallic compounds in the environment - general considerations. In: Craig, P. J., Ed. Organometallic Compounds in the Environment: Longmans; 1986: 1-64.
31. Craig, P. J. Organomercury compounds in the environment. In: Craig, P. J., Ed. Organometallic Compounds in the Environment: Longmans; 1986: 65-110.
32. Craig, P. J. Other organometallic compounds in the environment. In: Craig, P. J., Ed. Organometallic Compounds in the Environment: Longmans; 1986: 345-364.

33. De Jonghe, W. R. A. and Adams, F. C. Biogeochemical cycling of organic lead compounds. In: Nriagu, J. O. and Davidson, C. I., Eds. Toxic Metals in the Atmosphere: Wiley; 1986: 561-594.
34. De Jonghe, W. R. A. ans Adams, F. C. Measurement of organic lead in air: a review. **Talanta;** 1982; **29:** 1057-1067.
35. D'Itri, P. A, and D'Itri, F. M. Mercury Contamination: A Human Tragedy: Wiley-Interscience; 1977.
36. Domsch, K. H. and Gans, W. Fungi in Agricultural Soils. Trans. Edit. ed.: Longmans; 1972.
37. Fanchiang, Y.-T., Ridley, W. P.. and Wood, J. M. Kinetic and mechanistic studies on B_{12}-dependent methyl transfer to certain toxic metals ions. In: Brinckman, F. E. and Bellama, J. M, Eds. Organometals and Organometalloids: Occurrence and Fate in the Environment, ACS Symposium: Amer. Chem. Soc.; 1978: 54-64.
38. Fujiki, M. Methylmercury accumulation in plankton and fish. In: Baker, R. A., Ed. Contaminants and Sediments: Ann Arbor Science; 1980; 2: 485-491.
39. Gilman, J. C. A Manual of Soil Fungi. 2nd. ed.: Iowa State College Press; 1957.
40. Gosio, B. **Arch. Ital. Biol.;** 1901; **35:** 201.
41. Grandjean, P. and Nielson, T. Organolead compounds: environmental health aspects. **Residue Revs.;** 1979; **72:** 97-148.
42. Halpern, J. and Maher, J. P. Pentacyanobenzylcobaltate(III). A new series of stable organocobalt compounds. **J. Amer. Chem. Soc.;** 1964; **86:** 2311.
43. Harrison, R. M., Perry, R. and Slater, D. H. An adsorption technique for the determination of organic lead in street air. **Atmos. Environ.;** 1974; **8:** 1187-1194.
44. Harrison, R. M. and Perry, R. The analysis of tetraalkyl lead compounds and their significance as urban air pollutants. **Atmos. Environ.;** 1977; **11:** 847-852.
45. Harrison, R. M. and Laxen, D. P. H. Natural source of tetraalkyllead in air. **Nature;** 1978; **275:** 738-739.
46. Harrison, R. M. and Laxen, D. P. H. Organolead compounds absorbed upon atmospheric particulates: a minor component of urban air. **Atmos. Environ.;** 1977; **11:** 201-203.
47. Harrison, R. M. and Laxen, D. P. H. Tetraalkyllead in air: sources, sinks and concentrations. In: Management and Control of Heavy Metals in the Environment, International Conference, London: CEP Consultants; 1979: 257-261.
48. Hawker, L. E. and Linton, A. H. Microorganisms: Function Form and Organisms. 2nd Ed.; Arnold; 1979.
49. Hewitt, C. N. and Harrison, R. M. Organolead compounds in the environment. In: Craig, P. J., Ed. Organometallic Compounds in the Environment: Longmans; 1986: 160-197.

50. Ho, Y. B. Lead contamination in street dust in Hong Kong. **Bull. Environ. Contam. Toxicol.**; 1979; **21:** 639-647.
51. Huber, F., Schmidt, U. and Kirchmann, H. Aqueous chemistry of organolead and organothallium compounds in the presence of microorganisms. In: Brinckman, F. E. and Bellama, J. M, Eds. Organometals and Organometalloids: Occurrence and Fate in the Environment, ACS Symposium: Amer. Chem. Soc.; 1978: 65-81.
52. Huheey, J. E. Inorganic Chemistry. 2nd Ed.: Harper Row; 1978.
53. Jensen, S. and Jernelov, A. Behaviour of mercury in the environment. In: Mercury Contamination in Man and his Environment, Tech. Report: IAEA; 1972; No. 137: 43-47.
54. Jensen, S. and Jernelov, A. Biological methylation of mercury in aquatic organisms. **Nature;** 1969; **223:** 753-754.
55. Jernelöv, A. and Martin, A.-L. Ecological implications of metal metabolism by microorganisms. **Ann. Rev. Microbiol.;** 1975; **29:** 61-77.
56. Jewett, K. L., Brinckmann, F. E. and Bellama, J. M. Influence of environmental parameters on transmethylation between aquated metal ions. In: Brinckman, F. E. and Bellama, J. M, Eds. Organometals and Organometalloids: Occurrence and Fate in the Environment, ACS Symposium: Amer. Chem. Soc.; 1978: 158-187.
57. Jolly, W. L. Modern Inorganic Chemistry: McGraw Hill; 1984.
58. Jonasson, I. R. and Boyle R. W. Geochemistry of mercury and origins of natural contamination of the environment. **Canad. Mining Met. Bull.;** 1972; **65:** 32-39.
59. Kaiser, G. and Tolg, G. Mercury. In: Hutzinger, O., Ed. The Handbook of Environmental Chemistry: Springer Verlag; 1980; **3A:** 1-58.
60. Lechevalier, H. A. and Pramer, D. The Microbes: J. B. Lippincott Co.; 1971.
61. Maddock, B. G. and Taylor, D. The acute toxicity and bioaccumulation of some lead alkyl compounds in marine animals. In: Branica, M. and Konrad, Z., Eds. Lead in the Marine Environment; 1980: 233-261.
62. McBride, B. C. Merilees, H. Cullen, W. R. and Pickett, W. Anaerobic and aerobic alkylation of arsenic. In: Brinckman, F. E. and Bellama, J. M, Eds. Organometals and Organometalloids: Occurrence and Fate in the Environment, ACS Symposium: Amer. Chem. Soc.; 1978: 94-115.
63. Moore, J. W. and Ramamoorthy, S. Heavy metals in Natural Waters: Applied Monitoring and Impact Assessment: Springer-Verlag; 1984.
64. Olson, G. J. Microbial intervention in trace element containing process streams and waste products. In: Bernhard, M. Brinckman, F. E. and Sadler, P. J., Eds. The Importance of Chemical Speciation in Environmental Processes, Dahlem Konferezen: Springer Verlag; 1986: 493-512.
65. Rapomanikis, S. and Weber, J. H. Methyl transfer reactions of environmental significance involving naturally occurring and synthetic reagents. In: Craig, P. J., Ed. Organometallic Compounds in the Environment: Longmans; 1986: 279-307.

66. Ridley, W. P., Dizikes, L. J. and Wood, J. M. Biomethylation of toxic elements. **Science;** 1977; **197:** 329-332.
67. Schafer, M. L., James, U. R., Peeler, J. T., Hamilton, C. H. and Campbell, J. E. A method for the determination of methylmercuric compounds in fish. **J. Agric. Food Chem.;** 1975; **23:** 1079-1083.
68. Silver, S. Bacterial transformations of and resistence to heavy metals. In: Nriagu, J. O., Ed. Changing Metal Cycles and Human Health, Dahlem Konferenzen: Springer Verlag; 1984: 199-23.
69. Skerman, V. B. D. A Guide to the Identification of the Genera of Bacteria: Williams ans Wilkins Co.; 1959.
70. Stainer, R. Y., Adelberg, E. A. and Ingraham, J. L. General Microbiology. 4th Ed.: Prentice Hall; 1977.
71. Summers, A. O. Silver, S. Microbial transformations of metals. **Ann. rev Microbiol.;** 1978; **32:** 637-672.
72. Taylor, R. T. and Hanna, M. L. Methylcobalamin: methylation of platinum and demethylation with lead. **J. Environ. Sci. Health (Environ. Sci. Eng.);** 1976; **A11:** 201-211.
73. Thayer, J. S. and Brinckman, F. E. The biological methylation of metals and metalloids. In: Stone, F. G. A. and West, R, Eds. Advances in Organometallic Chememistry: Academic Press; 1982; 20: 313-356.
74. Thayer, J. S. Methylation of poorly soluble metal compounds: some environmental implications. In: Craig, P. J. and Glockling, F., Eds. The Biological Alkylation of Heavy Metals: Roy. Soc. Chem.; 1988: 201-214.
75. Tobias, R. S. The chemistry of organometallic cations in aqueous media. In: Brinckman, F. E. and Bellama, J. M, Eds. Organometals and Organometalloids: Occurrence and Fate in the Environment, ACS Symposium: Amer. Chem. Soc.; 1978: 130-148.
76. Vahter, M. and Marafante, E. In vivo methylation and detoxification of arsenic. In: Craig, P. J. and Glockling, F., Eds. The Biological Alkylation of Heavy Metals: Roy. Soc. Chem.; 1988: 105-119.
77. Westoo, G. Determination of methylmercury compounds in foodstuffs 1 methylmercury in fish, identification and determination. **Acta Chem. Scand.;** 1966; **20:** 2131-2137.
78. Williams, R. J. P. The transfer of methyl groups: a general introduction. In: Craig, P. J. and Glockling, F., Eds. The Biological Alkylation of Heavy Metals: Roy. Soc. Chem.; 1988: 5-19.
79. Windom, H. L. and Kendall, D. R. Accumulation and biotransformation of mercury in coastal and marine biota. In: Nriagu, J. O., Ed. The Biogeochemistry of Mercury in the Environment: Elsevier/Nth. Holland; 1979: 303-323.
80. Wong, P. T. S. and Chau, Y. K. Methylation and toxicity of lead in the aquatic environment. In: Management and Control of Heavy Metals in the Environment, International Conference, London: CEP Consultants; 1979: 131-134.

81. Wong, P. T. S., Chau, Y. K. and Luxon, P. L. Methylation of lead in the environment. **Nature**; 1975; **253**: 263-264.
82. Wood, J. M. Biological methylation of mercury can occur by many routes. **Chem. Eng. News**; 1971; **July 5**: 24-25.
83. Wood, J. M. Biological cycles for toxic elements in the environment. **Science**; 1974; **183**: 1049-1052.
84. Wood, J. M. and Goldberg, E. D. Impact of metals on the biosphere. In: Global Chemical Cycles and Their Alteration by Man.: Dahlem Workshop; 1977: 137-153.
85. Wood, J. M. Lead in the marine environment some biochemical considerations. In: Branica, M. and Konrad, Z., Eds. Lead in the Marine Environment: Pergamon; 1980: 299-303.
86. Wood, J. M. Mechanism for B12-dependent methyl transfer to heavy elements. In: Craig, P. J. and Glockling, F., Eds. The Biological Alkylation of Heavy Metals: Roy. Soc. Chem.; 1988: 62-76.
87. Wood, J. M., Segall, H. J., Ridley, W. P., Cheh, A., Cdudzk, W. and Thayer, J. S. Metabolic cycles for toxic metals in the environment. In: Int. Conf. Heavy Metals in the Environment, Toronto; 1975; 1: 49-68.
88. Wood, J. M. and Hong-Kang Wang. Microbial resistance to heavy metals. **Environ. Sci. Techn.**; 1983; **17**: 582A-590A.
89. Wood, J. M., Kennedy, F. S. and Rosen, C. G. Synthesis of methyl-mercury compounds by extracts of methanogenic bacterium. **Nature**; 1968; **220**: 173-174.
90. Wood, J. M. and Wang, H. K. Strategies for microbial resistance to heavy metals. In: Stumm, W., Ed. Chemical Processes in Lakes: Wiley; 1985: 81-98.

PART IV

HEAVY ELEMENTS IN HUMAN BEINGS

CHAPTER 13

THE HEAVY ELEMENTS IN HUMAN BEINGS

In this last section we will deal with three aspects of the heavy elements in human beings. In this chapter we will review the levels and distribution of the heavy metals in human tissues and fluids, and relate these where possible to sources. In Chapter 14 we will consider the amounts and the ways the heavy elements are taken into the body. In Chapters 15 and 16 we will review the health effects of the heavy elements, their effect on general health in Chapter 15, and the effect of lead on the neuropsychological functioning of children in Chapter 16. Because of the wealth of data available only a brief introduction to the influence of the heavy elements on human beings can be presented.

INTRODUCTION

Measurement of the concentrations of the heavy elements in the human body is achieved, either by acquiring, with a persons consent, material from their body, or post-mortem samples. Samples in the first case include blood, urine, faeces, saliva, sweat, hair, nails, milk and shed teeth. Post-mortem samples can be any part of the body, but the sections most often analysed are; kidney, liver, heart, muscle, brain, lung, bone and pancreas.

There are a number of reasons for monitoring the concentrations of the trace elements in human beings [19]. The most obvious is to estimate the health hazard of a particular element. Relating the concentration of the heavy elements in humans to environmental and occupational levels, is crucial in order to determine areas of health risk. It is therefore useful to have reference values of the heavy metals in human beings, as a base line to estimate the significance of elevated levels. The care in obtaining base line values has been described in Chapter 6. Knowledge of both, body levels of the heavy elements and human intakes of the elements provides data for estimating dose/response relationships. This information is valuable in monitoring health effects, in relation to peoples' intakes. It is essential, for the improvement of public health, to know the concentrations of the heavy elements in human beings in order to gauge the quality of our environment, or the changing habits of people.

In order to achieve some of the above goals it is best to analyse a body fluid or tissue which is, either sensitive to, or accumulates the particular element. For example arsenic is accumulated in hair and nails, cadmium in kidney (up to 40-50 years old), mercury in hair and kidney, lead in bone, teeth and hair. Some of the fluids and tissues that are suitable to use to monitor for the heavy elements are listed in Table 13.1. Also materials that give a reasonable estimate of the body burden of the element are valuable for analysis.

In order to compare data from different studies it is convenient to have the same units, but more particularly the same form of the sample, i.e. wet, dry or ash [24,106]. In wet material the fluids associated with the sample are still present, when dry the fluids have been removed by heating or freeze drying, and ashed material has been obtained by ignition, leaving only the mineral components. To convert concentration data from one basis to another a conversion factor is necessary, which unfortunately, is not a very precise figure. The conversion factors listed in Table 13.2 come from two sources, and there is reasonable agreement between them. There has been no attempt to convert the concentration units used for the reported data except for blood. Concentrations in blood are given as $\mu g\ dl^{-1}$ (µg per 100 ml of blood) and the density of 1.06 g ml^{-1} has been used to convert concentrations by mass to volume.

LEVELS OF THE HEAVY ELEMENTS IN HUMAN BEINGS

We will survey the levels of the heavy elements in the human body, and discuss briefly possible sources. Most work has been done on lead, followed by mercury, cadmium and arsenic. A number of compilations of the elemental content of human beings are available [10,106,121,207].

TABLE 13.1 Useful Tissues and Fluids for use in Biological Monitoring of the Heavy Elements

Material	As	Cd	Pb	Inorg. Hg	Org. Hg	Sb	Tl	Se
Blood	√	√	√	√	√	√	√	√
Bone			√					
Brain			√	√	√			
Faeces		√						√
Hair	√		√	√		√		√
Kidney		√	√	√	√			
Liver		√	√					
Milk								√
Placenta		√	√	√	√			
Teeth			√					
Urine	√	√	√	√	√	√	√	√

TABLE 13.2 Conversion Factors for Wet/Dry/Ash Weights for Some Human Tissues

Tissue	WW from DW		DW from WW		WW from AW	
	a	b	a	b	a	b
Brain	0.21	0.22	4.7	4.6	0.015	0.017
Kidney	0.23	0.25	4.4	4.0	0.011	0.013*
Liver	0.28	0.29	3.6	3.4	0.013	0.015
Lung	0.22	0.21	4.5	4.8	0.011	0.013
Muscle	0.21	0.27	4.7	3.6	0.012	0.013

*Cortex 0.018. Sources; a, Bowen, 1979 [24], b, Hamilton et al., 1972/73 [106].

Lead

Some concentrations of lead in various human tissues and fluids are listed in Tables 13.3, 13.4 and 13.5 at the end of the chapter. There is an abundance of published data, and only a few figures are presented to indicate typical levels.

A feature that stands out is the wide range of concentrations reported for lead in human materials, such as blood, teeth and hair. This is a result of the high contamination of our environment by lead, and that many groups of people are subject to varying exposures, depending on their distance from lead emitting sources, and the extent of contamination of their food. A lot of effort has been put into establishing the external sources of the lead in the human body. Most work has been done on lead in blood, and interpretation of the results with respect to lead sources.

The three materials mentioned above each have their particular analytical problems, which may influence the validity of the data. Blood lead analysis should, in the hands of competent analyst, be no problem. The main difficulty is sampling free of contamination, and in this regard finger prick samples are most likely to suffer from contamination unless extreme care is taken. The results at best give the current exposure, and this could be different a few weeks later. Hence one sampling is inadequate for an investigation of sources of the heavy metal. This becomes a particular problem when attempting to relate blood lead levels, a transitory measure, to neuropsychological functioning of children, which has developed over a long time (see Chapter 16).

Investigation of lead in teeth overcomes, to some extent, this last problem as they provide an historical record of lead intakes for the child. Teeth however, have a matrix in which the lead is far from homogeneously distributed (see Chapter 4) [59,83,84,162,170,204]. Therefore it is difficult to know which tooth, and what part of the tooth to analyse. This may be seen from the data in Table 13.6a for permanent teeth, indicating the range of lead levels that may be obtained, depending on the tooth section or tooth type [170]. The data for the tooth type relates closely to the age of the teeth (either formation or eruption age), i.e. the older teeth have more lead in them ($r = 0.92$, $p < 0.01$ for formation age, $r = 0.96$, $p < 0.001$ for eruption age) (Table 13.6b).

TABLE 13.6 Levels of Lead in Permanent Teeth

a Tooth section

Section	Level μg g^{-1}
Enamel*	7.9
Dentine*	20.4
Root dentine*	24.2
Coronal dentine*	20.7
Circumpulpal dentine*	99
Surface enamel*	1100
Incisor enamel**	
top labial	14.3
neck labial	20.7
top lingual	20.9
neck lingual	29.6

* Seven teeth, ** Seventeen teeth

b Tooth type Ratio of lead concentration in a tooth to the mean lead of a set*

Tooth	Ratio	Eruption age (y)	Tooth	Ratio	Eruption age (y)
Incisor central L**	1.57	6-7	Premolar first	0.90	10-12
Incisor central U**	1.17	7-8	Premolar second	0.80	10-12
Incisor lateral L	1.11	7-8	First molar L	1.36	6-7
Incisor lateral U	0.93	8-9	First molar U	1.74	6-7
Canine L	0.88	9-10	Second molar	0.80	11-13
Canine U	0.84	11-12	Third molar	0.40	17-21

$*\text{Ratio} = \frac{[\text{Pb}]\text{ in tooth}}{[\text{Pb}]\text{ in set}}$ ** L = lower jaw, U = upper jaw. Source of data; Purchase and Fergusson, 1986 [170].

There are even more difficulties in the analysis and interpretation of lead concentrations in hair. This is because hair has two sources of most trace metals internal (endogenous) and external (exogenous), and the separation of these by washing is not reliable. Also results depend on what part of the head the hair comes from, and how far from the scalp. For example, hair-cut samples (the end of the fibre) often have higher concentrations of lead than the hair sampled close to the scalp. The reason for this is that the exogenous contribution increases the longer the hair has been around.

The four major routes for entry of lead into the human body, are food, drink, air and dust. These materials contain both natural and pollution lead. The main sources of the latter are industrial (occupational, family members, and passive recipient of fall-out) lead in petrol, lead in paint, lead glazed vessels and lead solder.

Debate exists as to which lead sources are relevant to human beings. The debate centres particularly around the significance of petrol lead compared with other intakes of lead. The three methods of investigating this problem are; epidemiological, i.e. relating lead levels in humans to lead levels in the environment, lead isotope tracer studies (stable or radioactive), and lead intake versus lead in the body.

Sources of body lead The majority of work has been carried out on the relation between lead in air (PbA) and lead in blood (PbB). Both significant and non-significant correlations have been reported between these two quantities. Discrepancy arises from the confounding effect of the intake of air lead, from routes other than directly from the air [29,206]. Also, the relationship between PbA and PbB is curvelinear as shown in Fig 13.1 The quantity $\alpha = \Delta PbB/\Delta PbA$ originally thought to be constant, varies with both PbA and PbB as shown in Fig. 13.2. It should be noted however, that the data is for rather high values of PbA [36,99,109,206,207,213,230].

The parameter α is better defined as the quantity representing the slope of the PbB/PbA curve at a particular point, as the curve shape depends on the initial value of both PbB and PbA (Fig. 13.2). A second parameter β has been defined as the contribution to PbB from an exposure of PbA of 1 $\mu g\ m^{-3}$ [36]. The two parameters would be equal if the $\Delta PbB/\Delta PbA$ relationship was linear and passed through the origin, otherwise $\beta > \alpha$. The three experimental approaches mentioned above have been used to arrive at values of α (or β). Results from epidemiological studies are difficult to interpret due to confounding factors, such as other sources of environmental lead and the amount ingested [28,29,36,203]. Values of α reported, span <1 to >10, and it appears that figures of 2 for adults and 3-5 for children are reasonable. The value tends to increase the more intense the air lead level [16,28,29,36,39,40,99,101,104,109,175,181, 185,205,206,207,213,214,230]. Lead isotope studies provide estimates of PbB at a particular lead exposure, and therefore estimate β, for which a value of 3 has been suggested [36]. Cigarette smoking is a confounding factor as it is also a source of lead, and can raise the blood lead level [138,167,171,226].

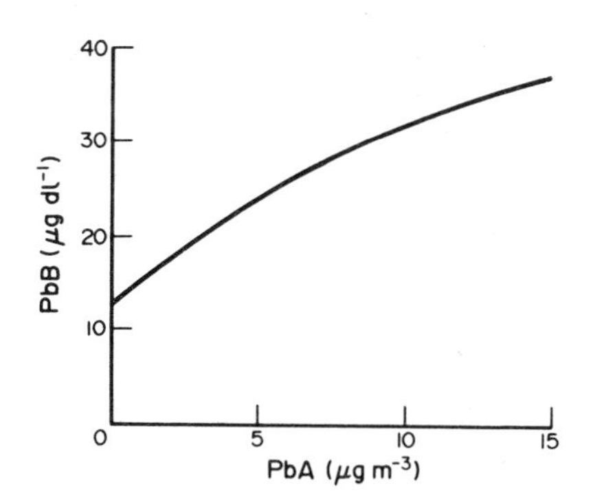

Fig. 13.1 The relationship between PbB and PbA. Sources of data; references 36,99,109,206,207,213,230.

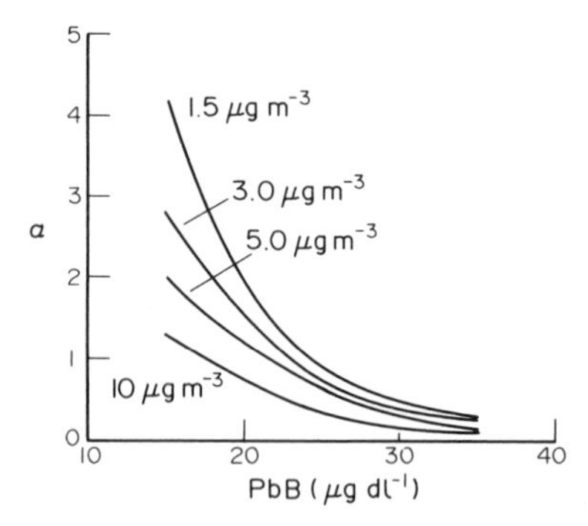

Fig.13.2 Variation of α with air lead and blood lead. Sources of data; references 36,99,109,206,207,213,230.

In some areas lead in water (PbW) is a significant problem, especially where lead plumbing has been used. A curvelinear relationship exists between water lead and blood lead which fits an equation of the type;

$$PbB = A + B(\sqrt[3]{PbW}).$$

The equation is essentially linear up to PbW = 100 μg l^{-1} [72,74,75,150-155, 167,195].

For food lead the relationship between intake and PbB is more difficult to estimate and quantify, because of the heterogeneous nature of food and its lead content. We will be deal with this matter in more detail in the next chapter.

Attempts to estimate the contribution of lead in dust (PbD) to the blood lead of children have been made. Positive correlations exist between soil and dust lead levels and lead in blood. Lead in both materials contribute significantly to the variance in the PbB levels of children [26,27,28,41,42,46,54,97,135,136,137, 141,185,187,188,238]. Two groups of children, one who lived in homes cleaned by wet mopping, and whose hands were frequently washed, and one with no change from usual procedures, were studied [41]. The change in the blood lead levels were quite striking as shown in the Table 13.7. A drop of 6.9 μg dl^{-1} (17.9%) occurred for the experimental group after one year, whereas the drop for the

TABLE 13.7 Levels of Blood Lead Before and After Controlling Dust Intake of Children

Children	Blood lead levels μg dl^{-1}			
	6 mths. before	At start	6 mths. after	12 mths. after
Experimental	37.6	38.6	33.3	31.7
Control	37.5	38.5	38.7	37.8

Source of data; Charney et al., 1983 [41].

controls was 0.7 μg dl^{-1} (1.8%). A relationship between PbB and dust and water lead has been observed for children and is;

$$PbB = \sqrt{(53.26 + 1.03(PbW) + 0.0381(PbD))}.$$

Old paint, where it occurs, is a particularly intense source of lead in dust [141].

The ratio of the change in concentration of lead in blood to the change of the level of lead in dust, i.e.,

$$\delta = \frac{\Delta PbB}{\Delta PbD},$$

provides an estimate of the influence of dust lead. Most values of δ reported range from 0.6 to >10, with a mean around 5, i.e for a lead concentration of 1000 μg g^{-1} in dust or soil the blood lead could be raised by 5 μg dl^{-1} [4,28,47,65-68,132,141,213]. The relationship is probably also curvelinear, i.e. a reduced elevation of PbB at the higher dust lead levels. For example, an increase in soil lead from 1000 to 10,000 μg g^{-1} corresponded to an increase in PbB of 8 μg dl^{-1} [47], indicating a leveling off of the blood lead value.

An alternative approach is to compare the PbB values of children with those of their mothers, especially children in the <1 to 3 years old range, where mouthing activities are common and PbB_{child} levels reach their peak. The ratio.

$$CM = \frac{PbB_{child}}{PbB_{mother}},$$

is usually >1, indicating either an additional source of lead for the children [37,67,68,97], or greater metabolic activity and diet intake of lead per kg of body weight for the child [17,36]. It is likely that all aspects are important.

Industrial lead emissions have been identified, in many cases, as the source of high blood and hair lead levels for people living in the vicinity of the industry [41,44,102,108,115,118,133,134,169,179,183,199,223]. The transfer of the metal, through dust in clothes etc, to the homes of the employees, raises the blood and hair lead levels of the family members. The data in Table 13.4 for hair lead levels shows clearly the difference between employees, their families and the general population [85,115,132].

The result of living in industrialized urban areas compared with remote or rural areas, is an elevation of the blood lead, hair lead and teeth lead levels [88,108,110,159,165,168,179,194]. Some features of urban living that have been associated with elevated levels of lead in the body are; the state of the housing, its age, use of lead paint, position in the city, and habits associated with different cultures or different living conditions [27,54,83,138,187,221].

Petrol lead and blood lead Investigation of the contribution of petrol lead to blood lead has received a great deal of attention. One of the more common methods, is to compare the PbB levels of groups of people in different lead environments, such as inner city with suburban, and urban with rural areas. It is difficult in such studies, to allow for confounding factors, so that the lead levels (environmental and blood) are the only significant variables. The general trend is that inner city people have higher PbB levels than suburban, and suburban higher than rural [7,9,27,58,75,99,132,147,159,190,202,219]. The trends are linked to both petrol and paint lead. The umbilical cord blood of 11,837 infants, born in Boston during a three month period, showed a positive correlation with petrol lead consumption in the city, maternal characteristics, soil and window-sill lead [174,176,177].

People whose work takes them into contact with petrol, e.g. petrol vendors, traffic policemen, taxi-drivers and toll-booth operators have been reported to have elevated blood lead levels relative to controls [16,172,224]. Some studies, however, did not find a significant relationship, and in some cases reverse relationships have been found [3,71,97]. A recent study of lead in deciduous teeth of 996 children [83] demonstrated a relationship between dentine lead and social background, residence in old wooden houses, the length of time of residence along busy streets and pica (Fig 13.3). All the variables were statistically significantly related to the tooth lead, though the factors were small.

Experiments using lead isotopes to identify the source of the blood lead have been carried out with varying success [60,144,145,192,220,222]. The stable isotopes of lead ^{206}Pb, ^{207}Pb and ^{208}Pb are produced from the decay of ^{238}U, ^{235}U and ^{232}Th respectively. The ratios of the concentrations of the three isotopes vary over the world's lead deposits. This is a consequence of the age of the deposits and their origin. Therefore, by using lead from a deposit with a particular lead isotope concentration ratio, it is possible to trace that source through the environment. An experiment in Turin, Italy, was carried out during 1974-1980 [82] where the petrol lead in the area, which had a $^{206}Pb/^{207}Pb$ ratio of 1.18, was replaced with lead from Broken Hill, Australia with the ratio 1.04. Changes in the lead isotope concentration ratio in petrol, aerosol, human blood, soil and

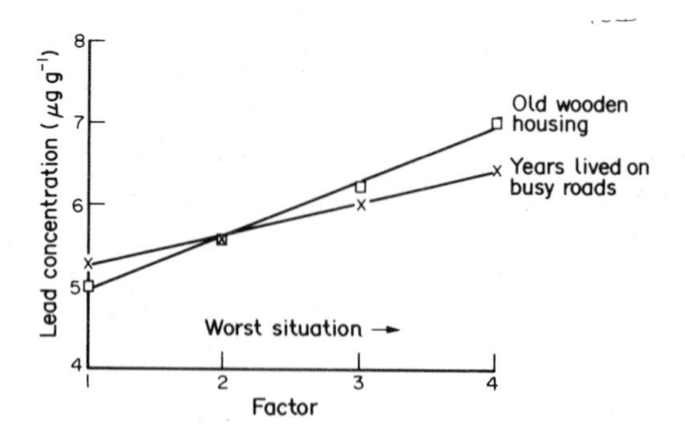

Fig.13.3 The relationship between dentine lead, residence time in old houses, length of time living on busy streets and social advantage. Source of data; Fergusson et al., 1988 [83].

plants was then monitored. Assuming equilibrium had been reached, the average petrol lead contribution to PbB for 35 adults (12 in Turin and 23 in the country) was calculated at 24% in Turin, 11-13% in the country. A three compartment model of the distribution of lead in the body was used to re-estimate the contributions [50] raising them to 26% and 14-17% respectively. Criticisms have been leveled at the experiment, particularly in its execution [38,77,184,230]. A similar experiment in Belgium has given comparable results [33].

A number of nation wide, world wide and large blood lead surveys have been carried out in different parts of the world. The results have been interpreted in a variety of ways [7,8,9,15,21,22,23,27,48,61,62,76,93,114,116,119,124, 138,142,143,160,161,172,185,191,200,201,202,226]. The clear result is that, world wide, there has been a steady drop in blood lead levels from the mid 1970's, see Table 13.8. The fall of around 5-7% yr^{-1} is similar in different parts of the world. In some cases, this fall has been despite no change in petrol lead levels. Some reasons for the change are; improved analytical procedures especially care over contamination, increased awareness by people of the dangers of lead, reduction in the use of lead in materials such as solder in food cans, lead paint and lead in petrol. A study of bone lead in Bavaria [1024] has shown that from 1974 to 1984 the bone lead concentration has fallen especially in younger men. The ratio of the geometric means for the lead concentrations in 1984 and 1974 is around 0.5 for men aged 20 to 40 years.

Cadmium

Some of the levels of cadmium in various human tissues and fluids are presented in Tables 13.9 to 13.11 for blood (Table 13.9), kidney and liver (Table 13.10) and other tissues (Table 13.11). The tables are at the end of the chapter.

TABLE 13.8 The Fall in Blood Lead Levels with Time

Country	Dates	% Fall	% Fall y^{-1}	Comments	Refs.
USA	1970-76	28	5	NY children	21,22,23
	1976-80	37	9	NHANES II	7,8,142,145
	1979-81	22	7	Cord PbB	174,177,176
UK	1979-81	~15	5	National	61
	1970's	~18		Preemployment	160
	1985		6-9	Exposed people	172
	1985		2-5	Controls	172
	1986		9-10	Exposed/controls	172
Wales	1970-80's		5	Three studies	70,73,76
Wales	1984-85	22		Cord PbB	76
N.Z.	1978-85	42	~5	Adults	116,143
	1978-85	40-44	~5	Children	116,143
Sweden	1978-84	34	4	Children	202

The commonly found range for cadmium in whole blood is 0.05-0.5 μg dl^{-1}, with a mean value around 0.1-0.15 μg dl^{-1}. The organ with the highest concentration of cadmium is the kidney, especially the outer layer, called the kidney cortex [45]. For most non-exposed people, the concentration is around 10-200 μg g^{-1} on a dry weight basis and 10-30 μg g^{-1} on a wet weight basis. The cadmium concentration rises with the age of the person up to around 40-50 years old and then it falls [107,166]. This suggest that the kidney accumulates the element, and a typical plot of concentration against age is given in Fig. 13.4 [166].

It is evident that the levels of cadmium in a number of human tissues are influenced by the smoking habits of people, the cadmium being elevated with smoking. This can be seen from the data on cadmium in blood [45,93,215, 216,226], kidney and liver [91,107,215], placenta [52,232] and human milk [57].

Cadmium concentrations also respond to different exposures that people experience, such as occupational, near to cadmium sources, urban (non-occupational), and rural. Results on human blood [110,128,134,183,215,223] demonstrate this (see Table 13.9). However interestingly, in the high cadmium area of Shipham in England, levels are not too different to those of the control village [216]. Other tissues where the difference between exposed and unexposed is obvious are kidneys and liver [25,215], hair [44,48,149], lungs [25] and urine [183,215,216]. The high levels of cadmium in kidney cortex (56.2 μg g^{-1} wet weight) [226] reported for Japanese, reflects elevated environmental concentrations of the metal in Japan. The cadmium is somewhat higher in the blood of pregnant women compared with their new born children, reported means being 0.15 and 0.10 μg dl^{-1} respectively. The placenta appears to act as some barrier to the transfer of cadmium [140,182].

It appears from the study of renal cortices from 1897 to 1981, that contemporary people (1980-1981), have significantly higher levels of cadmium, (40.67 μg g^{-1}, wet weight) compared with the period 1897-1937 (0.79 μg g^{-1} (wet weight) [49]. This difference is not obvious from the concentrations of the metal in the liver.

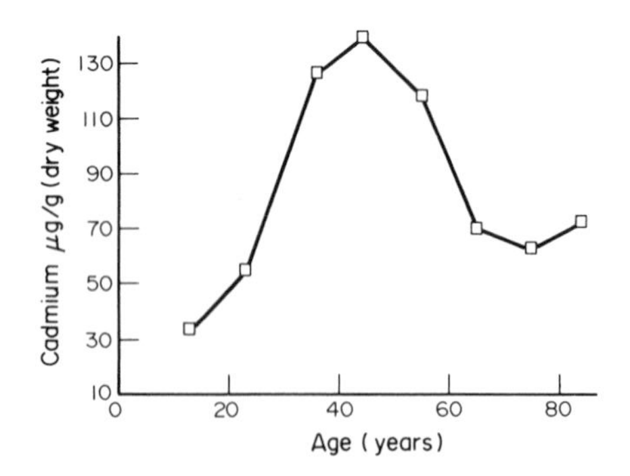

Fig.13.4 The variation of cadmium concentration in kidney with age. Source of data; Piscator and Lind, 1972 [166].

The concentrations of cadmium in the bones of stillbirths are elevated, compared with autopsy samples of infants that survived birth [30]. A wide range of cadmium concentrations are reported in teeth (Table 13.11), and it could be that some values are high due to contamination. It has been noted that the metal is highest in the tips of the teeth, the portion formed early in the life of the tooth [164]. Cadmium is not distributed homogeneously in lung tissue, suggesting that it originates from the intake of aerosol particles which have lodged in the lungs. Evidence for this is that the levels are higher in older people, and there is a fall off in concentration towards the upper parts of the lung [229].

Two groups who have studied cadmium in sweat differ in the concentrations found [180,211,212]. The difference may arise from contamination problems, or the different ways used to produce the sweat, such as thermally (e.g. sauna) or pharmacologically (e.g. use of the alkaloid pilocarpine). The rate at which sweat is produced may also be relevant.

Cadmium is trapped in soft tissue primarily by bonding to sulphur rich proteins, such as metallothionein. The protein metallothionein is a low molecular weight protein (RMW ~10,000), with about 9% sulphur content due to the high number of cysteine aminoacid residues in the protein. It is a suggested that the levels of cadmium are controlled by the extent of saturation of the bonding sites in the metallothionein. Only after prolonged exposure does the cadmium in urine start to reflect the external exposure. This is shown by the data plotted in Fig 13.5, where an increase in the air cadmium levels are reflected in increased cadmium urine levels, but only after 6 months exposure [12].

Some concentrations of cadmium, found in tissues of Japanese women with Itai-Itai disease, are shown in Table 13.12 [94]. The disease, to be discussed in Chapter 15, occurred in Japan, among older women who had long exposure to cadmium as well as other nutritional problems. Comparison of the data in the table with those in the previous tables suggests a high elevation of the metal. In this situation the liver had the highest concentrations of cadmium.

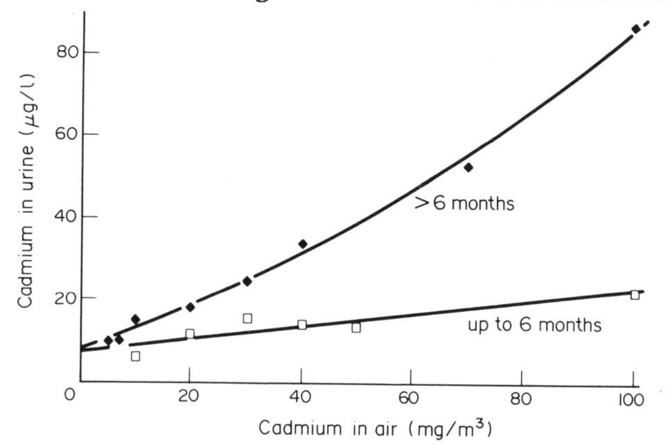

Fig. 13.5 A relationship between cadmium in air and urine. Source of data; Atessio and Bertelli, 1983 [12].

TABLE 13.12 Levels of Cadmium in Tissues of Women who had Itai-Itai Disease

Tissue	Concentration $\mu g\ g^{-1}$ (wet weight)
Bone	1.6-2.8
Brain	0.6
Liver	63.3-132
Lung	2.1-8.0
Muscle	14.1
Pancreas	5.2-64.7
Renal cortex	12-41.1
Skin	3.9-5.1

Source of data; Friberg et al., 1971 [94].

Mercury

Some reported levels of mercury in human blood, kidneys and lungs are listed in Table 13.13, in hair in Table 13.14 and in other tissues in Table 13.15, and are at the end of the chapter. The levels of mercury in the blood of pregnant mothers and their newborn babies indicate the placental barrier is weak towards mercury [182]. Up to 57% of mercury in the placenta has been estimated to be CH_3Hg^+ [53]. Whereas the concentrations of the metal in the blood of mothers and babies are similar, this is not reflected in their mercury hair levels. The concentration is less in the hair of the babies [197]. This may suggest either a significant amount of the mercury in the adult human hair is exogenous or metabolic processes for the child had not yet produced deposits in the hair.

The concentration of mercury in human tissues such as hair, kidney and liver were distinctly elevated for Minamata patients [146]. Progressive fall in the mercury levels occurred after the pollution into the bay had stopped, as shown by the concentration data for mercury in the hair of fishermen (see Table 13.14) [96]. The concentration of mercury in the brain of Minamata patients was very high, at least 60 times greater than for healthy people. A number of studies have demonstrated that the mercury concentrations found in human tissues of Japanese are elevated over those of people from many other parts of the world [157].

The elevated concentration of mercury in the blood of the Yanomamo Indians from South America compared with urban USA was unexpected, and suggests either a local or an imported food effect. The levels were highest in the males [110].

Hair has been frequently used as a means of estimating the mercury status of a person. Mercury levels in the hair appear to reflect eating habits, especially as regards fish consumption, for example the mercury in the hair of Eskimos [102] and Japanese (living in South America) [225] is high compared with USA males. Mercury concentrations either increase along the fibre or show little variation, depending on the circumstances [43,158]. Treatment of the hair with products containing mercury compounds (such as mercuric iodide) raise the levels of mercury to very high values [98]. The difference in mercury in hair concentrations between urban and rural populations [188] may reflect different environmental exposures, and also eating habits.

In some studies effort has been made to differentiate between inorganic and organic mercury in human tissues [53,90,217,228,235]. The reason for estimating this difference is that, organomercury is more toxic than inorganic mercury, hence knowledge of its concentration is important, in order to reveal the toxic potential of the element.

Arsenic

Some concentrations of arsenic in hair and other tissues are listed in Tables 13.16 and 13.17 respectively at the end of the chapter. Hair is the most frequently studied tissue for estimation of arsenic, as hair and nails are significant accumulator tissues for the element. For unexposed people a level within the range 0.02-2 $\mu g\ g^{-1}$ can be expected. Unless a person has been exposed to varying amounts of arsenic, there does not appear to be any significant change in the arsenic levels along hair fibres [43,98].

The arsenic in blood serum, responds to the arsenic content of food, especially fish intake [112]. The high fish intake of Eskimos may be the reason for them having five times more arsenic in their hair, compared with USA citizens [102]. The consumption of flat fish and crustacean, more than once a week, for Swedish people raised their arsenic levels to around 40 $\mu g\ g^{-1}$ (creatinine), compared with around 12 $\mu g\ g^{-1}$ (creatinine) for people who seldom ate seafood [227].

Both urine and hair have been used as bio-indicators (see Tables 13.16 and 13.17) for arsenic exposure in the work place and in the homes of workers [100,108,122,134,140]. The transfer of the element from the work-place to the home is seen in the elevated levels of arsenic in the urine of the children of workers in a copper metallurgy plant, i.e. 31.03 $\mu g\ l^{-1}$ (0-240 $\mu g\ l^{-1}$) compared with 6.41 $\mu g\ l^{-1}$ (0-54 $\mu g\ l^{-1}$) for children who are not exposed [100]. However, no difference was found for children living near a glass works which emit arsenic in Sweden, compared with controls [2]. On the other hand parental smoking had an effect on the children's urine arsenic levels, particularly when both parents smoked (Table 13.17) [2].

Selenium

Some reported levels of selenium in human tissues are listed in Table 13.18 at the end of the chapter. In a number of tissues, e.g. hair and blood, a wide range of values have been reported. There is evidence for a relationship between tissue levels and the natural concentrations found in the environment. In

Finland and New Zealand, where the natural levels of selenium are low the levels in the tissues of the citizens of the countries also tend to be low [1,20,117,233]. This is reasonable if the main source of selenium comes from food intake.

Unless hair has been treated by a preparation that contains selenium (for the treatment of dandruff), levels do not necessarily vary uniformly along the hair fibre [98,158]. Selenium appears to be high in various eye tissues, irrespective of wide variations between individuals, i.e. 0.48 to 40.4 $\mu g\, g^{-1}$ (dry weight) [237].

Antimony

Some levels of antimony in human tissues are listed in Table 13.19 at the end of the chapter. The concentrations found in hair can reflect the the distance of people from industrial operations [44]. The blood antimony concentrations of policemen employed as arms firing instructors were elevated compared with unexposed people [20]. On the whole, however, the levels of antimony in human tissues reflect food intakes, and only in special cases of exposure are the levels elevated.

Bismuth, Indium, Thallium and Tellurium

Some of the reported concentrations of the four elements bismuth, indium, thallium and tellurium are given in Table 13.20 at the end of the chapter. Available data on the elements, especially indium, thallium and tellurium are rather sparse.

HEAVY ELEMENTS IN A REPRESENTATIVE PERSON

It is clear from the above data, that it is not easy to choose concentrations of the heavy elements for a representative person, because of the wide range in the reported figures. It is also clear that the variations are often due to pollution. In the case of lead, a representative person is a modern person who has not been heavily exposed, but this does not represent the levels expected for a 'natural person'. Some suggested typical ranges for the ten elements in some human tissues, for a the person not unduly exposed to the elements, are listed in Table 13.21. The last column in the table gives the probable body burden for a person around 70 kg in weight.

A further aspect which influences the concentrations, is the length of time the elements remain in the body. This is expressed as an elements biological half-life. Some values are given in Table 13.22. Some elements are removed from the body in a matter of hours, such as arsenic, others including cadmium and lead, take many years to be removed. Also the removal time varies with the tissue, generally it takes longer from hard tissue compared with soft, and usually longer from soft tissue than body fluids. The removal is a function of the elements mobility in the body, which in turn is influenced by the chemical speciation of the element.

TABLE 13.21 Representative or Typical Levels of the Heavy Elements in Human Tissue (Representative Person)

Metal	Blood μg dl^{-1}	Kidney μg g^{-1} ww	Liver μg g^{-1} ww	Muscle μg g^{-1} ww	Hair μg g^{-1} dw	Bone μg g^{-1} dw	Body burden[a] mg/70 kg
Cd	0.05-0.5	10-30	2-3		0.5-2	0.5-2	50
Hg	0.2-2.0	0.1-2	0.01-1	0.02-0.5	0.5-2.0		13
In				0.004			
Tl	0.01-0.5	0.-0.1	0-0.2	0.017	0.02	0.002	8
Pb	1-25	0.2-1.5	0.2-1.5	0.01-0.5	1-20	0.2-10	40-120
As	0.1-1	0.005-0.3	0.02-0.2		0.02-2.0	0.08-1.6	18
Sb	0.3	0.005-0.1	0.005-0.3	0.005-0.2	0.05-1.0	0.01-1.5	8
Bi	<0.3	<0.1-1	<0.1-0.5	0.01		<0.2	
Se	5-25	0.2-1.5	0.24-0.4		0.5-1.0	1-9	13
Te	0.55?	0.018	0.003	0.004			

a Sources of data; references 24,25,49,51,63,64,89,91,189,193,203.

TABLE 13.22 Biological Half-Lives for Some Heavy Elements

Element	Body	Bone	Soft tissue
Cd	10-30 y		10-30 y
Hg	30-90 d		45-160 d
In	14-74 d		
Pb	5 y	20 y	21 d
As	10-30 h		

Sources of data; references 12,25,95,148,189,235.

TABLE 13.3 Some Levels of Lead in Human Blood(μg dl^{-1})

Place	Mean	Range	Comments	Ref.
Amazon	0.83	0-3.87	Amazon indians	110
Australia		13-23	City children	168
Belgium	26.4	15.7-45.8	<1 km from smelter	183
	13.6	9.4-25.4	2.5 km from smelter	183
	9.1	4.5-16.2	Rural children	183
	10.2	3.10-31.0	Pregnant mothers	182,140
	8.4	2.7-27.3	Newborn babies	182,140
	13.7		World survey	48
Brazil	58.9		Around smelter 1980	199
	36.7		Around smelter 1985	199
Britain	12.8		Inner city people	138
	11.0		Outer city people	138
	15.3	11.5-20.6	High lead/soft water	167
	12.8	7-27	Children	204
		7.9-13.2	Range of means	75
	31.8			106
	5.4	1.38-8.09	Controls	223
	7.49, 9.72		Steelworkers	223
	53.4, 50.3, 11.95		Non-ferrous workers	223
Canada	12.02*		Urban children	159
	9.95*		Suburban children	159
	8.91*		Rural children	159
China	41.8	24.0-63.0	Smelter workers	118
	10.21, 7.64		M/F non smokers	171
	12.43, 12.06		M/F smokers	171
Himalayas	3.4, 3.5		Adults, children	165
Holland	16.1*		Children	26
		7.9*-13.1*	Mean range, children	27
India	8.9*	2.9-31.2	Suburban children	128
	11.5*	2.9-47.7	Central city	128
	14.4	2.9-41.2	Adults	128
Island (Irish)	8.9, 7.9		Men and women	138
Malta	24.3		World survey	48
Mexico	19.5		World survey	48
New Zealand		17.0-9.7	Men, fall 1978-85	116
		11.4-6.4	Women, fall 1978-85	116
	10.2*	3.9-50.0	Children	198
Papua & NG	5.2	1.0-13.0	Highland children	168
Scotland	10.7	3.3-33.6	Children	141
Sweden	5.9		Part world survey	48
	5.47	1.4-25.0	Children	202
	6.0**	2.0-41.0		218
	3.44*	1.5-8.9	Near glass works	2
USA	14.6	<1-81	Urban people	110

Table 13.3 continued

Place	Mean	Range	Comments	Ref.
USA	15.0, 15.2		Children	165
	21.5	19.3-26.3	Range of means	6
		2-70	NHANES II survey	9
World		6.4-22.0	Mean range, males	226,93
		31.8-36.0	Exposed to traffic	51
	21			24
	12.7	3.7-47.7	Exposed to lead	215
	≥37.1	8.5-180.2	Occupat. exposed	215

*Geometric mean, ** Median. Other references 12,19,23,35,36,41,68,76,78,80,86,113, 115,132,133,134,143,151,152,169,172,185,187,193,195,216,221,231.

TABLE 13.4 Some Levels of Lead in Human Teeth, Bone and Hair ($\mu g\ g^{-1}$)

Place	Mean	Range	Comments	Ref.
Teeth				
Britain	4.30*, 4.34*		L jaw, 1st incisors	204
	5.68*, 5.58*		U jaw, 1st incisors	120
	4.1	1.0-10.3	Crowns	59
	5.9	1.4-26.9	Roots	59
Denmark	25.7	17.0-42.4	Sec. dentine	103
India	8.31**	2.0-28.8		127
	3.01**	0.22-14.4		127
Japan	7.20	1.12-64.7	Various areas	126
New Zealand	5.7**	1.4-27.1	Deciduous dentine	88
	6.24, 6.11		Deciduous dentine	87
	5.5		Whole enamel	56
Norway	3.73	1.60-6.98	Range of means, cities	92
		1.66-6.15	Range of means, rural	92
	5.6	1.5-13.4	Children	13
	2.4	0.9-7.8	Adult teeth	13
Scotland		1.04*-2.5*	Range of means	156
	9.12	0.6-89.7	Deciduous teeth	162
USA	92.3	2.8-313.6	Sec. dentine, sections	194
	19.6		Whole enamel	55
	3.1		Whole enamel	56
	16.6		Californians	131
	27.6		Hopi Indians	131
Mexico (Ind.)	4.3	0-35.3	Sec. dentine, sections	194
Alaska	56.0	9.6-97.8	Sec. dentine, reamed	194
Bone				
Asia	30*			51
Asia minor	26*			51
Britain	5.7	0.4-24.2	Stillbirths, ribs	30
	2.9	0.2-13.2	Stillbirths, vertebra	30

Table 13.4 continued

Place	Mean	Range	Comments	Ref.
Denmark	5.5	3.7-8.8		103
Japan	0.57**			130
Sth. Bavaria	5.59	0.95-35.22	Temporal bone, adults	63
	3.86	0.78-20.05	Mid femur, adults	63
	1.65	0.26-10.16	Pelvic bone, adults	63
		3.3-17.6	Temporal bone, children	64
		3.3-11.8	Mid femur, children	64
		2.6-6.3	Pelvic bone, children	64
USA	43*			51
	3.1			49
	72*	14-214	Blacks	107
	64*	11-134	Whites	107
	4.8*	0.2-32	Children	107
	4.1*	0.3-15	Children	107
		3.6-30		24
Other	29.0*, 34.5*		Hard & soft water areas	106
	31.0*			10
Hair				
Austria		0.97-44.9		18
Canada	10.1	0.5-25	Rural	44
	16.9	0.5-35	Urban	44
	45.2	10-350	Near smelter	44
	7.7		Children, exposed	149
	4.9, 4.1		Children, not exposed	149
	12.0		Adults, exposed	149
	3.4, 5.3		Adults, not exposed	149
Greenland	5.96			49
Japan	3.4			217
New Zealand	12.8	2.0-360	Levels relate to occup.	179
		1050-2410	Use hair preparations	179
	10.4	1.2-111	City survey	85
	363	124-1381	Lead workers	85
	67.0	7-313	Lead workers families	85
USA	16.2		Children	234
	6.55		Adults	234
		7.6-107.1	Inc. exposure areas	108
	13.4**	2.1-100	Children	54
	12.2**	2.0-155	Adults	54
	36		Maternal scalp hair	228
	14		Maternal pubic hair	228
	13		Neonatal hair	228
Other	1.0	0.05-15.0	Near roots	215
		3-70		24

* Ash weight, ** Geometric mean. Other references 84 and references therein, and 14,31,49,115,164,170,189,209210,236.

TABLE 13.5 Some Levels of Lead in Various Human Tissues and Fluids

Tissue	Mean	Range	Comments	Ref.
Blood		See Table 13.3		
Bone		See Table 13.4		
Brain $\mu g\ g^{-1}$		7.4-11.0w	Ethyllead poisoned	111
	0.3w			106
		5-14a	Range of means	51
Hair		See Table 13.4		
Heart $\mu g\ g^{-1}$		8.0, 9.4w		106
		5-24a	Range of means	51
Milk $\mu g\ l^{-1}$	1.04	<0.05-15.8		57
Kidney $\mu g\ g^{-1}$		1.2-6.8d		24
		36-98a	Range of means	51
		1-12d		166
	52a	15-246a	Black adults	107
	36a	15-204a	White adults	107
	26d	1.4-74a	Children	107
	0.29, 0.50w		Exposed people	25
	0.31w		Controls	25
		7.9-19.0w	Ethyllead poisoned	111
	1.4w		Whole kidney	106
	1.3w		Kidney cortex	106
	1.1w			10
Liver $\mu g\ g^{-1}$		3-12d		24
		59-160a	Range of means	51
	96a	15-517a	Black adults	107
	69a	15-788a	White adults	107
	40a	0.7-167a	Children	107
	0.41, 0.77w		Exposed people	25
	0.37w		Controls	25
		23.5-41w	Ethyllead popisoned	111
	2.3w			106
	1.5w			10
	1.46w	0.5-3.5w	N. Z.	163
Lungs $\mu g\ g^{-1}$		28-47a	Range of means	51
	0.22w	0.04-0.86w	Males	208
	0.19w	0.02-0.89w	Females	208
	0.17, 0.24w		Exposed people	25
	0.095w		Controls	25
	0.065w	0.0068-0.17w		229
	0.4w			106
	0.7w			10
Muscle $\mu g\ g^{-1}$		0.23-3.3d		24
		8.0w	Ethyllead poisoned	111
	0.05w			106
	0.2w			10
Nail $\mu g\ g^{-1}$		14-170d		24

Table 13.5 continued

Tissue	Mean	Range	Comments	Ref.
Nail μg g^{-1}		2.22-15.4d	Fingernail	18
		0.91-5.52d	Toenail	18
Pancreas μg g^{-1}		12.8w	Ethyllead poisoned	111
		24-49a	Range of means	51
Placenta μg g^{-1}	0.075w	0.011-0.395		182
Plasma μg dl^{-1}	13?			24
		0.002, 0.2		80
		0.54-4.42	Range of means	34,35
	4.6			189
Red cells μg dl^{-1}	46			24
		58.0-103.7	Occup. exposed	115
		30.1-24.9	Dec. 1974-81 F	113
		37.3-33.1	Dec. 1974-81 M	113
Sweat μg dl^{-1}	0.41	0.06-0.87	Males	211
	0.24	< 0.05-0.7	Females	211
	0.15		Males	212
		0.08-0.25	Females	212
Teeth		See Table 13.4		
Urine μg l^{-1}		20-150	Exposed people	215
	≤20	5-60	Unexposed people	215
	36			189
		23.9, 26.1	Taxi drivers	36
		18.7	Office workers	36
	84	20-250	Lead workers	118
Creatinine μg g^{-1}	27.4	5.2-70.6	<1 km from smelter	183
	8.15	3.3-20.6	2.5 km from smelter	183
	7.35	2.4-13.6	Rural	183
	52	19-115	Lead workers	118

a = ash weght, d = dry weight, w = wet weight

TABLE 13.9 Levels of Cadmium in Human Blood (μg dl^{-1})

Persons	Mean	Range	Comments	Ref.
Adults		0.06-15.9	Most <1μg dl^{-1}	91
	0.52			24
	<0.1	0.02-0.3	Non-smokers	215
	<0.2	0.02-0.5	Smokers	215
	≥15		Occup. exposed	215
		<0.05-0.12	Non-smokers	93,226
		<0.1-0.39	Smokers	93,226
		<0.05-0.12	Ten countries	93,226
	0.114	0.018-0.336	Controls	223
	0.05		Steel workers	223
	1.05		Non-ferrous work.	223

Table 13.9 continued

Persons	Mean	Range	Comments	Ref.
Adults	1.71	<0.1-9.6	Contemp. urban	110
	0.57	0.07-3.7	Amazon indian	110
	0.07		Shipham village	216
	0.10		Controls	216
	0.16		Smokers	216
	0.08		Non-smokers	216
	0.07		Never smoked	216
	0.19	0.05-0.91		128
	0.33*	<0.2-5.74		218
	0.15	0.01-1.01	Pregnant mothers	140,182
Children	0.10	0.01-1.03	New born children	140,182
	0.11	0.02-0.41	<1km from smelter	183
	0.06	0.01-0.27	2.5km from smelter	183
	0.07	0.01-0.24	Rural	183
	0.30	0.12-0.50	Near Zn smelter	134
	0.21	0.12-0.33	Near Pb smelter	134
	0.20	0.18-0.22	Control town	134
	0.21	0.03-4.1		128
	0.17	0.05-0.84	Suburban city	128
	0.26	0.03-4.1	Central city	128

All concentrations have been expressed as $\mu g\ dl^{-1}$ (μg in 100 ml of blood). CEC health hazard limit is 0.5 $\mu g\ dl^{-1}$ [19]. * Median.

TABLE 13.10 Levels of Cadmium in Kidney and Liver

Material	Mean	Range	Comments	Refs.
Kidney $\mu g\ g^{-1}$				
Kidney	4.16mg		Total Non-smokers	91
	10.28mg		Total smokers	91
		18-310d		24
	3d		Non-smokers	215
	6d		Smokers	215
		up to 30d	Occup exposed	215
		11-50d	Range of means	89
		11-22d		89
	13.9w			106
	35.1w			10
Renal cortex		10-130d	Unexposed	215
		20-500d	Exposed	215
		9-30.5w	Range of means	20,93
	56.2w		Japanese	20,93
	0.56w		Period 1897-1914	49
	0.79w		Period 1897-1937	49

Table 13.10 Continued

Material	Mean	Range	Comments	Refs.
Renal cortex	40.67w		Period 1980-1981	49
	50w			89
		33-118d	Inc. with age	166
		188-2696a	Inc. with age	107
	2068a		Non-cancer	107
	3237a		Cancer	107
	1286a	710-1823a	Never smoked	107
	2132a	1303-2891a	<1/2 pack d^{-1}	107
	2795a	1752-4771a	1/2-1 pack d^{-1}	107
	2812a	1562-4000a	>1 pack d^{-1}	107
	14.3w			106
	2.7w		Controls	25
	8.7w		Exposed, retired	25
	19.3w		Exposed	25
Liver μg g^{-1}				
	2.28mg		Total Non-smokers	91
	3.06mg		Total smokers	91
		2-22d		24
		1-3d	Unexp., non-smoker	215
		2-6d	Unexp., smoker	215
		5-15d	Exposed	215
		2-3w	Range of means	89
	1.33w		Period 1897-1914	49
	1.06w		Period 1897-1937	49
	1.15w		Period 1980-1981	49
		39-282a	Inc. with age	107
	7d			79
	0.1w			106
	0.27w		Controls	25
	0.82w		Exposed, retired	25
	1.45w	0.2-6.2w	N. Z.	163
	1.4w		Exposed	25
	4.30w			10

a = ash weight, d = dry weight, w = wet weight

TABLE 13.11 Levels of Cadmium in Tissue and Fluids of Human Beings

Material	Mean	Range	Comments	Refs.
Blood		See Table 13.9		
Bone μg g^{-1}d	1.8			24
	0.25			215
	5.2	0-31.5	Rib, stillbirth	30
	2.2	0-24.4	Vertebra, stillbirth	30
		0.03-0.95	Adults	30

Table 13.11 continued

Material	Mean	Range	Comments	Refs.
Bone μg g^{-1}d	3.6a		Hard water area	106
	0.06		Contemp. Japanese	130
	4.8a		Soft water area	106
Brain μg g^{-1}w	0.3			106
	<0.8			10
Eye tissue μg g^{-1}w		0.18-8.6, 72	Range of tissues	237
Faeces μg g^{-1}w		0.1-0.8		215
Hair μg g^{-1}w		0.24-2.7		24
	<0.5	0.1-2	Close to scalp	215
		0.5-2.5		228
	0.37		Greenlanders	49
	3.5, 2.0		Near Pb/Zn mining	108
	1.3, 0.9		Low exposure area	108
	4.1	0.45-8.2	Children, smelter	44
	2.0	0.32-3.4	Children, urban	44
	1.2	0.25-2.7	Children, rural	44
		0.8-1.8		189
	0.88*	0.14-6.90	Children	54
	0.76*	0.08-8.73	Adults	54
	1.11	0.60-1.45	Washed hair	233
	1.47	0.57-4.91	Unwashed hair	233
	2.98		Exposed adults	149
	0.45		Exposed children	149
	0.42, 0.24		Unexposed adults	149
	0.22, 0.35		Unexposed children	149
Human ng g^{-1}w	0.08	<0.002-4.05		57
Kidney		See Table 13 10		
Liver		See Table 13.10		
Lungs μg g^{-1}w	0.80	0.12-4.24	Adult males	208
	0.81	0.09-3.10	Adult females	208
	0.62		Exposed worker	25
	0.19		Exposed, retired	25
	0.019		Controls	25
	0.108	0.01-0.472	Not homogeneous	229
	2.3			106
Muscle μg g^{-1}d	0.03			106
		0.14-3.2		24
		0.02-0.3		215
	<0.9			10
Nail μg g^{-1}d		0.08-3.4		24
		0.02-0.48	Toe nail	18
		0.15-2.70	Finger nail	18
Pancreas μg g^{-1}d		0.01-0.80		215
Placenta μg g^{-1}d		0.02-0.16		215
	0.632	0.249-1.69	Smoking inc. Cd	232

Table 13.11 continued

Material	Mean	Range	Comments	Refs.
Placenta $\mu g\ g^{-1}d$	49	25-91	Non-smokers	52
Pancreas $\mu g\ g^{-1}d$	83	41-158	Smokers	52
$\mu g\ g^{-1}w$	0.011	0.0025-0.079		182
Saliva ng $g^{-1}w$	<2			215
	3.5	0.5-17	Cd and Zn high	139
Sweat $\mu g\ l^{-1}$	1.4	<0.5-10	Males	211
	2.6	<0.5-18	Females, oral contr.	211
	2.4	<0.5-5.5	Females, no contr.	211
	<3		Males and females	212
	84		Males	180
Teeth $\mu g\ g^{-1}d$	<0.5	0.05-2.5		215
		1.4-4.7	Urban	236
		3.1-4.0	Suburban	236
		1.8-1.9	Tooth crowns	209
		1.5-4	Tooth roots	209
	0.12	0.01-1.25	Japanese	126
	0.36, 0.40		Controls & retarded	164
		1.8, 1.4, 2.5	Tooth tips	164
	3.3		NZ, enamel	56
	0.99		USA, enamel	56
		1.9-4.7	Enamel, stillbirths	210
		0.9-3.7	Dentine, stillbirths	210
	0.2, 0.4		Ancient Norwegians	14
	1.87	0-27.0	Enamel	55
	0.10	0.04-0.24	Deciduous teeth	13
	0.10	0.03-0.51	Permanent teeth	13
Urine $\mu g\ l^{-1}$	<0.5	0.1-3.0	Unexposed people	215
		≥100	Exposed people	215
	1.2	0-4.5	Amazon Indians	110
	12, 44			189
	5		Health hazard limit	19
	0.59		Shipham, adults	216
	0.40		Controls, adults	216
	0.43		Shipham, children	216
	0.50		Controls, children	216
Creatinine	1.06	0.51-2.25	<1 km from smelter	183
$\mu g\ g^{-1}$	0.76	0.25-1.60	2.5 km from smelter	183
	0.23	0.05-0.79	Rural	183
	0.70		Shipham, adults	216
	0.62		Controls, adults	216
	0.44		Shipham, children	216
	0.86		Controls, children	216

d = dry weight, w = wet weight, * geometric mean.

TABLE 13.13 Some Levels of Mercury in Human Blood, Kidneys and Liver

Material	Mean	Range	Comments	Refs.
Whole blood	0.78			24
$\mu g\ dl^{-1}$		0.9-1.1		20
		0.5-2.0		125
		0.4-65	Exposed to $MeHg^{+}$	235
		4-550	Exposed to $MeHg^{+}$	235
	0.95	0.5-2.05	Urban adults USA	110
	1.43	0.3-5.7	Amazon Indians	110
	1.26	0.01-4.70	Pregnant women	140,182
	1.42	0.01-7.05	Newborn babies	140,182
	0.45*	0.18-2.73	Swedish people	218
	0.49*	0.19-2.58	Swedish people	218
	0.6			106
	0.5			10
Kidney		0.3-12d		24
$\mu g\ g^{-1}$		1.26-1.31w		20
		0.31, 1.55w	Exposed people	25
	0.12w		Controls	25
	0.09w			25
	2.75w	0-26.3w		123
	1.3w		6% organo Hg	53
	1.08d	0.16-4.42	Swedish people	157
	0.22w	0.036-0.98w	Swedish people	157
		0.091-2.1w	Range of means	157
	40.6w	3.1-144w	Minamata patients	146
		0-26w	A number of studies	95
Liver		0.018-3.7d		24
$\mu g\ g^{-1}$		0.027-0.037w		20
		0.005-3.7d		125
	0.10w		Controls	25
		0.06, 1.65w	Exposed people	25
		0.055-0.106w		25
	0.30w	0-0.9w		123
		0.14-0.24w	40% organo Hg	53
	0.34d	0.41-1.01d	Swedish people	149
	0.096w	0.0087-0.20w	Swedish people	157
	0.26w	<0.02-1.35w	N. Z.	163
	24.3w	0.3-70.5w	Minamata patients	146
		0-4.0w	A number of studies	95

*Median, d = dry weight, w = wet weight

TABLE 13.14 Some Levels of Mercury in Human Hair (μg g^{-1})

Mean	Range	Comments	Refs.
	1.2-7.6		24,125
0.28			43
	1.80-4.8	Along fibre	69
	1.71-5.2	Along fibre	69
	1.17-2.48	Along fibre	235
1.0	0.3-1.4	Amazon Indians	110
2.8		Females, 1972	49
9.8		Contemp. Greenlanders	49
	1-325	Long exposure to $MeHg^+$	235
9.2	2.6-73.8	Minamata fishermen, 1968	96
5.5		Minamata fishermen, 1969	96
3.7	1.2-9.5	Minamata fishermen, 1970	96
2.1		Inorganic mercury	217
3.1		Organic mercury	217
1.55, 1.08			102
	10.4-8870	Along fibre, treated hair	98
6			189
0.67	0.048-11.3	Children	54
0.77	0.05-14.0	Adults	54
3.31	1.0-8.07		233
1.88	0.02-40.6	Maternal scalp hair	197
1.01	ND-31.86	Maternal pubic hair	197
0.11	ND-0.62	Neonatal scalp hair	197
	1.57-2.40	Sth. American Japanese	225
	2.32-6.22	Sth. Amer. Japanese fish eating	225
2.01	0.42-6.08		196
5.7		Total Hg, Singapore	90
2.7		Inorg Hg, Singapore	90
3.0		Org Hg, Singapore	90
1.26	0.20-4.29	Swedish people	157
	1.26-4.2	A number of countries	157
	0.5-2.0	Frequent range	157
	0.84-3.27		18
1.88	0.37-9.4	Korea	129
1.73	0.17-8.80	Pakistan	173
4.34		Rural Malaysia	186
8.98		Urban Malaysia	186
138.2	2.46-705	Minamata patients	146
2.29	0.14-7.49	Outside Minamata	146
	1.8-8.8	Range of countries	95
	0.76-3.0		
	1.0-2.3	Alkyl Hg	228

TABLE 13.15 Some Levels of Mercury in Human Tissues and Fluids

Material	Mean	Range	Comments	Refs.
Blood		See Table 13.13		
Bone $\mu g\ g^{-1}$	<0.7			106
	0.45			24,125
Brain $\mu g\ g^{-1}$		0.29-0.38w		20
		0.005-2.94d		125
	0.10w	0-0.6w		178
		0.1-0.4w	39% $MeHg^+$	53
	6.2w	0-24.8w	Minamata patient	146
		0-3.0w	Range of countries	95
Hair		See Table 13.14		
Heart $\mu g\ g^{-1}$		0.013-0.015w		20
		0.005-0.15d		125
		0.14-0.24d	40% $MeHg^+$	53
Human $ng\ g^{-1}$	8.8			32
Kidney		See Table 13.13		
Liver		See Table 13.13		
Lungs $\mu g\ g^{-1}$		0.13-0.15w		20
		0.01-0.25d		125
	0.035w		Controls	25
		0.05, 0.56w	Exposed people	25
		0.046-0.094w	Range of means	25
	0.010w	0-1.0w		123
	<0.13w			229
		0.05-0.28w		229
Muscle $\mu g\ g^{-1}$		0.02-0.7d		24
		0.036-0.039w		20
		0.004-0.71d		125
	0.15w	0-1.0w		123
Nail $\mu g\ g^{-1}$		0.07-7		24
		0.40-1.46	Finger nail	18
		0.32-4.61	Toe nail	18
Pancreas $\mu g\ g^{-1}$		0.05-1.14d		125
	0.05w	0-0.7w		123
Placenta $\mu g\ g^{-1}$		0.06-0.12d		125
	0.011w	0.001-0.10w		182
	0.051d	0.021-0.099d		232
		0.02-0.04w	57% $MeHg^+$	53
Plasma $\mu g\ dl^{-1}$	0.65			24
		0.2-1.0		125
	0.3			189
		0.13-0.32	Range of means	95
Red cells $\mu g\ dl^{-1}$		0.7		24
		0.38-1.2	Range of means	95
Serum $\mu g\ dl^{-1}$	1.2			24,125

Table 13.15 continued

Material	Mean	Range	Comments	Refs.
Skin $\mu g\ g^{-1}$		0.003-3.34w		125
Sweat $\mu g\ day^{-1}$	0.9			189
Teeth $\mu g\ g^{-1}$	5.6		Dentine	49
	5.3		Dentine	131
Urine $\mu g\ l^{-1}$	7.2	2-26.0	Amazon Indians	110
	10			189
	<0.5		79% of sample	95

d = dry weight, w = wet weight

TABLE 13.16 Some Levels of Arsenic in Human Hair ($\mu g\ g^{-1}$)

Mean	Range	Comments	Refs.
	0.06-3.7		24
0.62		Normal hair	43
0.65		Sulphur deficient hair	43
2.5		Pre 1880 AD	49
1.5		1890-1910 AD	49
1.2		1910-1935 AD	49
0.04		1972 AD	49
	0.14, 0.12	USA males	102
10.6		Copper smelting area	108
5.2		Pb/Zn smelting area	108
1.7		Pb/Zn smelting & mining	108
0.8		Not exposed	108
0.4		Not exposed	108
	7.9-28	Different parts of head	98
0.68	0.47-1.7	Rural	44
0.75	0.40-2.1	Urban	44
1.9	0.63-4.9	Near smelter	44
32.5*		Mill and mine workers	122
3.0*		Indian children	122
0.38*, 0.23*		Controls	122
0.158	0.037-0.625	Bulgaria, washed hair	233
0.074	0.026-0.159	Washed hair	233
0.525	0.101-2.410	England	233
0.728	0.279-1.049	NZ	233
7.85	3.91-36.57	Workers in Cu plant	100
3.42	0-16.37	Other employees	100
0.25	0-1.85	Controls	100
0.275	0.015-0.74	Korea	129
0.26	0.04-1.41	Pakistan	173
0.27		Rural Malaysia	186
0.83		Urban Malaysia	186
	0.3-0.7		228

* Median

TABLE 13.17 Some Levels of Arsenic in Human Tissues and Fluids

Material	Mean	Range	Comments	Refs.
Blood $\mu g\ dl^{-1}$	20			106
		0.17-9		24
		3.5-7.5		20
	<1		Unexposed	20
	30			10
Bone $\mu g\ g^{-1}$		0.08-1.6d		24
Brain $\mu g\ g^{-1}$	0.10w			106
		0.024-0.037w		20
Hair		See Table 13.16		
Heart $\mu g\ g^{-1}$		0.024-0.037w		20
		0.00097-0.012w	20	
Milk $\mu g\ g^{-1}$	0.0032	0.0016-0.0060		105
	0.0012			32
Kidney $\mu g\ g^{-1}$	0.3w			106
		0.026-0.037w		20
		0.007-1.5d		24
		0.00043-0.009w	Exposed people	25
		0.011w		25
	0.005w		Controls	25
	0.3w			25
	0.11w			25
Lungs $\mu g\ g^{-1}$		0.0056-0.36w	Exposed people	25
	0.11w		Controls	25
	0.02w			106
	0.02w			25
	0.12w			25
	0.009w			25
Liver $\mu g\ g^{-1}$		0.006-0.023w	Exposed people	25
	0.006w		Controls	25
	0.46w			106
		0.002-0.46w	Range of values	25
		0.03-0.039w	USA	20
		0.35-1.02w	USA	20
	0.021w	<0.005-0.086w	NZ	163
		0.023-1.6d		24
Muscle $\mu g\ g^{-1}$		0.009-0.65d		24
		0.031-0.058w		20
Nail $\mu g\ g^{-1}$		0.2-3		24
Placenta $\mu g\ g^{-1}$	0.55d	0.27-1.08d		232
Plasma $\mu g\ dl^{-1}$	0.24			24
	19.0			189
Red cells $\mu g\ dl^{-1}$	0.27			24
Serum $\mu g\ dl^{-1}$	3			24
		0.86-6.84	Variation over 24 h	112

Table 13.17 continued

Material	Mean	Range	Comments	Refs.
Serum µg dl^{-1}		1.35-7.71	Variation over 24 h	112
		3.90-12.92	Variation over 42 h	112
		0.16-18.38	Variation over 42 h	112
Urine	18.7		Children, Cu smelter	134
µg l^{-1}	5.8		Controls	134
	170			189
	31.03	0-240	Children, Cu smelter	100
	6.41	0-54	Controls	100
	12.4	2.3-53.4	Inorganic Arsenic	227
	9.7	1.7-40.3	Inorganic Arsenic	227
	60.5	2.3-353.7	Organic Arsenic	227
	31.4	<1.0-202.7	Organic Arsenic	227
creatinine µg g^{-1}	4.4	1.6-9.7	Children, non-smoking parents	2
	5.6	2.5-13.5	Children, female parent smokes	2
	5.5	3.6-8.5	Children, male parent smokes	2
	13.0	10.5-16.1	Children, both parents smoke	2

d = dry weight, w = wet weight

TABLE 13.18 Some Levels of Selenium in Human Tissues and Fluids

Material	Mean	Range	Comments	Refs.
Blood µg dl^{-1}	17.1			24
		5.5-8.1	NZ	20
		4.2-10.9	Finland	20
	20	10-35	USA	1
	8.5			106
	6.4			106
	21.2			10
Bone µg g^{-1}		1-9		24
Brain µg g^{-1}	0.09w			106
	0.2w			10
Eye tissue µg g^{-1}		0.48-40.4d		237
Hair µg g^{-1}		0.6-6		24
	0.15		Normal hair	43
	0.21		S-deficient hair	43
	0.37		Modern Greenland	49
	1.05, 1.13		USA	102
		10.1-96.2	Along fibre Se treated	98
		4.3-87	Along fibre	158
		0.3-13		189

Table 13.18 continued

Material	Mean	Range	Comments	Refs.
Hair µg g^{-1}	0.32	0.025-1.65	Children	54
	0.30	0.025-1.58	Adults	54
	0.96	0.42-2.45	Bulgaria	233
	0.97	0.34-2.83	UK	233
	0.79	0.23-1.05	NZ	233
	0.42	0.21-0.63	Sweden	157
		0.42-1.4	Various countries	157
	1.03	0.42-1.91	Pakistan	173
	0.5, 0.58		Malaysia	186
Heart µg g^{-1}	0.16d		Infant	24
	0.27d		Adult	24
Milk µg g^{-1}	0.015	0.011-0.022		105
	0.107			32
	0.022			117
Kidney µg g^{-1}		0.4-3.5d		24
	0.92d		Infant	20
	0.63d		Adult	20
		0.54-0.60w	Exposed workers	25
	0.79w		Controls	25
		0.2-0.7w	Range of means	25
	3.81d	1.79-5.46d	Sweden	157
	0.78	0.36-1.29w	Sweden	157
		0.2-1.5w	Various countries	157
	0.1w		Whole kidney	106
	0.2w		Kidney cortex	106
Liver µg g^{-1}		0.35-2.4d		24
		0.34,0.39d	Infants/adults	20
		0.18-0.66w	Canada	20
		0.20-0.40w	England	20
	0.31w	0.13-0.83w	NZ	163
		0.18-0.27w	Exposed	25
	0.29w		Controls	25
		0.22-0.7w	Range of means	25
	1.19d	0.32-1.92d	Sweden	157
	0.33w	0.082-0.64w	Sweden	157
		0.25-2.30w	various countries	157
	0.3w			106
Lungs µg g^{-1}	0.17,0.21d		Infants/adults	20
		0.12-0.18w	Industrially exposed	25
	0.095w		Controls	25
	0.172w	0.042-0.52		229
	0.1w			106
	0.2w			10
Muscle µg g^{-1}		0.42-1.9d		24
	0.31,0.40d		Infants/adults	20

Table 13.18 continued

Material	Mean	Range	Comments	Refs.
Muscle μg g^{-1}	0.11w			106
	0.4w			10
Nail μg g^{-1}		1-8		24
Pancreas μg g^{-1}	0.05,0.13d		Infant/adult	20
		0.56-0.70d	Normal pancreas	20
		0.93-1.03d	Abnormal pancreas	20
Placenta μg g^{-1}	0.55d	0.27-1.08		232
Plasma μg dl^{-1}	<3			24
	15.3		Males	117
		4.8-21	Range of means	117
Red cells μg dl^{-1}	30.24		Males	117
		7.4-22	Range of means	117
	22			24
Saliva μg l^{-1}	2.3			117
Serum μg dl^{-1}	17.2			24
		64-108	Range over 24, 42 h	112
Sweat mg d^{-1}	0.34			189
Teeth μg g^{-1}	0.40,0.43		Enamel, NZ, USA	56
	1.47	0-18.1	Enamel	55
Urine μg l^{-1}	30			189
		50-1000	High Se environment	1
		20-100	Moderate Se environ.	1

d = dry weight, w = wet weight

TABLE 13.19 Some Levels of Antimony in Human Tissues and Fluids

Material	Mean	Range	Comments	Refs.
Blood μg dl^{-1}	0.33			24
	3.4	0-13	Firing instructors	20
	<1		Unexposed	20
	0.21			106
	0.50			10,106
Bone μg g^{-1}	1.3a		Hard water area	106
	1.7a		Soft water area	106
		0.01-0.6d		24
Brain μg g^{-1}	0.007w			106
Hair μg g^{-1}		0.09-3		24
	0.25		Normal hair	43
	0.12		S deficient hair	43
		0.33-0.72	Along hair fibre	43
	0.084			49
		0.025-0.067	Along hair fibre	158
		0.017-0.036	no trend	158
	7.9*	1.3-24	Rural	44

Table 13.19 continued

Material	Mean	Range	Comments	Refs.
Hair μg g^{-1}	9.7*	1.5-33	Urban	44
	14.6*	1.8-47	Near smelter	44
	6.5			189
	0.13	0.058-0.26	Bulgaria	233
	0.12	0.004-0.80	Sweden	157
	1.60	0.07-6.6	Korea	129
	0.15	0.03-0.70	Pakistan	173
	0.27		Urban malaysia	186
	0.37		Rural malaysia	186
Milk ng g^{-1}	<2			32
Kidney μg g^{-1}		0.026-0.22d		24
		0.010-0.20w	Exposed people	25
	0.013w		Controls	25
	0.005w			25
	0.006w		Whole kidney	106
	0.005w		Kidney cortex	106
Liver μg g^{-1}		0.011-0.42d		24
		0.024-0.028w	Exposed people	25
	0.013w		Controls	25
		0.003-0.020w		25
	0.01w			106
	0.01w			10
Lungs μg g^{-1}	0.06w			106
	1.0w			10
	0.0155w	0.0035-0.048w		229
		0.29-0.56w	Exposed workers	25
	0.033w		Controls	25
Muscle μg g^{-1}		0.042-0.19d		24
	0.009w			106
Nail μg g^{-1}		<0.03-0.75		24
Placenta μg g^{-1}	0.014d	0.008-0.024d		232
Plasma μg dl^{-1}	0.32			24
Red cells μg dl^{-1}	0.46			24
Serum μg dl^{-1}	0.26			24
Sweat mg d^{-1}	0.011			189
Teeth μg g^{-1}	0.09, 0.13		Enamel, NZ, USA	56
	0.20	0-3.0	Enamel	55
Urine μg l^{-1}	<0.05			18

*Median, a = ash weight, d = dry weight, w = wet weight

TABLE 13.20 Some Levels of Bismuth, Indium, Thallium and Tellurium in Human Tissues and Fluids

Material	Mean	Range	Comments	Refs.
Bismuth				
Blood μg dl^{-1}	1.6			24
	<0.32			106
Bone μg g^{-1}	<0.2d			24
	<0.2a			106
Brain μg g^{-1}	0.01w			106
	<0.03w			10
Hair μg g^{-1}	2?			24
Kidney μg g^{-1}		<0.1-2d	Whole kidney	106
	0.4w		Kidney cortex	106
	0.5w			10
Liver μg g^{-1}		0.015-0.33w		24
	0.004w			106
	0.09w			10
Lung μg g^{-1}	0.01w			106
	<0.05			106
Muscle μg g^{-1}	0.032d			24
	0.007w			106
	<0.02w			10
Nail μg g^{-1}	1.3			24
Plasma μg dl^{-1}	0.06			24
Indium				
Muscle μg g^{-1}	0.016d			24
Thallium				
Blood μg dl^{-1}	0.3	0-8		203
	0.55			24
	0.0005			106
		0-33	Range, exposed people	20
Bone μg g^{-1}	<0.6			106
	0.002			24
Brain μg g^{-1}		0.00042-0.0015w		203
		0-0.02w		20
	<0.001w			106
Hair μg g^{-1}		0.0048-0.016		203
		0.007-0.65		203
	<2			24
Kidney μg g^{-1}	0.013d			24
		0-0.08w		20
	<0.003w			106
		0.0014-0.0041w		203
Liver μg g^{-1}		0.00056-0.003w		203
		0.004-0.033d		24
		0-0.19w		20

Table 13.20 continued

Material	Mean	Range	Comments	Refs.
Liver μg g^{-1}	0.009w			106
Muscle μg g^{-1}	0.007w			203
	0.07d			24
Nail μg g^{-1}	0.0022			24
		0.0007-0.0049		203
Plasma μg dl^{-1}	<0.25			24
Urine μg l^{-1}		0-1		203
		0-1240	Exposed people	20
	0.22	0.06-0.61	Not exposed	11
	0.38	0.08-1.22	Cement workers	11
	0.33	0.06-1.04	Iron foundry workers	11
		0.2-0.5	Range of means	81
		<0.1-1.2	Range of values	81
Tellurium				
Blood μg dl^{-1}	0.55?			24
Hair μg g^{-1}	0.016			24
Kidney μg g^{-1}	0.07?d			24
Liver μg g^{-1}	0.014?d			24
Muscle μg g^{-1}	0.017?d			24
Plasma μg dl^{-1}	<3			24

a = ash weight, d = dry weight, w = wet weight

REFERENCES

1. Allaway, W H. Control of the environmental levels of selenium. **Trace Subst. in Environ. Health**; 1968; **2**: 181-206.
2. Andrén, P., Schütz, A. Vahter, M., Attewell, R., Johansson, L., Willers, S. and Skerfving, S. Environmental exposure to lead and arsenic among children living near a glassworks. **The Sci. Total Environ.**; 1988; **77**: 25-34.
3. Angle, C. R., McIntire, M. S. and Vest. G. Blood lead of Omaha school children: topographic correlation with industry, traffic and housing. **Nebr. Med. J.**; 1975; **60**: 91-102.
4. Angle, C. R. and McIntire, M. S. Environmental lead and children: the Omaha study. **J. Tocicol. Environ. Health**; 1979; **5**: 855-870.
5. Angle, C. R. and McIntire, M. S. Lead: environmental sources and red cell toxicity in urban children. EPA Report 650/1-75-003: EPA; 1975.
6. Angle, C. R., Marcus, A., Cheng, I-H. and McIntire, M. S. Omaha childhood blood lead and environmental lead: a linear total exposure model. **Environ. Res.**; 1984; **35**: 160-170.
7. Annest, J. L., Mahaffey, K. R., Cox, D. H. and Roberts, J. Blood lead levels for persons 6 months - 74 years of age: United States 1976-80. Adavance-data, DHSS Pub. No (PHS) 82-1250 No 79: PHS; 1982: 1-24.

8. Annest., J. L., Pirkle, J. L., Makuc, D., Neese, J. W., Bayse, D. D. and Kovar, M. G. Chronological trend in blood lead levels between 1976 and 1980. **N. Engl. J. Med.;** 1983; **308:** 1373-1377.
9. Annest, J. L. Trends in the blood lead levels of the US population: the second National Health and Nutrition Examination Survey (NHANES II) 1976-1980. In: Rutter, M. and Russel Jones, R., Eds. Lead Versus Health: Wiley; 1983: 33-58.
10. Anspaugh, L. R., Robinson, W. C. Martin, W. H. and Lowe, O. A. Compilation of Published Information on Element Concentrations in Human Organs in Both Normal and Diseased States: Univ. California; 1971; **II:** UCAL 51013 Pt. 2.
11. Apostoli, P., Maranelli, G., Mincia, C., Massola, A., Baldi, C. and Marchiori, L. Urinary thallium: critical problems, reference values and preliminary results of an investigation in workers with suspected industrial exposure. **The Sci. Total Environ.;** 1988; **71:** 513-518.
12. Atessio, L. and Bertelli, G. Examples of biological indicators of metals in occupational health. In: Facchetti, S, Ed. Analytical Techniques for Heavy Metals in Biological Fluids: Elsevier; 1983: 41-63.
13. Attramadal, A. and Jonsen, J. The content of lead, cadmium, zinc and copper in deciduous and permanent human teeth. **Acta Odont. Scand.;** 1976; **34:** 127-131.
14. Attrramadal, A. and Jonsen, J. Heavy trace elemnts in ancient Nowegian teeth. **Acta Odontol. Scand.;** 1978; **36:** 97-101.
15. Award, L., Huel, G., Lazer, P. and Boudene, C. Factors of inter-individual variations of blood lead levels. **Rev. Epidemiol. Sante Publique;** 1981; **29:** 113-124.
16. Azar, A., Snee, R. D. and Habibi, K. An epidemiologic approach to community air lead exposure using personal air samplers. In: Griffen, T. B. and Knelson, J. H., Eds. **Environmental Quality and Saftey, Suppl.;** 1975; **II:** 254-290.
17. Barltrop, D., Strehlow, C. D., Thornton, I. and Webb, J. S. Absorption of lead from dust and soil. **Postgrad. Med. J.;** 1975; **51:** 801-804.
18. Benescheck-Huber, I. and Benescheck, F. Correlation of trace metals in hair and nails. In: Health Related Monitoring of Trace Element Pollutants Using Nuclear Techniques, IAEA TECHDOC-330: IAEA; 1985: 21-31.
19. Berlin, A. Biological monitoring in environmental and occupational health in the European Community. In: Facchetti, S, Ed. Analytical Techniques for Heavy Metals in Biological Fluids: Elsevier; 1983: 17-39.
20. Berman, E. Toxic Metals and their Analysis: Heyden; 1980.
21. Billick, I. H., Curran, A. S. and Shier, D. R. Analysis of pediatric blood lead levels in New York city for 1970-76. **Environ. Health Perspec.;** 1979; **31:** 183-190.
22. Billick, I. H., Curran, A. S. and Shier, D. R. Relation of pediatric blood lead levels to lead in gasoline. **Environ. Health Perspec.;** 1980; **34:** 213-217.

23. Billick, I. H. Sources of lead in the environment. In: Rutter, M. and Russel Jones, R., Eds. Lead Versus Health: Wiley; 1983: 59-77.
24. Bowen, H. J. M. Environmental Chemistry of the Elements: Academic Press; 1979.
25. Brune, D., Nordberg, G. F., Webster, P. O. and Bivered, B. Accumulation of heavy metals in tissues of industrially exposed workers. In: Nuclear Activation Techniques in the Life Sciences: IAEA; 1974: 643-655.
26. Brunekreef, B., Veenstra, S. J., Biersteker, K. and Boleij, J. J. M. The Arnhem lead study I lead uptake by 1 to 3 year old children living in the vicinity of a secondary lead smelter in Arnhem, The Netherlands. **Environ. Res.;** 1981; **25:** 441-448.
27. Brunekreef, B., Noy, D., Biersteker, K. and Boleij, J. Blood lead levels of Dutch city children and their relationship to lead in the environment. **J. Air Pollut. Control Ass.;** 1983; **33:** 872-876.
28. Brunekreef, B. Childhood Exposure to Environmental Lead. Technical Report: MARC; 1986; No 34.
29. Brunekreef, B. The relationship between air lead and blood lead in children: a critical review. **The Sci. Total Environ.;** 1984; **38:** 79-123.
30. Bryce-Smith, D., Deshpande, R. R., Hughes, J. and Waldron, H. A. Lead and cadmium levels in stillbirths. **Lancet;** 1977; **May:** 1159.
31. Burkitt, A. J., Nickless, G. and Stack, M. V. Lead in teeth during the prenatal period. **Postgrad. Med. J.;** 1975; **51:** 778-779.
32. Byrne, A. R., Kosta, L. Ravnik, V. Stupar, J. and Hudnik, V. A study of certain trace elements in milk. In: Nuclear Activation techniques in the Life Sciences: IAEA; 1979: 255-269.
33. Caplun, E., Petit, D. and Picciotto, E. Lead in petrol. **Endeavour;** 1984; **8:** 135-144.
34. Cavalleri, A, Minoia, C., Pozzoli, L. and Baruffini, A. Determination of plasma lead levels in normal subjects and in lead exposed workers. **Brit. J. Ind. Med.;** 1978; **35:** 21-26.
35. Cavalleri, A., Minoia, C. and Capodaglio, E. Lead in plasma: kinetics and biological effects. In: Facchetti, S, Ed. Analytical Techniques for Heavy Metals in Biological Fluids: Elsevier; 1983: 65-74.
36. Chamberlain, A. C. Efect of airborne lead on blood lead. **Atmos. Environ.;** 1983; **17:** 677-692.
37. Chamberlain, A. C. Effect of airborne lead on blood lead. **Atmos. Environ.;** 1983; **17:** 2366-2367.
38. Chamberlain, A.C. Fallout of lead and uptake by crops. **Atmos. Environ.;** 1983; **17:** 693-706.
39. Chamberlain, A. C., Clough, W. S., Heard, M. J., Newton, D., Stott, A. N. B. and Wells, A. C. Uptake of inhaled lead from motor exhaust. **Postgrad. Med. J.;** 1975; **51:** 790-794.

40. Chamberlain, A. C., Clough, W. S., Heard, M. J., Newton, D., Stott, A. N. B. and Wells, A. C. Uptake of lead by inhalation of motor exhaust. **Proc. Roy. Soc. London;** 1975; **192:** 77-110.
41. Charney, E., Kessler, B., Farfel, M. and Jackson, D. Child lead poisoning: a controlled trial of the effect of dust control measures on blood lead levels. **New Eng. J. Med.;** 1983; **309:** 1089-1093.
42. Charney, E., Sayre, J. and Coulter, M. Increased lead absorption in inner city children: where does the lead come from? **Pediatrics;** 1980; **65:** 226-231.
43. Chatt, A., Saijad, M. DeSilva, K. N. and Secord, C. A. Human scalp hair as an epidemiologic monitor of environmental exposure to elemental pollutants. In: Health related monitoring of trace element pollution using nuclear techniques. IAEA TECDOC 330.: IAEA; 1985: 33-50.
44. Chattopadhyay, A., Roberts, T. M. and Jervis, R. E. Scalp hair as a monitor of community exposure to lead. **Arch. Environ. Health;** 1977; **32:** 226-236.
45. Cherry, W. H. Distribution of cadmium in human tissues. In: Nriagu, J. O., Ed. Cadmium in the Environment. Part II Health Effects: Wiley; 1981.
46. Chisholm, J. J. Current status of lead exposure and poisoning of children. **South. Med. J.;** 1976; **69:** 529-531.
47. Chisholm, J. J. and Barltrop, D. Recognition and management of children with increased lead absorption. **Arch. Dis. Child;** 1979; **54:** 249-262.
48. Clayes-Thoreau, F., Thiessen, L. Bruaux, P. Ducoffre, G. and Verduyn, G. Assessment and comparison of human exposure to lead between Belgium, Malta, Mexico and Sweden. **Int. Arch. Occup. Environ. Health;** 1987; **59:** 31-41.
49. Coleman, D. O. Human remains. In: Historical Monitoring: MARC Report No. 31; 1985: 282-315.
50. Colombo, A. and Fantechi, R. Isotope lead experiment: A dynamic analysis of the isotopic lead experiment results. Comm. European Commun. EUR 8760 EN; 1983.
51. Committee on Biological Effects of Atmospheric Pollution. Lead: Airborne Lead in Perspective: Nat. Acad. Sci.; 1972.
52. Copius Peereboom, J. W., de Voogt, P., van Hattum, B., Velde, W. v. d., Copius Peereboom-Stegeman, J. H. J. The use of human placenta as a biological indicator for cadmium exposure. In: Management ond Control of Heavy Metals in the Environmnet, Int. Conf. London: CEC Consultants; 1979: 8-10.
53. Craig, P. J. Organomercury compounds in the environment. In: Craig, P. J., Ed. Organometallic Compounds in the Environment: Longmans; 1986: 65-110.
54. Creason, J. P., Hinners, T. A., Bumgarner, J. E. and Pinkerton, C. Trace elements in hair as related to exposure in Metropolitan New York. **Clin. Chem.;** 1975; **21:** 603-612.

55. Curzon, M. E. J. and Crocker, D. C. Relationships of trace elements in human teeth enamel to dental caries. **Arch. Oral Biol.**; 1978; **23:** 647-653.
56. Curzon, M. E. J., Losee, F. L. and Macalister, A. D. Trace elements in the enamel of teeth from New Zealand and the USA. **N. Z. Dent. J.**; 1975; **71:** 80-83.
57. Dabiko, R. W., Karpinski, K. F., McKenzie, A. D. and Bodjik, C. D. Survey of lead and cadmium in human milk and correlation of levels with environmental and food factors. **The Sci. Total Environ.**; 1988; **71:** 65-66.
58. De Silva P. E. and Donnan, M. B. Blood lead levels in Victorian children. **Med. J. Aust.**; 1980; **2:** 315-318.
59. Delves, H. T., Clayton, B. E., Carmichael, A., Bubear, M. and Smith, M. An appraisal of the analytical signifiance of tooth-lead measurements as possible indices of environmental exposure of children to lead. **Ann. Clin. Biochem.**; 1982; **19:** 329-337.
60. Delves, H. T. Biomedical applications of ICP-MS. **Chem. Brit.**; 1988; **24:** 1009-1012.
61. Department of the Environment (DOE) UK. European screening programme for lead, United Kingdom results 1979-1980. Pollution report No 10; 1981.
62. Department of the Environment (DOE) UK. European screening programme for lead, United Kingdom results for 1981. Pollution report No 18; 1983.
63. Drasch, G. A., Böhm, J. and Baur, C. Lead in human bones, investigations of an occupationally non-exposed population in Southern Bavaria (FGR) I Adults. **The Sci. Total Environ.**; 1987; **64:** 303-315.
64. Drasch, G. A., Böhm, J. and Baur, C. Lead in human bones, investigations of an occupationally non-exposed population in Southern Bavaria (FGR) II Children. **The Sci. Total Environ.**; 1988; **68:** 61-69.
65. Duggan, M. J. Childhood exposure to lead in the surface dust and soil: a community health problem. **Pub. Health Rev.**; 1985; **13:** 1-54.
66. Duggan, M. J. Contribution of lead in dust to children's blood lead. **Environ. Health Perspec.**; 1983; **50:** 371-381.
67. Duggan, M. J. Lead in urban dust: an assessment. **Water, Air Soil Pollut.**; 1980; **14:** 309-321.
68. Duggan, M. J. Lead in dust as a source of children's body lead. In: Rutter, M. and Russel Jones, R., Eds. **Lead Versus Health:** Wiley; 1983: 115-139.
69. Ellis, J. B. and Revitt, D. M. Incidence of heavy metals in street surface sediments: solubility and grain size. **Water, Air Soil Pollut.**; 1982; **17:** 87-100.
70. Elwood, P. C. Blood lead and petrol lead. **Brit. Med. J.**; 1983; **286:** 1515.
71. Elwood, P. C., Essex-Cater, A. and Robb, R. C. Blood lead levels on islands. **Lancet**; 1984; **2:** 355.

72. Elwood, P. C., Thomas, H. and Sheltawy, M. Blood-lead levels in mother's and their children. **Lancet;** 1978; **1:** 1363-1364.
73. Elwood, P. C. Changes in blood lead concentrations in women in Wales 1972-82. **Brit. Med. J.;** 1983; **286:** 1553-1555.
74. Elwood, P. C., St. Leger, A. S. and Morton, M. Dependence of blood-lead on domestic water lead. **Lancet;** 1976; **1:** 1295.
75. Elwood, P. C., Gallacher, J. E. J., Phillips, K. M., Davies, B. E. and Toothill, C. Greater contribution to blood lead from water than air. **Nature;** 1984; **310:** 138-140.
76. Elwood, P. C. The source of lead in blood: a critical review. **The Sci. Total Environ.;** 1986; **52:** 1-23.
77. Elwood, P. C. Turin isotopic lead experiment. **Lancet;** 1983; **1:** 869.
78. Environmental Health Criteria: Lead: UNEP/WHO; 1977; **3.**
79. Evenson, M. A. and Anderson, C. T. Ultramicro analysis for copper, cadmium and zinc in human tissue by use of atomic absorption spectrophotometry and the heated graphite tube atomizer. **Clin. Chem.;** 1975; **21:** 537-543.
80. Everson, J. and Patterson, C. C. Ultra-clean isotope dilution/mass spectrometric analysis for lead in human blood plasma indicate that most reported results are artifically high. **Clin. Chem.;** 1980; **26:** 1603-1607.
81. Ewers, U. Environmental exposure to thallium. **The Sci. Total Environ.;** 1988; **71:** 285-292.
82. Facchetti, S., Geiss, F., Gaglione, P., Colombo, A., Garibaldi, G., Spallanzani, G. and Gilli, G. Isotope Lead Experiment - Status Report EUR 8352 EN; 1982.
83. Fergusson, D. M., Fergusson, J. E., Horwood, L. T. and Kinzett, N. G. A longintudinal study of dentine lead levels, intelligence, school performance and behaviour: Part I dentine lead levels and exposure to environmental risk factors. **J. Child Psychol. Psychiat.;** 1988: **29:** 781-792.
84. Fergusson, J. E. and Purchase, N. G. The analysis and levels of lead in human teeth - a review. **Environ. Pollut.;** 1987; **46:** 11-44.
85. Fergusson, J. E., Hibbard, K. A. and Lau, R. H. T. Lead in human hair: general survey, battery factory employees and their family members. **Environ. Pollut.;** 1981; **B12:** 235-248.
86. Fergusson, J. E. Lead: petrol lead in the environment and its contributions to human blood lead levels. **The Sci. Total Environ.;** 1986; **50:** 1-54.
87. Fergusson, J. E., Kinzett, N., Fergusson, D. M. and Horwood,. A longintudinal study of dentine lead levels, intelligence, school performance and behaviour: the measurement of dentine lead. **The Sci. Total Environ.;;** 1989; **80:** 229-241.
88. Fergusson J. E., Jansen M. L. & Sheat A. W. Lead in deciduous teeth in relation to environmental lead. **Environ. Techn. Lett.;** 1980; **1:** 376-383.

89. Fleisher, M., Sarofim, A. F., Fassett, D. W., Hammond, P., Shacklette, H. T., Nisbet, I. C. T. and Epstein, S. Environmental impact of cadmium: a review by the panel on hazardous trace substances. **Environ. Health Perspec.**; 1974: 253-323.
90. Foo., S. C., Ngim, C. H. Phoon, W. O. and Lee, J. Mercury in scalp hair of healthy Singapore residents. **The Sci. Total Environ.**; 1988; **72:** 113-122.
91. Forstner, U. Cadmium. In: Huntzinger, O., Ed. Handbook of Environmental Chemistry: Springer Verlag; 1980; **3A:** 59-107.
92. Fosse G. & Justesen N-P.B. Lead in deciduous teeth of Norwegian children. **Arch. Environ. Health**; 1978; **33:** 166-175.
93. Friberg, L. and Vahter, M. Assessment of exposure to lead and cadmium through biological monitoring results of a UNEP/WHO global study. **Environ. Res.**; 1983; **30:** 95-128.
94. Friberg, L., Piscator, M., Nordberg, G. F. and Kjellstrom, T. Cadmiun in the Environment: CRC Press; 1971.
95. Friberg, L. and Vostal, J. (Eds.). Mercury in the Environment: CRC Press; 1972.
96. Fujiki, M. The pollution of Minamata Bay by mercury and Minamata disease. In: Baker, R.A., Ed. Contaminants and Sediments: Ann Arbor; 1980; 2: 493-500.
97. Gallacher, J. E. J., Elwood, P. C., Phillips, K. M. Davies, B. E. and Jones, D. T. Relation between pica and blood lead in areas of differing lead exposure. **Arch. Dis. Child**; 1984; **50:** 40-44.
98. Gangadharan, S., Lakshmi, V. V. and Das, M. S. The growth of hair and the trace elemnt profile: a study of sectional analysis. **J. Radioanal. Chem.**; 1973; **15:** 287-304.
99. Garnys, V. P., Freeman, R. and Smythe L. E. Lead Burden of Sydney School Children: Univ. New South Wales; 1979.
100. Ghelberg, N. W. and Boder, E. Arsenic levels in the environment and in the human body in a copper metallurgy plant area. In: Management ond Control of Heavy Metals in the Environmnet, Int. Conf. London: CEC Consultants; 1979: 163-165.
101. Goldsmith, J. R. Effect of airborne lead on blood lead. **Atmos. Environ.**; 1983; **17:** 2365-2366.
102. Gordus, A. Factors affecting the trace metal content of human hair. **J. Radioanal. Chem.**; 1973; **15:** 229-243.
103. Grandjean, P., Nielson, O. V. and Shapiro, I. M. Lead retention in ancient Nubian and contemporary populations. **J. Environ. Path. Toxicol.**; 1979; **2:** 781-787.
104. Griffen, T. B., Coulston, F., Wills, H., Russell, J. C. and Knelson, J. H. Clinical studies of men continously exposed to airborne particulate lead. In: Griffen, T. B. and Knelson, J. H., Eds. **Environ, Qual. Saftey, Suppl.**; 1975; **II:** 221-240.

105. Grimanis, A. P. Vassilaki-Grimani, M. Alexion, D. and Papadatos, C. Determination of seven trace elements in human milk, powdered cow's milk and infant foods by neutron activation analysis. In: Nuclear Activation techniques in the Life Sciences: IAEA; 1979: 241-253.
106. Hamilton, E. I., Minski, M. J. and Cleary, J. J. The concentration and distribution of some stable elements in healthy tissues from the United Kingdom. An environmental study. **The Sci. Total Environ.**; 1972/73; **1:** 341-374.
107. Hammer, D. I., Calocci, A. V., Hasselblad, V., Williams, M. E. and Pinkerson, C. Cadmium and lead in autopsy samples. **J. Occup. Med.**; 1973; **15:** 956-963.
108. Hammer, D. I., Finklea, J. F., Hendricks, R. H. and Shy, C. M. Hair trace metal levels and environmental exposure. **Amer. J. Epidemiol.**; 1971; **23:** 84-92.
109. Hammond, P. B., O'Flaherty, E. J. and Gartside, P. S. The impact of air lead on blood lead in man - critique of the recent literature. **Food Cosmet. Toxicol.**; 1981; **19:** 631-638.
110. Hecker, L. H., Allen, H. E., Dinman, B. D. and Neel, J. V. Heavy metal levels in acculturated and unacculturated populations. **Arch. Environ. Health**; 1974; **29:** 181-185.
111. Hewitt, C. N. and Harrison, R. M. Organolead compounds in the environment. In: Craig, P. J., Ed. Organometallic Compounds in the Environment: Longmans; 1986: 160-197.
112. Heydorn, K., Damsgaard, E., Larsen, N. A. and Nielsen, B. Sources of variability of trace element concentrations in human serum. In: Nuclear Activation Techniques in the Life Sciences. IAEA; 1974: 129-142.
113. Hinton, D., Cresswell, B. C. L., Janus, E. D. and Malpress, W. A. Industrial lead exposure: a review of blood lead levels in South Island industries 1974-83. **N. Z. J. Med.**; 1984; **14 nov.:** 769-773.
114. Hinton, D., Walmsley, T., Frampton, C. and Malpress, W. A. Industrial lead exposure in the South Island 1986-88. **N. Z. J. Med.**; 1988: 214-215.
115. Hinton, D., Malpress, W. A., Cresswell, B. C. and Ussher, K. E. Teamwork can improve the health of lead process workers. **N. Z. J. Med.**; 1985; **8 May:** 336-339.
116. Hinton, D., Coope, P. A., Malpress, W. A. and Janus, E. D. Trends in blood lead levels in Christchurch (NZ) and environs 1978-85. **J. Epidem. Commun. Health**; 1986; **40:** 244-248.
117. Hojo, Y. Selenium and glutathione peroxidase in human saliva and other human body fluids. **The Sci. Total Environ.**; 1987; **65:** 85-94.
118. Huang, J., He, F., Wu, Y. and Zhang, S. Observation on renal function in workers exposed to lead. **The Sci. Total Environ.**; 1988; **71:** 535-537.
119. ICF Incorporated. The relationship between gasoline lead emissions and lead poisonings in Americans. Contract No 68-01-5845; 1982: 1-42.
120. Irving, H. M. N. H. The Analytical Applications of Dithizone: CRC; 1980.

121. Iyengar, G. V., Kollmer, W. E. and Bowen, H. J. M. The Elemental Composition of Human Tissue and Body Fluids: Verlag Chemie; 1978.
122. Jervis, R. E. and Tiefenbach, B. Arsenic accumulation in people working with and living near a gold smelter. In: Nuclear Activation Techniques in the Life Sciences: IAEA; 1974: 627-642.
123. Joselow, M. M., Goldwater, L. J. and Weinberg, S. B. Absorption and excretion of mercury in man. **Arch. Environ. Health**; 1967; **15:** 64-66.
124. Jost, D. and Sartorius, R. Improved ambient air quality due to lead in petrol regulation. **Atmos. Environ.**; 1979; **13:** 1463-1465.
125. Kaiser, G. and Tolg, G. Mercury. In: Hutzinger, O., Ed. The handbook of Environmental Chemistry: Springer Verlag; 1980; **3A:** 1-58.
126. Kaneko, Y. Inamori, I. and Nishimura, M. Zinc, lead, copper and cadmium in human teeth from different geographical areas of Japan. **Bull. Tokyo Dent. Coll.**; 1974; **15:** 233-243.
127. Khandekar, R. N., Raghunath, R. and Mishra, U. S. Lead levels in teeth of an urban Indian population. **The Sci. Total Environ.**; 1986; **58:** 231-236.
128. Khandekar, R. N., Raghunath, R. and Mishra, U. C. Levels of lead, cadmium, zinc and copper in the blood of an urban population. **The Sci. Total Environ.**; 1987; **66:** 185-191.
129. Kim, N. B., Chung, H. W. and Lee, K. Y. Trace element analysis of human hair by neutron activation analysis. In: Health Related Monitoring of Trace Element Pollutants Using Nuclear Techniques, IAEA TECH-DOC-330: IAEA; 1985: 169-174.
130. Kosudi, H., Hanihara, ., Suzuki, T., Hongo, T., Yoshinaga, J. and Morita, M. Elevated lead concentrations in Japense ribs of the Edo Era (300-120 BP). **The Sci. Total Environ.**; 1988; **76:** 109-115.
131. Kuhnlein, H. V. and Calloway, D. H. Minerals in human teeth: differences between preindustrial and contemporary Hopi Indians. **Amer. J. Clin. Nutr.**; 1977; **30:** 883-886.
132. Landrigan, P. J. and Bridbord, K. Additional exposure routes. In: Nriagu, J. O., Ed. Changing Metal Cycles and Human Health: Springer Verlag; 1984: 169-186.
133. Landrigan, P. J., Stephen, H. Gehlbach, M. D., Rosenblum, B. F., Shoults, J. M., Candelaria, R. M., Barthel, W. F., Liddle, J. A., Smrek, A. L., Staehling, N. W. and Sanders, J. F. Epidemic lead absorption near an ore smelter: the role of particulate lead. **New Engl J. Med.**; 1975; **292:** 123-129.
134. Landrigan, P. J. and Baker, E. L. Exposure of children to heavy metals from smelters: epidemiology and toxic cosequences. **Environ. Res.**; 1981; **25:** 204-224.
135. Landrigan, P. J., Baker, E. L., Feldman, R. G. Cox, D. H. Eden, K. V. Orenstein, W. A. Mather, J. A. Yankel, A. T. and Von Lindern, I. H. Increased lead absorption with anaemia and slowed nerve conduction in children near a lead smelter. **J. Pediatr.**; 1976; **89:** 904-910.

136. Landrigan, P. J., Baker, E. L., Feldman, R. G. Cox, D. H. Eden, K. V. Orenstein, W. A. Mather, J. A. Yankel, A. T. and Von Lindern, I. H. Increased lead absorption with anaemia and slowed nerve conduction in children near a lead smelter. Shoshone Lead Health Project: Idaho Dept. Health and Welfare; 1976: 90-115.

137. Landrigan, P. J., Baker, E. L., Whitworth, R. H. and Feldman, R. G. Neuroepidemiologic evolutions of children with chronic increased lead absorption. In: Needleman H. L., Ed. Low Lead Level Exposure: Raven Press; 1980: 17-33.

138. Landsdown, R. and Yule, W. The Lead Debate: The Environment, Toxicology and Child Health: Croom Helm; 1986.

139. Langmyhr, F. J. and Eyde, B. Determination of the total content and distribution of cadmium, copper and zinc in human parotid saliva. **Anal. Chim. Acta**; 1979; **107:** 211-218.

140. Lauwerys, R., Buchet, J-P., Roels, H. and Hubermont, G. Placental transfer of lead, mercury, cadmium and carbon monoxide in women. **Environ. Res.**; 1978; **15:** 278-289.

141. Laxen, D. P. H., Raals, G. M. and Fulton, M. Children's blood lead and exposure to lead in household dust and water - a basis for an environmental standard for lead in dust. **The Sci. Total Environ.**; 1987; **66:** 235-244.

142. Mahaffey, K. R., Annest, J. L., Roberts, J. and Murphy, R. S. National estimates of blood lead levels: United States, 1976-1980: association with selected demographic and socioeconomic factors. **N. Engl. J. Med.**; 1982; **307:** 573-579.

143. Malpress, W. A., Janus, E. d. and Hinton, D. Blood lead levels in the New Zealand population: preliminary communication. **N. Z. J. Med.**; 1984; **12 Dec.:** 868-869.

144. Manton. W. I. Significance of lead isotope composition of blood. **Nature**; 1973; **244:** 165-167.

145. Manton. W. I. Source of lead in blood. **Arch. Environ. Health**; 1977; **32:** 149-159.

146. Masazumi, H. and Smith, A. M. Minamata disease: a medical report. In: Smith, W. E. and Smith, A. M. Minamata: Chatto and Windus; 1975: 180-192.

147. McBride, W. G., Black, B. P. and English, B. J. Blood lead levels and behaviour of 400 preschool children. **Med. J. Aust.**; 1982; **2:** 26-29.

148. Miettinen, J. K. The accumulation and excretion of heavy metals in organisms. In: Krenkel, P. A., Ed. Heavy Metals in the Aquatic Environment: Pergamon Press; 1975: 155-162.

149. Moon, J., Smith, T. J. Tamaro, S., Enarson, D., Fadl, S. Davidson, A. J. and Weldon, L. Trace metal in scalp hair of children and adults in three Alberta Indian villages. **The Sci. Total Environ.**; 1986; **54:** 107-125.

150. Moore, M. R., Meredith, P. A. Campbell, B. C., Goldberg, A. and Pocock, S. J. Contributions of lead in drinking water to blood lead. **Lancet;** 1977; **2:** 661-662.
151. Moore, M. R., Goldberg, A., Meredith, P. A., Lees, R., Law, R. A. and Pocock, S. J. The contribution of drinking water lead to maternal blood lead concentrations. **Clin. Chem. Acta;** 1979; **95:** 129-133.
152. Moore, M. R. Lead exposure and water plumbosolvency. In: Rutter, M. and Russel Jones, R., Eds. **Lead Versus Health:** Wiley; 1983: 76-106.
153. Moore, M. R. Maternal foetal lead relationships - a population study in Glasgow. Proc. Symp. Conservation Soc. London; 1979: 15-32.
154. Moore, M. R. Prenatal exposure to lead and mental retardation. In: Needleman, H. L., Ed. Low Level Lead Exposure: Raven Press; 1980: 53-65.
155. Moore, M. R., Goldberg, A., Pocock, S. J., Meredith, P. A. Steward, I. M., MacAnespie, H., Lees, R. and Low, A. Some studies of maternal and infant lead exposure in Glasgow. **Scott. Med. J.;** 1982; **27:** 113-122.
156. Moore M.R., Campbell B.C., Meredith P.A., Beattie A.D., Goldberg A. & Campbell. The association between lead concentrations in teeth and domestic water lead. **Clin. Chim. Acta,;** 1978; **87:** 77-83.
157. Muramatsu, Y. and Parr, R. M. Concentrations of some trace elements in hair, liver and kidney from autopsy subjects - relationship between hair and internal organs. **The Sci. Total Environ.;** 1988; **76:** 21-40.
158. Obrusnik, I, Gislason, J. Maes, D., McMillan, D. K., D'Aurea, J. and Pate, B. D. The variation of trace elemnt concentrations in single human head hairs. **J. Radioanal. Chem.;** 1973; **15:** 115-134.
159. O'Heany, J., Kusiak, R., Duncan, C. E., Smith, J. F., Smith, L. F. and Spielberg, L. Blood lead and associated drink factors in Ontario Children. **The Sci. Total Environ.;** 1988; **71:** 477-483.
160. Oxley, G. R. Blood lead concentrations: apparent reduction over approximately one decade. **Int. Arch. Occup. Environ. Health;** 1982; **49:** 341-343.
161. Pallotti, G., Consolino, A., Bencivenga, B., Iacoponi, V., Morisi, G., and Taggi, F. Lead levels in whole blood of a population group from Rome. **The Sci. Total Environ.;** 1983; **31:** 81-87.
162. Paterson, L. J., Raals, G. M. Hunter, R., laxen, D. P. H. Fulton, G. S., Halls, D. J. and Sutcliffe, P. Factors influencing lead concentrations in shed deciduous teeth. **The Sci. Total Environ.;** 1988; **74:** 219-234.
163. Pickston, L., Lewin, J. F., Drysdale, J. M. Smith, J. M. and Bruce, J. determination of potentially toxic metals in human livers in New Zealand. **J. Anal. Toxicol.;** 1983; **7:** 2-6.
164. Pinchin, M. J., Newham, J. and Thompson, R. P. J. Lead, copper and cadmium in the teeth of normal and mentally retarded children. **Clinica Chim. Acta;** 1978; **85:** 89-94.

165. Piomelli, S., Corash, L., Corash, M. B., Seaman, C., Mushak, P., Glover, B. and Padgett, R. Blood lead concentrations in a remote Himalayan population. **Science**; 1980; **210:** 1135-1137.

166. Piscator, M. and Lind, B. Cadmium, zinc, copper and lead in human renal cortex. **Arch. Environ. Health**; 1972; **24:** 426-431.

167. Pocock, S. J. Shaper, A. G., Walker, M., Wale, C. J. Clayton, B., Delves, T., Lacey, R. F., Packham, R. F. and Powell, P. Effects of tap water lead, water hardness, alcohol and cigarettes on blood lead concentrations. **J. Epidemol. Commun. Health**; 1983; **37:** 1-7.

168. Poole, C., Smythe, L. E. and Alpers, M. Blood lead levels in Papua New Guinea children living in a remote area. **The Sci. Total Environ.**; 1980; **15:** 17-24.

169. Prpi´c-Maji´c, D., Kersanc, A., Pongracic, J. and Fugas, M. Biological indices of lead absorption and residential distance from lead emitting source. In: Management ond Control of Heavy Metals in the Environmnet, Int. Conf. London: CEC Consultants; 1979: 89-92.

170. Purchase, N. G. and Fergusson, J. E. Lead in teeth: the influence of tooth type and the sample within a tooth on lead levels. **The Sci. Total Environ.**; 1986; **52:** 239-250.

171. Qu, J-B., Jin, C, Liu, Yu-T., Yin, S-N. Watanabe, T., Nakatsuka, H., Seiji, K., Inoue, O. and Ikeda, M. Blood lead levels of the general population of three Chinese cities. **The Sci. Total Environ.**; 1988; **77:** 35-44.

172. Quinn, M. J. and Delves, H. T. Investigations of air pollution standing conference, preliminary results for 1986 from blood lead monitoring programme. IAPSC; 1987; 3/4.

173. Qureshi, I. H., Chaudhry, M. S., Ahmad, S. and Mannon, A. Measurement of trace elements in human head hair, tobacco, coal and food articles of Pakistan. In: Health Related Monitoring of Trace Element Pollutants Using Nuclear Techniques, IAEA TECHDOC-330: IAEA; 1985: 195-203.

174. Rabinowitz, M. B. and Needleman, H. L. Environmental demographic and medical factors related to cord blood lead levels. **Biol. Trace Element Res.**; 1984; **6:** 57-67.

175. Rabinowitz, M. B., Wetherill, G. W. and Kapple, J. D. Magnitude of lead intake from respiration by normal man. **J. Lab. Clin. Med.**; 1977; **90:** 238-248.

176. Rabinowitz, M. B. and Needleman, H. L. Petrol lead sales and umbilical cord blood lead levels in Boston, Massachusetts. **Lancet**; 1983; **1:** 63.

177. Rabinowitz, M. B. and Needleman, H. L. Temporal trends in the lead concentrations of umbilical cord blood. **Science**; 1982; **216:** 1429-1431.

178. Raptis, S. and Mueller, K. New methods for determination of cadmium in blood serum. **Clin. Chim. Acta**; 1978; **88:** 393-402.

179. Reeves, R. D., Jolley, K. W. and Buckley, P. D. Lead in human hair: relation to age, sex and environmental factors. **Bull. Environ. Contam. Toxicol.**; 1975; **14:** 579-587.
180. Robinson, J. W. and Weiss, S. The direct determination of cadmium in urine and perspiration using a carbon atomizer for atomic absorption spectroscopy. **J. Environ. Sci. Health**; 1980; **A15:** 635-662.
181. Roels, H. A., Buchet, J-P., Lauwerys, R. R., Braux, P., Clayes-Thoreau, F., Lafontaine, A. and Verduyn, G. Exposure to lead by the oral and pulmonary routes of children living in the vicinity of a primary lead smelter. **Environ. Res.**; 1098; **22:** 81-94.
182. Roels, H. A., Buchet, J-P., Bernard, A., Hubermont, G., Lauwerys, R. R. and Masson, P. Investigation of factors influencing exposure and response to lead, mercury and cadmium in man and animals. **Environ. Health Perspec.**; 1978; **25:** 91-96.
183. Roels, H. A., Buchet, J-P., Lauwerys, R., Braux, P., Claeys-Thoreau, F., Lafontaine, A., van Overschelde, J. and Verduyn, G. Lead and cadmium among children near a non-ferrous metal plant. **Environ. Res.**; 1978; **15:** 290-308.
184. Russel Jones, R. The contribution of petrol lead to blood lead via air, dust and food. **Atmos. Environ.**; 1983; **17:** 2367-2370.
185. Russel Jones, R. and Stephens, R. The contribution of lead in petrol to human lead uptake. In: Rutter, M. and Russel Jones, R., Eds. Lead Versus Health: Wiley; 1983: 141-177.
186. Sarmani, S., Koshy, T. and Zakaria, Z. Scalp hair as an indicator of environmental pollution in Malaysia. In: Health Related Monitoring of Trace Element Pollutants Using Nuclear Techniques, IAEA TECH-DOC-330: IAEA; 1985: 205-209.
187. Sayre, J. Dust lead contribution to lead in children. In: Lynam, D. R., Piananida, L. G. and Cole, J. F., Eds. Environmental Lead: Academic press; 1981: 23-40.
188. Schmitt, M., Philion, J. J., Larsen, A. A., Harnadek, M. and Lynch, A. J. Surface soils as a potential source of lead exposure for young children. **Canad. Med. Assoc. J.**; 1979; **121:** 1474-1478.
189. Schroder, H. A. and Nason, A. P. Trace element analysis in clinical chemistry. **Clin. Chem.**; 1971; **17:** 461-474.
190. Schutz, A., Ranstam, J. Skerfving, S. and Tejning. Blood lead levels in school children in relation to industrial emission and automobile exhausts. **Ambio**; 1984; **13:** 115-117.
191. Schwartz, J. The benefits of reducing lead in gasoline. Int. Conf. Air pollut. Control Assoc. Kansas; 1984.
192. Servant, J. and Delapart, M. Blood lead and lead-210 origins in residents of Toulouse. **Health Phys.**; 1981; **41:** 483-487.

193. Servant, S. Airborne lead in the environment in France. In: Nriagu, J. O. and Davidson, C. I., Eds. Toxic metals in the Atmosphere: Wiley; 1986: 595-619.

194. Shapiro I.M., Mitchell G., Davidson I. and Katz S.H. The lead content of teeth: evidence establishing new minimal levels of exposure. **Arch. Environ. Health**; 1975; **30:** 483-486.

195. Sherlock, J., Smart, G. Forbes, G. I., Moore, M. R., Patterson, W. J., Richards, W. N. and Wilson, T.S. Assessment of lead intakes and dose response for a population in Ayr exposed to a plumbosolvent water supply. **Human Toxicol.**; 1982; **1:** 115-123.

196. Shrestha, K. P. and Fornerino, I. Hair mercury content among residents of Cumaná, Venezuela. **The Sci. Total Environ.**; 1987; **63:** 7-81.

197. Sikorski, R. Paszkowski, T. and Szprengier-Juszkiewicz, T. Mercury in neonatal scalp hair. **The Sci. Total Environ.**; 1986; **57:** 105-110.

198. Silva, P. A., Hughes, P. and Faed, J. M. Blood lead levels in 11 year old children. **N. Z. J. Med.**; 1986; **99:** 179-183.

199. Silvany-Neto, A. M. Carvalho, F. M., Chaves, M. E. C., Brandão, A. M. and Tavares, T. M. Repeated surveillance of lead poisoning among children. **The Sci. Total Environ.**; 1989; **78:** 179-186.

200. Sinn, W. Relationship between lead concentration in the air and blood lead levels of people living and working in the centre of a city (Frankfurt blood lead study). I Experimental method and and examination of differences. **Int. Arch. Occup Environ, Health**; 1980; **47:** 93-118.

201. Sinn, W. Relationship between lead concentration in the air and blood lead levels of people living and working in the centre of a city (Frankfurt blood lead study). II Correlations and conclusions. **Int. Arch. Occup Environ, Health**; 1981; **48:** 1-23.

202. Skerfving, S., Schutz, A. and Ranstam, J. Decreasing lead exposure in Swedish children. **The Sci. Total Environ.**; 1986; **58:** 225-229.

203. Smith, I. C. and Carson, B. L. Trace Metals in the Environment Vol. I Thallium: Ann Arbor Science; 1977.

204. Smith, M., Delves, T. Lansdown, R. Clayton, B. and Graham, P. The effects of lead exposure on urban children : the Institute of Child Health/ Southampton study. **Dev. Med. Child Neurol., Suppl. 47**; 1983; **25:** 1-54.

205. Snee, R. D. Evalution of studies of the relationship between blood lead and air lead. **Int. Arch. Occup. Environ. Health**; 1981; **48:** 219-242.

206. Snee, R. D. Models for the relationship between blood lead and air lead. **Int. Arch. Occup. Environ. Health**; 1982; **50:** 303-319.

207. Snyder, W. S. (Chairman). Report of the Task Group on Reference Man. Int. Commission on Radiological Protection: PergamonPress; 1975; No. **23.**

208. Spickett, J. T. and Razner, I. Trace metals in lung tissue. **Int. Clean Air Conf. Brisbane:** Ann Arbor; 1978.

209. Stack, M. V., Burkitt, A. J. and Nickless, G. Characterization of teeth by trace elements. **Int. J. Forensic Dent.**; 1974; **2:** 62-65.
210. Stack, M. W. Burkitt, A. J. and Nickless, G. Trace metals in teeth at birth (1957-1963) and (1973-1973). **Bull. Environ. Contam. Toxicol.**; 1976; **16:** 764-766.
211. Stauber, J. L. and Florence, T. M. A comparative study of copper, lead, cadmium and zinc in human sweat and blood. **The Sci. Total Environ.**; 1988; **74:** 235-247.
212. Stauber, J. L. and Florence, T. M. The determination of trace metals in sweat by anodic stripping voltammetry. **The Sci. Total Environ.**; 1987; **60:** 263-271.
213. Stephens, R. Human exposure to lead from motor emissions. **Int. J. Environ. Studies**; 1981; **17:** 73-83.
214. Stephens, R. The total relationship between airborne lead and body lead burden. Proc. Symp. Conservation Soc. London; 1978: 1-12.
215. Stoeppler. M. General analytical aspects of the determination of lead, cadmium and nickel in biological fluids. in: Facchetti, S., Ed. Analytical Techniques for Heavy Metals in Biological Fluids: Elsevier; 1983.
216. Strehlow, C. D. and Barltrop, D. Health studies in The Shipham Report. **The Sci. Total Environ.**; 1988; **75:** 101-133.
217. Suzuki, T., Hongo, T., Morita, M. and Yamamoto, R. Elemental contamination of Japanese women's hair from historical samples. **The Sci. Total Environ.**; 1984; **39:** 81-91.
218. Svenson, B-G., Björnham, Å., Schütz, A., Lettevall, U., Nilsson, A. and Skerfving, S. Acid deposition and human exposure to toxic metals. **The Sci. Total Environ.**; 1987; **67:** 101-115.
219. Taskinen, H., Nordman, H., Hernberg, S. and Engstron. K. Blood lead levels in Finnish preschool children. **The Sci. Total Environ.**; 1981; **20:** 117-129.
220. Ter Haar, G. and Aronow, R. New information on lead in dirt and dust as related to childhood lead problem. **Environ. Health Perspec.**; 1974; **7:** 83-89.
221. Ter Haar, G. Sources of lead in children. In: Management ond Control of Heavy Metals in the Environmnet, Int. Conf. London: CEC Consultants; 1979: 70-76.
222. Ter Haar, G. and Aronow, R. Tracer studies of ingestion of dust by urban children. In: Griffin, T. B. and Knelson, J. H., Eds. **Environ. Qual. Saftey Suppl.**; 1975; **II:** 197-201.
223. Triger, D. R., Crowe, W., Ellis, M. J., Herbert, J. P., McDonnell, C. E. and Argent, B. B. Trace elemnt levels in blood of workers in two steel works and a non ferrous plant handling lead and cadmium compared with a non-exposed population. **The Sci. Total Environ.**; 1989; **78:** 241-261.

224. Tsuchiya, K., Sugita, M., Seki, Y., Kobayashi, Y., Hori, M. and Park C. B. Study of lead concentrations in atmosphere and population in Japan. In: Griffin, T. B. and Knelson, J. H., Eds. **Environ. Qual. Saftey Suppl;** 1975; **II:** 95-146.
225. Tusgane, S. and Kondo, H. The mercury content of hair of Japanese immigrants in various locations in Sth. America. **The Sci. Total Environ.;** 1987; **63:** 69-76.
226. Vahter, M. (Ed.). Assessment of human exposure to lead and cadmium through biological monitoring: UNEP/WHO; 1982.
227. Vahter, M. and Lind, B. Concentrations of arsenic in urine of the general population of Sweden. **The Sci. Total Environ.;** 1986; **54:** 1-12.
228. Valkovic, V. Trace Elements in Human Hair: Garland STPM Press; 1977.
229. Vanoeteren, C. and Cornelis, R. Evaluation of trace elements in human lung tissue I concentration and distribution. **The Sci. Total Environ.;** 1986; **54:** 217-230.
230. Volpe, R. H. Assessment of the health aspects of lead in petrol. Int. Lead and Zinc Research Organisation, Committee on Evaluation of Regulations, Global Approach (ERGA) of EEC; 1983.
231. Ward, N. I., Stephens, R. and Ryan, D. E. Comparison of three analytical methods for the determination of trace elements in whole blood. **Anal. Chim. Acta;** 1979; **110:** 9-19.
232. Ward, N. I., MacMahon, T. D. and Mason, J. A. Elemental analysis of human placenta by neutron irradiation and gamma ray spectrometry (standard, prompt and fast neutron). **J. Radioanal. Nucl. Chem., Art.;** 1987; **113:** 501-514.
233. Ward, N. I., Spyrou, N. M. and Damyanova, A. A. Study of hair element content from an urban Bulgarian population using NAA, assessment of environmental status. **J. Radioanal. Nucl. Chem., Art.;** 1987; **114:** 125-135.
234. Weiss, D., Whitten, B. and Leddy, D. Lead content of human hair (1871-1971). **Science;** 1972; **178:** 69-70.
235. WHO. Environmental Health Criteria 1 Mercury: WHO; 1976.
236. Wilkinson, D. R. and Palmer, W. Lead in teeth as a function of age. **Amer. Lab.;** 1975: 67-70.
237. Yamaguchi, T., Bando, M., Nakajima, A., Teral, M. and Suzuki-Yasumoto, M. An application of neutron activation analysis to biological materials IV approach to simultaneous determination of trace elements in human eye tissue with non-destructive neutron activation analysis. **J.Radioanal. Chem.;** 1980; **57:** 169-183.
238. Yankel, A. J., von Lindern, I. H. and Walter, S. D. The Silver Valley lead study: the relationship between chilhood blood lead levels and environmental exposure. **Air. Pollut. Control Assoc. J.;** 1977; **27:** 763-767.

CHAPTER 14

THE UPTAKE OF HEAVY ELEMENTS BY HUMAN BEINGS

In this chapter we will consider the ways the heavy elements enter human beings. The process may be broken down into four major steps: (1) the pathways for the transfer of the heavy elements to humans, (2) the intake of the heavy elements, (3) the uptake or absorption of the heavy elements into the blood stream, and (4) the distribution of the elements in the body, and their removal from the body. A simplified diagram for these steps is given in Fig. 14.1

The processes and materials associated with step 1 have been described in Chapters 7 to 11, and will not be discussed further. Much of the data in step 4 has been given in the previous chapter, therefore will will be concerned mainly with steps 2 and 3. We will review first the metabolic processes of respiration, digestion and absorption, and then the impact of the heavy metals.

INTAKE AND ABSORPTION

The three modes of intake of external materials for human beings are inhalation of air into the lungs, ingestion of food, water and at times non-food items into the gastrointestinal system and transfer through the skin. Other routes such as intravenous, intramuscular and vaginal will not be considered in any detail.

Digestion and Gastrointestinal Absorption

Food is digested in the stomach, by enzyme hydrolysis to produce small molecules which are absorbable. The pH of the stomach fluids are around 1 to 3 due to the presence of hydrochloric acid. Hence the heavy metals will become cationic or anionic (e.g. chloro-complex anions) in chemical form. The digested material moves into the duodenum and the small intestine, where the pH is around 6 to 7. Much of the absorption occurs in the duodenum, jejunum and ileum. The materials on the surface of the gastrointestinal tract may be absorbed

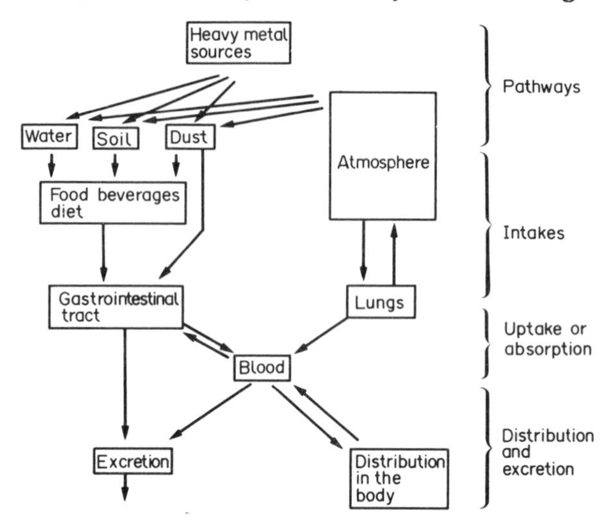

Fig. 14.1 The transfer of heavy elements into the human body.

into the walls and hence into the the blood stream. The transfer may be passive involving diffusion and osmosis, and this is the primary mechanism, or active involving consumption of energy. Transfer can also occur back from the blood into the intestine, called endogenous excretion. Liquids move more rapidly through the stomach into the intestine, than do solids.

Numerous factors influence the absorption process and hence the fraction of a metal intake that is absorbed. These include, the chemistry of the metal, especially the speciation of the metal at the time of absorption and the solubility of the metal species. Solubility is controlled by the pH at the segment of the intestine where the absorption takes place. The time taken for the digestion and the rate of movement in the intestine influence the kinetics of the conversion of the cations or anions at pH 1-3 (in the stomach) to other species in the intestine at pH 6-7. The amount of food eaten and the type of diet can also influence the absorption of metals. Microbial influences are important, as are the presence of organic chelating agents and other metals that compete for absorption sites [39,41,54,55,79]. Because of these various considerations it is not possible to generalize over the fraction of metals absorbed. The percentage of each metal absorbed from the gastrointestinal tract, into the blood, is given in Table 14.1.

Inhalation

The absorption of the heavy elements through the respiratory system is complex, as a number of processes are involved [13,14,31,39,54,55,79]. The first process is the inhalation of the air containing the heavy metal aerosol. The amount inhaled will depend on the breathing cycle (breaths per minute) and the ventilation rate (volume breathed per day). For an adult an average breathing

TABLE 14.1 Absorption Factor (%) for the Heavy Elements from the Gastrointestinal Tract into Blood

Element	Absorption factor %	Suggested value	Comments/conditions
Cd	3-8, 4.7-7	6	Influenced by $Ca^{2+}$and Zn^{2+} protein content and solubility of Cd compounds.
Hg Inorg.	7, 15, 8-20	10	Vaues for Hg^{2+}. The Hg^{+} is poorly absorbed because of low solubility.
Hg Org.	>80, 95, ~100	100	Almost completely absorbed because soluble in fats.
In	<5		
Tl	>70, 33, ~100	80	Rapid and complete absorption for soluble species.
Pb adult	7, 10	10	Influenced by diet and fasting.
Pb infant	25-53, 50	40	Influenced by diet and fasting.
As	~100, 5-25		High absorption for anionic and soluble species, low for insoluble species.
Sb	3-80, 50, 80		Similar to arsenic.
Bi	<5, 8	6	Soluble compounds more readily absorbed.
Se	>70, 33-80	80	Value for anionic soluble species, insoluble compounds poorly absorbed.
Te	25, 47, 88		For water soluble species.

Sources of data; references2,5,6,8,9,13,16,18,19,24,27,29,31-35,37,45,46,47,50,51, 54,55,60, 61,63,64,67,69,70,74,75,76,80,81,83,91.

cycle is around 5.0-5.3 sec. and the ventilation rate will vary with the amount of exertion, and figures in the literature range from 15 to 23 $m^3 d^{-1}$ [2,18,19,31,63, 74,75,76,83]. It has been estimated that for an adult working 8 hours of light activity 9.6 m^3 of air are breathed, for 8 hours of non-occupational activity the same amount and for 8 hours of resting 3.6 m^3 are breathed. This gives a total of 23 $m^3 d^{-1}$. A value of 22 $m^3 d^{-1}$ will be used in calculations, though the values of 20 and 15 $m^3 d^{-1}$ have also been used. The data for children is less well documented and values between 4-10 $m^3 d^{-1}$ are reported [18,19,24,31,74,75,76]. The second process is the amount of aerosol that is deposited in the lungs. This depends on the particle size, shape and charge (formal and/or electrostatic). Some particles penetrate deep into the pulmonary system, whereas others are deposited closer to the point of entry. In general larger particles, >2 μm, are trapped in the mucus of the upper respiratory system, and removed by ciliary activity either into the mouth or via. the glottis into the stomach. The very small, gaseous like particles, <0.01 μm, may penetrate deeply, but many will also be removed from the lungs by exhalation. The particles in the size range 0.1-2 μm can move deep into the the pulmonary and alveolar system, and eventually some will be absorbed into the the blood stream. The data listed in Table 14.2 is the estimated penetration depth of particles of different sizes [54].

The third process is absorption of the deposited material into the blood stream. This is more efficient (approximately 10 times) than the absorption process in the gastrointestinal tract. The amount of a metal absorbed via. the respiratory system is given by the equation;

$$A_a = A_i \times f_d \times f_a,$$

where A_a is the amount absorbed, A_i is the amount inhaled, f_d is the fraction deposited in the lungs, and f_a the fraction of the deposited material absorbed into the blood. For a particular metal the value of A_i is a function of the concentration of the element in the air and the respiratory characteristics of the person. The fraction f_d depends on the distribution of the metal over the range of particle sizes, and f_a is a function of the elements speciation and solubility of the particles. Some estimates of the two fractions are given in Table 14.3 for the elements cadmium, mercury, thallium, lead and arsenic. Except for mercury, and provided attention is given to particle size, the fraction deposited for many of the metals is around 0.4-0.5, and the fraction absorbed, depending on the solubility of the metal species, is around 0.6-1.0. Similar figures probably apply to the other five elements.

TABLE 14.2 Penetration Depth of Particulate Material into the Human Respiration System (% Retained)

Respiratory system		Particle size		
		20 μm	2 μm	0.2 μm
Nasopharyngeal	Mouth/nose	15		
	Pharynx	8		
Pulmonary	Trachea	10		
	Pulmonary bronchi	12		
	Secondary bronchi	19	1	
	Tertiary bronchi	17	2	
	Quartenary bronchi	6	2	1
	Terminal bronchioles	6	6	6
	Respiratory bronchioles	0	5	4
Alveolar	Alveolar ducts	0	25	11
	Alveolar sacs	0	0	0
	Retained	93	41	22
	Expelled	7	59	78

Source of data; Luckey and Venugopal, 1977 [54].

TABLE 14.3 Estimates of the Deposition in the Lungs and Absorption into the Blood of Heavy Elements in Aerosols

Element	f_d	f_a	Suggested $f_d f_a$
Cd	0.25, 0.1-0.5	0.4, 0.5, 1.0, 0.25-0.5	0.2
Hg metal	0.8, 0.75-0.85	0.8, 1.0	0.8
Hg inorg.	0.8, 0.1-0.5	0.8, 1.0	0.5
Hg org.	0.75-0.85	1.0	0.8
Tl		1.0	
Pb	0.3-0.8, 0.4, 0.5, 0.6	0.9, 0.5, 0.4, 0.9-1.0	0.4
As	0.3-0.6	1.0	0.4

f_d = fraction deposited in the lungs, f_a = fraction of deposited material absorbed into the blood stream. Sources of data; references 2,6,11,14,15,16,18,19,24,29,31-35,38,45,47,49,50,51,54,55, 60,61 ,63,64,67,69,74,75,76,81,91.

Skin Absorption of the Heavy Elements

The skin is also a route by which elements may enter the body. The skin is efficient at keeping out water, particles, ionic inorganic species and materials of high molecular weight. But it does not repel substances which are lipid soluble. Skin is therefore susceptible to the absorption of organometallic compounds and compounds soluble in some organic solvents. A summary of the potential for skin absorption of compounds of the heavy elements is given in Table 14.4. Lipid soluble materials are absorbed, but the situation is less clear for the ionic forms of the heavy elements.

INTAKE AND ABSORPTION OF THE HEAVY ELEMENTS

There are three principal ways of estimating the intakes of the heavy elements by people [6,17,29,30,42,57]. The first is called the duplicate diet method, whereby two identical meals are prepared, one is eaten and the other is analysed. The second is called the market basket method. In this case the concentrations of the heavy elements in food stuffs are obtained and, since there is usually a wide range of concentrations, an average figure is calculated. Then a typical or average food consumption is estimated so the average intake of a particular element can be determined. The last method makes use of the estimated fraction of an element absorbed into the body, and then measures the amount excreted in the urine and/or fæces.

For the first two methods the concentrations of the elements in some foods are at times below the detection limit. In this situation it is not obvious what concentration to use in the calculations, to use zero under estimates the concentration and to use the detection limit over estimates the concentration. At

TABLE 14.4 Absorption of the Heavy Elements Through Skin

Element	Species	Absorption
Cd	Cd^{2+}	Limited
Hg	Hg^{2+}, Hg	Some compound absorbed if dispersed in a suitable medium
	Organo.	Absorbed
Tl	Tl^{+}	Can be absorbed
Pb	Pb^{2+}	Some evidence for absorption
	Organo.	Readily absorbed
As	Fat soluble	Absorbed

Sources of data; references 35,38,45,53,54,60,91.

best, the methods are guides to the order of magnitude of heavy metal intakes. The third method depends on the assumption that urinary and fæcal metal levels relate to the amount absorbed, some endogenous excretion does occur (from the blood back to the intestine) but this is likely to be small, so that fæcal concentration data are probably reasonably reliable. The main problem is the accumulation of the elements in tissues of the body reducing the urinary levels.

We will now examine the intakes and uptakes of the elements, based on the absorption factors discussed. The elements most frequently studied are lead, cadmium and mercury, and to a lesser extent arsenic, selenium and thallium. The principal routes of entry for each of the elements are listed in Table 14.5.

Lead

Aerosol lead As described above the transfer of trace elements in particulate material from the air into the blood stream is a three stage process, inhalation, deposition and absorption. Using an adult ventilation rate of 22 $m^3 d^{-1}$, and a rate of 6.5 $m^3 d^{-1}$ for a child, the intake of lead aerosol into the lungs at different lead aerosol concentrations are given in column 3 of Table 14.6 [2,19,50,62,74,75,81].

The proportion of lead aerosol deposited in the lungs ranges from 30 to 85% [11,13,18,29,31,49,60,67]. A figure of 50% seems reasonable, however, the value of 40% is frequently used, and has been used in Table 14.6 (column 4). The proportion of the deposited lead absorbed into the blood is reported as 40-100% [2,11,13,19,31,50,60,63,74,75,76,81]. The proportion does depend on the particle size, but it seems that for the small lead particles almost 100% is absorbed. Many reports do not distinguish between the two stages and give a total absorption from air to blood as around 0.4. Similar figures are also used for the absorption of tetraalkyllead compounds [38].

From this information the results in the last column of Table 14.6, indicates that lead absorbed for an adult, may range from <0.9-8.8 $\mu g\ d^{-1}$, depending on the air lead concentration, and for children <0.26-2.6 $\mu g\ d^{-1}$. For a natural air lead

TABLE 14.5 Main Intake Routes of the Heavy Elements for Human Beings

Element	High source		Medium to low source				
	Occup.	Special popul.	Air	Food	Water	Smoke	Dust
Pb	√	√	√	√	√	(√)	√
Cd	√	√		√		√*	√
Hg	√	√	(√)	√			
As	√	√		√	(√)	(√)	(√)
Se	√	√		√			
Tl	√	√		√		(√)*	
Sb	√	√		√			
Bi	√	√		√			
Te	√	√		√			
In	√	√		√			

√ major source, (√) minor source, * cigarette smoking.

level of 0.00004 μg m^{-3} [62,81] the absorbed lead is much less, around 0.0004 μg d^{-1} for adults and 0.0001 μg d^{-1} for children.

An additional aerosol source of lead comes from cigarette smoking. A smoking rate of 30 cigarettes per day may add an aerosol lead concentration of 0.5-0.8 μg per cigarette. Taking the lower value, the lead intake per day will be 15 μg [19,29,63]. Using the same fractions of deposited and absorbed lead means around 6μg d^{-1} extra lead is absorbed by an adult smoker. This is of the same order that would come from air with a lead concentration of 0.5-1.0 μg m^{-3}.

TABLE 14.6 Intakes and Uptake of Lead from the Air by Adults and Children

Lead aerosol μg m^{-3}	Air intake m^3 d^{-1}	Lead intake μg d^{-1}	Deposit. factor	Deposit. intake μg d^{-1}	Absorp. factor	Lead absorbed μg d^{-1}
Adults						
1*	22	22	0.4	8.8	1	8.8
0.5**	22	11	0.4	4.4	1	4.4
0.1***	22	2.2	0.4	0.9	1	0.9
0.00004¶	22	0.00088	0.4	0.0004	1	0.0004
Children						
1*	6.5	6.5	0.4	2.6	1	2.6
0.5**	6.5	3.3	0.4	1.3	1	1.3
0.1***	6.5	0.65	0.4	0.26	1	0.26
0.00004¶	6.5	0.00026	0.4	0.0001	1	0.0001

* high urban value, ** urban, *** rural ¶ natural

Food lead Calculations of food and water lead intakes have been carried out by numerous workers [1,2,7,8,12,19,28,29,31,42,62,65,74,75,82,83]. The estimated intake depends on the level of consumption assumed, the concentration taken for the lead in the food and the quality of the analytical data. For example, two estimates of lead intake in a village found 85-157 μg d^{-1}, mean 101 μg d^{-1} using diary estimates (i.e. a record of what was eaten), and 20-171 μg d^{-1}, mean 60 μg d^{-1} using duplicate diets [28,57]. Two things stand out from these results, the wide variation found in any one of the methods, and the difference in the values obtained by the two methods. Hence trying to arrive at an result for an average person is therefore difficult and especially when children are involved.

Average estimates for lead in food used in such calculations are 0.05-0.2 μg g^{-1}, and for water 10-20 μg l^{-1}. The consumption of food and drink by adults span 1.5-2.0 kg d^{-1} and 1.0-2.0 l d^{-1}, whereas for children the quantities reported are 0.8-1.0 kg d^{-1} and 1.0-1.4 l d^{-1} respectively. Hence estimates of daily lead intake for adults cover a wide spread from 5 to 1700 μg d^{-1}, with most estimates around 120-150 μg d^{-1}. For children the range is 6-210 μg d^{-1} with most estimates around 40-60 μg d^{-1}. For breast fed babies an intake of 2-4 μg d^{-1} has been calculated, and Austrian babies were assessed to take in 6-42 μg d^{-1} [65]. An example of the type of calculation carried out is given in Table 14.7 for two concentrations of lead in food and lead in water.

TABLE 14.7 Intakes and Uptakes of Lead from Food and Water by Adults and Children

(a) Food

Conc. in food μg g^{-1}	Food intake kg d^{-1}	Lead Intake μg d^{-1}	Absorption factor	Lead uptake μg d^{-1}
Adults				
0.1	1.5	150	0.1	15
0.05	1.5	75	0.1	7.5
Children				
0.1	1.0	100	0.5	50
0.05	1.0	50	0.5	25

(b) Water

Conc. in water μg l^{-1}	Water intake l d^{-1}	Lead Intake μg d^{-1}	Absorption factor	Lead uptake μg d^{-1}
Adults				
20	1.5	30	0.1	3
10	1.5	15	0.1	1.5
Children				
20	1.0	20	0.5	10
10	1.0	10	0.5	5

Evidence is accumulating that the intake of lead from food is dropping [28]. One obvious reason for this is the reduction in the use of lead soldered cans, however, the amount of lead from other sources, that get into food, is also decreasing. One decrease is due to the reduction of lead in petrol and its contribution to dust. Also reported lead levels in food may be falling because of better contamination control during sampling and analysis.

The FAO/WHO recommended tolerable daily intake of lead from food and water is set at 430 $\mu g\ d^{-1}$, and a level suggested for children is 300 $\mu g\ d^{-1}$ (USPHS). This last figure however, assumes an absorption figure of 0.1 for children rather than the usual value of 0.4-0.5 (see below). This means the intake for children should be more like 60-75 $\mu g\ d^{-1}$. For a 70 kg person the FAO/WHO figure becomes 6.1 $\mu g\ kg^{-1}\ d^{-1}$, therefore for a 20 kg child the total daily intake would be around 120 $\mu g\ d^{-1}$.

The estimated levels of intakes based on the lead levels in fæces, assuming a 10% absorption factor (see below), for people living in four different countries, and assuming no endogenous excretion, are summarized in Table 14.8. The results for Sweden and Belgium are in agreement with duplicate meal measures of a total lead intake of 90 and 27-30 $\mu g\ d^{-1}$ respectively [6,17]. The data shows a strong curvelinear relationship between fæces lead and blood lead.

The absorption factor of lead from the gastrointestinal tract to the blood system is generally taken as 0.1, for both food and drink lead for adults, though it is likely that a higher absorption occurs for water compared with solid food. Values reported span 0.05-0.17, and depend to some extent on the age of the person, fasting and whether lead intakes occur between meals. A high calcium and iron diet decreases lead absorption, and when the elements are deficient lead absorption increases. For children a wide range of absorption factors are reported, 0.05-0.99, with most values around 0.4-0.5 [2,6,8,17,18,19,24,29,31, 46,50,55,60,62,63,67,74,75,76,78,81,83]. Assuming values of 0.1 and 0.5 for adults and children respectively, the absorption of lead intakes given in Table 14.7 have been estimated. The data for air (Table 14.6), and food and water (Table 14.7) are collected together in Table 14.9, from which the relative importance of each intake may be seen. Food lead is the most significant source

TABLE 14.8 Estimated Levels of Lead Intake in Four Countries

Country	Lead in fæces $\mu g\ g^{-1}$ (dw)	Estimated lead intake $\mu g\ d^{-1}$*	Total lead intake $\mu g\ d^{-1}$**	Blood lead $\mu g\ dl^{-1}$
Malta	11.1	361	401	24.7
Mexico	4.7	159	177	18.8
Belgium	3.2	82	91	13.2
Sweden	0.6	22	24	5.3

* Estimated on the basis of the fæces lead, ** Corrected for a 10% absorption factor. Sources of data; Bruaux and Svartengren, 1985 [6], Clayes-Thoreau et al., 1987 [17].

TABLE 14.9 Intake and Absorption of Lead from Air, Water and Food by Adults and Children

Source*	Intake	Lead intake μg d^{-1}	Absorpt. factor	Lead uptake μg d^{-1}	% of total	
					No smoking	Include smoking
Adults						
Air	22 m^3	11	0.4	4.4	21.1	16.4
Food	1.5 kg	150	0.1	15.0	71.8	55.8
Water	1.5 l	15	0.1	1.5	7.2	5.6
Smoking	30 cigs.	15	0.4	6.0		22.3
Total (including smoking)		191		26.9		
Total (excluding smoking)		176		20.9		
Children					No dust	Include dust
Air	6.5 m^3	3.3	0.4	1.3	2.3	1.6
Food	1.0 kg	100	0.5	50	88.8	61.5
Water	1.0 l	10	0.5	5	8.9	6.2
Dust	50 mg	50	0.5	25		30.8
Total (including dust)		163.3		81.3		
Total (excluding dust)		113.3		56.3		

* Concentrations: air 0.5 μg m^{-3}, food 0.1 μg g^{-1}, water 10 μg l^{-1}, smoking 0.5 μg cig^{-1}, dust 1000 μg g^{-1}.

for adults and children, and aerosol lead is more important for adults than for children. The high absorption factor of 0.5 for food for children is the reason why children absorb more lead into their bodies than adults even though their intake is less. The contribution from cigarette smoking is quite marked for adults, and 30 cigarettes a day can add about 22% of a persons uptake of lead.

Dust can be a significant source of lead, especially for children. Some attempts to quantify the intake of dust lead are summarized in Table 14.10 [3,17,19,21,24-27,52,59,75,76,84,85,86]. The results suggest a daily intake spanning 10 to 500 μg d^{-1}, with an average value between 50-100 μg d^{-1}. Dust easily accumulates on the hands of children. The amount of lead after play on the hands of a group of children was 13-86 μg, which was reduced to 3-19 μg after washing their hands. In a separate experiment the washed hands of children had a lead content in the remaining dust of 2-22 μg, and this increased to 10-46 μg after play [27]. The routes by which dust lead (street and house) can get into children is shown in the diagram in Fig. 14.2 [77,78]. There are a number of routes, and the main ones appear to be hand-to-mouth activities (1), pica for paint (2), and other non-food items (3).

The absorption of the dust lead from the gastrointestinal tract may be greater than for lead in food, as the extraction of lead in the dust in HCl at the pH of the stomach fluids is around 80-95% [22,23]. Lead on printed paper is mostly

TABLE 14.10 Lead Intake Through Dust on Hands of Children

Lead conc. in dust $\mu g\ g^{-1}$	Dust intake $mg\ d^{-1}$	Frequency of action	Lead intake $\mu g\ d^{-1}$	Reference
1200	20		48	85,86
1000	5-50 per sweet	2-20 sweets	100	21
1000	100		100	3
1000	20	10 sucks	10-100 mean 50	25,26
	10 on fingers	10 sucks	60	84
2400	10 on fingers	10 mouthings	240	52
1000	35		35	76
1400 urban	100		140	75
800 rural	100		80	75
3400	20		68	59
500	1000		500	19
			20-200	24

Other references; 7,31,27,48.

extracted in the pH range 1-2 [4]. Addition of the dust lead intake for children has been included in Table 14.9, next to food lead it is a significant source. The contribution from dust is probably underestimated because of the contribution of dust lead to the food and water intakes. It has been estimated that in the UK, about 13 $\mu g\ d^{-1}$ of food lead intake comes from dust on the food [12].

Alkyllead compounds are absorbed th-ough the skin, therefore cleaning skin with petrol containing tetraalkyllead is dangerous. It is likely that inor-

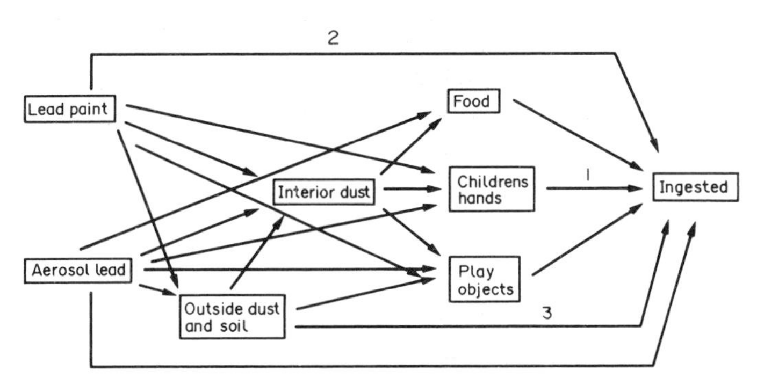

Fig. 14.2 Routes for the entry of dust lead into children.

ganic compounds are also absorbed [38,53,60], but it is unclear whether or not the inorganic lead gets into the blood, but is excreted in sweat and saliva [53].

The distribution of lead in the body is portrayed in Fig 14.3, based on a three compartment model [2,7,50,61,72,73]. The amounts in the various compartments and the transfer rate between compartments will vary, depending on the levels of intake assumed in the model.

Cadmium

The human intakes of cadmium are much less than for lead. But on a weight basis cadmium is more toxic, therefore it is possible, in the urban environment, to have too high an intake, especially if a person smokes.

Estimates of the total fraction of cadmium absorbed ($f_d f_a$) range from 0.16 to 0.5, with f_d around 0.1 to 0.5 and f_a 0.25 to 0.6. A value of 0.4 for $f_d f_a$ is probably reasonable, but this does depend on the speciation and solubility of the deposited cadmium compounds and metal [32,33,34,51,55,58,59,67,69,80,83]. For an aerosol cadmium concentration of 0.001-0.01 μg m^{-3}, and for an air intake of 22 m^3 d^{-1}, the intake of cadmium would be around 0.022-0.22 μg d^{-1}, with an absorption of 0.009-0.09 μg d^{-1}. The amount absorbed increases significantly if a person smokes. Cigarettes can contain up to 1-2 μg g^{-1} of cadmium in the tobacco. Therefore an additional 2-4 μg of the element is inhaled per day from the smoking of 20 cigarettes [32,33,34,36,44,69], which means 0.8 to 1.6 μg d^{-1} of cadmium is absorbed into the blood. This is much more than comes from typical air cadmium levels.

Several estimates have been made of the daily oral intake of cadmium for different countries, and the values span 10-120 μg d^{-1} [1,5,32,33,34,37,42,44,46, 51,55,57,58,59,65,69,71,80,83,89,90]. Intakes at the high end of the scale, occur in countries with high cadmium pollution, such as Japan. A median value, on a

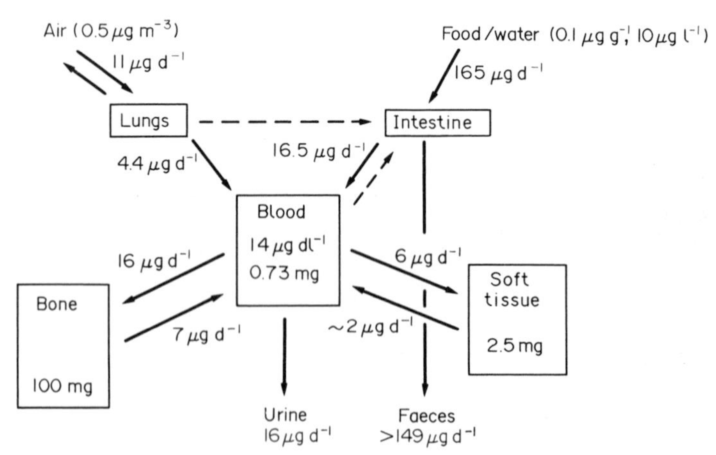

Fig. 14.3 The distribution of lead in the human body.
Sources of data; references 2,7,50,61,72,73

world basis, is around 35 μg d^{-1}. The FAO/WHO tolerable weekly intake of 400-500 μg corresponds to 57-71 μg d^{-1}. Hence it appears that a number of people are close to, or above, this limit. Many foodstuffs have low levels of cadmium which are frequently below the analytical detection limit, therefore the problem arises as to how to incorporate these into diet estimates of the element [44]. In a high cadmium environment, in Shipham village in England, the estimates of cadmium intakes were 33 μg d^{-1} (diary estimates), and 26 μg d^{-1} (duplicate diets), which were higher than the national average of 14 μg d^{-1} [57].

The absorption of cadmium from the gut has been estimated to be 4.7 to 7%, and the most common value used is 6% for adults. The absorption increases to around 10%, if the person is suffering from calcium and protein deficiency. Hence an intake of 35 μg d^{-1} would give rise to an uptake of 2.1 μg d^{-1}. The data in Table 14.11 is estimated intakes and absorptions for an adult, for different concentrations of cadmium in air, food and water. Food cadmium comes out as the major route, however, cadmium from smoking can match the food source.

Dust containing cadmium may also be an important source for children [32,59]. For a dust intake of 50-100 mg d^{-1} and a cadmium concentration of 5 μg g^{-1} the intake would be 0.25-0.5 μg d^{-1}, and an absorption of 0.015-0.03 μg d^{-1}. These figures are low, compared with a permissible food intake of around 15 μg d^{-1} (i.e. 0.9 μg d^{-1} absorbed) for a child. However, in high cadmium polluted areas, dust may become a significant source [59].

TABLE 14.11 Intake and Absorption of Cadmium from Air, Water and Food by Adults

Air				
Conc. μg m^{-3}	Air intake m^3 d^{-1}	Cd intake μg d^{-1}	Absorption	Cd uptake μg d^{-1}
0.001	22	0.022	0.4	0.009
0.01	22	0.22	0.4	0.09
20 cigs.	-	2-4	0.4	0.8-1.6
Food				
Conc. μg g^{-1}	Food intake kg	Cd intake μg d^{-1}	Absorption	Cd uptake μg d^{-1}
0.04	1.5	60	0.06	3.6
0.01	1.5	15	0.06	0.9
Water				
Conc. μg l^{-1}	Water intake l	Cd intake μg d^{-1}	Absorption	Cd uptake μg d^{-1}
2	1.5	3	0.06	0.2
10	1.5	15	0.06	0.9

Mercury

The intake and uptake of mercury is species sensitive, more so than for the other heavy elements. The three main species are elemental mercury (Hg^0), divalent 'inorganic' mercury (Hg^{2+}) and organomercury, especially methylmercury (CH_3Hg^+). A summary of the absorption factors for the three species is presented in Table 14.12 [5,9,16,35,45,46,55,67,80,83,88,91]. The main feature of the data is the high absorption of the mercury species.

Reported food intakes of mercury cover the range 3.0 to 25 μg d^{-1} [1,20,42,83], with a median value around 15 μg d^{-1}. The FAO/WHO permissible tolerable level has been set at 43 μg d^{-1} (0.3 mg $week^{-1}$) of which no more than 2/3 i.e. 29 μg d^{-1}, is methylmercury. Because of the bioaccumulation of mercury by fish and shellfish these food items can be high source of the metal [10,16,20,43,68,87]. The interrelation between fish consumption and mercury levels in fish muscle is given in Table 14.13 [68]. The figures in bold are ≤29 μg g^{-1}, the FAO/WHO limit for CH_3Hg^+.

Excluding fish, it is possible to make approximate calculations of the air and food intakes and uptakes of mercury by human beings using average concentrations in food and air [16,90,91]. An arbitrary assumption is made that 1/3 of the mercury in the food is inorganic and the rest is organic mercury. The results listed in Table 14.14 reveal that the organomercury component of food is the major source (>90%).

TABLE 14.12 Absorption Factors for Mercury Species in Human Beings

Species	Inhalation			Gut absorption	Skin absorption
	f_d	f_a	$f_d f_a$		
Hg(0)	0.8	~1.0	0.8	<0.0001	High
Hg(II)	0.5	~1.0	0.5	0.15	Debatable
CH_3Hg^+	0.8	~1.0	0.8	~1.0	High

Sources of data; references 5,9,16,35,45,46,55,67,80,83,91.

TABLE 14.13 Fish Consumption, the Level of Mercury in the Fish and Mercury Intake (μg d^{-1})

Fish consumption g d^{-1}	Mercury concentration in fish ng g^{-1}				
	100	200	500	1000	2000
20, average	**2**	**4**	**10**	**20**	40
100, elevated	**10**	**20**	50	100	200
300, high	**30**	60	150	300	600

Source of data; Piotrowski and Inskip, 1981 [68].

TABLE 14.14 Intake and Absorption of Mercury from Air, Water and Food by Adults

Material	Mercury concentration	Intake per day	Mercury intake $\mu g\ d^{-1}$	Chem. form $\mu g\ d^{-1}$	Absorp. factor	Uptake $\mu g\ d^{-1}$	Percent
Food	20 ng g^{-1}	1.5 kg	30	10 inorg	0.15	1.5	6.8
				20 org	1.0	20	90.5
Water	200 ng l^{-1}	1.5 l	0.3	0.1 inorg	0.15	0.015	0.1
				0.2 org	1.0	0.2	0.9
Air*	20 ng m^{-3}	22 m^3	0.44		0.8	0.35	1.6

* Mostly Hg(0) or CH_3Hg^+

Arsenic

Intakes of arsenic by inhalation and ingestion vary, depending on the source of the air, water and food. In relatively uncontaminated environments around 5 $\mu g\ d^{-1}$ of arsenic can be taken by inhalation and drink. Cigarette smoking however, can add a lot more if arsenic pesticides have been used on the tobacco plants [47,64]. The intakes of food arsenic span a wide range <10-1000 $\mu g\ d^{-1}$ [1,55,58,65,83,92] and depends on sea food consumption, and whether or not arsenic pesticides have been used. A typical arsenic intake through food, for an adult, is 10-40 $\mu g\ d^{-1}$, and 1-5 $\mu g\ d^{-1}$ has been suggested for babies [65].

The deposition of arsenic in the lungs is around 32-62% depending, on the particle size. Absorption is close to 100% for soluble arsenic species, and the gaseous arsenic compounds AsH_3 and $(CH_3)_3As$ [47,64,83]. Absorption of inorganic compounds from the gut is subject to their solubility, and varies from 5 to 25 % and even up to 100%. Organoarsenic compounds, as found in fish, are probably completely absorbed from the gastrointestinal tract [5,46,47,64,80]. The absorbed arsenic is rapidly distributed around the body, and except for accumulation in tissues such as hair and nails, it is also readily excreted from the body.

Selenium

Selenium is unique among the ten elements, in that it is probably an essential element for human beings On the other hand it is also toxic and the concentration difference between the levels which are deficient and toxic is small (see Chapter 15). An intake of <6-45 $\mu g\ d^{-1}$ maybe considered as deficient [5,55], whereas an intake of >5000 $\mu g\ d^{-1}$ is toxic [5].

Estimates of selenium intakes vary from low, i.e. <10, 25 and 30 $\mu g\ d^{-1}$, for areas in the world deficient in selenium, (parts of China, New Zealand and Finland respectively) to 500 to 5000 $\mu g\ d^{-1}$ for selenium rich areas [1,56,66,83]. Most diet intakes are 60-200 $\mu g\ d^{-1}$ with a median around 60-70 $\mu g\ d^{-1}$. Detailed analysis of hospital diets in Canada found 98.0 and 90.5 $\mu g\ d^{-1}$ selenium intake with the main sources being the meat and cereals [56]. An adequate and safe intake is given as 50-200 μg d^{-1} whereas WHO places the lower limit at 130 μg d^{-1} [46,55]. Babies being fed human or cows' milk had reported intakes from 5 to 21 μg d^{-1}. The lower end of the range occurs in countries with a deficiency of selenium in the soil (e.g. New Zealand and Belgium). The higher end of the range is the recommended level for milk-fed infants [40,83]. Absorption of selenium from the gastrointestinal tract is rapid and efficient, being around 30-80% depending on the selenium speciation [5,55].

Thallium

The main source of thallium in human intakes is through food and estimates suggest that an average intake is <2 μg d^{-1} [30,83]. Air inhalation seems to be unimportant unless smoking is a significant factor, as thallium levels may be elevated in tobacco e.g. 100 ng g^{-1} [83]. Based on studies with rats, and making appropriate adjustments, it has been suggested that 37 μg d^{-1} is an acceptable upper limit for thallium intake, another calculation suggests 14 μg d^{-1} [30]. Either way, the intakes for most people are well below these figures.

The absorption of thallium from the gastrointestinal tract is rapid and efficient, especially when present as the Tl^{+} ion in a soluble form [5,55]. Overall the absorption factor based on the estimate from levels in urine and fæces is around 33%. Thallium is also absorbed through the skin.

Remaining Elements

Much less is known about the remaining four elements, indium, antimony, bismuth and tellurium as regards their intakes and absorption into the human body [5,27,55,83]. The absorption of indium and bismuth are probably quite low <5%, whereas antimony and tellurium are more readily absorbed (see Table 14.1). Intakes for all of the four elements are probably low, but depend on the local environment and industrial uses.

REFERENCES

1. Adriano, D C. Trace Elements in the Terrestial Environment: Springer Verlag; 1986.
2. Australian Academy of Science. Health and environmental lead in Australia: Aust. Acad. Sci.; 1981.
3. Bloom, H. and Smythe, L. E. Environmental lead and its control in Australia. **Search**; 1983/84; **14**: 315-319.
4. Bogden, J. D., Joselow, M. M. and Singh, N. P. Extraction of lead from printed matter at physiological values of pH. **Arch. Environ. Health**; 1975; **30**: 442-444.
5. Bowen, H. J. M. Environmental Chemistry of the Elements: Academic Press; 1979.

6. Bruaux, P. and Svartengren, M. Assessment of human exposure to lead: comparison between Belgium, Mexico and Sweden: UNEP/WHO; 1985.
7. Brunekreef, B. Childhood Exposure to Environmental Lead. Technical Report: MARC; 1986; No 34.
8. Bryce-Smith, D. and Waldron, H. A. Lead in food - are todays regulations sufficient? **Chem. Brit.**; 1974; **10:** 202-206.
9. Butler, G. C. Exposure to mercury. In: Trace Metals: Exposure and Health Effects: CEC and Pergamon Press; 1979: 65-72.
10. Buzina, R., Suboticanec, K., Vukusi´c, T. Sapunar, J., Antoni´c, K. and Zorica, M. Effect of industrial pollution on seafood content and dietary intake of total and methylmercury. **The Sci. Total Environ.**; 1989; **78:** 45-57.
11. Chamberlain, A. C. Effect of airborne lead on blood lead. **Atmos. Environ.**; 1983; **17:** 677-692.
12. Chamberlain, A.C. Fallout of lead and uptake by crops. **Atmos. Environ.**; 1983; **17:** 693-706.
13. Chamberlain, A. C., Heard, M. J., Little, P. Newton, D., Wells, A. C. and Wiffen, R. D. Investigations into lead from motor vehicles. Rept. AERE Harwell, AERE-R 9198; 1978.
14. Chamberlain, A. C., Clough, W. S., Heard, M. J., Newton, D., Stott, A. N. B. and Wells, A. C. Uptake of inhaled lead from motor exhaust. **Postgrad. Med. J.**; 1975; **51:** 790-794.
15. Chamberlain, A. C., Clough, W. S., Heard, M. J., Newton, D., Stott, A. N. B. and Wells, A. C. Uptake of lead by inhalation of motor exhaust. **Proc. Roy. Soc. London**; 1975; **192:** 77-110.
16. Clarkson, T.W., Hamada, R. and Amin-Zaki, L. Mercury. In: Nriagu, J. O., Ed. Changing Metal Cycles and Human Health: Dahlem Konferenzen Springer Verlag; 1984: 285-309.
17. Clayes-Thoreau, F., Thiessen, L. Bruaux, P. Ducoffre, G. and Verduyn, G. Assessment and comparison of human exposure to lead between Belgium, Malta, Mexico and Sweden. **Int. Arch. Occup. Environ. Health**; 1987; **59:** 31-41.
18. Committee on Biological Effects of Atmospheric Pollution. Lead: Airborne Lead in Perspective: Nat. Acad. Sci.; 1972.
19. Committee on Lead in the Human Environment. Lead in the Human Environment: Nat. Acad. Sci.; 1980.
20. Craig, P. J. Environmental aspects of organometallic chemistry. In: Wilkinson, G., Stone, F. G. A. and Abel, E. W., Eds. Comprehensive Organometallic Chemistry; 1982; Vol. 2.
21. Day, J. P., Hart, M. and Robinson, M. S. Lead in urban street dust. **Nature**; 1975; **253:** 343-345.
22. Day, J. P. Lead pollution in Christchurch. **N. Z. J. Sci.**; 1977; **20:** 395-406.

23. Day, J. P., Fergusson, J. E. and Tay Ming Chee. Solubility and potential toxicity of lead in urban dust. **Bull. Environ. Contam. Toxicol.;** 1979; **23:** 497-502.
24. Duggan, M. J. Childhood exposure to lead in the surface dust and soil: a community health problem. **Pub. Health Rev.;** 1985; **13:** 1-54.
25. Duggan, M. J. and Williams, S. Lead in dust in city streets. **The Sci. Total Environ.;** 1977; **7:** 91-97.
26. Duggan, M. J. Lead in urban dust: an assessment. **Water, Air Soil Pollut.;** 1980; **14:** 309-321.
27. Duggan, M. J., Inskip, M. J., Rundle, S. A. and Moorcroft, J. S. Lead in playground dust and on the hands of school children. **The Sci. Total Environ.;** 1985; **44:** 65-79.
28. Elwood, P. C. The source of lead in blood: a critical review. **The Sci. Total Environ.;** 1986; **52:** 1-23.
29. Environmental Health Criteria: Lead: UNEP/WHO; 1977; **3.**
30. Ewers, U. Environmental exposure to thallium. **The Sci. Total Environ.;** 1988; **71:** 285-292.
31. Fergusson, J. E. Lead: petrol lead in the environment and its contributions to human blood lead levels. **The Sci. Total Environ.;** 1986; **50:** 1-54.
32. Fleisher, M., Sarofim, A. F., Fassett, D. W., Hammond, P., Shacklette, H. T., Nisbet, I. C. T. and Epstein, S. Environmental impact of cadmium: a review by the panel on hazardous trace substances. **Environ. Health Perspec.;** 1974: 253-323.
33. Forstner, U. Cadmium. In: Huntzinger, O., Ed. Handbook of Environmental Chemistry: Springer Verlag; 1980; **3A:** 59-107.
34. Friberg, L., Piscator, M., Nordberg, G. F. and Kjellstrom, T. Cadmium in the Environment: CRC Press; 1971.
35. Friberg, L. and Vostal, J. (Eds.). Mercury in the Environment: CRC Press; 1972.
36. Hammer, D. I., Calocci, A. V., Hasselblad, V., Williams, M. E. and Pinkerson, C. Cadmium and lead in autopsy samples. **J. Occup. Med.;** 1973; **15:** 956-963.
37. Heitanen, E. Gastrointestinal absorption of cadmium. In: Nriagu, J. O., Ed. Cadmium in the Environment. Part II Health Effects: Wiley; 1981: 55-68.
38. Hewitt, C. N. and Harrison, R. M. Organolead compounds in the environment. In: Craig, P. J., Ed. Organometallic Compounds in the Environment: Longmans; 1986: 160-197.
39. Hodgson, E. and Guthrie, F. E. (Eds.). Introduction to Biochemical Toxicology: Elsevier; 1980.
40. Hoji, Y. Selenium in Japanese baby foods. **The Sci. Total Environ.;** 1986; **57:** 151-159.

41. Houtman, J. P. W. and van den Hamer, C. J. A. (Eds.). Physiological and Biochemical Aspects of Heavy Elements in our Environment: Delft Univ. Pres; 1975.
42. Hubbard, A. W. and Lindsay, D. G. Dietary intakes of heavy metals by consumers in the United Kingdom. In: Management and Control of Heavy Metals in the Environment, Int. Conf. London: CEP Consultants; 1979: 52-55.
43. Huckabee, J. W., Elwood, J. W. and Hildebrand, S. G. Accumulation of mercury in fresh water biota. In: Nriagu, J. O, Ed. The Biogeochemistry of Mercury in the Environment: Elsevier/Nth. Holland; 1979: 277-302.
44. Hutton, M. Cadmium exposure and indicators of kidney function. MARC Tech. Rept. 29: MARC; 1983.
45. Kaiser, G. and Tolg, G. Mercury. In: Hutzinger, O., Ed. The handbook of Environmental Chemistry: Springer Verlag; 1980; **3A:** 1-58.
46. Kirk, P.W. W. and Lester, J. N. Significance and behaviour of heavy metals in waste water treatment processes IV water quality standards and criteria. **Sci. Total Environ.**; 1984; **40:** 1-44.
47. Lafontaine, A. Health effects of arsenic. In: Trace Metals: Exposure and Health Effects: CEC and Pergamon Press; 1979: 107-116.
48. Lag, J. and Steinnes, E. Regional distribution of selenium and arsenic in humus layers of Norwegian forest soils. **Geoderma**; 1978; **20:** 3-14.
49. Landrigan, P. J. and Bridbord, K. Additional exposure routes. In: Nriagu, J. O., Ed. Changing Metal Cycles and Human Health: Springer Verlag; 1984: 169-186.
50. Landsdown, R. and Yule, W. The Lead Debate: The Environment, Toxicology and Child Health: Croom Helm; 1986.
51. Lauwerys, R. R. Health effects of cadmium. In: Trace Metals: Exposure and Health Effects: CEC and Pergamon Press; 1979.
52. Lepow, M. L., Bruckman, L., Gillette, M., Markowitz, S., Robino, R. and Kapish, J. Investigations into sources of lead in the environment of urban children. **Environ. Res.**; 1975; **10:** 415-426.
53. Lilley, S. G., Florence, T. M. and Stauber, J. L. The use of sweat to monitor lead absorption through the skin. **The Sci. Total Environ.**; 1988; **76:** 267-278.
54. Luckey, T. D. and Venugopal, B. Metal Toxicity in Mammals: Plenum Press; 1977; Vol 1.
55. Luckey, T. D. and Venugopal, B. Metal Toxicity in Mammals: Plenum Press; 1978; Vol 2.
56. Mc Dowell, L. S. Griffen, P. R. and Chatt, A. Determination of selenium in individual food items using short lived nuclide Se -77m. **J. Radioanal Nucl. Chem. Art.**; 1987; **110:** 519-529.
57. Morgan, H., Smart, G. A. and Sherlock, J. C. Intakes of metal: In The Shipham report. **The Sci. Total Environ.**; 1988; **75:** 71-100.

58. Murti, C.R.K. The cycling of arsenic, cadmium, lead and mercury in India. In: Hutchinson, T.C. and Meena, K.M., Eds. Lead, Mercury, Cadmium and Arsenic in the Environment. Scope 31: Wiley; 1987: 315-333.
59. Muskett, C. J., Roberts, L. H. and Page, B. J. Cadmium and lead pollution from secondary metal refining operations. **The Sci. Total Environ.;** 1979; **11:** 73-87.
60. Newland, L. W. and Dawn, K. A. Lead. In: Hutzinger, O., Ed. Handbook of Environmental Chemistry: Springer Verlag; 1982; **3B:** 1-26.
61. O'Brien, B. J. The exposure commitment method with application to exposure of man to lead pollution. MARC Tech. Rept. 13: MARC; 1979.
62. Patterson, C. C. An alternative perspective - lead pollution in the human environment: origin extent and significance. In: Lead in the Human Environment: NAS; 1980: 265-349.
63. Patterson, C. C. Contaminated and natural lead environments of man. **Arch. Environ. Health;** 1965; **11:** 344-360.
64. Pershagen, G. Exposure to arsenic. In: Trace Metals: Exposure and Health Effects: CEC and Pergamon Press; 1979: 99-106.
65. Pfannhauser, W. and Widich, H. Source and distribution of heavy metals in food. In: Management and Control of Heavy Metals in the Environment, Int. Conf. London: CEP Consultants; 1979: 48-51.
66. Pickston, L., Lewin, J. F., Drysdale, J. M. Smith, J. M. and Bruce J. Determination of potentially toxic metals in human liver in New Zealand. **J. Anal. Toxicol.;** 1983; **7:** 2-6.
67. Piotrowski, J. K. and Coleman, D. O. Environmental hazards of heavy metals: summary evaluation of lead, cadmium and mercury. MARC Report No 20.; 1980.
68. Piotrowski, J. K. and Inskip, M. J. Health effects of methylmercury. MARC Tech. Rept. 24: MARC; 1981.
69. Piscator, M. Exposure to cadmium. In: Trace Metals: Exposure and Health Effects: CEC and Pergamon Press; 1979: 35-41.
70. Piscator, M. Metabolism and effects of cadmium. In: Management and Control of Heavy Metals in the Environment, Int. Conf. London: CEP Consultants; 1979: 1-7.
71. Probst, G. S. Cadmium: absorption, distribution and excretion in mammals. In: Mennear, J. E., Ed. Cadmium Toxicity: M. Dekker; 1979: 29-59.
72. Rabinowitz, M., Wetherill, G. W. and Kapple, J. D. Lead metabolism in the normal human stable isotope studies. **Science;** 1973; **182:** 725-727.
73. Rabinowitz, M., Wetherill, G. W. and Kapple, J. D. Studies of human lead metabolism by the use of stable isotope tracers. **Environ. Health Perspec.;** 1974; **7:** 145-153.
74. Report of a DHSS Working Party on Lead in the Environment. Lead and Health: DHSS; 1980.
75. Royal Commission on Environmental Pollution. Lead in the Environment: HMSO; 1983; 9th. Report.

76. Russel Jones, R. and Stephens, R. The contribution of lead in petrol to human lead uptake. In: Rutter, M. and Russel Jones, R., Eds. Lead Versus Health: Wiley; 1983: 141-177.
77. Sayre, J. Dust lead contribution to lead in children. In: Lynam, D. R., Piananida, L. G. and Cole, J. F., Eds. Environmental Lead: Academic press; 1981: 23-40.
78. Sayre, J. W., Charney, E., Vostal, J. and Pless, J. B. House and hand dust as a potential source of childhood lead exposure. **Amer. J. Dis. Child.;** 1974; **127:** 167-170.
79. Schmidt, R. F. and Thiers, G. (Eds.). Human Physiology: Springer Verlag; 1983.
80. Schroder, H. A. and Nason, A. P. Trace element analysis in clinical chemistry. **Clin. Chem.;** 1971; **17:** 461-474.
81. Settle, D. M. and Patterson, C. C. Lead in Albacore: guide to lead pollution in Americans. **Science;** 1980; **207:** 1167-1176.
82. Sherlock, J., Smart, G. Forbes, G. I., Moore, M. R., Patterson, W. J., Richards, W. N. and Wilson, T.S. Assessment of lead intakes and dose response for a population in Ayr exposed to a plumbosolvent water supply. **Human Toxicol.;** 1982; **1:** 115-123.
83. Snyder, W. S. (Chairman). Report of the Task Group on Reference Man. Int. Commission on Radiological Protection: Pergamon Press; 1975; No. 23.
84. Solomon, R. L. and Hartford, J. W. Lead and cadmium in dust and soils in a small urban community. **Environ. Sci. Techn.;** 1976; **10:** 773-777.
85. Stephens, R. Human exposure to lead from motor emissions. **Int. J. Environ. Studies;** 1981; **17:** 73-83.
86. Stephens, R. The total relationship between airborne lead and body lead burden. Proc. Symp. Conservation Soc. London; 1978: 1-12.
87. Stokes, P. M. and Wren, C. D. Bioaccumulation of mercury by aquatic biota in hydroelectric reservoirs: a review and consideration of mechanisms. In: Hutchinson, T. C. and Maema, K. M., Eds. Lead, Cadmium, Mercury and Arsenic in the Environment: Wiley, SCOPE; 1987: 255-277.
88. Suzuki, T. Metabolism of mercurial compounds. In: Goyer, R. A. and Mehlman, M. A., Eds. Toxicology of Trace Elements: Wiley; 1977: 1-37.
89. Thomas, B., Roughhan, J. A. and Watters, E. D. Lead and cadmium content of some vegetable foodstuffs. **J. Sci. Food Agric.;** 1972; **23:** 1493-1498.
90. Waldron, H. A. Health standards for heavy metals. **Chem. Brit.;** 1975; **11:** 354-357.
91. WHO. Environmental Health Criteria 1 Mercury: WHO; 1976.
92. Zielhuis, R. L. and Wibowo, A. A. F. Standard setting and metal speciation: arsenic. In: Nriagu, J. O., Ed. Changing Metal Cycles and Human Health, Dahlem Konferenzen: Springer Verlag; 1984: 323-344.

CHAPTER 15

THE TOXICITY OF HEAVY ELEMENTS TO HUMAN BEINGS

INTRODUCTION

A toxic material is a substance that has an adverse effect on health. Many chemicals could be classed as toxic, but some more so than others. The level of toxicity of a substance, relates to the amount that cause an adverse effect and, to some extent the type of effect [44,45,46,55,61,77,79,110]

Many chemical elements are also essential, or at least beneficial to human health, but they also can become toxic when taken in excess. Two curves which express the relationship between the amount and effect or response are given in Figs. 15.1a, b. The situation portrayed in Fig 15.1a applies only to selenium of the ten heavy elements, the rest are explained by the curve in Fig 5.1b. This curve is not meant to suggest that at low concentrations the toxic element is essential, but rather, that it does not appear to have any obvious adverse effect. There is debate over at what level an effect can be considered as adverse. The situation continually changes, as the sensitivity of measurements increase, and very small effects are recognized [61,79].

In some circumstances elements may have either antagonistic or synergic effects on the biological properties of other elements. A toxic element may be either helpful in reducing the toxic effects of another, or add to its toxic effect. For example selenium is antagonistic to mercury and reduces its toxicity. Also some toxic elements are used therapeutically for specific purposes, despite the toxic risks. A list of some of the therapeutic uses of the heavy metals is given in Table 15.1. Some of these applications have been replaced with organic antibiotics and bactericidal agents [61].

The toxic effects of an element are measured by its dose-response relationship, where the response is the sign of an adverse effect. Doses are either acute or chronic. An acute dose is a large amount of a toxic material which produces a rapid onset of effects, often intense and can result in death. A chronic dose is usually a lesser amount but continued over a longer period of time. Therefore

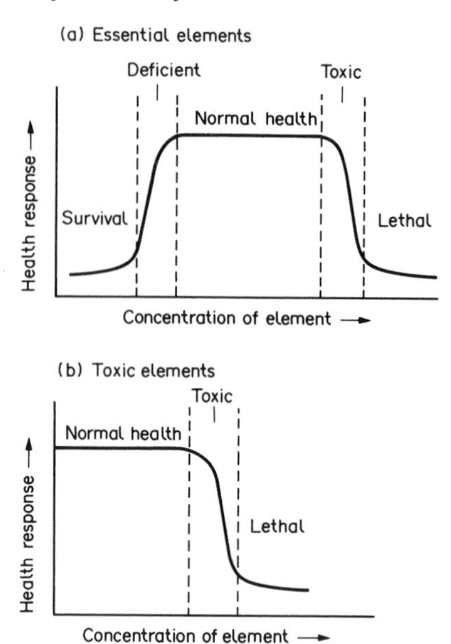

Fig. 15.1 The relationship between health response and the concentrations of the elements, (a) essential elements, (b) toxic elements.

the toxic material has a chance to build up in the body and its adverse effects are seen as a gradual onset of symptoms. At times the symptoms of acute and chronic exposures are different, and the effects may also differ [61].

Another feature which affects the response to a toxic metal is that some people are more sensitive, and therefore more at risk than others. For example

TABLE 15.1 Some Therapeutic Uses of Heavy Metal Salts for Human Beings

Element	Salt	Uses
Hg	Chloride, iodide, salicylate	Antiseptics, disinfectants, fumigants
	Oleate	Parasiticide
In	Radioactive	Locate tumors
Pb	Acetate	Topical astringent
As		Nutritional diseases
		Skin diseases
		Leukemia
Sb	Tartrate	Tropical diseases
Bi	Subcarbonate	Antacid
		Protects stomach ulcers

Source of data; Luckey and Venugopal, 1978 [61].

children are more at risk from damage to the central nervous system by CH_3Hg^+ and lead, older women are more at risk from the damaging effects of cadmium on bone. The S-shaped curve in Fig. 15.2 represents the situation. Some people are very sensitive (lower end of the curve) and some are not, i.e. they are tolerant to the toxin (higher end of the curve). The majority of people lie between these extremes.

For the heavy metals the critical organs affected differ from metal to metal. For example the critical organ for Cd^{2+} and Hg^{2+} ions is the kidney, for CH_3Hg^+ the brain, for lead the haematological system and the brain, and arsenic appears to be non-specific, though the skin is significantly affected. This does not mean to imply however, that other organs or tissues are not also affected. Some of the more affected organs for six of the heavy elements are listed in Table 15.2.

Attempts have been made to derive a scale of toxicity. There is debate over the units to use, either mass (or volume), or moles, or mass per kg of body weight. The scale given in Table 15.3 is for mass or volume per kg of body weight [61]. The problem with such scales is that they ignore other important factors, such as the particular the organ affected and the accessibility of the element.

A number of the heavy metals are toxic because of their interaction with sulphur containing biochemicals, such as enzymes and proteins. Some of the metals are soft acids and interact strongly with soft bases, of which sulphur is an example. Softness is related to the size of the atoms. Larger atoms, which have a lot of electrons, are polarizable and therefore are able to form strong interactions with other polarizable atoms. The interaction also has stereochemical requirements, namely does the bonding situation fit the normal or preferred stereochemistry of the element. Another factor to consider is the kinetics of toxic reactions. Some factors affecting the kinetics are, solubility of the metallic species, bond energies, and accessibility of the metal to a nucleophile [110].

The stability constants for metal organic systems is a useful guide to the biological influence of a metal. The apparent and approximate stability constants at a pH of 7 for the three metals mercury, lead and cadmium bonded to

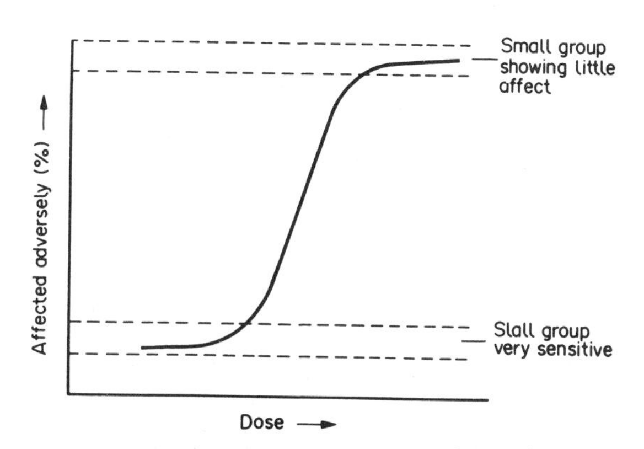

Fig. 15.2 Toxic doses in relation to the number of people affected.

TABLE 15.2 Organs Adversely Affected by the Heavy Elements

Organ/area	Element	Broad health effects
CNS*	CH_3Hg^+, Hg	Brain damage
	Pb^{2+}	Reduced neuropsylogical functioning
	Tl	Brain tumors
PNS**	CH_3Hg^+, Hg	Abnormal movement and reflexes
	Pb^{2+}	Peripheral neurological effects
	As	Peripheral neuropathy
	Tl	Polyneuritis
Renal system	Cd	Tubular, glomerular damage, proteinuria
	Hg^{2+}	Tubular nephrosis
	As	Tubular dysfunction
Liver	As	Cirrhosis
Blood system	Pb	Inhibits biosynthesis of haem
	Cd	Slight anemia
	As	Anemia
Oral, Nasal mucosa	Hg^{2+}	Stomatities
	As	Ulcers
Hair	Tl	Alopecia (loss of hair)
Respiratory tract	Cd	Emphysema
	As	Emphysema and fibrosis
	Hg	Bronchial effects
	Se	Respiratory inflammation
Skeleton	Cd	Osteomalacia
	Se	Tooth caries
Cardiovascular system	Cd	
	As	
Reproductive system	CH_3Hg^+, Hg	Spontaneous abortion
	As	
Teratogenesis	CH_3Hg^+	Deformed brain and body
	Tl	Deformed babies
Cancer	Cd	Prostate gland, lung
	As	Skin, lung
Chromosomal aberrations	Cd	
	As	

* CNS central nervous system, ** PNS peripheral nervous system. Te, Sb, Bi and In are not included. Source of data; Zielhuis, 1979 [111].

different groups are given in Table 15.4 [55]. This data clearly shows the affinity of the metals for sulphur compared with the other donor systems (-O, -N, PO_4^{3-} and Cl^-). The low molecular weight protein metallothionein (MW ~600-7000) which contains 61 or 62 amino acids is produced in the liver and is an important protection against some heavy metals such as cadmium, mercury and lead. One third of the amino acid residues in the protein are cysteine, giving a high sulphur donor atom content to the protein. The cysteine residues are so

TABLE 15.3 A Scale for Estimating the Degree of Toxicity

Degree of toxicity	Oral*	Inhalation**
Practically non-toxic	5000-15,000	10,000-100,000
Slightly toxic	500-5000	1000-10,000
Moderately toxic	50-500	100-1000
Highly toxic	5-50	10-100
Extremely toxic	1-5	5-10
Super toxic	<1	<5

* mg kg^{-1} of body weight to give LD_{50}, ** number of particles m^{-3} and exposure for four hours to give 33-66% mortality.
Source of data; Luckey and Venugopal, 1978 [61].

TABLE 15.4 Apparent Stability Constants for Mercury, Cadmium, and Lead for Different Biochemical Groups

Group	Log K (pH = 7)		
	Hg	Cd	Pb
Carboxyl	5.6	1.8	1.9
Amino	6.0	0.3	-0.5
Imidazole	3.7	2.7	2.2
Thiol	10.2	5.6	4.9
Phosphate	-	2.7	3.1
Chloride	7.3	2.0	1.6

Source of data; Kägi and Hapke, 1984 [55].

arranged that when a metal is coordinated cage structures are produced, such as M_3S_3 and M_4S_6 [57,78,86].

LEAD

Lead has two quite distinct toxic effects on human beings, physiological and neurological. We will discuss these two effects separately, and review the neurological effects in the next chapter.

Physiological Toxicity of Lead

The relatively immediate effects of acute lead poisoning are ill defined symptoms, which include nausea, vomiting, abdominal pains, anorexia, constipation, insomnia, anaemia, irritability, mood disturbance and coordination loss [6,9,21,29,31,32,43,48,62,82,92,95]. In more severe situations neurological effects such as restlessness, hyperactivity, confusion and impairment of memory [43] can result, as well as coma and death. For a group of men cutting lead painted steel, and who were inhaling and ingesting lead, the effects were excessive weakness and fatigue (100%), anaemia (90%), abdominal pains (80%), constipation (50%), pleuritic pains (50%) and anorexia (50%). The blood lead levels of the men were very high at 104-139 μg dl^{-1} [31].

Inhibition of the biosynthesis of haem The first signs of low level lead exposure are effects on the biosynthesis of haem (the iron porphyrin component of haemoglobin). The biosynthetic pathway for the haem group, from glycine and succinyl CoA, is given in Fig 15.3, and the sites where lead interferes are marked. There are five positions for possible interference, and two of these, 2 and 5, are well defined sites where lead inhibits the biosynthesis [5,6,9,21,22, 29,31,32,42,52,62,68,70,76,81,82,87,90,92].

The enzyme δ-aminolevulinic acid dehydrase (ALAD) is very sensitive to lead inhibition. The result is that δ-aminolevulinic acid (ALA) appears in the urine and serum of the person and ALAD accumulates in the serum. The estimation of ALA in urine however, is not sensitive to lead toxicity as it is only detectable at relatively high levels of the metal in blood, around 30-40 μg dl^{-1}. On the other hand increased ALAD activity in the blood is very sensitive to lead and is apparent at blood lead levels around 9 μg dl^{-1} and as low as 4-6 μg dl^{-1} [77]. Hence the effect is discernible for the general population. It has been claimed by some, that the first signs of increased ALAD activity does not represent a health risk, whereas others claim it represents interference by lead in a natural process and therefore is a health problem. A potential risk of ALA elevation is that it

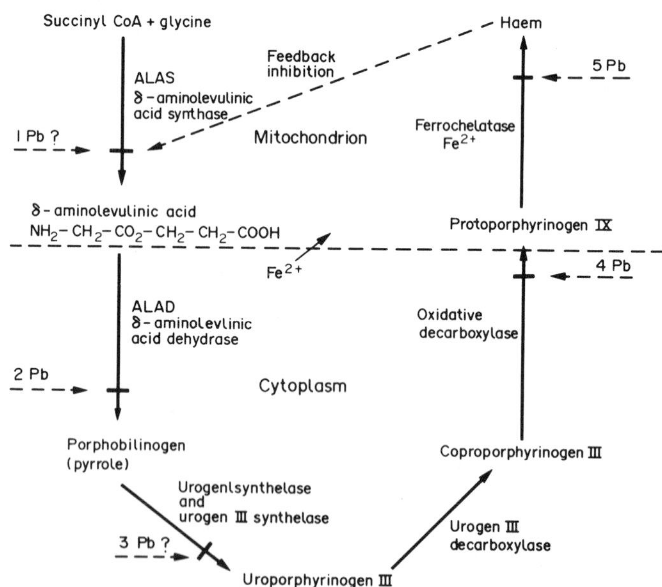

Fig. 15.3 The biosynthesis of haem and sites where lead inhibits the process. Sources; references 5,6,9,21,22, 29,31,32,42,52,62,68,70,76,81,82,87,90,92.

may be involved in neuro-toxic effects, and may get into the brain [52]. The level of lead exposure as measured by blood lead, and the health effects on the haematological system are given in Table 15.5 [9,21,29,31,68, 77,81,89].

The other well defined interference by lead, is the inhibition of ferrochelatase, the enzyme that enables the Fe^{2+} ion to be incorporated into the protoporphyrin IX. The effect of this is that protoporphyrin IX accumulates in the erythrocyte (red blood cells). This has been given various names viz. free erythrocyte protoporphyrin (FEP), erythrocyte protoporphyrin (EP) and zinc protoporphyrin (ZnPP) [9,21,29,31,52,81]. The first name is incorrect as the protoporphyrin is not free, as zinc takes over the place of iron when it is not available [9]. A rise in the erythrocyte protoporphyrin in blood is apparent at blood lead levels around 12-15 $\mu g\ dl^{-1}$. Its analytical detection is assisted by the strong fluorescence of ZnPP. Because the iron does not go into the protoporphyrin it accumulates in the blood (e.g. ferritin), and in tissues that store iron [21,29,31,81]. The appearance of the erythrocyte protoporphyrin in the blood, is said to be the first sign of an adverse effect of lead, and also indicates some impairment of certain mitochondrial functions, where the processes are occurring [2,9,21,29,31,52,77,81,89].

The other sites in the biosynthesis of the haem group, where lead interferes are marked 1, 3 and 4 in Fig. 15.3. In these cases the inhibition is not so well characterized, for example haem in blood also has an inhibitory feedback on the activity of ALAS. The effects of inhibition at sites 3 and 4 are to increase the amounts of porphobilinogen, uroporphyrins and coproporphyrinogen III (CP) in urine, especially in cases of severe lead poisoning [9,21,29,31,81].

Estimates of the dose (blood lead concentration), and response (species in blood) relationships for adults and children calculated by probit analysis are given in Fig 15.4 [81]. These results, some of which are tentative, give the thresholds for blood lead concentrations as: ALAD activity 9 and 11 $\mu g\ dl^{-1}$ for

TABLE 15.5 Blood Lead Levels and Health Effects on the Haematological System

PbB $\mu g\ dl^{-1}$	Health effect
4-6	Inhibits ALAD
15-25	Rise in blood porphyrin (children)
12-31	Rise in blood porphyrin (female adults)
19-35	Rise in blood porphyrin (male adults)
25	Anemia in Children
31-39	Inhibits red cell ATPase
34-38	Increase in urinary ALA and coproporphyrin
50	Decreased blood haemoglobin (adults)
63	Anemia in adults

Sources; references 9,21,29,31,68,77,81,89.

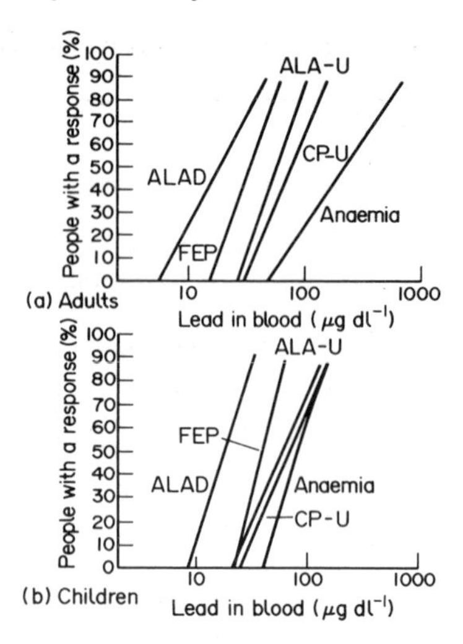

Fig. 15.4 Lead dose response relationship for (a) adults and (b) children. Source of data; Piotrowski and O'Brien, 1980 [81].

adult men and children respectively, signs of EP 19 and 21 μg dl^{-1}, ALA in urine 34 and 38 μg dl^{-1} CP in urine 37 and 41 μg dl^{-1}, anaemia 63 and 25 μg dl^{-1} [81].

Other physiological health effects In the renal system the initial damage occurs to the peripheral tubular system, resulting in reduced re-absorption of glucose and α-aminoacids, and the appearance of materials in the urine. The effects are aminoaciduria, glucosuria, hyperphosphaturia and also hyperphosphatanemia. Chronic lead nephropathy may develop where the kidney contracts with anteriosclerotic changes, and glomerular atrophy. Sometimes renal failure can occur [6,9,21,22,29,32,42,52,62,70,76,82,92]. These effects occur for persons occupationally exposed, or people who have consumed lead over a period from unsuspected sources, such as liquids kept in lead glazed vessels.

Other health effects where lead has been implicated are cot deaths (sudden infant death syndrome) [9,30], and still births [9,14,82]. Experimental results indicate elevated tissue lead levels for the infants, compared with infants who died for other reasons. For pregnant women elevated lead concentrations increase the risk of hypertension and birth defects [89]. It is clear from the effect of lead on the biosynthesis on haem that the metal influences the activity of enzymes [31,92]. The functions of the two endocrine organs, the thyroid and the adrenal, are impaired in some cases by lead, interfering with the biosynthesis of tryptophan [29,70,76]. Some human reproduction functions may also be affected by lead [22], and in some countries women are not allowed to work in a lead contaminated environment. A list of some of the health effects on children and adults of lead in relation to lead exposure as measured by blood lead levels is given in Table 15.6 [22].

TABLE 15.6 Effects of Lead at Different Blood Lead Levels

Blood lead level $\mu g\ dl^{-1}$	Effect
4-6	Inhibits ALAD
15	Elevation of EP in blood
20-25	Chromosomal abnormalities
30	Toxicity to fetus
30-40	Reduced fertility to women
30-40	Altered spermatogensis for men
40	Reduced peripheral nerve conduction
40	Reduced hemoglobin synthesis
40-60	Psychological sensory and behaviour changes
50	Impaired kidney function
>50	Anemia
50-60	Peripheral neuropathy
100-120	Encephalopathy

Sources of data; references 22,77,89

There is no evidence available that lead has carcinogenic or teratogenic effects on human beings [29,75,82,92]. Lead is stored in bone, and whereas no health effect has been associated with the metal in bone, the lead can become remobilized under stressful circumstances, in the same way that calcium is mobilized. Therefore lead in bone is a potential source for other tissues [22,95].

Some of the relationships between lead and other metals are summarized in Table 15.7. The general effect is that lead competes with other elements in the body and therefore produces an apparent deficiency of the essential element [1,11,62,94].

TABLE 15.7 The Interaction Between Lead and Other Elements in Biological Systems

Element	Effect of lead
Fe	Lead competes with iron in the intestine, inhibits the incorporation of iron into protoporphyrin IX, gives the effect of iron deficiency and anemia.
Ca	Lead increases calcium deficiency, on the other hand calcium can alleviate lead toxicity, hence the reason why lead workers are asked to drink milk.
Zn	Lead interfers with zinc enzymes and added zinc can alleviate the effects of lead.
Cu	Lead increases copper deficiency.

Sources; references 1,11,62,94.

Treatment Because of the widespread occurrence of lead poisoning among workers in lead industries, and accidental poisoning among the general population, methods have been developed to remove lead from the human system. The main method is to add a reagent to the blood stream, intravenously or subcutaneously, which complexes with lead and removes it from the body [9,13,17]. The reagent $CaNa_2EDTA$ chelates lead in the extracellular fluid, and is excreted in urine, whereas 2,3-dimercaptopropanol, $HSCH_2$-C(SH)-CH_2OH, (British Anti-Lewisite reagent, BAL) removes lead directly from intracellular space and is excreted in urine and fæces. D-penicillamine, HS-$C(CH_3)_2$-$CHNH_2COOH$, can also be used to coordinate lead and remove it from the body. In severe situations a number of treatments are necessary, because remobilization of lead from storage tissues (e.g. bone) produces the toxicity again.

MERCURY

Of the ten heavy elements mercury has featured in the most serious outbreaks of heavy metal poisoning among the general population, especially in Japan, Iraq and Nth. America. We will discuss some of these outbreaks in more detail below.

The toxicity of mercury occurs at three levels, depending on the chemical form of the element. The order of decreasing toxicity is [56]; Alkyl Hg (esp. CH_3Hg^+) > Hg metal vapour > Hg^{2+} salts and phenyl and methoxy mercury salts. The last two organic species have weak C-Hg bonds, which are easily broken (see Chapter 13). Hence the species soon transform to Hg^{2+} in the body. This does not apply to the C-Hg bond in CH_3Hg^+, and its persists in the body.

Mercury has well characterized toxic effects on both the physiological and the neurological systems of the body. The most destructive is the effects of CH_3Hg^+ on the central nervous system.

Mercury Metal

The volatility of elemental mercury, and its use in a number of circumstances, means it is a serious toxin, especially in the occupational environment. The critical organs are the lungs, kidneys and the brain. The vapour crosses the blood-brain barrier, and the metal is oxidized in the brain and in the blood stream [39,76,84,103,108]. Once oxidized the mercury remains in the brain, whereas the free metal may move out again.

The effects of mercury vapour on the respiratory tract are coughing, acute bronchial inflammation, chest pains and in severe cases respiratory arrest [39,56,108]. Studies of workers exposed to mercury vapour have indicated a range of effects including loss of appetite, tremors, insomnia, shyness, diarrhea, vomiting and soreness in the oral cavity [12,39,40,76,108]. Russian workers have reported an asthemic-vegetative syndrome, which, when associated with mercury is called micromercurialism. Symptoms are decreased productivity, fatigue, loss of self confidence, muscular weakness and depression. Other symptoms are enlargement of the thyroid gland, haematological changes and excretion of mercury in the urine [39,76,108]. Some of these effects indicate disorders of the central and peripheral nervous systems. Renal effects from high levels of mercury vapour exposure are proteinuria [76,103].

Another form of metallic mercury that many people have contact with is the mercury/tin amalgam in teeth [56,60,98,106]. It is regarded that the amalgam has a very low solubility in saliva and gastrointestinal fluids. However, dentists and dental assistants are exposed to the mercury vapour, and elevated mercury levels have been found among dental workers. Levels of mercury have been found elevated in the blood of people for a few days after they have had a filling inserted. This could arise from either mercury vapour while the amalgam is hardening, or from absorption from amalgam particles that get into the gastro-intestinal tract or both. Whether or not abrasion over the years continually adds mercury to the body is not really known, though people with fillings do have higher mercury levels in their breath than those with no fillings. Guinea pigs, who wear their teeth down rapidly, have been found with elevated mercury levels in their brain, kidney, liver and heart when they had amalgam fillings [37]. Some people have an allergic reaction to mercury on their lips and the oral mucus membrane [56,60].

Inorganic Mercury Salts

Inorganic mercury salts are absorbed into the body less than mercury vapour, but the amount depends on the solubility of the species. The critical organ from intake by inhalation or ingestion is the kidney and the effect has been called sublimate nephrosis. Inorganic mercury(II) also has an adverse effect on the central nervous system.

Acute ingestion of inorganic mercury can cause precipitation of protein in the gastrointestinal tract and produce gastric pain, vomiting and bloody diarrhea. Renal damage can occur, including oliguria, severe anuria with azotemia and in severe cases renal failure. Fatal effects have been associated with levels of mercury in the kidney of around 16 $\mu g\ g^{-1}$ [66]. For chronic exposure, the renal effects recorded are proteinuria, albuminuria and œdema. The renal lesions produced by mercury are mainly on the tubular epithelium. Damage also occurs to the oral cavity [8,39,56,62,66,76,84,108].

The neurological effects are much the same as reported for other chemical forms of mercury. These include tremors, erethism - a psychiatric disturbance where a person is very quickly aroused to irritability and excitability, as well as changes in personality and behaviour - fatigue, loss of memory and self confidence, and development of idiosyncrasy. In severe cases delirium with hallucinations and manic-depressive disorders have been reported [39,56].

Methylmercury

The most serious mercury toxin is methylmercury both in its effects and its availability in the past. This form of mercury is now less available, however, mercury in sediments will continue to be methylated and bioaccumulated by fish for years to come. Hence there is a need for caution over the exposure of people to CH_3Hg^+, especially from the eating of fish. We will discuss the neurological health effects of methylmercury and then some of the more dramatic and serious outbreaks of this form of mercury poisoning.

Neurological effects of methylmercury Methylmercury readily crosses the placental and blood-brain barriers and it causes disintegration (lysis) of cells within the brain. This may involve a CH_3Hg-S interaction. Of the methylmercury ingested and absorbed into the body 90-95% becomes associated with the red cells and 5-10% with the plasma. It is the CH_3Hg^+ in the plasma that gets into the brain.

The main areas of brain function that are damaged by methylmercury are those that control sensory, visual, auditory and coordination. Therefore it appears that methylmercury is selective in its influence on the brain. The pathological changes observed, are in the cerebrum, where cortical damage to visual areas of the occipital cortex occurs, some damage, of varying intensity, to the granular layer in the cerebellum, and some significant changes to white matter in the brain [8,18,20,24,39,40,52,54,56,62,67,76,83,84,88,103,105,108].

The effects observed on human beings are, initially loss of sensation at the extremities and around the mouth (paresthesia) followed by loss of coordination in movement, e.g. walking (ataxia), slurred speech (dysarthia), loss of hearing, restricted visual field, blindness, coma and death. When a person suffering these effects is removed from the source of mercury some improvement occurs, but this does not represent reversibility of the damage but rather that some other brain function takes over part of the loss. Deafness is not restorable at all. Since the brain damage is not reversible it is therefore cumulative [39,40,54,56,62,76,83,88,105,108]. The frequency percent of some of the symptoms found among people in Minamata are listed out in Table 15.8 [105].

Pregnant women, and unborn children, are most sensitive to to CH_3Hg^+ toxicity, especially the developing brain system of the child [18,52,67]. Infants born to mothers with high intakes of mercury have had serious mental disturbances, including retardation of mental and physical development. Studies of the brains of such infants, who died, indicated signs of severe brain damage, which was not just restricted to focal neuronal damage but also faulty development of brain structure caused by abnormal migration of neurons to the cerebellar and cerebral cortices. A reduction in brain mass of 26-55% was reported and up to 50% or more loss of neurons was discovered [54,83].

There appears to be a latent period, between the intake of methylmercury and the appearance of the symptoms. For example for people poisoned at Niigata (see below) the sensory disturbance at the extremities started first and

TABLE 15.8 Frequency of Symptoms Experienced by People at Minamata Suffering from Methylmercury Poisonong

Symptom	Congenital %	Post-natal Children %	Post-natal Adults %
Mental disturbance	100	100	71
Disturbed coordination	100	100	80-90
Disturbed gait	100	100	82
Disturbed speech	100	94	88
Disturbed hearing	5	67	85
Constricted visual field	?	100	100
Impaired chewing & swallowing	100	89	94
Enhanced tendon reflexes	82	72	38
Pathological reflexes	54	50	12
Involuntary movements	73	0	0
Salivation	77	56	24
Forced laughing	27	29	-

Sources of data; references 54,83,105

was prevalent after 3 years, whereas disturbance of the perioral area, and ataxia, both began after 1 year and were prevalent by 4-5 years, and the constricted visual field was delayed until 4 years [54,83].

Outbreaks of methylmercury poisoning There have been a number of outbreaks of mercury poisoning throughout the centuries, some of which have been well documented. Mention has already been made (Chapter 1) of the term 'mad as a hatter' which may have derived from the effects on people from the use of $Hg(NO_3)_2$ in the felting of hats [26,60]. Some big and severe outbreaks of mercury poisoning have occurred in recent times and these are listed in Table 15.9. Though the source and chemical form of mercury differs from event to event, the common toxin is CH_3Hg^+. The sources are; methylmercury fungicide seed dressing, the effluent from factories making vinyl chloride and acetaldehyde, in which Hg^{2+} is a catalyst and which is discharged along with methylated mercury formed during the process, and metallic mercury from the chloroalkali cells using mercury as the cathode. In the latter two cases methylation of Hg and Hg^{2+} occur in the aquatic sediments [19,20,26,39,52,54,60,65,66,83,105,108].

Minamata disease, or poisoning, was first reported in 1953 when a child showed signs of brain damage, and by 1956 78 cases had been reported, of which seven were fatal. The cause originally considered as infectious was diagnosed as heavy metal poisoning and finally as mercury poisoning in 1959. The source of the mercury for the people (which was CH_3Hg^+) was the fish and shellfish living in Minamata Bay. The fish obtained the mercury from the

TABLE 15.9 Outbreaks of Methylmercury Poisoning in Recent Times

Date	Place	Effect	Source of mercury
1953-60's	Minamata Japan	>3000 cases 100 deaths	CH_3Hg^+ and other mercury products from a plastics factory
1956	Iraq	100 cases	From eating CH_3Hg^+ treated grain
1960	Iraq	1000 cases	From eating CH_3Hg^+ treated grain
1963-65	Guatemala	45 cases	From eating CH_3Hg^+ treated grain
1965	Niigata Japan	669 cases 55 deaths	CH_3Hg^+ and other mercury products from a plastics factory
1967	Ghana	150 cases	From eating CH_3Hg^+ treated grain
1969	Pakistan	100 cases	From eating CH_3Hg^+ treated grain
1970's	Canada		Mercury from chloroalkali cells in paper manufacture
1971-72	Iraq	6350 cases 459 deaths	From eating CH_3Hg^+ treated grain

Sources of data; references 19,20,26,39,52,54,60,65,66,83,105,108.

discharges from a nearby plastics factory. Incidences of the disease (poisoning) followed a seasonal trend with more in the summer when the fish catches were greater. The situation in Niigata was very similar to that of Minamata [54, 65,83,105].

The Iraqi episode in 1971-72 has been the biggest single event where close to 6350 people were hospitalized, of whom 459 died. Most people affected were in the age range 5-19 years. The source of the mercury was a shipment of CH_3Hg^+ treated wheat and barley seeds, meant for sowing, but was used for making flour for bread [105].

The Canadian Indians, in certain areas of Canada, have also suffered from mercury poisoning. It appears that the same syndrome is prevalent and arises from the eating of fish. The source of the mercury is from pulp and paper mills where caustic soda and chlorine are made using the mercury cathode cell for the electrolysis of aqueous sodium chloride. Methylmercury formed in the river sediments and consumed and bioaccumulated by the fish and then consumed by the local people [26,52,54,83].

Other areas where the likelihood of mercury toxicity from fish consumption have been identified are Peru, Papua-New Guinea and some coastal regions of the Mediterranean [54,83].

Thresholds The thresholds for methylmercury poisoning have been estimated for a number of tissues and for a particular symptom. A list of some of these are given in Table 15.10. There is broad agreement between the values, but differences can be expected when diverse experimental conditions are used. The onset of paresthesia occurs at a body burden of 0.4-0.5 mg kg^{-1}, which on the scale of toxicity given in Table 15.3 means the methylmercury is super toxic. A fatal body burden is around 4 mg kg^{-1}, which would be classed as extremely to

TABLE 15.10 Threshold Concentrations for Mercury in the Human Body

Tissue or fluid	Threshold level	Comments
Body burden	25 mg	Onset of paresthenia
51 kg person	55 mg	Onset of ataxia
	90 mg	Onset of dysarthia
	170-180 mg	Onset of deafness
	200 mg	Death
70 kg person	30 mg	Onset of paresthenia (Niigata)
	2-3 $\mu g\ g^{-1}$	Signs and symptoms (Iraq)
Blood	8 $\mu g\ dl^{-1}$	First signs and symptoms
	100 $\mu g\ dl^{-1}$	Toxic
	10 $\mu g\ dl^{-1}$	Toxic for pregnant women
	5 $\mu g\ dl^{-1}$	Toxic for fetus (maternal blood)
	20 $\mu g\ dl^{-1}$	Onset of symptoms (Sweden)
	20 $\mu g\ dl^{-1}$	Chromosome damage
	30-40 $\mu g\ dl^{-1}$	Fetal damage (brain level 3 $\mu g\ g^{-1}$)
	≥140 $\mu g\ dl^{-1}$	Fatal (brain level 12 $\mu g\ g^{-1}$)
Brain	5 $\mu g\ g^{-1}$	
	>6 $\mu g\ g^{-1}$	Death (Minamata)
	1.2-3.4 $\mu g\ g^{-1}$	Critical level (Japan/Iraq)
Hair	20 $\mu g\ g^{-1}$	First signs and symptoms
	30 $\mu g\ g^{-1}$	Toxic for pregnant women
	15-20 $\mu g\ g^{-1}$	Toxic for fetus (maternal hair)
Liver	>10 $\mu g\ g^{-1}$	Death (Iraq)

Sources of data; references 52,54,56,83,88,103,105.

super toxic. The threshold estimates, for the usual symptoms, taken from the Iraqi experience are shown in Fig 15.5 [7,103]. A mercury intake of 200 $\mu g\ d^{-1}$, will produce the onset of paresthesia with an 8% risk [18].

Treatment of Mercury Toxicity

The treatment of mercury poisoning entails an increase in the rate of elimination of mercury from the body, either by forming a species (e.g. a mercury complex) which allows the metal to be excreted or trapping the metal in the gastrointestinal tract. A number of reagents have been used to complex the mercury. Two promising reagents are N-acetyl d,l-penicillamine CH_3-CO(NH)-CH(C(CH_3)$_2$-SH)-CO_2H, and dimercaptosuccinic acid HO_2C-CH(SH)-CH(SH)-CO_2H. Other materials used for elimination include BAL, penicillamine, glutathione, and EDTA. A thiol resin has been used for trapping the metal in the gut [64,88,105,108]. Most of the reagents make use of the strong interaction between mercury and sulphur.

Diet selenium also protects against mercury toxicity and appears to release CH_3Hg^+ from sulphur, probably by the formation of a CH_3Hg-Se bond. This

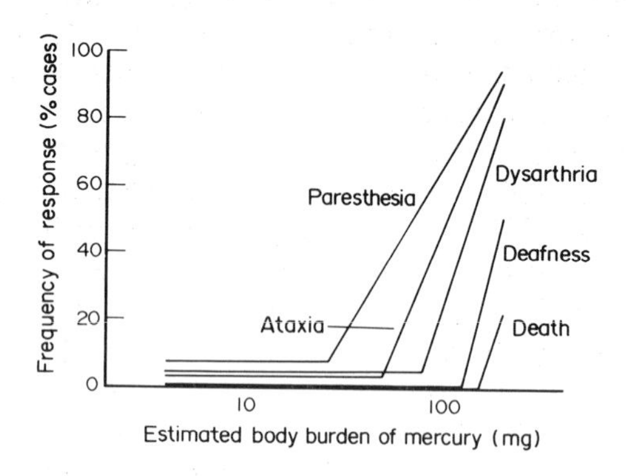

Fig. 15.5 Relationship between signs and symptoms of mercury toxicity and body burden of mercury at the onset of symptoms. Sources; Bakir et al., 1973 [7], Suzuki, 1979 [103].

does not eliminate the mercury from the body, but redistributes it. Also CH_3Hg^+, when interacting with selenium (selenite) produces $(CH_3)_2Se$, and the less toxic Hg^{2+} [24,40,54,56,62,64,76,83,106].

CADMIUM

As with mercury, cadmium is a very toxic metal, and has been responsible for a number of deaths. The most serious situation being the disease called Itai Itai disease. The major effects of cadmium poisoning are experienced in the lungs, kidneys and bones.

Acute effects of inhalation are bronchitis and pneumonitis, and toxemia in the liver [38,61,62,63,76,107]. Effects can be fatal if the dose, such as 8 mg m^{-3} of CdO fume, is inhaled for five hours [34]. The acute effects of oral intakes of cadmium are excess salivation, nausea, vomiting, abdominal pains, diarrhea, vertigo and for large doses, loss of consciousness [34,59,61,62,76]. A lethal dose of cadmium would be >350-500 mg [59].

Chronic inhalation of cadmium compounds as fumes or dust produce pulmonary emphysema, where the small air sacs of the lungs become distended and eventually destroyed reducing lung capacity. The effect becomes apparent after a number of years of occupational exposure [13,34,35,38,59,61,62,63,73, 76,85,107]. Cadmium induced emphysema was first discovered in 1948 [38] and appears to be less common since the late 1970's [59]. This is probably because of improved working conditions.

Both the chronic inhalation and oral intakes of cadmium affect the kidneys producing in the first instance proteinuria, similar to proximal tubule proteinuria [10,13,34,35,38,52,53,59,62,71,73,74,76,77,82,84,85,90,96,107]. The renal damage, which is first seen by the appearance of low molecular weight (MW

12,000-30,000) proteins in the urine, is caused by the impaired re-adsorption function of the proximal tubules. The tubules are part of the filtering system that separates waste materials, such as urea, creatinine and sulphate from necessary materials which are reabsorbed back into the blood stream. The low molecular weight proteins excreted are α_2-, β_2- and γ-globulins, such as β_2-microglobulin (MW ~11,8000), retinol binding protein (MW ~21,000), lysozyme and γ-globulin L-chains [10,38,52,53,71,74,84,90,96]. This type of proteinuria is not specific to cadmium, and care is needed in population studies to have adequate controls, and matched for age [53]. The most common measure of the renal effect is the amount of β_2-microglobulin in the urine, and its presence is said to represent an adverse health effect. Retinol binding protein can also be used as a measure of the toxicity. A later, and maybe independent effect, is the appearance of high molecular weight proteins, such as albumin and transferrin from a glomerular dysfunction of the kidneys [38,53,74,90]. Later effects of further renal damage is aminoaciduria, phosphaturia, glucosuria and calcium in urine [38,53,52, 71,85]. The appearance of Ca^{2+} and PO_4^{3-} indicates that, as a consequence of the damage, there is interference in the metabolism of the two ions [35,38,85], and this shows up in an increased incidence of kidney stones for people with cadmium poisoning [85,96]. The critical level of cadmium in the kidneys is 100-300 $\mu g\ g^{-1}$ (wet weight) and a mean of 200 $\mu g\ g^{-1}$ (ww). In addition to dose the exposure time is also important [38,59,74,82].

The dramatic toxic effect of cadmium is the development of Itai Itai disease, where the outcome is osteomalacia, which is a softening of bones usually produced by a deficiency of vitamin D. The early signs of the problem is pain in the joints, lumbago pains and pseudo-fracturing of the bones. It is an intensely painful disease leading to deformity of the bones [13,34,35,38,59,71, 72,73,82,85,107].

In the Jintsu area of Japan, where the disease was first recognized in the 1950's and 1960's, it was the older women, who had been through a number of pregnancies, and had nutritional and vitamin D deficiency, contracted the disease. The problem has been recognized elsewhere in the world, and cadmium is a contributing factor. Its intervention is probably adversely affecting the Ca^{2+}, PO_4^{3-} and vitamin D metabolism in the body [13,34,35,38,52,59,62, 71,72,73]. Administration of Vitamin D relieves the symptoms [72].

The estimated intake of cadmium by the women in the Jintsu area, was 600 $\mu g\ d^{-1}$ over periods as long as 20 years, more than 10 times typical intakes [34, 38,72,82]. Levels of cadmium in the rice were around 1 $\mu g\ g^{-1}$, and sometimes as high as 3 $\mu g\ g^{-1}$. The close match between the incidences of the disease and levels of cadmium in the soils, used to grow the rice, is quite striking [38,72]. Up to early 1981, the number of deaths reported numbered 66, with many more diagnosed or suspected to have the disease [4].

Other influences cadmium can have on human health are less well defined. Its effect on enzymes is to replace essential elements, such as zinc and therefore render the enzyme biologically ineffective. This takes place because of the competition between cadmium and zinc for the binding sites on the enzyme, and where these are sulphur cadmium has a distinct chemical advantage, being a softer acid than zinc [35,55,62,94].

Cases of high blood pressure or hypertension have been attributed to cadmium toxicity. The evidence is more conclusive for animals than human beings, and epidemiological studies are ambiguous. Some people with hypertension have been found with elevated cadmium in their kidneys and arterial walls and an elevated Cd/Zn concentration ratio, but whether this is causative is not known [10,34,35,38,61,62,95,101]. Slight anaemia has been associated with cadmium toxicity. This may occur because of competition between cadmium and iron in the body, giving the appearance of iron deficiency [35,38,62,85].

The carcinogenic effect of cadmium has been established in animals. The situation for human beings is less clear, but there has been found a relationship between cadmium exposure in industry and incidences of prostatic cancer. A large scale epidemiological study seems warranted [35,38,59,61,62,67,75, 82,102]. Cadmium sulphide may have some mutagenic effects [38]. Testicular destruction has been induced in animals by cadmium, but there is no evidence that this occurs for human beings. The cadmium levels in the bones of still births has been reported as being higher than in normal people. Whether cadmium is implicated is not clear [14].

The biological interaction of cadmium with other elements is summarized in Table 15.11 [11,34,62,82,85,94,96]. The principal effect is competition between cadmium and the other elements generating a deficiency of the other element.

The main protein that cadmium is associated with is metallothionein, which has a stoichiometry of seven cadmium atoms per protein molecule [13,55, 57,76,78,84,86,102]. Cadmium, like some other metals, induces the synthesis of metallothionein in the liver. It is likely that the metallothionein-cadmium complex is then transported to the kidney where it is absorbed by the proximal tubule cells. The acute effects of cadmium may arise from the metal not being bound to the metallothionein, and chronic effects when the capacity of the

TABLE 15.11 The Biochemical Interaction of Cadmium with Other Elements

Element	Effect of cadmium or other element
Zinc	Cadmium interfers with metabolism, replaces in enzymes
Copper	Cadmium interfers with metabolism, creates deficiency
Iron	Cadmium interfers with metabolism, creates deficiency
Calcium	Cadmium interfers with metabolism, creates deficiency
Selenium	Selenium protects against cadmium
Manganese	Manganese can protect agaist low level cadmium

Sources of data; references 11,34,62,82,85,94,96.

metallothionein to bind cadmium becomes exceeded [84]. It is at this stage that cadmium becomes more evident in the urine (see Fig 13.5 p471).

A summary of the principal sites where cadmium exhibits its toxic effect in human beings, and the resulting health effect is given in Table 15.12. Therapy for cadmium poisoning is not readily available. For example, removal of cadmium by a complexing agent takes the metal through the kidneys, which may add to the renal effects of the element [63,107].

ARSENIC

Of the ten heavy elements the toxicity of arsenic is probably the one people know most about. This is because of its use in murder, in crime fiction anyway, and because it was once available as a home garden herbicide.

The chemical forms of arsenic are important as regards its toxicity. The toxicity of the element decreases in the order, As(III) > As(V) > organoarsenic [3,15,36,62,84]. The gas arsine, AsH_3, is particularly toxic [36,62]. It appears that one of the bodies defences against arsenic is alkylation, producing the less toxic alkylarsenic(V) compounds $CH_3AsO(OH)_2$ and $(CH_3)_2AsO(OH)$, which are then excreted from the body [15,27].

The acute effects of arsenic poisoning by oral intake are, intense abdominal pains, nausea, vomiting, diarrhea from damage to the gastrointestinal tract,

TABLE 15.12 The Health Effects of Cadmium on Human Beings

Exposure	Site	Health effects	Comments
Acute	Lungs	Bronchitis, pneumonitis, can be fatal	
	Gastro-intestinal tract	Nausea, vomiting, abdominal pains, gastroententis, diarrhea, vertigo, can be fatal	
	Liver	Toxemia	
Chronic	Lungs	Emphysema	
	Kidneys	Proteinuria, aminoaciduria, phosphaturia, glucosuria, Cd^{2+} in urine, kidney stones	Due to proximal tubule and glomerular dysfunction
	Bones	Intense pain, lumber pains, pseudofracturing, osteomalacia	May also require vitamin D and nutrition deficiency
	Enzymes	Inactivate enzymes	Zn replaced by Cd
	Blood system	Hypertension	In animals
		Anemia	Cd competes with Fe
	Prostate gland	Cancer	Needs confirming
	Other metals	Creates a deficiency	

finally coma and death. Respiratory effects from inhalation are, irritation in the nose and throat. Also neurological effects occur such as headache, vertigo, restlessness and irritability [8,62].

Trivalent arsenic interferes with enzymes by bonding to HS- and HO-groups, especially when there are two adjacent HS- groups in the enzyme. Arsenic inhibits enzymes, such as the pyruvate oxidase system, δ-aminoacid oxidase, choline oxidase and transaminase [8,27,62]. The strong bond between As(III) and sulphur, may be the reason why arsenic accumulates in the keratin tissues hair and nail [62].

Arsenic affects the skin, and in its most severe form causes skin cancer. The signs of arsenic toxicity on the skin, are hyperpigmentation, especially in areas not exposed to the sun, and hyperkeratosis on the palms of hands and soles of feet [8,52,58]. In the early stages arsenic may give the appearance of a 'milk and roses' complexion [8], which has led to the mistaken idea that arsenic was good for the complexion. The black foot disease (peripheral arteriosclerosis and gangrene) is the result of long time exposure [8,58]. The ultimate outcome after prolonged exposure is skin cancer [27,49,52,58,60,61,62,77,84,112]. An estimated intake of 200 $\mu g\ l^{-1}$ of arsenic in drinking water over a long period, is calculated as the threshold for skin cancer as found in Taiwan and Chili [52,77]. The usual arsenic levels in drinking water are less, and more like 2 $\mu g\ l^{-1}$ [52]. Most examples of arsenic generated skin cancer are from occupational exposure, or long term intake from a source which was not recognized as containing arsenic. Arsenic has also been implicated in lung cancer, especially when the arsenic compound inhaled is of low solubility. However, the evidence for this effect is less strong [27,49,52,58,60,61,62,77,84,112].

Arsenic has an effect on the liver [8,36,52,58,60,62] producing cirrhosis of the liver, and a rare form of cancer called haemongioendothelioma. These effects have been observed when a solution of potassium arsenite and potassium carbonate, called Fowler's solution, was used medicinally [52,58,60]. Fowler's solution was used for a number of illnesses, including malaria, asthma, cancer, epilepsy and tuberculosis, to name a few [60]!

Whereas arsenic(III) is regarded the more toxic form of the element, arsenic(V), as arsenate, can be disruptive by competing with phosphate. For example arsenate uncouples oxidative phosphorylation [3,8,36,62]. Oxidative phosphorylation is the process by which adenosine-5'-triphosphate (ATP) is produced, while at the same time reduced nicotinamide adenine dinucleotide (NADH) is oxidized [23].

$$3ADP + 3H_3PO_4 \rightarrow 3ATP + 3H_2O$$

$$NADH + H^+ + 1/2O_2 \rightarrow NAD^+ + H_2O$$

The arsenate disrupts this process by producing an arsenate ester of ADP, which is unstable and undergos hydrolysis non-enzymatically. Hence, the energy metabolism is inhibited and glucose-6-arsenate is produced rather than glucose-6-phosphate [3]. Arsenic may also replace the phosphorus in DNA, and this appears to inhibit the DNA repair mechanism [36,58,62].

A number of other health effects of arsenic have been reported. The element can affect the blood system and interfere with the porphyrin biosynthesis, and affect the white blood cells [8,36,52,58]. Effects on the reproductive system, such as spontaneous abortions, have been attributed to arsenic [52]. Loss of hearing has also been associated with large scale arsenic poisonings. Incidences, such as contamination of milk in Japan and emissions from copper smelters, have caused hearing loss [52]. Similar to many of the other heavy elements, arsenic also affects the renal system, particularly the reabsorption process [8,36]

The peripheral nervous system is reported to be influenced by arsenic, such as peripheral neuritis and motor sensory paralysis [8,52]. Abnormal electromyographic effects i.e. nerve impulses to muscles, have been observed for people with arsenic concentrations in hair >1 $\mu g\ g^{-1}$ [49]. A summary of the health effects of arsenic are summarized in Table 15.13.

Arsenic and selenium are antagonistic to each other in the body, and each counteract the toxicity of the other. However, arsenic may also interfer with the essential role of selenium in human metabolism (see below) [36,62].

An organoarsenic compound called Lewisite, $ClCH=CH\text{-}AsCl_2$, was developed as a poison gas in 1918. It has broad inhibitory properties on metabolic processes. An antidote was developed called British-Anti-Lewisite (see above for lead therapy) [3].

SELENIUM

Selenium is the only element of the ten heavy elements which is both essential and toxic. Therefore selenium fits the dose-response curve (Fig. 15.1a) described in the beginning of this chapter. The interval between a deficient concentration (<0.04 $\mu g\ g^{-1}$) and a toxic concentration (>4 $\mu g\ g^{-1}$) in food is a factor of 100, which appears large, but in fact is not because both the concentrations are so low. A lethal dose in food is 0.1-0.5 μg g-1 The curve applies to animals, where it has been demonstrated that the element is essential, but also toxic in excess. The element is toxic for human beings, but increasing evidence suggests that it is also essential, in the same way as for animals. Therefore people may suffer either from a deficiency or from too much. The general population are more likely to suffer from the first problem, whereas an excess selenium intake is more likely to occur in industrial situations.

Some of the diseases of animals (e.g. cattle and sheep) which are responsive to selenium are muscular dystropy called white muscle disease, exudative diathesis i.e. seepage and accumulation of fluid throughout the body, and hepatosis dietetica which is destructive lesions on the liver [97]. In addition animals can become ill and die from too high a selenium intake. Animals may take in too much selenium by eating plants that are high in selenium either because the plant is a selenium accumulator (see Chapter 10), or selenium in the

TABLE 15.13 The Toxic Effects of Arsenic on Human Beings

Affected area	Toxic effects
Skin	Hyperpigmentation, hyper-keratosis, black foot disease, gangrene, skin cancer
Lungs	Lung cancer, needs confirmation
Liver	Cirrhosis, haemongioendothelioma
Kidneys	Renal re-adsorption problems
Blood system	Inhibits biosynthesis of porphyrin, affects white blood cells
Reproductive system	Spontaneous abortions
Peripheral nervous system	Peripheral neuropathy, paralysis, loss of hearing
Enzymes	Inhibits some enzymes
ATP	Inhibits oxidative phosphorylation
DNA	Inhibits DNA repair mechanism
Gastrointestinal tract	Damage to intestine, intense pain

soil is readily available because of the pH and redox conditions (see Chapter 9) [8,62,109]. Two typical examples of selenium toxicity in animals, are alkali disease and the blind staggers [62,109]. Blind staggers is characterized by anorexia, emaciation and collapse and can be produced by eating plants with 100-10,000 $\mu g\ g^{-1}$ selenium content. Less acute, alkali disease is produced from consuming plants with a selenium content 25-50 $\mu g\ g^{-1}$, and is typified by loss in weight, degeneration and fibrosis of the heart, liver and kidneys, and impaired reproductive capacity [62].

Selenium as an Essential Element

Selenium is an essential element for the enzyme glutathione peroxidase in animals, and it is also likely that the element has the same function in human beings. For example, if selenium is added to diet the glutathione peroxidase activity is increased in human erythocytes [23,41,50,51,62,66,97,100,109]. Strong relationships exist between the mean levels of glutathione peroxidase and selenium, and protein and selenium in human body fluids, as shown by the data plotted in Fig 15.6 [50,51]. This implies an essential function for human beings.

The enzyme glutathione peroxidase (GPO) protects tissue from oxidative damage arising from H_2O_2, organic peroxides and also hydroxy radicals. The reactions:

$$\underset{\text{glutathione}}{2GSH} + H_2O_2 \xrightarrow{\text{GPO}} \underset{\text{oxidized glutathione}}{GSSG} + 2H_2O$$

or

$$2GSH + ROOH \xrightarrow{GPO} GSSG + ROH + H_2O$$

summarize the process [100]. The enzyme has a relative molecular weight (RMW) of ~76,000-92,000 and consists of four subunits, with RMW's around 19,000-23,000. Each unit contains one selenocysteine residue, hence there are 4Se atoms per enzyme [60,100]. The selenocysteine has a lower redox potential, and higher nucleophilicity than the sulphur analogue, and its greater reactivity may be its specific role in the enzyme [100]. The selenocysteine may act as an intermediate in the electron transfer process given below [109].

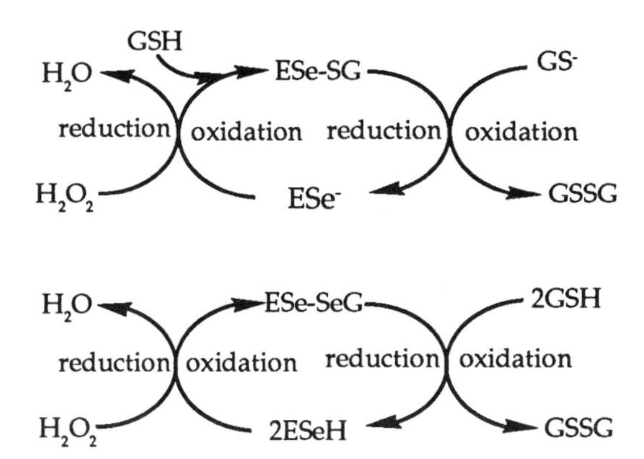

Selenium in diet appears to have the same nutritional effect as α-tocopherol, which is also an antioxidant. Selenium is involved in other enzyme systems, including glycine reductase, formate dehydrogenase, hydrogenase, nicotinate hydroxylase and xanthine dehydrogenase [62,100,109].

Epidemiological studies of cancer mortalities have revealed a negative relationship with selenium in human tissues and fluids, and with levels in soils,

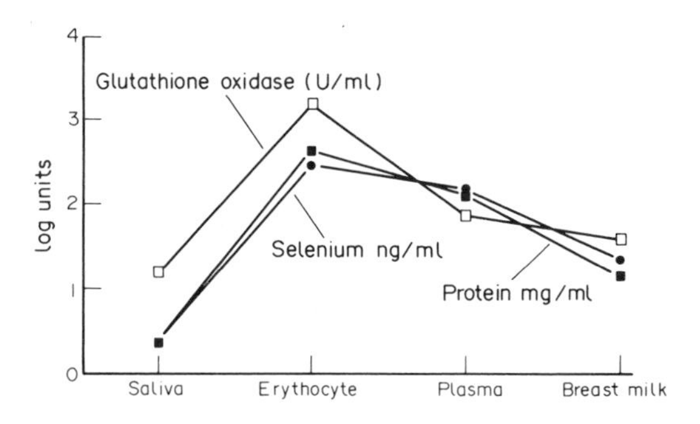

Fig. 15.6 The relationship of selenium to protein and glutathione oxidase. Sources of data; Hojo, 1986, 1987 [50,51].

suggesting that selenium may be helpful in preventing cancer [33,50,61,62,97]. The effect has been found in relation to deaths from cancer of digestive organs, peritoneum, respiratory system, breast, urinary organs and the lymphatic system [33]. People living in selenium rich areas have a low risk of certain cancers. On the other hand selenium has been shown to cause cancer in test animals [61].

There are also other areas of human health where a deficiency of selenium is a disadvantage. Some forms of childhood malnutrition, such as kwashiorker, may have a component of selenium deficiency [8,33,50]. Selenium deficiency can also occur when a person is on long term parenteral nutrition (e.g. intravenous) [50,51]. Also in certain situations selenium deficiency can occur, where it did not exist before, because of changed circumstances. Examples of this are in cases of severe burns and cancer [97]. Heart disease also responds to selenium, though it is probably not the only factor [51,60,97]. Keshwan disease, which is a congestive heart failure in children, was first found in 1935 in China and related to low selenium areas. The prevalence of the disease was dramatically reduced by supplementing the diet with selenium [60,97]. There seems clear evidence that trace amounts of selenium are essential, or at least helpful to human beings. A low intake of selenium in New Zealand does not relate to low levels in the liver [80]. Some of the ways in which selenium is essential for human beings are given in Table 15.14.

Selenium as a Toxic Element

The early symptoms of acute toxic effects of selenium are sore throat, fever, vomiting, irritation to eyes and nose, headache, gastrointestinal irritation, somnolence, drop in blood pressure, dermatitis and a garlic odour on the breath [8,33,62]. At a later stage pulmonary œdema, vascular disruption, hemorrhaging, laboured breathing and respiratory failure occur [33,62,69].

The speed and severity of the toxic effects depends on the chemical form of the selenium taken in by inhalation or ingestion [16,33,62,97]. For example, elemental selenium is not considered very toxic because of its insolubility, however, selenium dust can be dangerous in the lungs. Hydrogen selenide is more toxic than hydrogen sulphide [97]. Selenium, dioxide which could be a problem in industrial environments, causes localized irritation to the eyes, and pulmonary œdema. With water selenous acid is produced, and both SeO_2 and H_2SeO_3 are about as irritating as SO_2 and H_2SO_3 [33].

The documented effects of chronic exposure to selenium are numerous, especially for people having industrial exposure, or living in high selenium (seleniferous) areas. Some problems are respiratory inflammation, pulmonary œdema, dermatitis, gastrointestinal congestion and hemorrhage, depression, metallic taste, garlic odour to the breath and inflammation around the nail beds. For longer term exposure, the problems are jaundice, cirrhosis of the liver, kidney damage and porphinuria, spleen atrophy, loss of nails and hair and dental caries [33,62,69,97]. The greater incidence of dental caries in selenium

TABLE 15.14 The Toxic and Essential Effects of Selenium to Human Beings

Exposure	Affected area	Effects if deficient or in excess
Essential		
Trace	Tissues	Antioxidant, involved with glutathione oxidase
	Various organs	Cancer
	General growth	Nutritional deficiency
	Heart	Keshwan disease, congestive heart failure
Toxic		
Acute	Respiratory system	Sore throat, irritates nose, garlic breath, laboured breathing, pulmonary œdema, respiratory failure
	Gastrointestinal tract	Irritation, vomiting
	Skin	Dermatitis
	Blood system	Vascular disruption, hemorrhaging
Chronic	Respiratory system	Inflammation, pulmonary œdema, garlic breath
	Gastrointestinal tract	Congestion, hemorrhaging
	Skin	Dermatitis
	Nails	Inflammation, loss
	Hair	Loss
	Liver	Jaundice, cirrhosis
	Kidney	Porphinuria
	Spleen	Atrophy
	Teeth	Caries

rich areas may be because of the alteration of the protein component of tooth enamel [62]. The garlic odour of breath is probably due to the formation of $(CH_3)_2Se$ in the body, which being volatile is gradually expelled. This could be a detoxification process [33,62]. Some of the effects of selenium as a toxin are listed in Table 15.14.

One way in which selenium is toxic to human beings, is in the replacement of sulphur in the body. This disrupts sulphur metabolism, such as the uptake of SO_4^{2-}, and the production of methionine [16,62,97]. Alternatively, it has been suggested that selenium does not replace sulphur but bonds to sulphur compounds and interfere in this way [97].

Selenium has a moderating influence on the toxicity of other heavy metals. The toxicity of mercury, thallium, cadmium and tellurium are modified by selenium [62,76]. Also there is a small negative but significant correlation between lead in blood and plasma and selenium [47]. The strong interaction between selenium and some heavy metals may be the way selenium is effective. The toxic metal being bound to selenium is not available to bond to sulphur in the biological system.

THALLIUM, INDIUM, ANTIMONY, BISMUTH AND TELLURIUM

Much less is known about the toxic effects of the remaining five elements, thallium, indium, antimony, bismuth and tellurium. This is because, even in occupational situations, the elements are not common, and are not general environmental hazards. However, they may be encountered in everyday life.

Thallium

The element of the group for which most is known is thallium, because of its past use as a rodenticide and that for human beings could come into contact with it. Also, it was used therapeutically for the removal of head hair, regardless of its toxic effects. The hair follicle is destroyed by thallium. Thallium tends to accumulate in the body [8,25,28,99].

The main effects of acute toxicity are on the gastrointestinal tract, and the nervous system. Symptoms, including abdominal pains, vomiting, diarrhea, gastrointestinal hemorrhage, convulsions, coma, rapid heart beat, disturbed rhythm of the heart, pneumonia and respiratory failure have all been reported. Some of the neurological effects are encephalitis, brain tumour, epilepsy, polyneuritis, optic and nerve atrophy, tingling pain at the extremities [8,28,62,99,100]. People with thallium poisoning experience an extreme sensitivity of the extremities to touch [62]. The signs of thallium poisoning are not immediate, and take a few days to weeks to appear. Depending on the dose of the element the physiological symptoms can start to appear 12-48 hours later and the neurological symptoms 2-5 days later. The loss of hair (alopecia) can take up to 10 days [25,28,62,99].

The chronic effects of thallium are largely seen in its effects on the nervous system, producing peripheral polyneuritis, which is the inflammation of many nerves, and paralysis of the extremities. In addition liver, kidney (albuminuria and hematuria) and heart (hypertension and angina like pains) damage occurs [25,28,62,99]. There is also evidence of teratogenic effects, as babies were born deformed when thallium had been taken by the mother after the third month of pregnancy [62].

The toxic effects of thallium as Tl^+, are most likely due to its competition with K^+. The radii of the ions are 164 pm and 152 pm respectively, only a 7% difference. Therefore cell membranes cannot distinguish $Tl^+$and K^+, hence the thallous ion will be able to enter cells and inhibit the activity of potassium. Thallium also seems to interfere with sulphur metabolism and sulphur containing enzymes [8,62]. Increased intakes of potassium counteracts the effect of thallium and assists in its excretion. Selenium is found to reduce the toxic effects of thallium, and vice versa [62].

Treatment of thallium poisoning can be achieved using prussian blue, which is potassium ferroferricyanide. Potassium is replaced by thallium, and by forced duiresis the thallium salt of prussian blue is excreted. Active charcoal may also be used to remove thallium by absorption.

Indium

The more soluble the indium species ingested the more toxic they are. Some of the acute consequences are reduction in eating and drinking, neurological effects such as convulsive movements and limb paralysis, renal and hepatic damage and lowered lymphocyte count. For chronic exposure the outcome is more mild and degenerative changes occur in the urinogenital tract. Autopsy results on a person who died of indium poisoning showed evidence of damage to the spleen, adrenals, heart, and also liver necrosis and degenerative changes to the tubules that surround the glomeruli in the kidneys [62].

Antimony

Antimony has similar toxic properties to arsenic, and is considered the more toxic in both acute and chronic cases. Also like arsenic, the trivalent oxidation state is more toxic than the pentavalent. It interferes by bonding to HS- groups. Certain antimony compounds, such as the tartrate, causes irritation of the gastric mucosa and have been used as emetics. Some of the symptoms of antimony poisoning are, diarrhea, drop in body temperature and lowered respiratory rate. Chronic effects include albuminuria, jaundice and damage to the heart, liver and kidneys [62,104].

Bismuth

Bismuth also bonds to HS- groups in biological molecules. It has distinct neurological effects with symptoms such as confusion, drowsiness, hallucinations, severe headaches, peripheral neuritis and mycoclonic jerks. The usual effects on the ailimentary system are experienced, but some bismuth compounds can be used therapeutically for coating intestinal mucosa and protecting ulcers from attack by acid, allowing them to heal. Hepatitis, nephrotoxicity and damage to the renal tubules in the proximal region have been reported to be caused by bismuth poisoning [8,62].

Tellurium

In general tellurium is less toxic than selenium because its compounds tend to be less soluble at physiological pH's. Tellurium compounds are relatively readily reduced by bacteria to produce the less toxic tellurium metal. The order of toxicity is Te(IV) > Te(VI) > Te(0). There is no evidence to date that tellurium is an essential element, but it could well act in a similar manner to selenium in uncoupling oxidative phosphorylation. Tellurium may also cross the blood brain barrier and cause neurological outcomes.

Most exposures by inhalation or ingestion occur in occupational situations. In the lungs it is likely that tellurium metal is formed and deposited. The methyl compound $(CH_3)_2Te$, also forms, as indicated by the strong and lasting garlic odour to breath. Some of the other acute effects are metallic taste, nausea, restlessness, giddiness, sleeplessness, headache, diminished reflexes, tremors, convulsions and unconsciousness. Over long periods and at lower intakes chronic effects develop. These include beside the garlic odour to breath and sweat, dryness of the mouth, somnolence, loss of appetite and nausea [33,62,93].

HEAVY ELEMENT TOXICITY

It is clear from the above discussion that many of the toxic effects of the heavy elements are general and not specific for the element. Also many of the effects are similar for the different elements. This is not surprising as a number of the elements have similar chemistries, especially their ability to bond to sulphur. It does mean, however, that care is necessary in diagnosing a particular metal as the source of the problem. In general more information is necessary, such as investigating if the person has had access to the element, or is living in an environment where the element occurs. Some of the more important organs affected by the heavy elements are listed in Table 15.2. The effects of the heavy metals are probably more easily diagnosed in these organs.

A great number of people have been affected by the four elements lead, cadmium, mercury and arsenic and estimates on the extent of the problem have been made and are listed in Table 15.15 [77].

TABLE 15.15 Estimated Magnitude of the Extent of Heavy Element Poisonings

Element	Production 1985 1000 t	Global emissions 1000 t y^{-1}			People affected
		Air	Water	Soil	
Pb	4100	332	138	796	>1 billion[a]
Cd	14	7.6	9.4	22	0.25-0.5 million[b]
Hg	6	3.6	4.6	8.3	40,000-80,000[c]
As	50	18.8	41	82	Hundreds of thousands[d]

a PbB >20 μg dl^{-1}, b producing renal dysfunction, c certified mercury poisonings, d symptom of skin disorder and water arsenic at 2 μg l^{-1}. Sources of data; Nriagu, 1988 [77] and Table 2.1.

REFERENCES

1. Abdulla, M., Svensson, S. and Nordén, A. Antagonistic effect of zinc in heavy metal poisoning. In: Management and Control of Heavy Metals in the Environment, Int. Conf.: CEP Consultants; 1979: 179-182.
2. Alessio, L. Relationship between "chelatable lead" and the indicators of exposure and effect in current and past occupational exposure. **The Sci. Total Environ.**; 1988; **71:** 243-252.
3. Andreae, M. O. Organoarsenic compounds in the environment. In: Craig, P. J., Ed. Organometallic Compounds in the Environment: Longmans; 1986: 198-228.

4. Asami, T. Pollution of soils by cadmium. In: Nriagu, J. O., Ed. in: Changing Metal Cycles and Human Health, Dahlem Konferenzen; 1984: 95-111.
5. Atessio, L. and Bertelli, G. Examples of biological indicators of metals in occupational health. In: Facchetti, S, Ed. Analytical Techniques for Heavy Metals in Biological Fluids: Elsevier; 1983: 41-63.
6. Australian Academy of Science. Health and environmental lead in Australia: Aust. Acad. Sci.; 1981.
7. Bakir, F., Damluji, S. F., Amin-Zaki, L., Murtadha, M., Khalide, A., Al-Rawi, N. Y., Tikriti, S., Dhaher, H. I., Clarkson, T. W., Smith, J. C. and Doherty, R. A. Methyl mercury poisoning in Iraq. **Science;** 1973; **181:** 230-241.
8. Berman, E. Toxic Metals and their Analysis: Heyden; 1980.
9. Boeckx, R. L. Lead poisoning in children. **Anal. Chem.;** 1986; **58:** 274A-287A.
10. Bousquet, W. F. Cardiovascular and renal effects of cadmium. In: Mennear, J. E., Ed. Cadmium Toxicity: Dekker; 1979: 133-157.
11. Bremner, I. and Mills, C. F. Effects of diet on the toxicity of heavy metals. In: Management and Control of Heavy Metals in the Environment, Int. Conf. London: CEP Consultants; 1979: 139-146.
12. Bruaux, P. and Svartengren, M. Assessment of human exposure to lead: comparison between Belgium, Mexico and Sweden: UNEP/WHO; 1985.
13. Bryce-Smith, D. Heavy metals as contaminates of the human environment: Chem. Soc. Chem. Cassetts; 1975.
14. Bryce-Smith, D., Deshpande, R. R., Hughes, J. and Waldron, H. A. Lead and cadmium levels in stillbirths. **Lancet;** 1977; **May:** 1159.
15. Buchet, J. P. and Lauwerys, R. Evaluation of exposure to inorganic arsenic in man. In: Facchetti, S., Ed. Analytical Techniques for Heavy Metal in Biological Fluids: Elsevier; 1983: 75-90.
16. Chau, Y. K. and Wong, P. T. S. Organic Group VI elements in the environment. In: Craig, P. J., Ed. Organometallic Compounds in the Environment: Longmans; 1986: 254-278.
17. Cicchella, G., Arcangeli, G. and Rizzardini, L. Treatment of occupational lead poisoning by slow sulcutaneous administration of CaNa2EDTA. **The Sci. Total Environ.;** 1988; **71:** 551-552.
18. Clarkson, T.W., Hamada, R. and Amin-Zaki, L. Mercury. In: Nriagu, J. O., Ed. Changing Metal Cycles and Human Health: Dahlem Konferenzen Springer Verlag; 1984: 285-309.
19. Clarkson, T. W., Crispin Smith, J., Marsh, D. O. and Turner, M. D. A review of dose-response relationships resulting from human exposure to methylmercury compounds. In: Krenkel, P. A., Ed. Heavy Metals in the Aquatic Environment: Pergamon; 1975: 1-12.
20. Clarkson, T. W. and Marsh, D. O. The toxicity of methylmercury in man: dose-response relationships in adult populations. In: Nordberg, G. F., Ed. Effects and Dose-Response Relationships of Toxic Metals: Elsevier; 1976: 246-261.

21. Committee on Biological Effects of Atmospheric Pollution. Lead: Airborne Lead in Perspective: Nat. Acad. Sci.; 1972.
22. Committee on Lead in the Human Environment. Lead in the Human Environment: Nat. Acad. Sci.; 1980.
23. Conn, E. E. and Stumpf, P. K. Outlines of Biochemistry. 4th Ed.: Wiley; 1976.
24. Craig, P. J. Organomercury compounds in the environment. In: Craig, P. J., Ed. Organometallic Compounds in the Environment: Longmans; 1986: 65-110.
25. De Groot, G. and Heijst, A. N. P. Toxicokinetic aspects of thallium poisoning. Methods of treatment by toxin elimination. **The Sci. Total Environ.**; 1988; **71:** 411-418.
26. D'Itri, P. A, and D'Itri, F. M. Mercury Contamination: A Human Tragedy: Wiley-Interscience; 1977.
27. Elwood, P. C., Essex-Cater, A. and Robb, R. C. Blood lead levels on islands. **Lancet**; 1984; **2:** 355.
28. Emsley, J. E. The trouble with thallium. **New Sci.**; 1978: 392-394.
29. Environmental Health Criteria: Lead: UNEP/WHO; 1977; **3.**
30. Erickson, M. M., Poklis, A., Dickinson, A. W. and Hillman, L. S. Tissue mineral levels in victims of sudden infant death syndrome I. Toxic metals - lead and cadmium. **Pediat. Res.**; 1983; **17:** 779-784.
31. Fell , G. S. Lead toxicity problems of definition and laboratory evaluation. **Ann. Clin. Biochem.**; 1984; **21:** 453-460.
32. Ferrier, R. J. Lead in the environment in New Zealand. **Chem. N. Z.**; 1986; **50:** 107-111.
33. Fishbein, L. Toxicology of selenium and tellurium. In: Goyer, R. A. and Mehlman, M. A., Eds. Toxicology of Trace Elements: Wiley; 1975: 191-240.
34. Fleisher, M., Sarofim, A. F., Fassett, D. W., Hammond, P., Shacklette, H. T., Nisbet, I. C. T. and Epstein, S. Environmental impact of cadmium: a review by the panel on hazardous trace substances. **Environ. Health Perspec.**; 1974: 253-323.
35. Forstner, U. Cadmium. In: Huntzinger, O., Ed. Handbook of Environmental Chemistry: Springer Verlag; 1980; **3A:** 59-107.
36. Fowler, B. A. Toxicology of environmental arsenic. In: Goyer, R. A. and Mehlman, M. A., Eds. Toxicology of Trace Elements: Wiley; 1977: 75-90.
37. Fredin, B. The distribution of mercury in various tissues of guinea-pigs after application of dental amalgam fillings (a pilot study). **The Sci. Total Environ.**; 1987; **66:** 263-268.
38. Friberg, L., Piscator, M., Nordberg, G. F. and Kjellstrom, T. Cadmium in the Environment: CRC Press; 1971.
39. Friberg, L. and Vostal, J. (Eds.). Mercury in the Environment: CRC Press; 1972.

40. Gatti, G. L., Marcì, A. and Silano, V. Biological and health effects of mercury. In: CEC Trace Metals Exposure and Health Effects: CEC & Pergamon; 1979: 73-98.
41. Godwin, K. O. The role and metabolism of selenium in the animal. In: Nicholas, D. J. D. and Eagan, A. R., Eds. Trace Elements in Soil-Plant-Animal Systems: Academic Press; 1975: 259-270.
42. Goyer, R. A. and Mushak, P. Lead toxicity laboratory aspects. In: Goyer, R. A. and Mehlman, M. A., Eds. Toxicology of Trace Elements: Wiley; 1977.
43. Grandjean, P. and Nielson, T. Organolead compounds: environmental health aspects. **Residue Revs.**; 1979; **72:** 97-148.
44. Group Report. Perspectives and prospects on health effects of metals. In: Nriagu, J. O., Ed. Changing Metal Cycles and Human Health (Dahlem Konferenzen): Springer Verlag; 1984: 407-423.
45. Group Report. Routes of exposure to humans and bioavalibility. In: Nriagu, J. O., Ed. Changing Metal Cycles and Human Health (Dahlem Konferenzen): Springer Verlag; 1984: 375-388.
46. Group Report. Structure mechanism and toxicity. In: Nriagu, J. O., Ed. Changing Metal Cycles and Human Health (Dahlem Konferenzen): Springer Verlag; 1984: 391-404.
47. Gustafson, A., Schütz, A. Anderson, P. and Skerfving, S. Small effects on plasma selenium levels by occupational lead exposure. **The Sci. Total Environ.**; 1987; **66:** 39-43.
48. Hewitt, C. N. and Harrison, R. M. Organolead compounds in the environment. In: Craig, P. J., Ed. Organometallic Compounds in the Environment: Longmans; 1986: 160-197.
49. Hindmarsh, J. T., McLetchie, O. R., Heffernan, L. P. M., Hayne, O. A., Ellenberger, H. A., McCurdy, R. F. and Thiebaux, H. J. Electromyographic abnormalities in chronic environmental arsenicalism. **J. Anal. Toxicol.**; 1977; **1:** 270-276.
50. Hojo, Y. Selenium in Japanese baby foods. **The Sci. Total Environ.**; 1986; **57:** 151-159.
51. Hojo, Y. Selenium and glutathione peroxidase in human saliva and other human body fluids. **The Sci. Total Environ.**; 1987; **65:** 85-94.
52. Hutton. M. Human health concerns of lead, mercury, cadmium and arsenic. In: Hutchinson, T. C. and Meema, K. M., Eds. Lead, Mercury, Cadmium and arsenic in the Environment: Wiley, SCOPE; 1987: 53-68.
53. Hutton, M. Cadmium exposure and indicators of kidney function. MARC Tech. Rept. 29: MARC; 1983.
54. Inskip, M. J. and Piotrowski, J. K. Review of the health effects of methylmercury. **J. Appl. Toxicol.**; 1985; **5:** 113-133.
55. Kägi, J. H. R. and Hapke, H. -J. Biochemical interactions of mercury. cadmium and lead. In: Nriagu, J. O., Ed. Changing Metal Cycles and Human Health, Dahlem Konferenzen: Springer Verlag; 1984: 237-250.

56. Kaiser, G. and Tolg, G. Mercury. In: Hutzinger, O., Ed. The handbook of Environmental Chemistry: Springer Verlag; 1980; **3A:** 1-58.
57. Kotsonis, F. N. and Klaassen, C. D. Metallothionein and its interactions with cadmium. In: Nriagu, J. O., Ed. Cadmium in the Environment. Part II Health Effects: Wiley; 1981: 595-616.
58. Lafontaine, A. Health effects of arsenic. In: Trace Metals: Exposure and Health Effects: CEC and Pergamon Press; 1979: 107-116.
59. Lauwerys, R. R. Health effects of cadmium. In: Trace Metals: Exposure and Health Effects: CEC and Pergamon Press; 1979.
60. Lenihan, J. The Crumbs of Creation: Adam Hilger; 1988.
61. Luckey, T. D. and Venugopal, B. Metal Toxicity in Mammals: Plenum Press; 1977; Vol 1.
62. Luckey, T. D. and Venugopal, B. Metal Toxicity in Mammals: Plenum Press; 1978; Vol 2.
63. MacFarland, H. N. Pulmonary effects of cadmium. In: Mennear, J. E., Ed. Cadmium Toxicity: Dekker; 1979: 113-132.
64. Magos, L. and Webb, M. Synergism and antagonism in the toxicology of mercury. In: Nraigu, J. O., The Biogeochemistry of Mercury in the Environment: Elsevier/North Holland; 1979: 581-599.
65. Masazumi, H. and Smith, A. M. Minamata disease: a medical report. In: Smith, W. E. and Smith, A. M. Minamata: Chatto and Windus; 1975: 180-192.
66. Melton, E. K. Alfred E. Stock and the insidious "quecksilbervergiftung". **J. Chem. Ed.**; 1977; **54:** 211-213.
67. Moore, J. W. and Ramamoorthy, S. Heavy metals in Natural Waters: Applied Monitoring and Impact Assessment: Springer-Verlag; 1984.
68. Moore, M. R. Haematological effects of lead. **The Sci. Total Environ.**; 1988; **71:** 419-431.
69. Newland, L. W. Arsenic, beryllium, selenium and vanadium. In: Hutzinger, O., Ed. Handbook of Environmental Chemistry: Springer Verlag; 1982; **3B:** 27-68.
70. Newland, L. W. and Dawn, K. A. Lead. In: Hutzinger, O., Ed. Handbook of Environmental Chemistry: Springer Verlag; 1982; **3B:** 1-26.
71. Nogawa, K. Cadmium. In: Nriagu, J. O., Ed. Changing Metal Cycles and Human Health, Dahlem Konferenzen: Springer Verlag; 1984: 275-284.
72. Nogawa, K. Itai-Itai disease and follow up studies. In: Nriagu, J. O., Ed. Cadmium in the Environment. Part II Health Effects: Wiley; 1981: 1-37.
73. Nomiyama, K. Toxicity of cadmium - mechanism and diagnosis. In: Krenkal, P. A., Ed. Heavy Metals in the Aquatic Environment: Pergamon Press; 1975: 15-23.
74. Nomyama, K. Renal effects of cadmium. In: Nriagu, J. O., Ed. Cadmium in the Environment. Part II Health Effects: Wiley; 1981: 643-689.

75. Nordberg, G. F. Current concepts in the assessment of effects of metals in chronic low-level exposures. Consideration of experimental and epidemiological evidence. **The Sci. Total Environ.**; 1988; **71:** 243-252.
76. Nordberg, G. F. Effects and Dose-Response Relationships of Toxic Metals, Ed: Elsevier; 1976.
77. Nriagu, J. O. A silent epidemic of environmental metal poisoning? **Environ. Pollut.**; 1988; **50:** 139-161.
78. Otvos, J. D., Petering, D. H. and Shaw, C. F. Structure-reactivity relationship of metallothionein, a unique metal-binding protein. **Comm. Inorg. Chem.**; 1989; **9:** 1-35.
79. Pfitzer, E. A. General concepts and definitions for dose-response and dose-effect relationships of toxic metals. In: Nordberg, G. F., Ed. Effects and Dose-Response Relationships of Toxic Metals: Elsevier; 1976: 140-146.
80. Pickston, L., Lewin, J. F., Drysdale, J. M. Smith, J. M. and Bruce, J. Determination of potentially toxic metals in human livers in New Zealand. **J. Anal. Toxicol.**; 1983; **7:** 2-6.
81. Piotrowski, J. K. and O'Brien, B. J. Analysis of the effects of lead in tissue upon human health using dose-response relationships. MARC Tech. Rept. No. 17: MARC; 1980.
82. Piotrowski, J. K. and Coleman, D. O. Environmental hazards of heavy metals: summary evaluation of lead, cadmium and mercury. MARC Report No 20.; 1980.
83. Piotrowski, J. K. and Inskip, M. J. Health effects of methylmercury. MARC Tech. Rept. 24: MARC; 1981.
84. Piscator, M. The dependence of toxic reactions on the chemical species of the elements. In: Bernhard, M., Brinckman, F. E. and Sadler, P. J., Eds. The Importance of Chemical Speciation in Environmental Processes, Dahlem Konferenzen: Springer Verlag; 1986: 59-70.
85. Piscator, M. Metabolism and effects of cadmium. In: Management and Control of Heavy Metals in the Environment, Int. Conf. London: CEP Consultants; 1979: 1-7.
86. Probst, G. S. Cadmium: absorption, distribution and excretion in mammals. In: Mennear, J. E., Ed. Cadmium Toxicity: M. Dekker; 1979: 29-59.
87. Prpi´c-Maji´c, D., Kersanc, A., Pongracic, J. and Fugas, M. Biological indices of lead absorption and residential distance from lead emitting source. In: Management and Control of Heavy Metals in the Environment, Int. Conf. London: CEC Consultants; 1979: 89-92.
88. Rabenstein, D. L. The chemistry of methylmercury toxicology. J. Chem. Ed.; 1978; **55:** 292-296.
89. Rabinowitz, M. Lead and pregnancy. **Birth**; 1988; **15:** 236-241.
90. Roels, H. A., Buchet, J-P., Bernard, A., Hubermont, G., Lauwerys, R. R. and Masson, P. Investigation of factors influencing exposure and response to lead, mercury and cadmium in man and animals. **Environ. Health Perspec.**; 1978; **25:** 91-96.

91. Royal Commission on Environmental Pollution. Lead in the environment: HMSO; 1983; 9th. Report.

92. Royal Society of New Zealand. Lead in the Environment in New Zealand. Miscell. Series 14: Roy. Soc. N. Z.; 1986.

93. Sadeh, T. Biological and biochemical aspects of tellurium derivatives. In: Patai, S., Ed. The Chemistry of Organic Selenium and Tellurium Compounds Vol. 2: Wiley; 1987: 367-376.

94. Sandstead, H. H. Nutrient interactions with toxic elements. In: Goyer, R. A. and Mehlman, M. A., Eds. Toxicology of Trace Elements: Wiley; 1977: 241-256.

95. Schroder, H. A. and Nason, A. P. Trace element analysis in clinical chemistry. **Clin. Chem.;** 1971; **17:** 461-474.

96. Scott, R., Rundle, J., Fell, G. Cunningham, C., Ottaway, J. and Willemse, P. The urological significance of chronic cadmium poisoning. In: Management and Control of Heavy Metals in the Environment, Int. Conf. London: CEP Consultants; 1979: 27-30.

97. Shamberger, R. J. Selenium in the environment. **The Sci. Total Environ.;** 1981; **17:** 59-74.

98. Sinn, W. Relationship between lead concentration in the air and blood lead levels of people living and working in the centre of a city (Frankfurt blood lead study). I Experimental method and and examination of differences. **Int. Arch. Occup Environ, Health;** 1980; **47:** 93-118.

99. Smith, I. C. and Carson, B. L. Trace Metals in the Environment Vol. I Thallium: Ann Arbor Science; 1977.

100. Soda, K., Tanaka, H. and Esaki, N. Biochemistry of physiologically active selenium compounds. In: Patai, S., Ed. The Chemistry of Organic Selenium and Tellurium Compounds Vol. 2: Wiley; 1987: 349-365.

101. Strehlow, C. D. and Barltrop, D. Health studies in The Shipham Report. **The Sci. Total Environ.;** 1988; **75:** 101-133.

102. Sunderman, F. W. Metal carcinogenesis. In: Goyer, R. A. and Mehlman, M. A., Eds. Toxicology of Trace Elements: Wiley; 1977: 257-295.

103. Suzuki, T. Dose-effect and dose-response relationships of mercury and its derivatives. In: Nriagu, J. O., Ed. The Biogeochemistry of Mercury in the Environment: Elsevier/North Holland; 1979: 399-413.

104. Svenson, B-G., Björnham, Å., Schütz, A., Lettevall, U., Nilsson, A. and Skerfving, S. **The Sci. Total Environ.;** 1987; **67:** 101-115.

105. Takizawa, Y. Epidemiology of mercury poisoning. In: Nriagu, J. O., Ed. The Biogeochemistry of Mercury in the Environment: Elsevier/North Holland; 1979: 325-365.

106. Treptow, R. S. Amalgam dental filling, Part II The chemistry and a few problems. **Chemistry;** 1978; **51:** 15-19.

107. Tsuchiya, K. Clinical signs, symptoms, and prognosis of cadmium poisoning. In: Nriagu, J. O., Ed. Cadmium in the Environment. Part II Health Effects: Wiley; 1981: 39-54.

108. WHO. Environmental Health Criteria 1 Mercury: WHO; 1976.
109. Williams, R. J. P. A short note on selenium biochemistry. In: Williams, R. J. P. and Da Silva, J. R. R. F., Eds. New Trends in Bio-Inorganic Chemistry: Academic Press; 1978: 253-260.
110. Williams, R. J. P. Structural aspects of metal toxicity. In: Nriagu, J. O., Ed. Changing Metal Cycles and Human Health, Dahlem Konferenzen: Springer Verlag; 1984: 251-263.
111. Zielhuis, R. L. General report: health effects of trace metals. In: Di Ferrante, E., Ed. Trace Metals: Exposure and Health Effects: CEC/Pergamon; 1979: 239-247.
112. Zielhuis, R. L. and Wibowo, A. A. F. Standard setting and metal speciation: arsenic. In: Nriagu, J. O., Ed. Changing Metal Cycles and Human Health, Dahlem Konferenzen: Springer Verlag; 1984: 323-344.

CHAPTER 16

THE NEUROTOXICITY OF LEAD

INTRODUCTION

Most of the heavy elements have neurotoxic effects, either on the peripheral nervous system (PNS), or on the central nervous system (CNS). A brief list of the neurotoxic effects of the heavy elements is given in Table 16.1. The neurotoxicity of mercury and arsenic is not disputed, and there is no doubt that lead has neurotoxic effects on both children and adults when the dose is high. A lot of effort has been put in to investigating whether or not low level lead exposure impairs the neuropsychological functioning of children [28,65,79,100]. The weight of evidence does point to some effect on intelligence, cognitive functioning and behaviour. It is necessary in such an investigation to have sensitive and reliable measures of both lead levels and neuropsychological functioning, as well as statistical techniques to handle the data, and reliable estimates of confounding variables. Though lead is not as dangerous as mercury on a mass basis, there are other reasons why lead is more dangerous as a general population toxin. These are, its greater production, and its wide use in areas where people come into contact with it, especially lead solder, painted houses and lead additives in petrol. A comparison of some of these features for the four most common toxic heavy elements, are listed in Table 16.2.

ESTIMATES OF THE BODY LEAD BURDEN

The body lead burden can be estimated by measurement either of the lead content in materials such as lead in blood, teeth, hair, urine, fæces and bone, or the biochemical effect of lead, such as its effect on the haemopoeitic system.

Lead in Blood

The most common measure of lead is lead in blood, it is mainly estimated in the whole blood, but also in the plasma or in the red blood cells. Blood is difficult

TABLE 16.1 Neurotoxicity of the Heavy Elements

Element	Neurotoxic effects
Mercury	CNS: brain damage, visual, sensory, auditory and coordination dysfunction, PNS: e.g. tremors
Cadmium	Nothing reported
Lead	CNS: encephalitis, behaviour, inattention, IQ deficits PNS: reduced nerve conduction
Indium	PNS: paralysis of limbs, convulsive movements
Thallium	CNS: encephalitis, brain tumour, PNS: polyneuritis
Arsenic	CNS: vertigo, restless, irritability, hearing loss, PNS: motor paralysis, peripheral neuritis
Antimony	?
Bismuth	PNS: peripheral neuritis
Selenium	CNS: depression, irritability
Tellurium	PNS: tremors, diminished reflexes, convulsions

to sample accurately and reliably, and is sampled either as venous blood or capillary blood. Both methods are subject to contamination, but the problem is more significant in the capillary method as a small volume of blood is taken. However, with rigorous methodology capillary sampling should be satisfactory for small samples [43], and provided high results are checked with venous sampling.

A single PbB measurement is not adequate as a persons blood lead can fluctuate over time [36,38], and it represents recent exposure [33,75,77,100, 105,115]. A person could easily be incorrectly classified if a high lead exposure had occurred a long time before the blood was taken. Blood lead results typically have a variability of ±10 to ±15%.

Lead in Teeth

After blood, children's deciduous teeth have been the next most often used tissue for estimating body lead. The distinct advantage of teeth is that they provide a record of past lead exposure [5,19,29,36,42,77,90,106,109,126], there-

TABLE 16.2 Relative Toxicity and Availability of Mercury, Cadmium, Lead and Arsenic to Human Beings

Element	Crustal conc. $\mu g\ g^{-1}$	Production x 1000 t (1985)	Toxicity	Tolerable intake $\mu g\ d^{-1}$
Lead	13	4100	mod. to highly	430
Mercury	0.08	6	extremely	~40
Cadmium	0.2	14	highly	~60
Arsenic	1.8	50	highly	~60?

fore there is less of a chance of incorrectly classifying people in terms of their overall lead exposure [33,86,100,118]. Moderate association has been found between PbB and tooth lead (e.g. $r = 0.47$, $p<0.001$) [5,118,127], however, as the two materials reflect different temporal lead exposures, it is not necessary that they show a close correlation. We have discussed previously the problem of tooth analysis and the different matrices and type of teeth in Chapters 4 and 13. If agreement could be reached on what tooth material and tooth type to sample, it is likely that teeth would be the best tissue to use in epidemiological studies of the health effects of lead.

Lead in Hair

An accessible tissue for the measurement of lead is human scalp hair, which has been used in the study of neuropsychological effects of the metal [43,97,98,121]. As discussed in Chapter 13 hair presents more problems than teeth because of the endogenous and exogenous sources of lead in the fibre. The best hair sample is that taken close to the scalp, reflecting more closely internal lead levels. The best position to sample is probably the nape of the neck. Hair roots may be even better material to sample.

Lead in Other Materials

Other materials such as urine and fæces have not been widely used as a measure of the body lead burden [5,14,20,43,49,119,120]. Lead released from the body in urine and fæces by chelating agents, such as penicillamine and EDTA may be used to measure the body lead burden [14].

As outlined previously lead interferes with the biosynthesis of haem at a number of steps. Therefore various species and abnormal products accumulate, providing possible markers for the body lead burden [19]. The markers most frequently studied are, EP, the activity of δ-ALAD in blood and levels of ALA and CP in urine. Little use has been made of these methods.

NEUROPSYCHOLOGICAL METHODS

Cognitive and Behaviour Testing

A variety of neuropsychological measures are available which aim to measure functions such as: IQ (full scale, performance and verbal) fine and perceptual motor skills, verbal processes, visual perception processes, gross motor activity, behaviour ratings and nerve conduction. An area of difficulty is, finding tests sensitive enough to detect small changes in neuropsychological functioning [19,24,33,36,40,41,45,77,82,84,86,87,91,100,104,105,108,109,111]. The Wechsler Pre-school and Primary Scale of Intelligence, WPPSI, test which measures perceptual motor ability revealed significant differences between groups with PbB levels above and below 29-39 $\mu g\ dl^{-1}$, whereas the Denver Development Screening Test, DDST, test which measures language and fine motive adaptive ability revealed differences between groups with PbB levels above and below 39-60 mg dl^{-1} [91]. Therefore it is necessary to consider a combination of tests, perhaps a mixture of a standardized IQ test (e.g. Wechsler Intelligence Scale for Children - Revised, WISC-R) and specialized neuropsychological tests [108]. Measures in more than one setting may help distinguish between pervasive and situational effects [109]. Testing must be objective and blind with respect to the lead measure [19], and bias may be introduced by using more than one tester.

Increasing the battery of tests does not necessarily improve the results. At a 5% level of significance one test in 20 will show a significant effect by chance [34,86,108,109,111]. It is difficult to obtain objective parent and teacher ratings of child behaviour, and care is necessary in producing a questionnaire, and in the method of presentation to parents [72,126].

Confounding Variables

Confounding variables (CV's) are factors which associate with the outcome of interest, i.e. the neuropsychological function, or the lead exposure or with both together [19]. Where appropriate these variables have to be considered, and a list is given in Table 16.3. Two methods are available to handle CV's, either

TABLE 16.3 Confounding Variables

Socioeconomic status (SES)		Birth and medical factors
Parental:	IQ	Pre-natal problems
	Education level	Birth weight
	Attitude to child	Birth order
	Attitude to school	Number of pregnancies
	Interest	Mother's age at birth of child
	Restrictiveness	Mother's exposure to toxins
Father head of house		Mother's mental health
Father's occupation		Infections
Mother's occupation		Iron balance
Family income		Trauma
Care-giving environment		Medical problems
Nourishment		Pica
Social class		Genetic factors
Social disadvantage		Host sensitivity
Marital relationship		
Environmental factors		**Physical factors**
Type of residence		Age
Length of time in residence		Sex
Geographic location		Race
Home cleanliness		Number of siblings
Smoking habits		
Alcohol use		

matching exposed and control groups for the variables, which is difficult to do adequately, or controlling for the variables statistically [19,33,100,105]. In longitudinal studies (i.e. carried out over time) of the same sample, it is reasonable to expect most CV's to remain constant [37].

Extensive criticism of various studies centres around the attention, or the inattention, given to CV's [36,37,71,82,84,87,111]. The socio-economic status (SES) of the subjects is relevant, but is difficult to measure quantitatively. Social class, or father's occupation, or income have been used, but they are rather crude measures. Both, under-estimates and over-estimates of SES, can influence the final assessment of the effect of lead. Many of the factors listed in Table 16.3 are inter-dependent and if too many of these are used in an analysis this could conceal the influence of other factors [1108,110].

The biological variability of people is also a factor, to which little attention has been given [24]. Some people are sensitive to a certain level of a toxin, while others are not.

Statistical Methods

The lead concentrations and neuropsychological data obtained have to be handled by statistical techniques, in particular multivariate analysis. The ability to discriminate between various inter-active factors depends on the sample size [18,19,40,41,82,100, 109,111]. The sample size required is determined by, the level of significance desired, how small a difference one wishes to detect, and the frequency of the factor in the population [19]. To increase the certainty, and decrease the detectable difference, requires an increase in sample size. Samples too small tend to give inconclusive results.

The selection of exposed and control groups has to be random, but for a number of reasons some subjects will be eliminated from a study This may produce a sample bias which needs to be tested. When the sampling has been selective or stratified, care is necessary over generalizing the results [19].

Confounding variables may be determined by forward selection, where the experimenter makes the choice, or by backward selection, where a CV is eliminated statistically because its inclusion or exclusion does not change the lead exposure regression coefficient. This can mean that a CV may be required for one outcome measure, but not for another [37].

Outliers in the data need to be determined statistically and removed before further processing of the data [37]. In a material, such as teeth, because of the natural variability in the lead level an outlier may in fact be a real value and must be included [42]. Some statistical packages have constraints over the number of covariates that can be handled, and this could be artificially restrictive [37]. The statistical limit to the number of outcome measures (per subject) that are required to demonstrate group, as distinct from individual, differences is one third the sample size [13].

EPIDEMIOLOGICAL STUDIES

The steps in epidemiological studies include, determination of the sample size, sampling, measurement of the lead exposure and the outcome factors (i.e. neuropsychological functioning), deciding on confounding variables and statistical handling of the results. Epidemiological studies can be either retrospective or prospective. In a retrospective study people with a particular outcome (limited to one) are compared with a control group regarding their lead exposure. It is difficult, and may be impossible, to fully control for the CV's. In a prospective study, which is a better approach, groups with different lead exposure are studied over a range of outcome measures [7,16,23,24,33,37, 54,100,105]. For longitudinal prospective studies, lead exposure and outcome measures are estimated over time, and the CV's could well remain relatively constant and are less of a problem. An ideal study would start with pregnant women and then follow their children over a number of years.

Cause and Effect

It does not automatically follow that if a strong association exists between two factors that the relationship is causal [22,37,63,100,109,122]. Nevertheless, a strong association cannot be ignored. Requirements to establish a cause and effect relationship are: a consistent trend over a number of studies, a dose-response relationship, the effect is biologically plausible, that other traumata can be excluded or shown to not rule out the cause under consideration, the association is strong and specific, that the cause precedes the effect in time, that the evidence is coherent i.e. consistent with the natural history of the outcome and disease, and experimental evidence is available such as a study of accidental exposure [19,108]. A strict adherence to all these requirements could well rule out any cause and effect relationship being discovered. Eventually it is the trend in the available evidence, and the quality of the epidemiological study that are important in the decision.

Some evidence obtained does not distinguish clearly between either lead being the cause of the outcome or the outcome predisposing people to high lead exposure [37,109]. In such cases other information is required about the situation in order for one conclusion to be favoured over another [37].

Two models of cause and effect have been proposed [19]; (a) multiple independent causal factors,

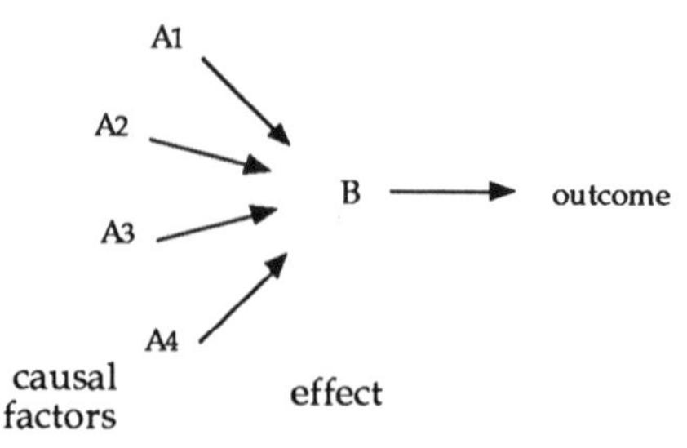

and (b) cumulative and sometimes independent causal factors

A1 + A2 + A3 + A4 → B → Outcome
causal factors effect

The second model is often used in analysis of results, but a more complex situation may well exist, i.e. a mixture of models (a) and (b),

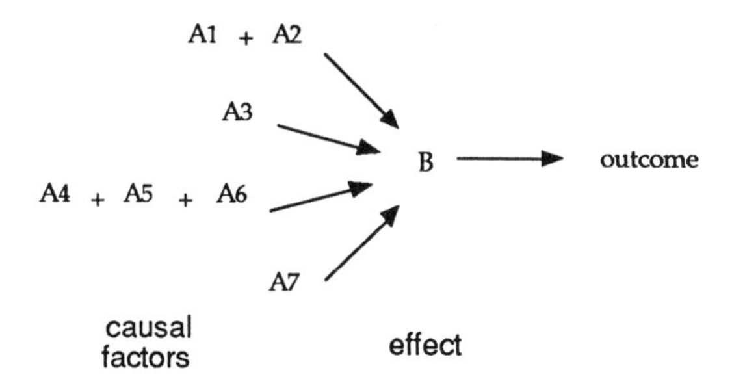

NEUROPSYCHOLOGICAL STUDIES

The various neuropsychological studies on lead can be classified in different ways, such as the degree of lead exposure, or in terms of the subject, either an adult or child, or the type of epidemiological study, i.e. prospective or retrospective, cross sectional or longitudinal. A method of classification used by Rutter [108-111] will be followed here. The classification is: (1) studies of children with mental retardation or behavioural deviance; (2) studies on the effect of reducing the body lead burden by chelation; (3) studies of people living close to smelters; (4) studies of small groups of people in a clinical environment and (5) studies of large populations selected in a random manner. Clearly the groups merge into each other, especially groups 3, 4 and 5.

Mental Retardation Studies

Studies of mental retardation in relation to lead are retrospective, in that an outcome has been determined and people with or without the outcome are compared as regards their lead exposure. The principal limitation, is the restriction in the number of neuropsychological outcomes. The sample is normally stratified and is generally small in size.

Outcomes that have been selected are; mental retardation [9,32,74,76], learning disability [97,98] and autism [16,17]. Studies in Glasgow [9,76] demonstrate good correlations between mental retardation, blood lead levels and water lead levels. The quality of the matching of the mentally retarded and control children, and the handling of CV's may be criticized, but even so for water lead levels >800 mg l^{-1} the chance of mental retardation increases by a factor of 1.7. In a Welsh study [32], no relationship between water lead levels and mental ability was observed, but in this case, the water lead levels were

<300 mg l^{-1}. Autistic children with a high incidence of pica were found to have elevated blood lead levels [17], but the significant relationship disappeared in a second study, though the trend persisted [16].

In other studies, children diagnosed with mental retardation and/or hyperactivity [20,21,23,24,129] have been grouped according to; those with a probable cause, those with a possible cause and those with an unknown cause. The groups were then compared according to their lead exposure, and in some cases [21,24], the blood lead levels were higher in the children with no known cause of their illness. Some criticisms directed at this work include; possible sample bias, little or no consideration given to pica and SES, poor matching of groups, small samples, over diagnosing probable cause and unknown cause, and no clear source of lead [15,45,82,109]. Even so, a distinct difference in the PbB levels between the groups with probable and unknown aetiologies existed. Whether or not a causal relationship occurs is less certain.

Chelation Studies

In principle the investigation of the neuropsychology of children with a high lead burden before and after chelation is a powerful method to demonstrate a causal relationship [109]. Hyperactive children, of no-known or minimal cause, or with a history of lead poisoning were treated with either penicillamine, which removes lead from the system, or with methylphenidate, which helps in hyperactivity, or with a placebo. Improvement in teaching ratings, parent ratings and global impressions occurred for children on both drugs, but no change was apparent for those who received the placebos, however, more dropouts occurred from the study for the placebo group. The children receiving penicillamine also had their blood lead levels reduced [25]. Though the quality of the sample has been questioned the results do point to a possible causal relationship, especially because of the use of contrasting interventions [108].

Smelter Studies

Studies of people living near or working in lead based industries, have one advantage, namely, the source of lead is clearly definable. A number of studies have been carried out on children and adults [2,4,7,28,44,46,47,48,52,58,59,60-64,69,70,73,95,104,114,115,116,123,130]. Most have used blood lead as the measure of lead exposure, and frequently only one measurement was made. Two PbB measurements at different times provides the basis for a better classification of the high and low lead groups. An Australian study addressed this problem, and a number of blood lead measurements have been taken at different times during the study [4,28].

A problem for smelter studies is obtaining a big enough sample, and maintaining it throughout the study. Another problem is determining the 'best time' to carry out the neuropsychological testing of outcome measures in relation to the lead measure. Finding matched pairs, increases difficulties and often the sample size can be drastically reduced because of this. Often age, sex and race have been matched, and estimates of SES have varied from, not considered to the use of the father's occupation and income.

In a large (592 children at 2 years old) prospective epidemiological study at Port Pirie, Australia [4,28], the children's lead exposure was tracked from the 16th. week of pregnancy. The increase in blood lead levels in the sample (especially for post-natal blood) was associated with a significantly reduced Bayley Mental Development Index (MDI) score. The MDI score drops ~2 units for every 10 $\mu g\ dl^{-1}$ lead in the blood. It is planned to carry the study on to age 7 years, and preliminary results at age 4, using the McCarthy scale of children's ability, also shows a significant negative association to blood lead. It was also noticed that preterm births are more likely when the blood lead level of the mother was elevated [4,28]. A similar result has been found in Glasgow [28].

A summary of some of the studies on children is given in Table 16.4. Summaries have disadvantages as much of the data has to be omitted, and it gives the impression that the studies are all of equal quality. A number of the investigations find a lead effect [4,57,59,60,130], whereas others find no effect [52,70,104]. The weight of evidence is suggestive that elevated blood lead levels are associated with some impairment of IQ and neurologic functioning. It is likely that a spectrum of effects occur [63], and what is observed depends on the cut-off point between the high and low lead groups, as well as the sensitivity of the neuropsychological tests. A cut-off point of 35-40 $\mu g\ dl^{-1}$ could be too high, so that the two groups differ little in their neuropsychologic effects. The Australian study indicates a lead effect where the blood lead levels are much less, around 11-20 $\mu g\ dl^{-1}$ [4,28].

Clinic-Type Studies

A number of clinic-type studies have been carried out, mostly in the U.S.A. The studies are generally prospective and cross-sectional. Frequently the the cut-off blood lead level is >30 mg dl^{-1} for the lead group and <30 mg dl^{-1} for the controls, which, as for the smelter studies, may be too high. However, at the time when many of the studies were carried out, a PbB level of 30 mg dl^{-1} was not considered too high. Teeth and hair have also been used as lead indicators.

The principal outcome measure used in the studies is mental ability as measured by IQ tests. Electroencephalographic (EEG) measurements have also been used. The CV's mostly considered were age, sex, race and SES. Parental IQ and/or education have been considered as important variables.

A majority of the studies (including the EEG studies) demonstrate a significant lead effect, and in some a small negative trend was observed, but not statistically significant. Comparative data, and the findings of the various investigations, are listed in Table 16.5. The results from studies which have not taken into account CV's must be considered with caution. A longitudinal study [30,31], with some methodological errors, indicated that over a 3-4 year period the damage initially observed was still obvious. Unfortunately, the blood lead

TABLE 16.4 Summary of Studies on Smelter Children

Location	Sample size	Blood lead µg dl^{-1}	Neuropsychol. tests	Authors' findings	Comments	Reference
U.K.						
London	215	Mean 33.1	IQ, behaviour	No effect	A weak PbB/distance correl., one PbB, SES not handled well.	2,64
Birmingham	851 1642	Near factory unpolluted areas	11+ Exam	No effect, closest to factory did slightly less well	No measure of lead exposure, matched somewhat for age, SES, birthrank, maternal age.	52
Manchester	24 23	>35 ≤35	Behaviour, general development	No effect, high lead did slightly less well	Matched for age, sex, SES, parent educ. area, length of residence; age & school accounted for most variation; 3 yr study.	104
London	34 48 49 35	7-10 11-12 13-16 17-32	IQ, educational attainment, teacher ratings	An effect for IQ and dose/response	Controlled for SES, sex, age; tested 9-12 months after one PbB measure; decided SES measure too crude.	130,63
U.S.A.						
El Paso	50 & 46 81 & 78	≥40 <40	IQ, behaviour neurologic tests	An effect (IQ (perform.) finger wrist tapping)	Some matching, age, sex, SES, language, time of residence.	123 62
	138 controls	mean 50 mean 20	IQ, teacher reports, neurologic tests	No effect	Some matching, age, sex, race, income; children away from area during investigation.	70,69
Shoshone (Idaho)	202	≥40, <40	Nerve conduction velocity (NCV)	An effect	Concentric sampling, matched, age, sex, SES, time of residence.	58,59,60
	34 + 20 matched pairs	>40, <40	IQ	No effect	Matched, sex, age, SES, length of residence, type of residence; variation between testers >samples.	47
	50 matched pairs	≥40, <40	IQ	An effect		
	168		IQ	An effect		

Table 16.4 continued

Location	Sample size	Blood lead $\mu g\ dl^{-1}$	Neuropsychol. tests	Authors' findings	Comments	Reference
Shoshone (Idaho)	50 matched pairs 6 with low NCV	≥40, <40	NCV	No effect	NCV on Gregory's group and 6 with low NCV in Landrigan's study	44
Belgium	42 (<1 km) 36 (2.5 km) 73 controls	27.9 16.0 11.5	EEG	No effect	Controls matched for age and SES	35
Australia (Pt. Pirie)	592	9.4 (mothers) at 16 months 10.4 (mothers) at birth	Bayley, MDI, PDI	An effect	Corrects for counfounding vcariables e.g. caretaking environment	4

TABLE 16.5 Summary of Clinic-type Studies of Children

Study	Sample	Blood lead $\mu g\ dl^{-1}$	Tests	Results	Comments
Boston (USA) (1972,1974) [102,103]	58 42 (1.5 yrs later)	>40 >35	IQ, neurologic	IQ 8 pt increase after 1.5 yr	Longitudinal study, no CV considered.
Boston (USA) (1973) [39]	24 22	>40 controls	NCV	A lead effect	No CV considered.
New york (1974) [1]	Variety	40-80	IQ, behaviour	IQ 22 pt difference low PbB & tooth lead and tooth lead poisoned	No CV considered.
New York (1972,1977) [55,56]	24, 25 31, 36	81, 38 80, 28	Neurologic, social, language	No effect	Matched; age, sex, race, SES, pica and neonatal environ . Cut-off PbB level high.
Virginia (USA) (1972, 1975) [30,31]	70, 72 67, 70 (restudy 3-4 yrs later)	58, - 202, 117 $\mu g\ g^{-1}$ tooth lead	IQ, neurologic, behaviour	1972 5 IQ, 1975 3 IQ points difference	Matched; race, sex, SES, housing density, no. children <6 yrs,pica in lead groups. Results suggest permanent damage, Longitudinal study.
New York (1974,1976,1981) [36,96,128]	30, 50 32, 31 27, 73	40-70, 10-30 27-49, ≤26 ≥38, <37	IQ, teacher ratings, behaviour	1981 12 IQ point difference	Controlled for age, sex, parental IQ and educ., birth wt., SES positive effect in both studies, but dismissed by authors (1981) due to inadequate control of SES.
Cleveland (1985,86) [28]	132 Mother/infant pairs	6.5 mothers 5.8 cord	Behaviour	Some effect	Three out of 17 outcomes associate with maternal or cord blood lead.
Cincinnati, (1975) [6]	27 matched pairs	>50, <30	IQ, neurologic, behaviour	IQ -1 to 2 points difference	Matched; age, sex, race, SES; parent and teacher assessments weak.
Rhode Is (1974, 1979) [107]	45 45	≥40 <40	IQ, neurologic behaviour	IQ 16 points between controls & lead poisoned and 5 points controls & long exposed.	Matched; sex, age, race, SES; appears to show a dose/response relationship.

Table 16.5 continued

Study	Sample	Blood lead $\mu g\ dl^{-1}$	Tests	Results	Comments
Boston (USA) (1977) [88]	41 35	>50 <30	IQ, neurologic	Weak positive effect	Matched SES, birth weight; big time gap between PbB and outcome measures.
New York (1978,1979) [26,27]	579	5-44	Behaviour, teacher ratings, academic achievement	Positive result	Controlled for age, sex, culture group; SES assumed same for all sample.
Nth Carolina (1980,81) [72]	40 43	≥31 <31	Behaviour, neurologic	No effect	Matched; sex, age, race, SES; well conducted, hyperactivity in home environment, no school.
Chicago (USA) (1981) [112]	Variety	<40 to >200	IQ, neurologic, NCV, EEG.	Negligible effect IQ 4 pts difference <40 to >200 mg dl^{-1}	No control of CV, except case of siblings.
Maryland (1981) [121]	149	Hair lead	IQ, neurologic	Approx. 30 IQ pts over entire range	Controlled for sex, race, SES, urbanicity. Used hair lead, could have sample bias. Cd also has an association
Sydney (Aust.) (1982) [68]	72-108 72-108	>19 <9	IQ, neurologic, behaviour	No clear difference	Handles CV by correlation, controlled for age, sex.
Boston (USA) (1980,1981) [13,89]	19 22	Dentine lead	Psychol. measures EEG	Positive effect	Combined EEG and psychological testing and obtained improved explanation of variance.
Nth Carolina (USA) (1981,1983) [11,92,93]	69 100 (63 studied) 28 (reasses)	6-59 7-59 14-39	EEG IQ, EEG EEG	All give a positive effect for lead	Results suggest effect occur at PbB <15 mg dl^{-1}.

levels of the controls were not measured. The SES factor was carefully considered, making use of parent education, occupation and family income.

Some studies [36,96,128] have aroused a lot of comment, probably because the authors reversed the conclusions of their earlier study. Initially a positive effect was found to exist between a lead group with PbB of 40-70 μg dl^{-1} and controls 10-30 μg dl^{-1} [96]. A restudy of the same group [36], five years later, demonstrated only a small positive effect which was not statistically significant. The authors' suggested there were methodological problems in the first study. In a third prospective study 132 mother/infant pairs were studied using maternal blood at the time of delivery and cord blood lead levels for estimating the lead exposure. Significant relationships were found between the cord blood lead and abnormal reflexes and neurological soft signs, and also between maternal blood lead and muscle tonus [28].

The four EEG studies [11,13,89,92,93] all showed an effect with increasing PbB levels. Also an effect was discernible at PbB levels <15 μg dl^{-1}, though no threshold was apparent over the PbB range 7-59 μg dl^{-1}. The combined use of EEG and psychological measures [13,89] appears to be a more powerful predictor of PbB values and neuropsychological impairment than the two separate measures.

The results for most of the clinic-type studies are weighted towards a neuropsychological effect of lead. Issues, such as the danger level of PbB, and satisfactory cut-off levels for study, are not resolved.

Population Studies

Provided the sample in a population study is representative, the findings may be extrapolated to the general population. Hence there is a lot of interest in this type of epidemiological study. Over recent years a number of studies have been conducted [28,40,41,43,51,66,77,89,91,100,101,113,117,118,126,127].

There are difficult areas in general population studies. Locating the source of lead can be difficult. The sample needs to be representative, but in the end the experimenter is dependent on the cooperation of the participants. A range of confounding variables can be expected, making it necessary to take care over establishing and controlling for these factors. Finally, it is best to find a measure of lead which is more than one single measurement. A summary of some of the studies is given in Table 16.6. Many of the studies involve a great deal of work involving a number of disciplines.

Studies in the USA The investigation that created interest in population studies and has evoked most comment and criticism, was published in 1979 by Needleman's group [77]. From a population of 3329 eligible children two groups, one of 100 children with dentine lead <6 μg g^{-1} and one of 58 children with dentine lead >24 μg g^{-1} were selected. The two groups were then tested with a number of neuropsychological tests and the results statistically analysed, controlling for mother's age at time of subject's birth, mother's educational level, father's SES, number of pregnancies and parental IQ. The IQ's found for the low and high lead groups were: full scale 106.6, 102.1; verbal 103.9, 99.3; and the performance 108.7, 104.9, i.e. a difference in the region of 4-5 points. Also ten of eleven teacher behaviour ratings were significantly different between the two lead groups. In addition 2146 children were given teacher behaviour ratings and the results considered in terms of six dentine lead groups. The high lead groups performed less well on nine of the eleven behaviour ratings.

Numerous criticisms of the study include; the use of dentine lead as a lead marker, omission of the source of lead, sample selection, handling of the confounding variables, the large number (66) of neuropsychologic outcomes tested, the testing technique, the statistical treatment of the data, and the methods of obtaining the teacher ratings [3,8,35,36,37,45,67,70,100,106,122]. The work has also had favourably criticism, such as the planning, good reasons for excluding certain people and a sound choice of psychological tests [3,99,106,109]. Needleman and coworkers have responded to most of the above points [78,80,83-86] and despite the criticisms the study does demonstrate an effect of lead on neuropsychological functioning of children.

The data from the study has been analysed in other ways [10], which indicates that lead disrupts the child/maternal IQ relationship. For the low lead group maternal IQ accounted for 24.7% of the child's IQ variance and the addition of lead did not alter the value, whereas for the high lead group, maternal IQ accounted for 16.1% of the variance, which rose to 27.1% when lead was included.

Two further studies [86,89] have replicated the teacher behaviour rating study on a group of 1273 children. A group of the children studied in the 1979 report were re-studied 4 years later [86] with respect to behaviour in school. Some of the tests indicated a dose-response relationship.

A 4-7 point deficit on the Bayley MDI scale was observed for 216 children at age 6 months, grouped into high, medium and low lead exposures based on maternal cord blood lead levels. The lower score occurred for the higher lead group [28]. The deficit persisted when tested at ages 12, 18 and 24 months. The primary source of lead was identified as coming from the mother during pregnancy.

A study of 242 mothers and 280 infants in Cincinnati [28] found that the prenatal blood lead (mothers) showed a significant negative relationship with MDI and PDI (Bayley Psychological Motor Development Index), but not with the child's post natal blood lead levels.

A group of 218 children in Chicago were divided into three lead groups based on their blood lead levels [91]. A number of blood lead measurements were carried out on each child over 3 years. A significant difference was found between the high and low lead groups in verbal productivity, perceptual and visual motor functioning, however no control of CV's was attempted, except that the sample was said to be homogeneous.

TABLE 16.6 Summary of Population Studies of Children

Study	Lead Measure	Sample	Tests	Confounding variables	Results
Boston (1979) [77]	Dentine	3329 teeth from 2335 a)high lead >24 μg g^{-1} 100 low lead <6 mg g^{-1} 58	WISC-R Seashore rhythm Token Sentence repetition, reaction time	Family size, mother's age at time of birth, mother's educ., SES parental IQ	Significant differences between high and low lead groups IQ full scale 4.5, verbal 4.6, performance 3.8.
		b)Six groups, <5.1, 5.1-8.1, 8.2-11.8, 11.9-17.1, 17.2-27.0, >27.0 μg g^{-1} 2146 children	Eleven teacher behaviour ratings	None on whole sample, same as above on high and low lead grups	Significant trend observed indicating a dose-response relationship.
Lowell, Mass. (1981) [89]	Dentine	1273 children Five groups: ≤6.4, 6.5-8.7, 8.71-12.01-18.1, ≥18.2 μg g^{-1} Selected 215 children from 447	Eleven teacher behaviour ratings grouped into 5 clusters	SES, birthweight, mother's education	Significant trend for behaviours; distractable, disorbanised and frustrated, a dose-response relationship.
West German (1983) [126]	Whole teeth	26 matched pairs low lead <3 μg g^{-1} high lead > μg g^{-1} Original sample 1238 children	WISC-R visual motor integration gross motor integration	Matched for age, sex, father's occupation (other factors similar for two groups)	Near significant difference in IQ, performance. 6 points (p = 0.08), full scale 7 points (p = 0.09); visual motor integration significantly different (GFT test); verbal IQ (5 pts) and gross motor integration not significant.
(1983) [127]	Whole teeth	Three groups <4.2 μg g^{-1}, 36 5.8-7.2 μg g^{-1}, 56 >9.8 μg g^{-1}, 23 selected from 317 children out of 3669	WISC-R visual motor integration Vienna reaction time; behaviour ratings	Sex, age, labour duration, socio-hereditory background (school type & father's occupation)	A difference in IQ's but not significant, visual motor integration (GFT) significantly different, some parent behaviour ratings significantly different, not for teacher ratings.

Table 16.6 continued

Study	Lead Measure	Sample	Tests	Confounding variables	Results
Australia (Sydney) (1979) [43]	Blood Hair	Four school areas 1200 children	Questionnaire to parents and teachers	None, pica not associated ith lead	Significant correlations in behavioural and anti-social behaviour.
London (1983) [118]	Whole tooth crown	Three groups <2.5 $\mu g\ g^{-1}$ 5.0-5.5 $\mu g\ g^{-1}$ >8.0 $\mu g\ g^{-1}$ 403 from 3890 from a possible 6875 children	WISC-R, word reading, maths., seashore rhythm, visual seuqential memory subset, sentence memory, shape copying, behaviour ratings	Mother's IQ, parental educ., interest, family characteristics, social background, birth condition, developmental delay, marital relation-ship, mother's mental state, sex, age, tester, - controlled when required.	Non-significant IQ differences (verbal 2.3, performance 1.8, full scale 2.3, word reading 4.2) after control for C.V. Show a consistent trend in decrease with increase in lead.
Chicago (1983) [91]	Blood	Three groups <25 $\mu g\ g^{-1}$ 30-40 $\mu g\ g^{-1}$ 41-60 $\mu g\ g^{-1}$ 218 children	WPPSI	None, groups relatively homogenous	Statistical significant differences for high lead with other two groups, in verbal productivity, perceptual and visual motor function.
New Zealand (Dunedin) (1988) [117]	Blood	579, 11 yr. four groups <7, 7-10, 10-13, >13 $\mu g\ dl^{-1}$	WISC-R, spelling, reading, behaviour	SES, no correl. with PhB, mother's ability	IQ not significantly associated with PhB, but reading, spelling and behaviour were.
London (1985,87) [100,101]	Tooth crowns	<1.0-34 $\mu g\ g^{-1}$ 403 from 3890	Tests as for ref. 118	Parental and social factors	Non-signif. IQ, mothers IQ main determinant Boys IQ/lead assoc. remained signif.
USA (1987) [113]	Blood	4519 from NHANES II survey	Hearing threshold	Various ear conditions SES, urbanization	Signif. increase in hearing threshold with lead also early infant activities, e.g. sit walk.

Table 16.6 continued

Study	Lead Measure	Sample	Tests	Confounding variables	Results
New Zealand (Christchurch) (1988) [40,41]	Dentine	<1.0->20.0 µg g⁻¹ 664-888 children 8-9 yrs. behaviour	IQ, Burt reading test, teacher ratings of school performance, teeth sampling	Parent educ. level, SES family environ., perinatal history, child's educ. exper.,	IQ/lead assoc., but not significant with confounders but stat. signif. school performance, and attentive/restless behaviour.
Birmingham (UK) (1984) [50]	Blood	187 pre-school 15.6 µg dl⁻¹	Child behaviour, psychomotor, cognitive	SES, parents educ., marital situation, age child's activities	Non-signif. IQ/lead,after controlling for confounders.
Birmingham (UK) (1988) [51]	Blood	201 5.5 yr. olds vigilance	Reaction times, motor skills, behaviour, IQ mother's IQ	SES, parents mental health, life events marital situation	Non-signif. IQ/lead assoc., marginally based on father's occupation, some motor skills relate to lead levels.
London (UK) (1986) [66]	Blood	194, 7-24 µg dl⁻¹	IQ, educ., attain., teacher ratings of performance	SES (occupation), age	Non-signif lead/IQ relationship after use of confounders.
Boston (USA) (1984) [cf. 28]	Cord blood	249; 1.8,6.5,14.6 µg dl⁻¹ three groups	Bayley MDI	Yes	Signif association, cord blood lead and reduced MDI measure.
Cincinnati (USA) (1986) [cf. 28]	Blood mothers infants	8.0 µg dl⁻¹ 4.5 µg dl⁻¹	Bayley MDI, PDI	Race, SES,, home environ., tobacco & alcohol use, natal problems	Signif. assoc. pre-natal blood lead and MDI measure.

The blood lead levels obtained in the USA during the NHANES II survey were studied in relation to the hearing threshold of 4519, 4-19 year olds. A significant association was discovered, which showed a higher threshold (frequency) existed for the people with higher lead levels. Also, using the same blood lead data, significant associations were found with the age a child first sat up, walked and spoke, as well as evidence for hyperactivity [113].

Studies in West Germany Two investigations have been carried out in West Germany [53,124-127], in which dentine lead was used as the lead marker. From a sample of 1238 children, 904 incisors were analysed from 604 children, eventually 26 pairs matched for age, sex and fathers' occupation (SES) were obtained. Lower IQ's were found for the high lead group, but the difference with the low lead group did not reach statistical significance. The small sample may have had something to do with this.

In a second study at Stolberg, in a mining area, [124,127] a sample of 3669 children were used. The dentine lead of 115 children were divided into three groups. Some blood lead levels were also measured. Verbal IQ decreased, 117, 115, 109 with increasing dentine lead ($p<0.1$), whereas other IQ measures were not significant, though the trend was in the same direction. However, after correcting for the CV's, the verbal IQ's differences became not significant.

Studies in the United Kingdom A large study carried out on London children [118], was both well planned and detailed. From a total population of 6875, 6-7 yr old children, from 168 London schools, 4105 children provided teeth of which the whole tooth crowns were analysed. A sub-sample of 403 children were then studied in detail. The groupings within this sample were: low lead ($<2.5\ \mu g\ g^{-1}$) 145, medium lead (5.0-$5.5\ \mu g\ g^{-1}$) 103, and high lead ($>8.0\ \mu g\ g^{-1}$)155. A number of neurologic and psychometric tests were carried out on the sub-sample. For the lead groups, unadjusted for covariates, the three IQ scores were significantly different for the high and low lead groups, as was a word reading test and one block of 'reaction time without delay' tests. In addition the results showed the trend for IQ; low lead < medium lead < high lead. Re-calculation of the data using the CV's removed the statistical significance. The variance in the tooth lead was accounted for by family cleanliness, pica, years the child sucked their thumb, mother's smoking habit, proximity to waste land, age of house and child's play space. If anything the study may have over compensated for the confounding factors.

A number of other studies have been completed in the UK. The work described above [118] has been reevaluated using the dentine lead as a continuous variable rather than grouped. Again no association was found between tooth lead and child's IQ. Parental IQ had the major influence on the the child IQ. However, there was a small significant association, after allowing for CV's, between tooth lead and the IQ of boys. This was an unexpected result, and needs to be replicated [100,101].

An earlier study of children in London, living near a smelter (Table 16.4) [63,130], found an association between blood lead and IQ. This was replicated [66], this time using children living near a major road system. The results this time did not reveal any association, though children of manual workers did have a non-significant trend of decreasing IQ and increasing blood lead. One reason may have been the small sample of 87 children. The difference between this result and the earlier study, may have been in the SES mix of the samples, as more children of middle class families were included in the second study [66].

Two studies have been carried out on Birmingham children, 187 preschoolers, and 201 children 5.5 years old. The results of both studies indicated no association with blood lead and IQ, though for the older group a non-significant trend was found. In some neuropsychological tests significant associations were observed [50,51].

Studies in the South Pacific Three studies have been carried out in the South Pacific. A sample of 1200 children, in Sydney, Australia, were studied using capillary blood and hair for lead measurement [43]. The principal finding was that proximal hair lead related to behavioural and antisocial problems most strongly. Blood lead and behavioural factors were also significantly related.

In Dunedin, New Zealand, as part of a Child Multidiscipline Health and Development Study, 574 eleven year old children were studied using blood lead as the lead marker [117]. Whereas IQ was not significantly associated with lead, a number of cognitive functions and attention/behavioural ratings were significantly associated, after controlling for the confounding variables.

A larger study of children (664 to 888 depending on the neuropsychological measure) has been carried out in Christchurch, New Zealand The children were aged 8 and 9 years and dentine lead was used as the lead measure [40,41]. The lead measure was treated as a continuous variable, and allowance was made for the intrinsic variability of the sample. As for many other studies a trend of decreasing IQ and increasing lead was observed, but after controlling for CV's the trend was not statistically significant. Teacher ratings, however, of school performance remained significant [40]. Maternal and teacher ratings of inattention and restlessness were found to have a small, but consistent, and significant association with the dentine lead variable [41].

Mechanism

As yet there is no clear picture of the mechanism by which the lead can influence the neurological functioning of children. One possibility [12,81] is that δ-ALA (δ-aminolevulinic acid), which is raised in the blood because of the inhibition of haem synthesis by lead, is a neurotoxin and can cross the blood-brain barrier. The δ-ALA could be a weak antagonist to the neuro-transmitter γ-aminobutyric acid (GABA), $HO_2C\text{-}CH_2\text{-}CH_2\text{-}CH_2\text{-}NH_2$. The chemical structure of GABA and δ-ALA, $HO_2C\text{-}CH_2\text{-}CH_2\text{-}CO\text{-}CH_2\text{-}NH_2$ are not unrelated, and the latter may mimic the GABA as a false neuro-transmitter. GABA is believed to be an important neuro-transmitter in the cerebral cortex and lead may inhibit its metabolism. Both GABA, and maybe the neurotoxic effects of δ-ALA, appear to be mainly associated with areas such as stress, anxiety and hyperactivity, and if this is so it tends to fit in with the results of the studies discussed above.

In conclusion it may be said that, overall from the large number of different studies the weight of evidence does suggest an association between low level lead exposure and neurological functioning of children. The lead may slightly affect the IQ of children, but usually not significantly. This may because IQ is not the best outcome to measure in relation to lead. It does seem however, that behaviour, restlessness, and attentiveness are more clearly and significantly associated with lead in children. This could be considered a serious situation, because such behaviour problems are not conducive to children learning irrespective of the IQ. The affect of lead on the human race has been with us for centuries [94] because of its wide use in industrialized societies. Whether or not the situation is any worse today has been debated, but it does appear from the work described above that that the urban population at least, is being influenced by the neurotoxicity of lead.

REFERENCES

1. Albert, R. E., Shore, R. E., Sayers, A. J., Strehlow, C., Kneip, T. J., Pasternack, B. S., Friedhoff, A. J., Covan, F. and Cimino, J. A. Follow up of children over-exposed to lead. **Environ. Health Perspec.**; 1974; **7:** 33-39.
2. Alexander, F. W. The uptake of lead by children in different environments. **Environ. Health Perspec.**; 1974; **7:** 155-159.
3. Australian Academy of Science. Health and environmental lead in Australia: Aust. Acad. Sci.; 1981.
4. Baghurst, P. A. Robertson, E. F., McMichael, A. J.., Vimpani, G. V., Wigg, N. R. and Roberts, R. R. The Port Pirie cohort study: lead effects on pregnancy outcome and early childhood development. **Neuroltoxicol.**; 1987; **8:** 395-402.
5. Baloh, R. W. Laboratory diagnosis of increased lead absorption. **Arch. Environ. Health**; 1974; **28:** 198-208.
6. Baloh, R. W., Sturm, R., Green, B. and Gleser, G. Neuropsychological effects of chronic asymptomatic increased lead absorption. **Arch. Neurol.**; 1975; **32:** 326-330.
7. Baloh, R. W., Spivey, G. H., Brown, C. P., Morgan, D., Campion, D. S., Browdy, B. L., Valentine, J. L. Gonick, H. C. Massey, F. J. and Culver, B. D. Subclinical effects of chronic increased lead absorption - a prospective study II, results of baseline neurologic testing. **J. Occup. Med.**; 1979; **21:** 490-496.
8. Barr, M., Meinrath, H. and Isherwood, R. Research into lead pollution. **Lancet**; 1979; **1:** 1289.

9. Beattie, A. D. Moore, M. R., Goldberg, A., Finlayson, M. J. W., Graham, J. F., Mackie, E. M. Main, J. C., Mclaren, D. A., Murdoch, R. M. and Stewart, G. T. Role of chronic low-level lead exposure in the aetiology of mental retardation. **Lancet**; 1975; **1:** 589-592.

10. Bellinger, D. C. and Needleman, H. L. Lead and the relationship between maternal and child intelligence. **J. Pediat.**; 1983; **102:** 523-527.

11. Benignus, V. A., Otto, D. A., Muller, K. E. and Seiple, K. J. Effects of age and body lead burdrn on CNS function in young children II EEG spectra. **Electroencep. Clin. Neurophysiol.**; 1981; **52:** 240-248.

12. Bryce-Smith, D. Environmental chemical influences on behaviour and mentation. **Chem. Soc. Revs.**; 1986; **15:** 93-123.

13. Burchfiel, J. L. Duffey, F. H. Bartels, P. H. and Needleman H. L. The combined discriminating power of quantitative electroencephalography and neuropsychologic measures in evaluating central nervous system effects of lead at low levels. In: Needleman H. L., Ed. Low Level Lead Exposure: Raven Press; 1980: 91-119.

14. Chisholm, J. J. Current status of lead exposure and poisoning of children. **South. Med. J.**; 1976; **69:** 529-531.

15. Christophers, A. J. Environmental lead question a point of view from Victoria. **Med. J. Aust.**; 1980; **2:** 300-304.

16. Cohen, D. J., Paul, R., Anderson, G. M. and Hareherik, D. F. Blood lead in autistic children. **Lancet**; 1982; **2:** 94-95.

17. Cohen, D. J.. Johnson, W. T. and Caparulo, B. K. Pica and elevated blood lead levels in autistic and atypical children. **Amer. J. Diseases Child.**; 1976; **130:** 47-48.

18. Corrigan, J. Occupational health experts hits out at anti-lead campaigners. **Med. News**; 1981; **13:** 6.

19. Cowan, L. D. and Leviton, A. Epidemiologic considerations in the study of the sequelæ of low level lead exposure. In: Needleman, H. L., Ed. Low Level Lead Exposure: Raven Press; 1980: 91-119.

20. David, O. J., Clark, J. and Voeller, K. Lead and hyperactivity. **Lancet**; 1972; **2:** 900-903.

21. David, O. J., Hoffman, S. P., Sverd, J. and Clark, J. Lead and hyperactivity. Lead levels among hyperactive children. **J. Abnorm. Child Psychol.**; 1977; **5:** 405-416.

22. David, O. J. Letter to the editor. **Arch. Environ. Health**; 1979; **34:** 379.

23. David, O. J., Hoffman, S. P., McGann, B., Sverd, J. and Clark, J. Low lead levels and mental retardation. **Lancet**; 1976; **2:** 1376-1379.

24. David, O. J., Grad, G., McGann, B. and Koltun, A. Mental retardation and "non-toxic" lead levels. **Amer. J. Psychiat.**; 1982; **139:** 806-609.

25. David, O. J., Hoffman, S. P., Clark, J., Grad, G. and Sverd, J. Penicillamine in the trearment of hyperactive children with moderately elevated lead levels. In: Rutter, M. and Jones, R. R., Eds. Lead versus Health: Wiley; 1983: 297-317.

26. David, O. J., Clark, J. and Hoffman, S. P. The subclinical effects of lead on children. Proc. Symp. Conservation Soc. London; 1978: 29-40.
27. David, O. J., Hoffman, S. P. and Kagey, B. Sub-clinical lead levels and behaviour in children. **Trace Subst. Environ. Health**; 1979; **13**: 52-58.
28. Davis, J. M. and Svendsgaard, D. J. Lead and child development. **Nature**; 1987; **329**: 297-300.
29. de la Burde, B. and Shapiro, I. M. Dental lead, blood lead and pica in urban children. **Arch. Environ. Health**; 1975; **30**: 281-284.
30. de la Burde, B. and Choate, M. S. Does asymptomatic lead exposure in children have latent sequelal. **J. Pediat.**; 1972; **81**: 1088-1091.
31. de la Burde, B. and Choate, M. S. Early asymptomatic lead exposure and development at school age. **J. Pediat.**; 1975; **87**: 638-642.
32. Elwood, P. C. Morton, M. and St Leger, A. S. Lead in water and mental retardation. **Lancet**; 1976; **1**: 590-591.
33. Epidemiological study protocol on biological indicators of lead neurotoxicity in children: WHO; 1983; ICP/RCE 903.
34. Ernhart, C. B. Lead in petrol. **Lancet**; 1972; **2**: 209-210.
35. Ernhart, C. B., Landa, B. and Schell, N. B. Lead levels and intelligence. **Pediat.**; 1981; **68**: 903-905.
36. Ernhart, C. B., Landa, B. and Schell, N. B. Subclinical levels of lead and development deficit - a multivariate followup reassessment. **Pediat.**; 1981; **67**: 911-919.
37. Expert Committee. Pediatric neurobehavioural evaluations. Appendix 12c, Air Quality Criteria for Lead: EPA; 1983.
38. Falk, H. L. Conclusons of the committee on human health consequences of lead exposure from automobile emissions. **Environ. Health Perspec.**; 1977; **19**: 243-246.
39. Feldman, R. G., Haddow, J. Kopito, L. and Schwachman, H. Altered peripheral nerve conduction velocity: chronic lead intoxication in children. **Amer. J. Diseases Child.**; 1973; **125**: 39-41.
40. Fergusson, D. M., Fergusson, J. E., Horwood, L. T. and Kinzett, N. G. A longintudinal study of dentine lead levels, intelligence, school performance and behaviour: Part II. Dentine lead and cognitive ability. **J. Child Psychol. Psychiat.**; 1988; **29**: 793-809.
41. Fergusson, D. M., Fergusson, J. E., Horwood, L. T. and Kinzett, N. G. A longintudinal study of dentine lead levels, intelligence, school performance and behaviour: Part III Dentine lead levels and attention/ activity. **J. Child Psychol. Psychiat.**; 1988; **29**: 811-824.
42. Fergusson, J. E., Kinzett, N. G., Fergusson, D. M. and Horwood, L. T. A longintudinal study of dentine lead levels, intelligence, school performance and behaviour: The measurement of dentine lead. **The Sci. Total Environ.**; 1989; **80**: 229-241
43. Garnys, V. P., Freeman, R. and Smythe L. E. Lead Burden of Sydney School Children: Univ. New South Wales; 1979.

44. Gartside, P. S. and Panke, R. K. A discussion concerning the significance of results for children tested in the Shoshone project. Shoshone Lead Health Project; 1976: 116-119.

45. Gloag, D. Is low level lead polution dangerous? **Brit. Med. J.**; 1980; **281:** 1622-1625.

46. Grandjean, P., Arnvig, E. and Beckmann, J. Psychological dysfunction in lead exposed workers. **Scand. J. Work Environ. Health**; 1978; **4:** 295-303.

47. Gregory, R. J. Lehman, R. E. and Mohan, P. J. Intelligence test results for children with and without undue lead absorption. Shoshone Lead Health Project; 1976: 120-149.

48. Haenninen, H., Hernberg, S., Mantere, P., Vesanto, R. and Jalkanen, M. Psychological performance of subjects with low exposure to lead. **J. Occup. Med.**; 1978; **20:** 683-689.

49. Hammond, P. B., Clark, C. S., Gartside, P. S., Berger, O., Walker, A. and Michael, L. W. Fecal lead excretion in young children as related to sources of lead in their environment. **Int. Arch. Occup. Environ. Health**; 1980; **46:** 191-202.

50. Harvey, P. G., Hamlin, M. W. Kumar, R. and Delves, H. T. Blood lead, behaviour and intelligence test performance in preschool children. **The Sci. Total Environ.**; 1984; **40:** 45-60.

51. Harvey, P. G., Hamlin, M. W., Kumar, R., Morgan, G., Spurgeon, A. and Delves, H. T. Relatioships between blood lead, behaviour, psychometric and neuropsychological test performance in young children. **Brot. J. Develop. Psychol.**; 1988; **6:** 145-156.

52. Hebel, J. R., Kinch, D. and Armstrong, E. Mental capability of children exposed to lead pollution. **Brit. J. Preven. Social Med.**; 1976; **30:** 170-174.

53. Hrdina, K. and Winneke, G. Neuropsychological research on children with increased tooth lead content. **Conf. German Assoc. Hygiene Microbiol.**; 1978.

54. Hutton. M. Human health concerns of lead, mercury, cadmium and arsenic. In: Hutchinson, T. C. and Meema, K. M., Eds. Lead, Mercury, Cadmium and Arsenic in the Environment: Wiley, SCOPE; 1987: 53-68.

55. Kotok, D., Kotok, R. and Heriot, J. J. Cognitive evaluation of children with elevated blood lead levels. **Amer. J. Diseases Child.**; 1977; **131:** 791-793.

56. Kotok, D. Development of children with elevated blood lead levels, a controlled study. **J. Pediat.**; 1972; **80:** 57-61.

57. Landrigan, P. J., Gehlbach, S. H., Rosenblum, B. F., Shoults, J. M., Candlearia, R. M., Bartle, W. F., Liddle, J. A., Smrek, A. L., Staehling, N. W. and Sanders, J-D. F. Epidemic lead absorption near an ore smelter. **New Engl. J. Med.**; 1975; **292:** 123-129.

58. Landrigan, P. J. and Baker, E. L. Exposure of children to heavy metals from smelters: epidemiology and toxic cosequences. **Environ. Res.**; 1981; **25:** 204-224.

59. Landrigan, P. J., Baker, E. L., Feldman, R. G. Cox, D. H. Eden, K. V. Orenstein, W. A. Mather, J. A. Yankel, A. T. and Von Lindern, I. H. Increased lead absorption with anaemia and slowed nerve conduction in children near a lead smelter. **J. Pediatr.**; 1976; **89:** 904-910.

60. Landrigan, P. J., Baker, E. L., Feldman, R. G. Cox, D. H. Eden, K. V. Orenstein, W. A. Mather, J. A. Yankel, A. T. and Von Lindern, I. H. Increased lead absorption with anaemia and slowed nerve conduction in children near a lead smelter. Shoshone Lead Health Project: Idaho Dept. Health and Welfare; 1976: 90-115.

61. Landrigan, P. J., Baker, E. L., Whitworth, R. H. and Feldman, R. G. Neuroepidemiologic evolutions of children with chronic increased lead absorption. In: Needleman H. L., Ed. Low Lead Level Exposure: Raven Press; 1980: 17-33.

62. Landrigan, P. J., Whitworth, R. H., Baloh, R. W., Staehling, N. W., Bartle, W. F. and Rosenblum, B. F. Neuropsychological dysfunction in children with chronic low-level lead absorption. **Lancet**; 1975; **1:** 708-712.

63. Lansdown, R., Yule, W. Urbanowicz, M.-A. and Millar I. B. Blood lead, itelligence, attainment and behaviour in school children: overview of a pilot study. In: Rutter, M. and Jones, R. R., Eds. Lead versus Health: Wiley; 1983: 267-296.

64. Lansdown, R. G., Shepherd, J., Clayton, B. E. Delves, H. T. Graham, P. J. and Turner, W. C. Blood-lead levels, behaviour and intelligence: a population study. **Lancet**; 1974; **1:** 538-541.

65. Lansdown, R. and Yule, W. The Lead Debate: The Environment, Toxicology and Child Health: Croom Helm; 1986.

66. Lansdown, R. Yule, W., Urbanowicz, M. A. and Hunter, J. The relationship between blood-lead concentrations, intelligence, attainment and behaviour in a school population: the second London study. **Int. Arch. Occup. Environ. Health**; 1986; **57:** 225-235.

67. Marshall, E. EPA faults classic lead poisoning study. **Science**; 1983; **222:** 906.

68. McBride, W. G., Black, B. P. and English, B. J. Blood lead levels and behaviour of 400 preschool children. **Med. J. Aust.**; 1982; **2:** 26-29.

69. McNeil, J. L. and Ptasnik, J. S. Epidemiological study of a lead contaminated area. Final Rept. Project LH-208: Int. Lead and Zinc Research Organisation; 1975.

70. McNeil, J. L., Ptasnik, J. S. and Croft, D. B. Evaluation of long-term effects of elevated blood lead concentrations in asymptomatic children. **Arch. Ind. Hyg. Toxicol.**; 1977; **26 (Suppl.):** 97-118.

71. Milar, C. R., Schroeder, S. R., Mushak, P., Dolcourt, J. L. and Grant, L. D. Contributions of the caregiving environment to increased lead burden of children. **Amer. J. Mental Defic.**; 1980; **84:** 339-344.

72. Milar, C. R., Schroeder, S. R., Mushak, P. and Boone, L. Failure to find hyperactivity in preschool children with moderately elevated lead burden. **J. Pediat. Psychol.**; 1981; **6:** 85-95.
73. Milburn, H., Mitran E. and Crockford, G. W. An investigation of lead workers for subclinical effects of lead using three performance tests. **Annals Occup. Hyg.**; 1976; **19:** 239-249.
74. Moncrieff, A. a. Koumides, O. P., Clayton, B. E. Patrick, A. D. Renwick, A. G. C. and Rberts, G. C. Lead poisoning in children. **Arch. Diseases Chid.**; 1964; **39:** 1-13.
75. Moore, M. R. Exposure to lead in childhood. **Nature**; 1980; **283:** 334-335.
76. Moore, M. R., Meredith, P. A. and Goldberg, A. A retrospective analysis of blood-lead in mentally retarded children. **Lancet**; 1977; **1:** 717-719.
77. Needleman H.L., Gunnoe C., Leviton A., Reed R., Persie H., Maher C. & Barret. Deficits in psychologic and classroom performance of children with elevated. **New Engl. J. Med.**; 1979; **300:** 689-695.
78. Needleman, H. L. Appendix to the ECAO critique; 1984.
79. Needleman, H. L. and Bellinger, D. The developmental consequences of childhood exposure to lead. In: Lahey, B. B. and Kazdin, A. E., Eds. Advances in Clinical Child Psychology: Plenum Press; 1984; 7: 195-220.
80. Needleman, H. L. EPA review of lead study. **Science**; 1984; **223:** 116.
81. Needleman, H. L. The hazard to health of lead exposure at low dose. In: Nriagu, J. O., Ed. Changing Metal Cycles and Human Health, Dahlem Konferenzen: Springer Verlag; 1984: 311-322.
82. Needleman, H. L. and Landrigan, P. J. The health effects of low level exposure to lead. **Ann. Rev. Pub. Health**; 1981; **2:** 277-298.
83. Needleman, H. L. and Leviton, A. Lead and neurobehavioural deficit in children. **Lancet**; 1979; **2:** 104.
84. Needleman, H. L. Lead and neuropsychological deficit: finding a threshold. In: Needleman, H. L., Ed. Low Level Lead Exposure: Raven Press; 1980: 43-51.
85. Needleman, H. L. and Verducci, J. Lead and child behaviour. **Lancet**; 1982; **2:** 605.
86. Needleman, H. L. Low level lead exposure and neurpsychological performance. In: Rutter, M. and Jones, R. R., Eds. Lead versus Health: Wiley; 1983: 229-248.
87. Needleman, H. L. Neuropsychological dysfunction and classroom behaviour in children with elevated dentine lead. Proc. Symp. The Conservation Soc., London; 1979: 53-69.
88. Needleman, H. L. Studies in subclinical lead exposure, environmental health effects research series. PB - 271649: US EPA; 1977.
89. Needleman, H. L. Studies in children exposed to low levels of lead. EPA 600/ 1 - 81-066: EPA; 1981.

90. Needleman, H. L., Davidson, I., Sewell, E. M. and Shapiro, I. M. Subclinical lead exposure in Philadelphia school children. **New Engl. J. Med.**; 1874; **290**: 245-248.
91. Odenbro, A., Greenberg, N., Vroegh, K. and Bederka, J. Functional disturbance in lead exposed children. **Ambio**; 1983; **12**: 40-44.
92. Otto, D. A., Benignus, V. A., Muller, K. E. and Barton, C. N. Effects of age and body lead burden on CNS function in young children I slow cortical potentials. **Electroenceph. Clin. Neurophysiol.**; 1981; **52**: 229-239.
93. Otto, D. A., Benignus, V. A., Muller, K. E. and Barton, C. N. Electrophysiological evidence of changes in CNS function at low-to-moderate blood lead levels in children. In: Rutter, M. and Jones, R. R., Eds. Lead versus Health: Wiley; 1983: 319-331.
94. Patterson, C. C. Lead in ancient human bones and its relevance to historical developments of social problems with lead. **The Sci. Total Environ.**; 1987; **61**: 167-200.
95. Pepko, J. D., Corum, C. R., Jones, P. D. and Garcia, L. S. The effects of inorganic lead on behavioural and neurologic function. NIOSH Tech. Rept. PHS Pub. 78-128: US Dept. Health Educ. and Welfare; 1978: 1-92.
96. Perino, J. and Ernhart, C. B. The relation of subclinical lead level to cognitive and sensorimotor impairment in black preschoolers. **J. Learn. Disabil.**; 1974; **7**: 616-620.
97. Pihl, R. O., Drake, H. and Vrana, F. Hair analysis in learning and behaviour problems. In: Browm, A, C., and Crounse, R. C., Eds. Hair, Trace elements and Human Illness: Praeger; 1980: 128-143.
98. Pihl, R. O. and Parkes, M. Hair element content in learning disabled children. **Science**; 1977; **198**: 204-206.
99. Pitcher, H. M. Comments on issues raised in the analysis of the neuropsychological effects of low level lead exposure. Office of Policy Analysis EPA; 1984.
100. Pocock, S. J. and Ashby, D. Environmental lead and children's intelligence: a review of recent epidemiological studies. **The Statist.**; 1985; **34**: 31-44.
101. Pocock, S. J., Ashby, D. and Smith, M. Lead exposure and children's intellectual performance. **Int. J. Epidemiol.**; 1987; **16**: 57-67.
102. Pueschel, S. M., Kopito, L. and Schwachman, H. Children with an increased lead burden: a screening and follow-up study. **J. Amer. Med. Assoc.**; 1972; **222**: 462-466.
103. Pueschel, S. M. Neurological and psychomotor functions in children with an increased lead burden. **Environ. Health Perspec.**; 1974; **7**: 13-16.
104. Ratcliffe, J. M. Development and behavioural functions in young children with elevated blood lead levels. **Brit. J. Prevent. Social Med.**; 1977; **31**: 258-264.
105. Ratcliffe, J. M. Neurological and behavioural toxicology of moderate lead exposure in children. In: Management and Control of Heavy Metals in the Environment, Int. Conf. London: CEP Consultants; 1979: 81-84.

106. Report of a DHSS Working Party on Lead in the Environment. Lead and health: DHSS; 1980.
107. Rummo, J. H., Routh, D. K. Rummo, N. J. and Brown, J. F. Behavioural and neurological effects of symptomatic and asymptomatic lead exposure in children. **Arch. Environ. Health;** 1979; **34:** 120-124.
108. Rutter, M. Low level lead exposure: sources, effects and implications. In: Rutter, M. and Jones, R. R., Eds. Lead versus health: Wiley; 1983: 333-370.
109. Rutter, M. Raised lead levels and impaired cognitive/behavioural functioning: a review of the evidence. **Develop. Med. Child Neurol.;** 1980; **22 (Suppl. 42):** 1-26.
110. Rutter, M. The relatioship between science and policy making: the case of lead. **Clean Air;** 1983; **13:** 17-32.
111. Rutter, M. Scientific issues and the state of the art in 1980. In: Rutter, M. and Jones, R. R., Eds. Lead versus health: Wiley; 1983: 1-15.
112. Sachs, H. K. Prognosis for children with chronic lead exposure. In: Lynam, D. R., Piantanida, L. G. and Cole, J. F., Eds. Environmental Lead: Academic Press; 1981: 41-48.
113. Schwartz, J. and Otto, D. Blood lead, hearing thresholds and neurobehavioural development in children and youth. **Arch. Environ. Health;** 1987; **42:** 153-160.
114. Seppäläinen, A, M., Hernberg, S. and Kock, B. Relationship between blood lead levels and nerve conduction velocities. **Neurotoxicol.;** 1979; **1:** 313-332.
115. Seppäläinen, A, M., Tola, S., Hernberg, S. and Kock, B. Subclinical neuropathy a "safe" levels of lead exposure. **Arch. Environ. Health;** 1975; **30:** 180-183.
116. Seppäläinen, A, M., Hernberg, S. Subclinical lead neuropathy. **Amer. J. Indust. Med.;** 1980; **1:** 413-420.
117. Silva, P. A., Hughes, P., Williams, S. and Faed, J. M. Blood lead, intelligence, reading attainment and behaviour in eleven old children in Dunedin, New Zealand. **J. Child Psychol. Psychiat.;** 1988; **29:** 43-52.
118. Smith, M., Delves, T. Lansdown, R. Clayton, B. and Graham, P. The effects of lead exposure on urban children : the Institute of Child Health/ Southampton study. **Dev. Med. Child Neurol., Suppl. 47;** 1983; **25:** 1-54.
119. Ter Haar, G. and Aronow, R. New information on lead in dirt and dust as related to childhood lead problem. **Environ. Health Perspec.;** 1974; **7:** 83-89.
120. Ter Haar, G. and Aronow, R. Tracer studies of ingestion of dust\by urban children. In: Griffin, T. B. and Knelson, J. H., Eds. **Environ. Qual. Saftey Suppl.;** 1975; **II:** 197-201.

121. Thatcher, R. W., Lester, M. L., McAlaster, R. and Horst, R. Effects of low levels of cadmium and lead on cognitive functioning of children. **Arch. Environ. Health**; 1982; **37**: 159-166.
122. Volpe, R. H. Assessment of the health aspects of lead in petrol. Int. Lead and Zinc Research Organisation, Committee on Evaluation of Regulations, Global Approach (ERGA) of EEC; 1983.
123. Whitworth, R. H., Rosenblum, B. F., Dickerson, M. S. and Baloh, R. W. Epidemiologic notes and reports follow-up on human lead absorption. **Morbid. Mortal. Weekly Rept.**; 1974; **23**: 157-159.
124. Winneke, G. Neurobehavioural and neuropsychological effects of lead. In: Rutter, M. and Jones, R. R., Eds. **Lead versus Health**: Wiley; 1983: 249-265.
125. Winneke, G. Neuropsychological studies in children with elevated tooth-lead levels. Proc. Symp. Conservation Soc., London; 1979: 33-52.
126. Winneke, G., Hrdina, K-G. and Brockhaus, A. Neuropsychological studies in children with elevated tooth lead concentrations, Part I, a pilot study. **Int. Arch. Occup. Environ. Health**; 1983; **51**: 169-183.
127. Winneke, G., Kramer, U., Brockhaus, A., Ewers, U., Kujanek, G., Lechner, H. and Janke, W. Neuropsychological studies in children with elevated tooth lead concentrations, Part II, extended study. **Int. Arch. Occup. Environ. Health**; 1983; **51.**
128. Yamins, J. The relationship of subclinical lead intoxication to cognitive and language functioning in preschool children. PhD Dissertation, Hofstra Univ. Hempsted N.Y.; 1976.
129. Youroukos, S., Lyberatos, C., Philippidou, A., Gardikas, C. and Tsomi, A. Increased blood lead levels in mentally retarded children in Greece. **Arch. Environ. Health**; 1978; **33**: 297-300.
130. Yule, W., Lansdown, R., Millar, I. B. and Urbanowicz, M. A. The relationship between blood-lead concentrations, intelligence and attainment in a school population: a pilot study. **Develop. Med. Child Neurol.**; 1981; **33**: 567-576.

INDEX